MW01089093

Reliability Evaluation of Power Systems

Second Edition

Reliability Evaluation of Power Systems

Second Edition

Roy Billinton
University of Saskatchewan
College of Engineering
Saskatoon, Saskatchewan, Canada

and

Ronald N. Allan
University of Manchester
Institute of Science and Technology
Manchester, England

PLENUM PRESS • NEW YORK AND LONDON

Library of Congress Cataloging-in-Publication Data

Billinton, Roy.
Reliability evaluation of power systems / Roy Billinton and Ronald N. Allan. -- 2nd ed.
p. cm.
Includes bibliographical references and index.
ISBN 0-306-45259-6
1. Electric power systems--Reliability. I. Allan, Ronald N. (Ronald Norman) II. Title.
TK1010.B55 1996
621.31--dc20 96-27011
CIP

ISBN 0-306-45259-6

First published in England by Pitman Books Limited

A Division of Plenum Publishing Corporation
233 Spring Street, New York, N. Y. 10013

10 9 8 7 6 5 4 3 2 1

Printed in the United States of America

Preface to the first edition

This book is a sequel to *Reliability Evaluation of Engineering Systems: Concepts and Techniques*, written by the same authors and published by Pitman Books in January 1983.* As a sequel, this book is intended to be considered and read as the second of two volumes rather than as a text that stands on its own. For this reason, readers who are not familiar with basic reliability modelling and evaluation should either first read the companion volume or, at least, read the two volumes side by side. Those who are already familiar with the basic concepts and only require an extension of their knowledge into the power system problem area should be able to understand the present text with little or no reference to the earlier work. In order to assist readers, the present book refers frequently to the first volume at relevant points, citing it simply as *Engineering Systems.*

Reliability Evaluation of Power Systems has evolved from our deep interest in education and our long-standing involvement in quantitative reliability evaluation and application of probability techniques to power system problems. It could not have been written, however, without the active involvement of many students in our respective research programs. There have been too many to mention individually but most are recorded within the references at the ends of chapters.

The preparation of this volume has also been greatly assisted by our involvement with the IEEE Subcommittee on the Application of Probability Methods, IEE Committees, the Canadian Electrical Association and other organizations, as well as the many colleagues and other individuals with whom we have been involved.

Finally, we would like to record our gratitude to all the typists who helped in the preparation of the manuscript and, above all, to our respective wives, Joyce and Diane, for all their help and encouragement.

Roy Billinton
Ron Allan

*Second edition published by Plenum Press in 1994.

Preface to the second edition

We are both very pleased with the way the first edition has been received in academic and, particularly, industrial circles. We have received many commendations for not only the content but also our style and manner of presentation. This second edition has evolved after considerable usage of the first edition by ourselves and in response to suggestions made by other users of the book. We believe the extensions will ensure that the book retains its position of being the premier teaching text on power system reliability.

We have had regular discussions with our present publishers and it is a pleasure to know that they have sufficient confidence in us and in the concept of the book to have encouraged us to produce this second edition. As a background to this new edition, it is worth commenting a little on its recent chequered history. The first edition was initially published by Pitman, a United Kingdom company; the marketing rights for North America and Japan were vested in Plenum Publishing of New York. Pitman subsequently merged with Longman, following which, complete publishing and marketing rights were eventually transferred to Plenum, our current publishers. Since then we have deeply appreciated the constant interest and commitment shown by Plenum, and in particular Mr. L. S. Marchand. His encouragement has ensured that the present project has been transformed from conceptual ideas into the final product.

We have both used the first edition as the text in our own teaching programs and in a number of extramural courses which we have given in various places. Over the last decade since its publication, many changes have occurred in the development of techniques and their application to real problems, as well as the structure, planning, and operation of real power systems due to changing ownership, regulation, and access. These developments, together with our own teaching experience and the feedback from other sources, highlighted several areas which needed reviewing, updating, and extending. We have attempted to accommodate these new ideas without disturbing the general concept, structure, and style of the original text.

We have addressed the following specific points:

- A complete rewrite of the general introduction (Chapter 1) to reflect the changing scenes in power systems that have occurred since we wrote the first edition.

- Inclusion of a chapter on Monte Carlo simulation; the previous edition concentrated only on analytical techniques, but the simulation approach has become much more useful in recent times, mainly as a result of the great improvement in computers.
- Inclusion of a chapter on reliability economics that addresses the developing and very important area of reliability cost and reliability worth. This is proving to be of growing interest in planning, operation, and asset management.

We hope that these changes will be received as a positive step forward and that the confidence placed in us by our publishers is well founded.

Roy Billinton
Ron Allan

Contents

1 Introduction

1.1 Background

Electric power systems are extremely complex. This is due to many factors, some of which are sheer physical size, widely dispersed geography, national and international interconnections, flows that do not readily follow the transportation routes wished by operators but naturally follow physical laws, the fact that electrical energy cannot be stored efficiently or effectively in large quantities, unpredicted system behavior at one point of the system can have a major impact at large distances from the source of trouble, and many other reasons. These factors are well known to power system engineers and managers and therefore they are not discussed in depth in this book. The historical development of and current scenarios within power companies is, however, relevant to an appreciation of why and how to evaluate the reliability of complex electric power systems.

Power systems have evolved over decades. Their primary emphasis has been on providing a reliable and economic supply of electrical energy to their customers [1]. Spare or redundant capacities in generation and network facilities have been inbuilt in order to ensure adequate and acceptable continuity of supply in the event of failures and forced outages of plant, and the removal of facilities for regular scheduled maintenance. The degree of redundancy has had to be commensurate with the requirement that the supply should be as economic as possible. The main question has therefore been, "how much redundancy and at what cost?"

The probability of consumers being disconnected for any reason can be reduced by increased investment during the planning phase, operating phase, or both. Overinvestment can lead to excessive operating costs which must be reflected in the tariff structure. Consequently, the economic constraint can be violated although the system may be very reliable. On the other hand, underinvestment leads to the opposite situation. It is evident therefore that the economic and reliability constraints can be competitive, and this can lead to difficult managerial decisions at both the planning and operating phases.

These problems have always been widely recognized and understood, and design, planning, and operating criteria and techniques have been developed over many decades in an attempt to resolve and satisfy the dilemma between the economic and reliability constraints. The criteria and techniques first used in practical applications, however, were all deterministically based. Typical criteria are:

(a) Planning generating capacity—installed capacity equals the expected maximum demand plus a fixed percentage of the expected maximum demand;
(b) Operating capacity—spinning capacity equals expected load demand plus a reserve equal to one or more largest units;
(c) Planning network capacity—construct a minimum number of circuits to a load group (generally known as an $(n-1)$ or $(n-2)$ criterion depending on the amount of redundancy), the minimum number being dependent on the maximum demand of the group.

Although these and other similar criteria have been developed in order to account for randomly occurring failures, they are inherently deterministic. Their essential weakness is that they do not and cannot account for the probabilistic or stochastic nature of system behavior, of customer demands or of component failures.

Typical probabilistic aspects are:

(a) Forced outage rates of generating units are known to be a function of unit size and type and therefore a fixed percentage reserve cannot ensure a consistent risk.
(b) The failure rate of an overhead line is a function of length, design, location, and environment and therefore a consistent risk of supply interruption cannot be ensured by constructing a minimum number of circuits.
(c) All planning and operating decisions are based on load forecasting techniques. These techniques cannot predict loads precisely and uncertainties exist in the forecasts.

1.2 Changing scenario

Until the late 1980s and early 1990s, virtually all power systems either have been state controlled and hence regulated by governments directly or indirectly through agencies, or have been in the control of private companies which were highly regulated and therefore again controlled by government policies and regulations. This has created systems that have been centrally planned and operated, with energy transported from large-scale sources of generation through transmission and distribution systems to individual consumers.

Deregulation of private companies and privatization of state-controlled industries has now been actively implemented. The intention is to increase competition, to unbundle or disaggregate the various sectors, and to allow access to the system by an increased number of parties, not only consumers and generators but also traders of energy. The trend has therefore been toward the "market forces" concept, with trading taking place at various interfacing levels throughout the system. This has led to the concept of "customers" rather than "consumers" since some customers need not consume but resell the energy as a commodity. A consequence of these developments is that there is an increasing amount of energy generated at local distribution levels by independent nonutility generators and an increasing number of new types of energy sources, particularly renewables, and CHP (combined heat and power) schemes being developed.

Although this changing scenario has a very large impact on the way the system may be developed and operated and on the future reliability levels and standards, it does not obviate the need to assess the effect of system developments on customers and the fundamental bases of reliability evaluation. The need to assess the present performance and predict the future behavior of systems remains and is probably even more important given the increasing number of players in the electric energy market.

1.3 Probabilistic reliability criteria

System behavior is stochastic in nature, and therefore it is logical to consider that the assessment of such systems should be based on techniques that respond to this behavior (i.e., probabilistic techniques). This has been acknowledged since the 1930s [2–5], and there has been a wealth of publications dealing with the development of models, techniques, and applications of reliability assessment of power systems [6–11]. It remains a fact, however, that most of the present planning, design, and operational criteria are based on deterministic techniques. These have been used by utilities for decades, and it can be, and is, argued that they have served the industry extremely well in the past. However, the justification for using a probabilistic approach is that it instills more objective assessments into the decision-making process. In order to reflect on this concept it is useful to step back into history and recollect two quotes:

> A fundamental problem in system planning is the correct determination of reserve capacity. Too low a value means excessive interruption, while too high a value results in excessive costs. The greater the uncertainty regarding the actual reliability of any installation the greater the investment wasted.
>
> The complexity of the problem, in general, makes it difficult to find an answer to it by rules of thumb. The same complexity, on one side, and good engineering and sound economics, on the other, justify the use of methods of analysis permitting the systematic evaluations of all important factors involved. There are no exact methods available which permit the solution of reserve problems with the same exactness with which, say, circuit problems are solved by applying Ohm's law. However, a systematic attack of them can be made by "judicious" application of the probability theory.
>
> (GIUSEPPE CALABRESE (1947) [12]).

> The capacity benefits that result from the interconnection of two or more electric generating systems can best and most logically be evaluated by means of probability methods, and such benefits are most equitably allocated among the systems participating in the interconnection by means of "the mutual benefits method of allocation," since it is based on the benefits mutually contributed by the several systems. (CARL WATCHORN (1950) [13])

These eminent gentlemen identified some 50 years ago the need for "probabilistic evaluation," "relating economics to reliability," and the "assessment of benefits or worth," yet deterministic techniques and criteria still dominate the planning and operational phases.

The main reasons cited for this situation are lack of data, limitation of computational resources, lack of realistic reliability techniques, aversion to the use of probabilistic techniques, and a misunderstanding of the significance and meaning of probabilistic criteria and risk indices. These reasons are not valid today since most utilities have valid and applicable data, reliability evaluation techniques are very developed, and most engineers have a working understanding of probabilistic techniques. It is our intention in this book to illustrate the development of reliability evaluation techniques suitable for power system applications and to explain the significance of the various reliability indices that can be evaluated. This book clearly illustrates that there is no need to constrain artificially the inherent probabilistic or stochastic nature of a power system into a deterministic domain despite the fact that such a domain may feel more comfortable and secure.

1.4 Statistical and probabilistic measures

It is important to conjecture at this point on what can be done regarding reliability assessment and why it is necessary. Failures of components, plant, and systems occur randomly; the frequency, duration, and impact of failures vary from one year to the next. There is nothing novel or unexpected about this. Generally all utilities record details of the events as they occur and produce a set of performance measures. These can be limited or extensive in number and concept and include such items as:

- system availability;
- estimated unsupplied energy;
- number of incidents;
- number of hours of interruption;
- excursions beyond set voltage limits;
- excursions beyond set frequency limits.

These performance measures are valuable because they:

(a) identify weak areas needing reinforcement or modifications;
(b) establish chronological trends in reliability performance;
(c) establish existing indices which serve as a guide for acceptable values in future reliability assessments;
(d) enable previous predictions to be compared with actual operating experience;
(e) monitor the response to system design changes.

The important point to note is that these measures are statistical indices. They are not deterministic values but at best are average or expected values of a probability distribution.

The same basic principles apply if the future behavior of the system is being assessed. The assumption can be made that failures which occur randomly in the past will also occur randomly in the future and therefore the system behaves probabilistically, or more precisely, stochastically. Predicted measures that can be compared with past performance measures or indices can also be extremely beneficial in comparing the past history with the predicted future. These measures can only be predicted using probabilistic techniques and attempts to do so using deterministic approaches are delusory.

In order to apply deterministic techniques and criteria, the system must be artificially constrained into a fixed set of values which have no uncertainty or variability. Recognition of this restriction results in an extensive study of specified scenarios or "credible" events. The essential weakness is that likelihood is neglected and true risk cannot be assessed.

At this point, it is worth reviewing the difference between a hazard and risk and the way that these are assessed using deterministic and probabilistic approaches. A discussion of these concepts is given in *Engineering Systems* but is worth repeating here.

The two concepts, hazard and risk, are often confused; the perception of a risk is often weighed by emotion which can leave industry in an invidious position. A hazard is an event which, if it occurs, leads to a dangerous state or a system failure. In other words, it is an undesirable event, the severity of which can be ranked relative to other hazards. Deterministic analyses can only consider the outcome and ranking of hazards. However, a hazard, even if extremely undesirable, is of no consequence if it cannot occur or is so unlikely that it can be ignored. Risk, on the other hand, takes into account not only the hazardous events and their severity, but also their likelihood. The combination of severity and likelihood creates plant and system parameters that truly represent risk. This can only be done using probabilistic techniques.

1.5 Absolute and relative measures

It is possible to calculate reliability indices for a particular set of system data and conditions. These indices can be viewed as either absolute or as relative measures of system reliability.

Absolute indices are the values that a system is expected to exhibit. They can be monitored in terms of past performance because full knowledge of them is known. However, they are extremely difficult, if not impossible, to predict for the future with a very high degree of confidence. The reason for this is that future performance contains considerable uncertainties particularly associated with numerical data and predicted system requirements. The models used are also not entirely accurate representations of the plant or system behavior but are approximations. This poses considerable problems in some areas of application in which absolute values are very desirable. Care is therefore vital in these applications, particularly in situations in which system dependencies exist, such as common cause (mode) failures which tend to enhance system failures.

Relative reliability indices, on the other hand, are easier to interpret and considerable confidence can generally be placed in them. In these cases, system behavior is evaluated before and after the consideration of a design or operating change. The benefit of the change is obtained by evaluating the relative improvement. Indices are therefore compared with each other and not against specified targets. This tends to ensure that uncertainties in data and system requirements are embedded in all the indices and therefore reasonable confidence can be placed in the relative differences. In practice, a significant number of design or operating strategies or scenarios are compared, and a ranking of the benefits due to each is made. This helps in deciding the relative merits of each alternative, one of which is always to make no changes.

The following chapters of this book describe methods for evaluating these indices and measures. The stress throughout is on their use as relative measures.

The most important aspect to remember when evaluating these measures is that it is necessary to have a complete understanding of the engineering implications of the system. No amount of probability theory can circumvent this important engineering function. It is evident therefore that probability theory is only a tool that enables an engineer to transform knowledge of the system into a prediction of its likely future behavior. Only after this understanding has been achieved can a model be derived and the most appropriate evaluation technique chosen. Both the model and the technique must reflect and respond to the way the system operates and fails. Therefore the basic steps involved are:

- understand the ways in which components and system operate;
- identify the ways in which failures can occur;
- deduce the consequences of the failures;
- derive models to represent these characteristics;
- only then select the evaluation technique.

1.6 Methods of assessment

Power system reliability indices can be calculated using a variety of methods. The basic approaches are described in *Engineering Systems* and detailed applications are described in the following chapters.

The two main approaches are analytical and simulation. The vast majority of techniques have been analytically based and simulation techniques have taken a minor role in specialized applications. The main reason for this is because simulation generally requires large amounts of computing time, and analytical models and techniques have been sufficient to provide planners and designers with the results needed to make objective decisions. This is now changing, and increasing interest is being shown in modeling the system behavior more comprehensively and in evaluating a more informative set of system reliability indices. This implies the need to consider Monte Carlo simulation. (See *Engineering Systems*, Ref. 14, and many relevant papers in Refs. 6–10.)

Analytical techniques represent the system by a mathematical model and evaluate the reliability indices from this model using direct numerical solutions. They generally provide expectation indices in a relatively short computing time. Unfortunately, assumptions are frequently required in order to simplify the problem and produce an analytical model of the system. This is particularly the case when complex systems and complex operating procedures have to be modeled. The resulting analysis can therefore lose some or much of its significance. The use of simulation techniques is very important in the reliability evaluation of such situations.

Simulation methods estimate the reliability indices by simulating the actual process and random behavior of the system. The method therefore treats the problem as a series of real experiments. The techniques can theoretically take into account virtually all aspects and contingencies inherent in the planning, design, and operation of a power system. These include random events such as outages and repairs of elements represented by general probability distributions, dependent events and component behavior, queuing of failed components, load variations, variation of energy input such as that occurring in hydrogeneration, as well as all different types of operating policies.

If the operating life of the system is simulated over a long period of time, it is possible to study the behavior of the system and obtain a clear picture of the type of deficiencies that the system may suffer. This recorded information permits the expected values of reliability indices together with their frequency distributions to be evaluated. This comprehensive information gives a very detailed description, and hence understanding, of the system reliability.

The simulation process can follow one of two approaches:

(a) Random—this examines basic intervals of time in the simulated period after choosing these intervals in a random manner.

(b) Sequential—this examines each basic interval of time of the simulated period in chronological order.

The basic interval of time is selected according to the type of system under study, as well as the length of the period to be simulated in order to ensure a certain level of confidence in the estimated indices.

The choice of a particular simulation approach depends on whether the history of the system plays a role in its behavior. The random approach can be used if the history has no effect, but the sequential approach is required if the past history affects the present conditions. This is the case in a power system containing hydroplant in which the past use of energy resources (e.g., water) affects the ability to generate energy in subsequent time intervals.

It should be noted that irrespective of which approach is used, the predicted indices are only as good as the model derived for the system, the appropriateness of the technique, and the quality of the data used in the models and techniques.

1.7 Concepts of adequacy and security

Whenever a discussion of power system reliability occurs, it invariably involves a consideration of system states and whether they are adequate, secure, and can be ascribed an alert, emergency, or some other designated status [15]. This is particularly the case for transmission systems. It is therefore useful to discuss the significance and meaning of such states.

The concept of adequacy is generally considered [1] to be the existence of sufficient facilities within the system to satisfy the consumer demand. These facilities include those necessary to generate sufficient energy and the associated transmission and distribution networks required to transport the energy to the actual consumer load points. Adequacy is therefore considered to be associated with static conditions which do not include system disturbances.

Security, on the other hand, is considered [1] to relate to the ability of the system to respond to disturbances arising within that system. Security is therefore associated with the response of the system to whatever disturbances they are subjected. These are considered to include conditions causing local and widespread effects and the loss of major generation and transmission facilities.

The implication of this division is that the two aspects are different in both concept and evaluation. This can lead to a misunderstanding of the reasoning behind the division. In reality, it is not intended to indicate that there are two distinct processes involved in power system reliability, but is intended to ensure that reliability can be calculated in a simply structured and logical fashion. From a pragmatic point of view, adequacy, as defined, is far easier to calculate and provides valuable input to the decision-making process. Considerable work therefore has been done in this regard [6–10]. While some work has been done on the problem of "security," it is an exciting area for further development and research.

It is evident from the above definition that adequacy is used to describe a system state in which the actual entry to and departure from that state is ignored and is thus defined as a steady-state condition. The state is then analyzed and deemed adequate if all system requirements including the load, voltages, VAR requirements, etc., are all fully satisfied. The state is deemed inadequate if any of the power system constraints is violated. An additional consideration that may sometimes be included is that an otherwise adequate state is deemed to be adequate if and only if, on departure, it leads to another adequate state; it is deemed inadequate if it leads to a state which itself is inadequate in the sense that a network violation occurs. This consideration creates a buffer zone between the fully adequate states and the other obviously inadequate states. Such buffer zones are better known [14] as alert states, the adequate states outside of the buffer zone as normal states, and inadequate states as emergency states.

This concept of adequacy considers a state in complete isolation and neglects the actual entry transitions and the departure transitions as causes of problems. In reality, these transitions, particularly entry ones, are fundamental in determining whether a state can be static or whether the state is simply transitory and very temporary. This leads automatically to the consideration of security, and consequently it is evident that security and adequacy are interdependent and part of the same problem; the division is one of convenience rather than of practical experience.

Power system engineers tend to relate security to the dynamic process that occurs when the system transits between one state and another state. Both of these states may themselves be acceptable if viewed only from adequacy; i.e., they are both able to satisfy all system demands and all system constraints. However, this ignores the dynamic and transient behavior of the system in which it may not be possible for the system to reside in one of these states in a steady-state condition. If this is the case, then a subsequent transition takes the system from one of the so-called adequate states to another state, which itself may be adequate or inadequate. In the latter case, the state from which the transition occurred could be deemed adequate but insecure. Further complications can arise because the state from which the above transition can occur may be inadequate but secure in the sense that the system is in steady state; i.e., there is no transient or dynamic transition from the state. Finally the state may be inadequate and insecure.

If a state is inadequate, it implies that one or more system constraints, either in the network or the system demand, are not being satisfied. Remedial action is therefore required, such as redispatch, load shedding, or various alternative ways of controlling system parameters. All of these remedies require time to accomplish. If the dynamic process of the power system causes departure from this state before the remedial action can be accomplished, then the system state is clearly not only inadequate but also insecure. If, on the other hand, the remedial action can be accomplished in a shorter time than that taken by the dynamic process, the state is secure though inadequate. This leads to the conclusion that "time to perform" a remedial action is a fundamental parameter in determining whether a state is adequate and secure, adequate and insecure, inadequate and secure, or inadequate and insecure. Any state which can be defined as either inadequate or insecure is clearly a system failure state and contributes to system unreliability. Present reliability evaluation techniques generally relate to the assessment of adequacy. This is not of great significance in the case of generation systems or of distribution systems; however, it can be important when considering combined generation and transmission systems. The techniques described in this book are generally concerned with adequacy assessment.

1.8 System analysis

As discussed in Section 1.1, a modern power system is complex, highly integrated, and very large. Even large computer installations are not powerful enough to analyze in a completely realistic and exhaustive manner all of a power system as a single entity. This is not a problem, however, because the system can be divided into appropriate subsystems which can be analyzed separately. In fact it is unlikely that it will ever be necessary or even desirable to attempt to analyze a system as a whole; not only will the amount of computation be excessive, but the results are likely to be so vast that meaningful interpretation will be difficult, if not impossible.

The most convenient approach for dividing the system is to use its main functional zones. These are: generation systems, composite generation and transmission (or bulk power) systems, and distribution systems. These are therefore used as the basis for dividing the material, models, and techniques described in this book. Each of these primary functional zones can be subdivided in order to study a subset of the problem. Particular subzones include individual generating stations, substations, and protection systems, and these are also considered in the following chapters. The concept of hierarchical levels (HL) has been developed [1] in order to establish a consistent means of identifying and grouping these functional zones. These are illustrated in Fig. 1.1, in which the first level (HLI) refers to generation facilities and their ability on a pooled basis to satisfy the pooled system demand, the second level (HLII) refers to the composite generation and transmission (bulk power) system and its ability to deliver energy to the bulk supply points, and the third level (HLIII) refers to the complete system including distribution and its ability to satisfy the capacity and energy demands of individual consumers. Al-

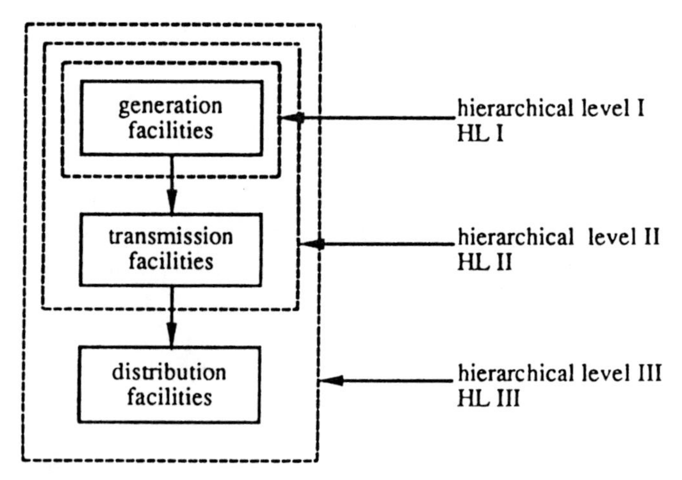

Fig. 1.1 Hierarchical Levels

though HLI and HLII studies are regularly performed, complete HLIII studies are usually impractical because of the scale of the problem. Instead the assessment, as described in this book, is generally done for the distribution functional zone only.

Based on the above concepts and system structure, the following main subsystems are described in this book:

(a) Generating stations—each station or each unit in the station is analyzed separately. This analysis creates an equivalent component, the indices of which can be used in the reliability evaluation of the overall generating capacity of the system and the reliability evaluation of composite systems. The components therefore form input to both HLI and HLII assessments. The concepts of this evaluation are described in Chapters 2 and 11.

(b) Generating capacity—the reliability of the generating capacity is evaluated by pooling all sources of generation and all loads (i.e., HLI assessment studies). This is the subject of Chapters 2 and 3 for planning studies and Chapter 5 for operational studies.

(c) Interconnected systems—in this case the generation of each system and the tie lines between systems (interconnections) are modeled, but the network in each system (intraconnections) is not considered. These assessments are still HLI studies and are the subject of Chapter 4.

(d) Composite generation/transmission—the network is limited to the bulk transmission, and the integrated effect of generation and transmission is assessed (i.e., HLII studies). This is the subject of Chapter 6.

(e) Distribution networks—the reliability of the distribution is evaluated by considering the ability of the network fed from bulk supply points or other local infeeds in supplying the load demands. This is the subject of Chapters 7–9. This considers the distribution functional zone only. The load point indices evaluated in the HLII assessments can be used as input values to the distribution zone if the overall HLIII indices are required.

(f) Substations and switching stations—these systems are often quite complicated in their own right and are frequently analyzed separately rather than including them as complete systems in network reliability evaluation. This creates equivalent components, the indices of which can be used either as measures of the substation performance itself or as input in evaluating the reliability of transmission (HLII) or distribution (HLIII) systems. This is the subject of Chapter 10.

(g) Protection systems—the reliability of protection systems is analyzed separately. The indices can be used to represent these systems as equivalent components in network (transmission and distribution) reliability evaluation or as an assessment of the substation itself. The concepts are discussed in Chapter 11.

The techniques described in Chapters 2–11 focus on the analytical approach, although the concepts and many of the models are equally applicable to the simulation approach. As simulation techniques are now of increasing importance and increasingly used, this approach and its application to all functional zones of a power system are described and discussed in Chapter 12.

1.9 Reliability cost and reliability worth

Due to the complex and integrated nature of a power system, failures in any part of the system can cause interruptions which range from inconveniencing a small number of local residents to a major and widespread catastrophic disruption of supply. The economic impact of these outages is not necessarily restricted to loss of revenue by the utility or loss of energy utilization by the customer but, in order to estimate true costs, should also include indirect costs imposed on customers, society, and the environment due to the outage. For instance, in the case of the 1977 New Year blackout, the total costs of the blackouts were attributed [16] as:

- Consolidated Edison direct costs 3.5%
- other direct costs 12.5%
- indirect costs 84.0%

As discussed in Section 1.1, in order to reduce the frequency and duration of these events and to ameliorate their effect, it is necessary to invest either in the design phase, the operating phase, or both. A whole series of questions emanating from this concept have been raised by the authors [17], including:

- How much should be spent?
- Is it worth spending any money?
- Should the reliability be increased, maintained at existing levels, or allowed to degrade?
- Who should decide—the utility, a regulator, the customer?
- On what basis should the decision be made?

The underlying trend in all these questions is the need to determine the worth of reliability in a power system, who should contribute to this worth, and who should decide the levels of reliability and investment required to achieve them.

The major discussion point regarding reliability is therefore, "Is it worth it?" [17]. As stated a number of times, costs and economics play a major role in the application of reliability concepts and its physical attainment. In this context, the question posed is: "Where or on what should the next pound, dollar, or franc be invested in the system to achieve the maximum reliability benefit?" This can be an extremely difficult question to answer, but it is a vital one and can only be attempted if consistent quantitative reliability indices are evaluated for each of the alternatives.

It is therefore evident that reliability and economics play a major integrated role in the decision-making process. The principles of this process are discussed in *Engineering Systems*. The first step in this process is illustrated in Fig. 1.2, which shows how the reliability of a product or system is related to investment cost; i.e., increased investment is required in order to improve reliability. This clearly shows

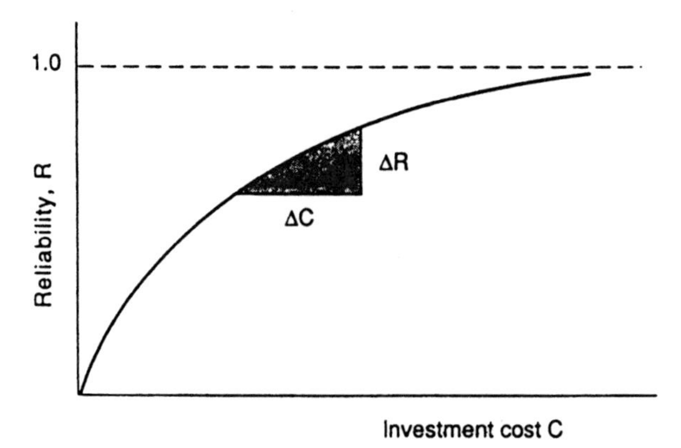

Fig. 1.2 Incremental cost of reliability

the general trend that the incremental cost ΔC to achieve a given increase in reliability ΔR increases as the reliability level increases, or, alternatively, a given increase in investment produces a decreasing increment in reliability as the reliability is increased. In either case, high reliability is expensive to achieve.

The incremental cost of reliability, $\Delta C/\Delta R$, shown in Fig. 1.2 is one way of deciding whether an investment in the system is worth it. However, it does not adequately reflect the benefits seen by the utility, the customer, or society. The two aspects of reliability and economics can be appraised more consistently by comparing reliability cost (the investment cost needed to achieve a certain level of reliability) with reliability worth (the benefit derived by the customer and society).

This extension of quantitative reliability analysis to the evaluation of service worth is a deceptively simple process fraught with potential misapplication. The basic concept of reliability-cost/reliability-worth evaluation is relatively simple and can be presented by the cost/reliability curves of Fig. 1.3. These curves show that the investment cost generally increases with higher reliability. On the other hand, the customer costs associated with failures decrease as the reliability increases. The total costs therefore are the sum of these two individual costs. This total cost exhibits a minimum, and so an "optimum" or target level of reliability is achieved. This concept is quite valid. Two difficulties arise in its assessment. First, the calculated indices are usually derived only from approximate models. Second, there are significant problems in assessing customer perceptions of system failure costs. A number of studies and surveys have been done including those conducted in Canada, United Kingdom, and Scandinavia. A review of these, together with a detailed discussion of the models and assessment techniques associated with reliability cost and worth evaluation, is the subject of Chapter 13.

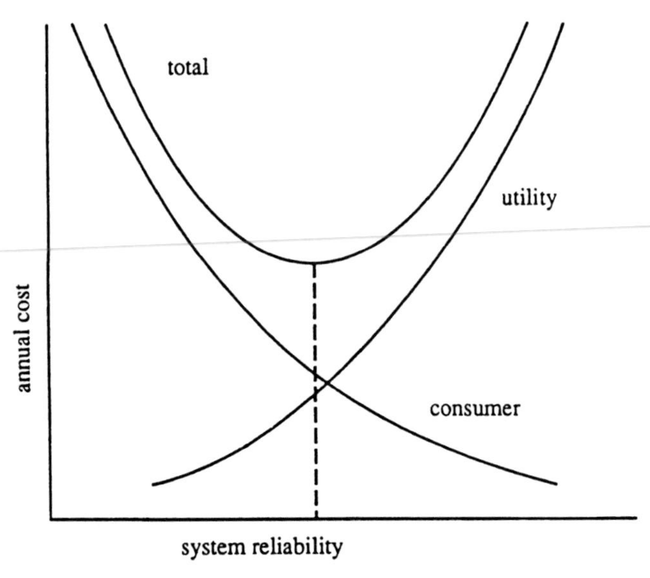

Fig. 1.3 Total reliability costs

1.10 Concepts of data

Meaningful reliability evaluation requires reasonable and acceptable data. These data are not always easy to obtain, and there is often a marked degree of uncertainty associated with the required input. This is one of the main reasons why relative assessments are more realistic than absolute ones. The concepts of data and the types of data needed for the analysis, modeling, and predictive assessments are discussed in Ref. 18. The following discussion is an overview of these concepts.

Although an unlimited amount of data can be collected, it is inefficient and undesirable to collect, analyze, and store more data than is required for the purpose intended. It is therefore essential to identify how and for what purposes it will be used. In deciding which data is needed, a utility must make its own decisions since no rigid rules can be predefined that are relevant to all utilities. The factors that must be identified are those that have an impact on the utility's own planning, design, and asset management policies.

The processing of this data occurs in two distinct stages. Field data is first obtained by documenting the details of failures as they occur and the various outage durations associated with these failures. This field data is then analyzed to create statistical indices. These indices are updated by the entry of subsequent new data. The quality of this data depends on two important factors: confidence and relevance. The quality of the data, and thus the confidence that can be placed in it, is clearly dependent on the accuracy and completeness of the compiled information.

It is therefore essential that the future use to which the data will be put and the importance it will play in later developments are stressed. The quality of the statistical indices is also dependent on how the data is processed, on how much pooling is done, and on the age of the data currently stored. These factors affect the relevance of the indices in their future use.

There is a wide range of data which can be collected and most utilities collect some, not usually all, of this data in one form or another. There are many different data collection schemes around the world, and a detailed review of some of these is presented in Ref. 18. It is worth indicating that, although considerable similarities exist between different schemes, particularly in terms of concepts, considerable differences also exist, particularly in the details of the individual schemes. It was also concluded that no one scheme could be said to be the "right" scheme, just that they are all different.

The review [18] also identified that there are two main bases for collecting data: the component approach and the unit approach. The latter is considered useful for assessing the chronological changes in reliability of existing systems but is less amenable to the predictive assessment of future system performance, the effect of various alternative reinforcements schemes, and the reliability characteristics of individual pieces of equipment. The component approach is preferable in these cases, and therefore data collected using this approach is more convenient for such applications.

1.11 Concluding comments

One point not considered in this book is how reliable the system and its various subsystems should be. This is a vitally important requirement and one which individual utilities must consider before deciding on any expansion or reinforcement scheme. It cannot be considered generally, however, because different systems, different utilities, and different customers all have differing requirements and expectations. Some of the factors which should be included in this decision-making consideration, however, are:

(a) There should be some conformity between the reliability of various parts of the system. It is pointless to reinforce quite arbitrarily a strong part of the system when weak areas still exist. Consequently a balance is required between generation, transmission, and distribution. This does not mean that the reliability of each should be equal; in fact, with present systems this is far from the case. Reasons for differing levels of reliability are justified, for example, because of the importance of a particular load, because generation and transmission failures can cause widespread outages while distribution failures are very localized.

(b) There should be some benefit gained by an improvement in reliability. The technique often utilized for assessing this benefit is to equate the incremental or marginal investment cost to the customer's incremental or marginal valuation of the improved reliability. The problem with such a method is the uncertainty in the customer's valuation. As discussed in Section 1.9, this problem is being actively studied. In the meantime it is important for individual utilities to arrive at some consistent criteria by which they can assess the benefits of expansion and reinforcement schemes.

It should be noted that the evaluation of system reliability cannot dictate the answer to the above requirements or others similar to them. These are managerial decisions. They cannot be answered at all, however, without the application of quantitative reliability analysis as this forms one of the most important input parameters to the decision-making process.

In conclusion, this book illustrates some methods by which the reliability of various parts of a power system can be evaluated and the types of indices that can be obtained. It does not purport to cover every known and available technique, as this would require a text of almost infinite length. It will, however, place the reader in a position to appreciate most of the problems and provide a wider and deeper appreciation of the material that has been published [6–11] and of that which will, no doubt, be published in the future.

1.12 References

1. Billinton, R., Allan, R. N., 'Power system reliability in perspective', *IEE J. Electronics Power*, **30** (1984), pp. 231–6.
2. Lyman, W. J., 'Fundamental consideration in preparing master system plan', *Electrical World*, **101**(24) (1933), pp. 788–92.
3. Smith, S. A., Jr., Spare capacity fixed by probabilities of outage. *Electrical World*, **103** (1934), pp. 222–5.
4. Benner, P. E., 'The use of theory of probability to determine spare capacity', *General Electric Review*, **37**(7) (1934), pp. 345–8.
5. Dean, S. M., 'Considerations involved in making system investments for improved service reliability', *EEI Bulletin*, **6** (1938), pp. 491–96.
6. Billinton, R., 'Bibliography on the application of probability methods in power system reliability evaluation', *IEEE Transactions*, **PAS-91** (1972), pp. 649–60.
7. IEEE Subcommittee Report, 'Bibliography on the application of probability methods in power system reliability evaluation, 1971–1977', *IEEE Transactions*, **PAS-97** (1978), pp. 2235–42.
8. Allan, R. N., Billinton, R., Lee, S. H., 'Bibliography on the application of probability methods in power system reliability evaluation, 1977–1982', *IEEE Transactions*, **PAS-103** (1984), pp. 275–82.

9. Allan, R. N., Billinton, R., Shahidehpour, S. M., Singh, C., 'Bibliography on the application of probability methods in power system reliability evaluation, 1982–1987', *IEEE Trans. Power Systems*, **3** (1988), pp. 1555–64.
10. Allan, R. N., Billinton, R., Briepohl, A. M., Grigg, C. H., 'Bibliography on the application of probability methods in power system reliability evaluation, 1987–1991', *IEEE Trans. Power Systems*, **PWRS-9**(1) (1994).
11. Billinton, R., Allan, R. N., Salvaderi, L. (eds.), *Applied Reliability Assessment in Electric Power Systems*, IEEE Press, New York (1991).
12. Calabrese, G., 'Generating reserve capability determined by the probability method', *AIEE Trans. Power Apparatus Systems*, **66** (1947), 1439–50.
13. Watchorn, C. W., 'The determination and allocation of the capacity benefits resulting from interconnecting two or more generating systems', *AIEE Trans. Power Apparatus Systems*, **69** (1950), pp. 1180–6.
14. Billinton, R., Li, W., *Reliability Assessment of Electric Power Systems Using Monte Carlo Methods*, Plenum Press, New York (1994).
15. EPRI Report, 'Composite system reliability evaluation: Phase 1—scoping study', Final Report, EPRI EL-5290, Dec. 1987.
16. Sugarman, R., 'New York City's blackout: a $350 million drain', *IEEE Spectrum Compendium, Power Failure Analysis and Prevention*, 1979, pp. 48–50.
17. Allan, R. N., Billinton, R., 'Probabilistic methods applied to electric power systems—are they worth it', *IEE Power Engineering Journal*, **May** (1992), 121–9.
18. CIGRE Working Group 38.03, '*Power System Reliability Analysis—Application Guide*, CIGRE Publications, Paris (1988).

2 Generating capacity—basic probability methods

2.1 Introduction

The determination of the required amount of system generating capacity to ensure an adequate supply is an important aspect of power system planning and operation. The total problem can be divided into two conceptually different areas designated as static and operating capacity requirements. The static capacity area relates to the long-term evaluation of this overall system requirement. The operating capacity area relates to the short-term evaluation of the actual capacity required to meet a given load level. Both these areas must be examined at the planning level in evaluating alternative facilities; however, once the decision has been made, the short-term requirement becomes an operating problem. The assessment of operating capacity reserves is illustrated in Chapter 5.

The static requirement can be considered as the installed capacity that must be planned and constructed in advance of the system requirements. The static reserve must be sufficient to provide for the overhaul of generating equipment, outages that are not planned or scheduled and load growth requirements in excess of the estimates. A practice that has developed over many years is to measure the adequacy of both the planned and installed capacity in terms of a percentage reserve. An important objection to the use of the percentage reserve requirement criterion is the tendency to compare the relative adequacy of capacity requirements provided for totally different systems on the basis of peak loads experienced over the same time period for each system. Large differences in capacity requirements to provide the same assurance of service continuity may be required in two different systems with peak loads of the same magnitude. This situation arises when the two systems being compared have different load characteristics and different types and sizes of installed or planned generating capacity.

The percentage reserve criterion also attaches no penalty to a unit because of size unless this quantity exceeds the total capacity reserve. The requirement that a reserve should be maintained equivalent to the capacity of the largest unit on the system plus a fixed percentage of the total system capacity is a more valid adequacy criterion and calls for larger reserve requirements with the addition of larger units to the system. This characteristic is usually found when probability techniques are used. The application of probability methods to the static capacity problem provides an analytical basis for capacity planning which can be extended to cover partial or complete integration of systems, capacity of interconnections, effects of unit size and design, effects of maintenance schedules and other system parameters. The economic aspects associated with different standards of reliability can be compared only by using probability techniques. Section 2.2.3 illustrates the inconsistencies which can arise when fixed criteria such as percentage reserves or loss of the largest unit are used in system capacity evaluation.

A large number of papers which apply probability techniques to generating capacity reliability evaluation have been published in the last 40 years. These publications have been documented in three comprehensive bibliographies published in 1966, 1971, and 1978 which include over 160 individual references [1–3]. The historical development of the techniques used at the present time is extremely interesting and although it is rather difficult to determine just when the first published material appeared, it was almost fifty years ago. Interest in the application of probability methods to the evaluation of capacity requirements became evident about 1933. The first large group of papers was published in 1947. These papers by Calabrese [4], Lyman [5], Seelye [6] and Loane and Watchorn [7] proposed the basic concepts upon which some of the methods in use at the present time are based. The 1947 group of papers proposed the methods which with some modifications are now generally known as the 'loss of load method', and the 'frequency and duration approach'.

Several excellent papers appeared each year until in 1958 a second large group of papers was published. This group of papers modified and extended the methods proposed by the 1947 group and also introduced a more sophisticated approach to the problem using 'game theory' or 'simulation' techniques [8–10]. Additional material in this area appeared in 1961 and 1962 but since that time interest in this approach appears to have declined.

A third group of significant papers was published in 1968/69 by Ringlee, Wood *et al.* [11–15]. These publications extended the frequency and duration approach by developing a recursive technique for model building. The basic concepts of frequency and duration evaluation are described in *Engineering Systems*.

It should not be assumed that the three groups of papers noted above are the only significant publications on this subject. This is not the case. They do, however, form the basis or starting point for many of the developments outlined in further work. Many other excellent papers have also been published and are listed in the three bibliographies [1–3] referred to earlier.

The fundamental difference between static and operating capacity evaluation is in the time period considered. There are therefore basic differences in the data used in each area of application. Reference [16] contains some fundamental definitions which are necessary for consistent and comprehensive generating unit reliability, availability, and productivity. At the present time it appears that the loss of load probability or expectation method is the most widely used probabilistic technique for evaluating the adequacy of a given generation configuration. There

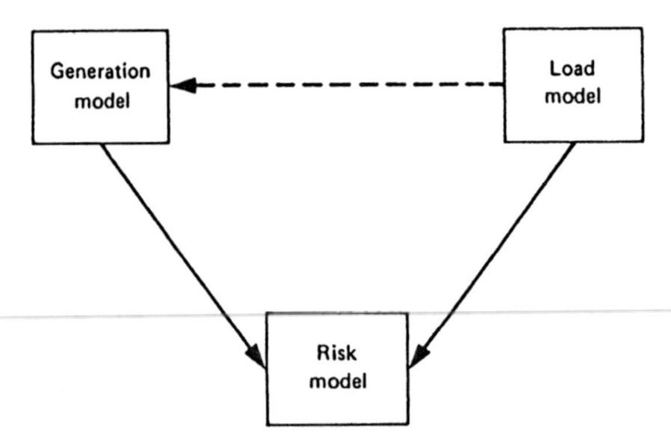

Fig. 2.1 Conceptual tasks in generating capacity reliability evaluation

are, however, many variations in the approach used and in the factors considered. The main elements are considered in this chapter. The loss of energy expectation can also be decided using a similar approach, and it is therefore also included in this chapter. Chapter 3 presents the basic concepts associated with the frequency and duration technique, and both the loss of load and frequency and duration methods are detailed in Chapter 4 which deals with interconnected system reliability evaluation.

The basic approach to evaluating the adequacy of a particular generation configuration is fundamentally the same for any technique. It consists of three parts as shown in Fig. 2.1.

The generation and load models shown in Fig. 2.1 are combined (convolved) to form the appropriate risk model. The calculated indices do not normally include transmission constraints, although it has been shown [39] how these constraints can be included, nor do they include transmission reliabilities; they are therefore overall system adequacy indices. The system representation in a conventional study is shown in Fig. 2.2.

The calculated indices in this case do not reflect generation deficiencies at any particular customer load point but measure the overall adequacy of the generation system. Specific load point evaluation is illustrated later in Chapter 6 under the designation of composite system reliability evaluation.

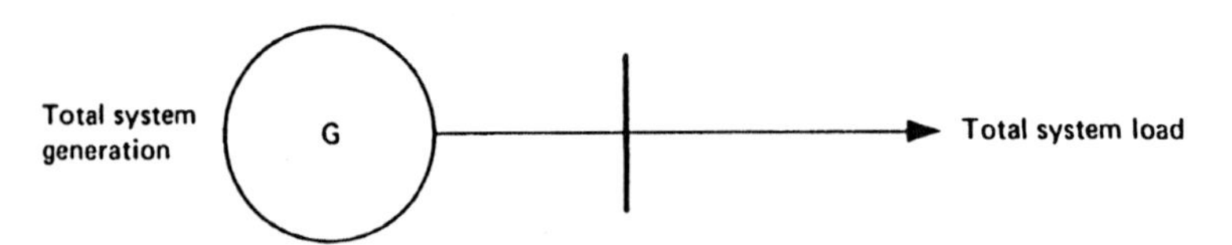

Fig. 2.2 Conventional system model

2.2 The generation system model

2.2.1 Generating unit unavailability

The basic generating unit parameter used in static capacity evaluation is the probability of finding the unit on forced outage at some distant time in the future. This probability was defined in *Engineering Systems* as the unit unavailability, and historically in power system applications it is known as the unit forced outage rate (FOR). It is not a rate in modern reliability terms as it is the ratio of two time values. As shown in Chapter 9 of *Engineering Systems*,

$$\text{Unavailability (FOR)} = U = \frac{\lambda}{\lambda + \mu} = \frac{r}{m + r} = \frac{r}{T} = \frac{f}{\mu}$$

$$= \frac{\Sigma[\text{down time}]}{\Sigma[\text{down time}] + \Sigma[\text{up time}]} \qquad 2.1(a)$$

$$\text{Availability} = A = \frac{\mu}{\lambda + \mu} = \frac{m}{m + r} = \frac{m}{T} = \frac{f}{\lambda}$$

$$= \frac{\Sigma[\text{up time}]}{\Sigma[\text{down time}] + \Sigma[\text{up time}]} \qquad 2.1(b)$$

where λ = expected failure rate
μ = expected repair rate
m = mean time to failure = MTTF = $1/\lambda$
r = mean time to repair = MTTR = $1/\mu$
$m + r$ = mean time between failures = MTBF = $1/f$
f = cycle frequency = $1/T$
T = cycle time = $1/f$.

The concepts of availability and unavailability as illustrated in Equations 2.1(a) and (b) are associated with the simple two-state model shown in Fig. 2.3(a). This model is directly applicable to a base load generating unit which is either operating or forced out of service. Scheduled outages must be considered separately as shown later in this chapter.

In the case of generating equipment with relatively long operating cycles, the unavailability (FOR) is an adequate estimator of the probability that the unit under similar conditions will not be available for service in the future. The formula does not, however, provide an adequate estimate when the demand cycle, as in the case of a peaking or intermittent operating unit, is relatively short. In addition to this, the most critical period in the operation of a unit is the start-up period, and in comparison with a base load unit, a peaking unit will have fewer operating hours and many more start-ups and shut-downs. These aspects must also be included in arriving at an estimate of unit unavailabilities at some time in the future. A working

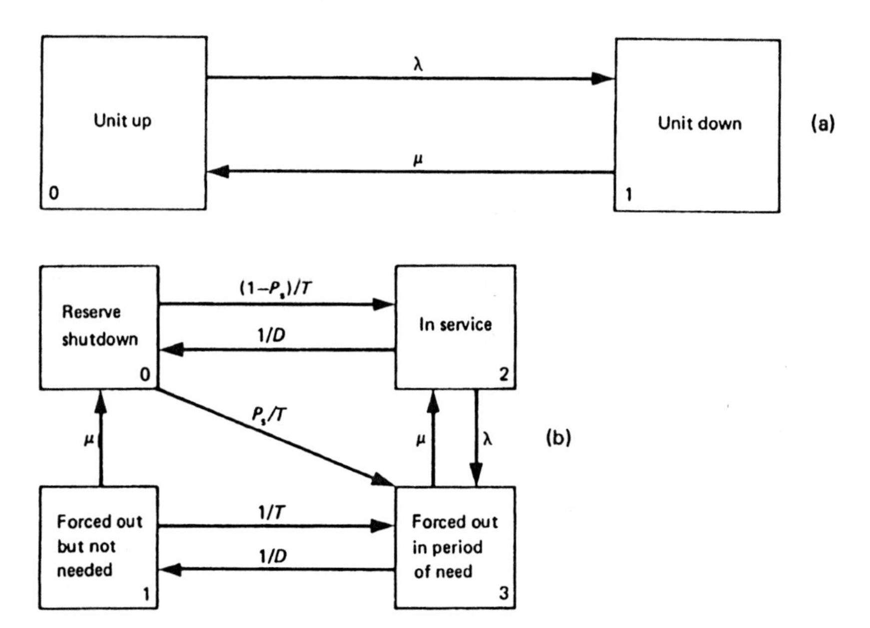

Fig. 2.3 (a) Two-state model for a base load unit
(b) Four-state model for planning studies
T Average reserve shut-down time between periods of need
D Average in-service time per occasion of demand
P_s Probability of starting failure

group of the IEEE Subcommittee on the Application of Probability Methods proposed the four-state model shown in Fig. 2.3(b) and developed an equation which permitted these factors to be considered while utilizing data collected under the conventional definitions [17].

The difference between Figs 2.3(a) and 2.3(b) is in the inclusion of the 'reserve shutdown' and 'forced out but not needed' states in Fig. 2.3(b). In the four-state model, the 'two-state' model is represented by States 2 and 3 and the two additional states are included to model the effect of the relatively short duty cycle. The failure to start condition is represented by the transition rate from State 0 to State 3.

This system can be represented as a Markov process and equations developed for the probabilities of residing in each state in terms of the state transition rates. These equations are as follows:

$$P_0 = \frac{\mu T[D\lambda + 1 + D(\mu + 1/T)]}{A}$$

where

$$A = (D\lambda + P_s)\left[(\mu T + 1) + \left(\mu + \frac{1}{T}\right)D\right] + \left[(1 - P_s) + D\left(\mu + \frac{1}{T}\right)\right](\mu(T + D))$$

$$P_1 = \frac{D\lambda + P_s}{A}$$

$$P_2 = \frac{D\mu(1 - P_s + \mu D + D/T)}{A}$$

$$P_3 = \frac{D(\mu + 1/T)(D\lambda + P_s)}{A}$$

$$\text{The conventional FOR} = \frac{P_1 + P_3}{P_1 + P_2 + P_3}$$

i.e. the 'reserve shutdown' state is eliminated.

In the case of an intermittently operated unit, the conditional probability that the unit will not be available given that a demand occurs is P, where

$$P = \frac{P_3}{P_2 + P_3}$$

$$P = \frac{(\mu + 1/T)(D\lambda + P_s)}{\mu[D(\mu + 1/T) + 1] + D\lambda(\mu + 1/T) + P_s/T}$$

The conditional forced outage rate P can therefore be found from the generic data shown in the model of Fig. 2.3(b). A convenient estimate of P can be made from the basic data for the unit.

Over a relatively long period of time,

$$\hat{P}_2 = \frac{\text{service time}}{\text{available time} + \text{forced outage time}} = \frac{\text{ST}}{\text{AT} + \text{FOT}}$$

$$(\hat{P}_1 + \hat{P}_3) = \frac{\text{FOT}}{\text{AT} + \text{FOT}}$$

Defining

$$f = \frac{P_3}{P_1 + P_3} = \frac{(\mu + 1/T)}{(1/D + \mu + 1/T)} = \frac{(1/r + 1/T)}{(1/D + 1/r + 1/T)}$$

where $r = 1/\mu$.

The conditional forced outage rate P can be expressed as

$$P = \frac{f(P_1 + P_3)}{P_2 + f(P_1 + P_3)} = \frac{f(\text{FOT})}{ST + f(\text{FOT})}$$

The factor f serves to weight the forced outage time FOT to reflect the time the unit was actually on forced outage when in demand by the system. The effect of this modification can be seen in the following example, taken from Reference [17].

Average unit data

$$\begin{aligned}\text{Service time . ST} &= 640.73 \text{ hours}\\ \text{Available time} &= 6403.54 \text{ hours}\\ \text{No. of starts} &= 38.07\\ \text{No. of outages} &= 3.87\\ \text{Forced outage time FOT} &= 205.03 \text{ hours}\end{aligned}$$

Assume that the starting failure probability $P_s = 0$

$$\hat{D} = \frac{640.73}{38.07} = 16.8 \text{ hours}$$

$$\hat{D} + \hat{T} = \frac{6403.54}{38.07} = 168 \text{ hours}$$

$$\hat{r} = \frac{205.03}{3.87} = 53 \text{ hours}$$

$$\hat{m} = \frac{640.73}{3.87} = 166 \text{ hours}$$

Using these values

$$f = \left(\frac{1}{53} + \frac{1}{155.2}\right) \Big/ \left(\frac{1}{16.8} + \frac{1}{53} + \frac{1}{151.2}\right) = 0.3$$

$$\text{The conventional forced outage rate } = \frac{205.03}{640.73 + 205.03} \times 100$$

$$= \underline{24.24\%}$$

$$\text{The conditional probability } P = \frac{0.3(205.03)}{640.73 + 0.3(205.03)} \times 100$$

$$= \underline{8.76\%}$$

The conditional probability P is clearly dependent on the demand placed upon the unit. The demand placed upon it in the past may not be the same as the demand which may exist in the future, particularly under conditions of generation system inadequacy. It has been suggested [18] that the demand should be determined from the load model as the capacity table is created sequentially, and the conditional probability then determined prior to adding the unit to the capacity model.

2.2.2 Capacity outage probability tables

The generation model required in the loss of load approach is sometimes known as a capacity outage probability table. As the name suggests, it is a simple array of capacity levels and the associated probabilities of existence. If all the units in the system are identical, the capacity outage probability table can be easily obtained using the binomial distribution as described in Sections 3.3.7 and 3.3.8 of *Engineering Systems*. It is extremely unlikely, however, that all the units in a practical

Table 2.1

Capacity out of service	*Probability*
0 MW	0.9604
3 MW	0.0392
6 MW	0.0004
	1.0000

system will be identical, and therefore the binomial distribution has limited application. The units can be combined using basic probability concepts and this approach can be extended to a simple but powerful recursive technique in which units are added sequentially to produce the final model. These concepts can be illustrated by a simple numerical example.

A system consists of two 3 MW units and one 5 MW unit with forced outage rates of 0.02. The two identical units can be combined to give the capacity outage probability table shown as Table 2.1.

The 5 MW generating unit can be added to this table by considering that it can exist in two states. It can be in service with probability $1-0.02=0.98$ or it can be out of service with probability 0.02. The two resulting tables (Tables 2.2, 2.3) are therefore conditional upon the assumed states of the unit. This approach can be extended to any number of unit states.

The two tables can now be combined and re-ordered (Table 2.4). The probability value in the table is the probability of exactly the indicated amount of capacity being out of service. An additional column can be added which gives the cumulative probability. This is the probability of finding a quantity of capacity on outage equal to or greater than the indicated amount.

The cumulative probability values decrease as the capacity on outage increases: Although this is not completely true for the individual probability table, the same general trend is followed. For instance, in the above table the probability of losing 8 MW is higher than the probability of losing 6 MW. In each case only two units are involved. The difference is due to the fact that in the 8 MW case, the 3 MW loss contribution can occur in two ways. In a practical system the probability of having a large quantity of capacity forced out of service is usually quite small,

Table 2.2 5 MW unit in service

Capacity out	*Probability*
0 + 0 = 0 MW	(0.9604) (0.98) = 0.941192
3 + 0 = 3 MW	(0.0392) (0.98) = 0.038416
6 + 0 = 6 MW	(0.0004) (0.98) = 0.000392
	0.980000

Table 2.3 5 MW unit out of service

Capacity out	*Probability*
0 + 5 = 5 MW	(0.9604) (0.02) = 0.019208
3 + 5 = 8 MW	(0.0392) (0.02) = 0.000784
6 + 5 = 11 MW	(0.0004) (0.02) = 0.000008
	0.020000

as this condition requires the outage of several units. Theoretically the capacity outage probability table incorporates all the system capacity. The table can be truncated by omitting all capacity outages for which the cumulative probability is less than a specified amount, e.g. 10^{-8}. This also results in a considerable saving in computer time as the table is truncated progressively with each unit addition. The capacity outage probabilities can be summated as units are added, or calculated directly as cumulative values and therefore no error need result from the truncation process. This is illustrated in Section 2.2.4. In a practical system containing a large number of units of different capacities, the table will contain several hundred possible discrete capacity outage levels. This number can be reduced by grouping the units into identical capacity groups prior to combining or by rounding the table to discrete levels after combining. Unit grouping prior to building the table introduces unnecessary approximations which can be avoided by the table rounding approach. The capacity rounding increment used depends upon the accuracy desired. The final rounded table contains capacity outage magnitudes that are multiples of the rounding increment. The number of capacity levels decreases as the rounding increment increases, with a corresponding decrease in accuracy. The procedure for rounding a table is shown in the following example.

Two 3 MW units and one 5 MW unit with forced outage rates of 0.02 were combined to form the generation model shown in Table 2.4. This table, when

Table 2.4 Capacity outage probability table for the three-unit system

Capacity out of service	*Individual probability*	*Cumulative probability*
0	0.941192	1.000000
3	0.038416	0.058808
5	0.019208	0.020392
6	0.000392	0.001184
8	0.000784	0.000792
11	0.000008	0.000008
	1.000000	

Table 2.5

Capacity on outage (MW)	*Individual probability*	
0	$0.941192 + \frac{2}{5}(0.038416)$	$= 0.9565584$
5	$0.019208 + \frac{3}{5}(0.038416)$	
	$+ \frac{4}{5}(0.000392) + \frac{2}{5}(0.000784)$	$= 0.0428848$
10	$\frac{1}{5}(0.000392) + \frac{3}{5}(0.000784)$	
	$+ \frac{4}{5}(0.000008)$	$= 0.0005552$
15	$\frac{1}{5}(0.000008)$	$= \underline{0.0000016}$
		1.0000000

rounded at 5 MW increments, will contain only capacity outage magnitudes of 0, 5, 10 and 15 MW. The rounded table is obtained as shown in Table 2.5.

The general expression for this rounding process is

$$P(C_j) = \frac{C_k - C_i}{C_k - C_j} P(C_i)$$

$$P(C_k) = \frac{C_i - C_j}{C_k - C_j} P(C_i)$$

for all states i falling between the required rounding states j and k.

The use of a rounded table in combination with the load model to calculate the risk level introduces certain inaccuracies. The error depends upon the rounding increment used and on the slope of the load characteristic. The error decreases with increasing slope of the load characteristic and for a given load characteristic the error increases with increased rounding increment. The rounding increment used should be related to the system size and composition. Also the first non-zero capacity-on-outage state should not be less than the capacity of the smallest unit.

The generation system model in the form shown in Table 2.4 can be used to illustrate the basic inadequacies of the conventional deterministic approaches to capacity evaluation.

2.2.3 Comparison of deterministic and probabilistic criteria

It was noted in Section 2.1 that deterministic risk criteria such as 'percentage reserve' and 'loss of largest unit' do not define consistently the true risk in the system. In order to illustrate this objectively, consider the following four systems:

—system 1, 24 × 10 MW units each having a FOR of 0.01
—system 2, 12 × 20 MW units each having a FOR of 0.01
—system 3, 12 × 20 MW units each having a FOR of 0.03
—system 4, 22 × 10 MW units each having a FOR of 0.01

All four systems are very similar but not identical. In each system, the units are identical and therefore the capacity outage probability table can be easily constructed using the binomial distribution. These arrays are shown in Table 2.6

Table 2.6 Capacity Outage Probability Tables for systems 1–4

System 1 Capacity (MW)		*Probability*	
Out	*In*	*Individual*	*Cumulative*
0	240	0.785678	1.000000
10	230	0.190467	0.214322
20	220	0.022125	0.023855
30	210	0.001639	0.001730
40	200	0.000087	0.000091
50	190	0.000004	0.000004
System 2 Capacity (MW)		*Probability*	
Out	*In*	*Individual*	*Cumulative*
0	240	0.886384	1.000000
20	220	0.107441	0.113616
40	200	0.005969	0.006175
60	180	0.000201	0.000206
80	160	0.000005	0.000005
System 3 Capacity (MW)		*Probability*	
Out	*In*	*Individual*	*Cumulative*
0	240	0.693841	1.000000
20	220	0.257509	0.306159
40	200	0.043803	0.048650
60	180	0.004516	0.004847
80	160	0.000314	0.000331
100	140	0.000016	0.000017
120	120	0.000001	0.000001
System 4 Capacity (MW)		*Probability*	
Out	*In*	*Individual*	*Cumulative*
0	220	0.801631	1.000000
10	210	0.178140	0.198369
20	200	0.018894	0.020229
30	190	0.001272	0.001335
40	180	0.000061	0.000063
50	170	0.000002	0.000002

and have been truncated to a cumulative probability of 10^{-6}. It can be seen that a considerable number of capacity outage states have been deleted using this truncation technique.

The load level or demand on the system is assumed to be constant. If the risk in the system is defined as the probability of not meeting the load, then the true risk in the system is given by the value of cumulative probability corresponding to the outage state one increment below that which satisfies the load on the system. The two deterministic risk criteria can now be compared with this probabilistic risk as in Sections (a) and (b) following.

(a) *Percentage reserve margin*

Assume that the expected load demands in systems 1, 2, 3 and 4 are 200, 200, 200 and 183 MW respectively. The installed capacity in all four cases is such that there is a 20% reserve margin, i.e. a constant for all four systems. The probabilistic or true risks in each of the four systems can be found from Table 2.6 and are:

risk in system 1 = 0.000004
risk in system 2 = 0.000206
risk in system 3 = 0.004847
risk in system 4 = 0.000063

These values of risk show that the true risk in system 3 is 1000 times greater than that in system 1. A detailed analysis of the four systems will show that the variation in true risk depends upon the forced outage rate, number of units and load demand. The percentage reserve method cannot account for these factors and therefore, although using a 'constant' risk criterion, does not give a consistent risk assessment of the system.

(b) *Largest unit reserve*

Assume now that the expected load demands in systems 1, 2, 3 and 4 are 230, 220, 220 and 210 MW respectively. The installed capacity in all four cases is such that the reserve is equal to the largest unit which again is a constant for all the systems. In this case the probabilistic risks are:

risk in system 1 = 0.023855
risk in system 2 = 0.006175
risk in system 3 = 0.048650
risk in system 4 = 0.020229

The variation in risk is much smaller in this case, which gives some credence to the criterion. The ratio between the smallest and greatest risk levels is now 8:1 and the risk merit order has changed from system 3–2–4–1 in the case of 'percentage reserve' to 3–1–4–2 in the case of the 'largest unit' criterion.

It is seen from these comparisons that the use of deterministic or 'rule-of-thumb' criteria can lead to very divergent probabilistic risks even for systems that are very similar. They are therefore inconsistent, unreliable and subjective methods for reserve margin planning.

2.2.4 A recursive algorithm for capacity model building

The capacity model can be created using a simple algorithm which can also be used to remove a unit from the model [19]. This approach can also be used for a multi-state unit, i.e. a unit which can exist in one or more derated or partial output states as well as in the fully up and fully down states. The technique is illustrated for a two-state unit addition followed by the more general case of a multi-state unit.

Case 1 No derated states

The cumulative probability of a particular capacity outage state of X MW after a unit of capacity C MW and forced outage rate U is added is given by

$$P(X) = (1 - U)P'(X) + (U)P'(X - C) \tag{2.2}$$

where $P'(X)$ and $P(X)$ denote the cumulative probabilities of the capacity outage state of X MW before and after the unit is added. The above expression is initialized by setting $P'(X) = 1.0$ for $X \leq 0$ and $P'(X) = 0$ otherwise.

Equation (2.2) is illustrated using the simple system shown in Table 2.7. Each unit in Table 2.7 has an availability and unavailability of 0.98 and 0.02 respectively (Equation 2.1).

The system capacity outage probability is created sequentially as follows:

Step 1 Add the first unit

$$\begin{aligned}
P(0) &= (1-0.02)(1.0) + (0.02)(1.0) = 1.0\\
P(25) &= (1-0.02)(0) + (0.02)(1.0) = 0.02
\end{aligned}$$

Step 2 Add the second unit

$$\begin{aligned}
P(0) &= (1-0.02)(1.0) + (0.02)(1.0) = 1.0\\
P(25) &= (1-0.02)(0.02) + (0.02)(1.0) = 0.0396\\
P(50) &= (1-0.02)(0) + (0.02)(0.02) = 0.0004
\end{aligned}$$

Step 3 Add the third unit

$$\begin{aligned}
P(0) &= (1-0.02)(1.0) + (0.02)(1.0) = 1.0\\
P(25) &= (1-0.02)(0.0396) + (0.02)(1.0) = 0.058808\\
P(50) &= (1-0.02)(0.0004) + (0.02)(1.0) = 0.020392\\
P(75) &= (1-0.02)(0) + (0.02)(0.0396) = 0.000792\\
P(100) &= (1-0.02)(0) + (0.02)(0.0004) = 0.000008
\end{aligned}$$

The reader should utilize this approach to obtain Table 2.4.

Table 2.7 System data

Unit no.	*Capacity (MW)*	*Failure rate (f / day)*	*Repair rate (r / day)*
1	25	0.01	0.49
2	25	0.01	0.49
3	50	0.01	0.49

Table 2.8 50 MW unit—three-state representation

State	*Capacity out*	*State probability (p_i)*
1	0	0.960
2	20	0.033
3	50	0.007

Case 2 Derated states included

Equation (2.2) can be modified as follows to include multi-state unit representations.

$$P(X) = \sum_{i=1}^{n} p_i P'(X - C_i) \tag{2.3}$$

where n = number of unit states
C_i = capacity outage of state i for the unit being added
p_i = probability of existence of the unit state i.
when n = 2, Equation (2.3) reduces to Equation (2.2).

Equation (2.3) is illustrated using the 50 MW unit representation shown in Table 2.8.

If the two-state 50 MW unit in the previous example is replaced by the three-state unit shown in Table 2.8, Step 3 becomes

$$P(0) = (0.96)(1.0) + (0.033)(1.0) + (0.007)(1.0) = 1.0$$
$$P(20) = (0.96)(0.0396) + (0.033)(1.0) + (0.007)(1.0) = 0.078016$$
$$P(25) = (0.96)(0.0396) + (0.033)(0.0396) + (0.007)(1.0) = 0.0463228$$
$$P(45) = (0.96)(0.0004) + (0.033)(0.0396) + (0.007)(1.0) = 0.0086908$$
$$P(50) = (0.96)(0.0004) + (0.033)(0.0004) + (0.007)(1.0) = 0.0073972$$
$$P(70) = (0.96)(0) + (0.033)(0.0004) + (0.007)(0.0396) = 0.0002904$$
$$P(75) = (0.96)(0) + (0.033)(0) + (0.007)(0.0396) = 0.0002772$$
$$P(100) = (0.96)(0) + (0.033)(0) + (0.007)(0.0004) = 0.0000028$$

2.2.5 Recursive algorithm for unit removal

Generating units are periodically scheduled for unit overhaul and preventive maintenance. During these scheduled outages, the unit is available neither for service nor for failure. This situation requires a new capacity model which does not include the unit on scheduled outage. The new model could be created by simply building it from the beginning using Equations (2.2) or (2.3). In the case of a large system this requires considerable computer time if there are a number of discrete periods to consider. Equations (2.2) and (2.3) can be used in reverse, however, to find the capacity model after unit removal.

Consider Equation (2.2):

$$P(X) = (1 - U)P'(X) + (U)P'(X - C) \tag{2.2}$$

$$P'(X) = \frac{P(X) - (U)P'(X - C)}{(1 - U)} \tag{2.4}$$

In Equation (2.4) $P'(X - C) = 1.0$ for $X \le C$. This procedure can be illustrated using the example of case 1 in Section 2.2.4. The 50 MW unit is removed from the capacity outage probability table as follows:

$$P(0) = \frac{(1.0) - (0.02)(1.0)}{0.98} = 1.0$$

$$P(25) = \frac{(0.058808) - (0.02)(1.0)}{0.98} = 0.0396$$

$$P(50) = \frac{(0.020392) - (0.02)(1.0)}{0.98} = 0.0004$$

This is the capacity model shown in Step 2. The reader can remove a 25 MW unit to obtain the values in Step 1.

The equation for removal of a multi-state unit is obtained from Equation (2.3):

$$P(X) = \sum_{i=1}^{n} p_i P'(X - C_i) \tag{2.3}$$

$$P'(X) = \frac{P(X) - \Sigma_{i=2}^{n} p_i P'(X - C_i)}{p_1} \tag{2.5}$$

It is left to the reader to apply Equation (2.5) to the previous case in which the unit shown in Table 2.8 was added to the two 25 MW units. The direct application of Equation (2.5) requires that all the derated states and full outage state of the unit being removed be multiples of the rounding increment used in the capacity outage probability table. In practice, the derated states chosen to model the unit are not the entire set of derated states but a selected representative set of states. It is therefore logical to make the derated states identical to a multiple of the rounding increment. The total capacity of the unit may also not be a multiple of the rounding increment. In this case the removal can be accomplished by removing separately from the existing table two hypothetical units, one having a capacity less than and the other having a capacity greater than the unit to be removed, both being equal to a multiple of the rounding increment. This produces two individual tables which can then be averaged to form the required table.

2.2.6 Alternative model-building techniques

Generating system capacity outages have a discrete distribution and their probabilities are normally evaluated using the well-known recursive technique previously described. It is found that if the system is very large the discrete distribution of system capacity outages can be approximated by a continuous distribution [20]. Such a distribution approaches the normal distribution as the system size increases. If the assumption is made that the distribution of capacity on forced outage is a normal distribution, then the development of the capacity outage probability table is relatively simple. A single entry in the table can be obtained using only the mean and variance of the distribution. The results obtained using this continuous model of system capacity outages are found [37] to be not sufficiently accurate when compared to those obtained using the recursive technique. Schenk and Rau [21] have therefore proposed a Fourier transform method based on the Gram–Charlier expansion of a distribution to improve the accuracy of the continuous model. The complete mathematical description of the proposed method is given in Reference [21]. The step-by-step procedure is summarized as follows.

Let

$C_i =$ capacity of unit i in MW
$q_i =$ forced outage rate of unit i
$n =$ number of generating units

Step 1 Calculate the following quantities for each unit in the system.

$$m_1(i) = C_i q_i$$

$$m_2(i) = C_i^2 q_i$$

$$m_3(i) = C_i^3 q_i$$

$$m_4(i) = C_i^4 q_i$$

$$V_i^2 = m_2(i) - m_1^2(i)$$

$$M_3(i) = m_3(i) - 3m_1(i)m_2(i) + 2m_1^3(i)$$

$$M_4(i) = m_4(i) - 4m_1(i)m_3(i) + 6m_1^2(i)m_2(i) - 2m_1^4(i)$$

Step 2 From the results of Step 1, calculate the following parameters.

$$M = \sum_{i=1}^{n} m_1(i)$$

$$V^2 = \sum_{i=1}^{n} V_i^2$$

$$M_3 = \sum_{i=1}^{n} M_3(i)$$

$$M_4 = \sum_{i=1}^{n} (M_4(i) - 3V_i^4) + 3V^4$$

$$G_1 = M_3/V^3$$

$$G_2 = (M_4/V^4) - 3$$

Step 3 From the results of Step 2 and for any desired capacity outage of x MW, calculate

$$Z_1 = \frac{x - M}{V}$$

$$Z_2 = \frac{x + M}{V}$$

According to the numerical value of Z_2, three cases are considered.

Case 1 If $Z_2 \leq 2.0$

Calculate two areas, Area 1 and Area 2, under the standard normal density function either from tables for the standard Gaussian distribution $N(Z)$ or from the equations given in Section 6.7.3 of *Engineering Systems*. The normal density function can be expressed as

$$N(Z) = \frac{1}{\sqrt{(2\pi)}} e^{-\frac{1}{2}Z^2}, \quad -\infty < Z < \infty$$

and the two areas are defined as

$$\text{Area 1} = \int_{Z_1}^{\infty} N(Z)\, dZ$$

$$\text{Area 2} = \int_{-\infty}^{-Z_2} N(Z)\, dZ = \int_{Z_2}^{\infty} N(Z)\, dZ$$

The probability of a capacity outage of x MW or more is given by

Prob[capacity outage $\geq x$ MW] = Area 1 + Area 2

Case 2 ***If*** **$2 < Z_2 \leq 5.0$**

Calculate Area 1 and Area 2 as in Case 1. In addition, calculate the following expressions

$$N^{(2)}(Z_i) = (Z_i^2 - 1)N(Z_i)$$

$$N^{(3)}(Z_i) = (-Z_i^3 + 3Z_i)N(Z_i)$$

$$N^{(5)}(Z_i) = -Z_i^5 + 10Z_i^3 - 15Z_i)N(Z_i)$$

$$K_i = G_1 - \tfrac{1}{6}N^{(2)}(Z_i) - \frac{G_2}{24}N^{(3)}(Z_i) - \frac{G_1^2}{72}N^{(5)}(Z_i)$$

where i takes on values of 1 and 2.

The probability of a capacity outage of x MW or more is given by

Prob[capacity outage $\geq x$ MW] = Area 1 + Area 2+ $K_1 + K_2$

Case 3 ***If*** $Z_2 > 5.0$

For this case only Area 1 of Case 1 is used as well as K_1 of Case 2. Area 2 and K_2 can be neglected since their numerical values are very small in this range. The required probability for a given x MW is given by

Prob[capacity outage $\geq x$ MW] = Area 1 + K_1

The technique described above for developing a capacity outage probability table of a given system has utilized the two-state representation of a generating unit. In situations in which a system has some derated units, the step-by-step procedure is still applicable with the first four expressions in Step 1 taking the form

$$m_1(i) = c_i q_i + \sum_{k=1}^{r} c_{ik} q_{ik}$$

$$m_2(i) = c_i^2 q_i + \sum_{k=1}^{r} c_{ik}^2 q_{ik}$$

$$m_3(i) = c_i^3 q_i + \sum_{k=1}^{r} c_{ik}^3 q_{ik}$$

$$m_4(i) = c_i^4 q_i + \sum_{k=1}^{r} c_{ik}^4 q_{ik}$$

where

q_i = FOR for a full capacity outage
q_{ik} = FORs for partial capacity states
c_{ik} = capacities of partial capacity states
r = number of derated states.

When a unit is added or removed from a capacity outage probability table, the new table can be developed using the same procedure after the new parameters M, V, G_1 and G_2 are obtained.

The Fourier transform method for developing the capacity outage probability table is illustrated using the five generator system given in Table 2.11. The numerical results are as follows.

Step 1 The quantities associated with this step are evaluated once, since the units are identical and the values are given below

m_1	m_2	m_3	m_4	V^2	M_3	M_4
0.40	16.0	640.0	25600.0	15.84	620.928	24591.3088

Step 2 The values of the different parameters associated with this step are shown below

M	V^2	M_3	M_4	G_1	G_2
2.0	79.2	3104.64	138010.88	4.4047724	19.0020406

Step 3 For a capacity outage of 40 MW, the values of Z_1 and Z_2 become 4.269932 and 4.719399 respectively. Since $2 < Z_2 \leq 5.0$, Case 2 applies. The values associated with the different expressions in this step are given below.

Area 1	$N(Z_1)$	$N^{(2)}(Z_1)$
0.9774×10^{-5}	0.43834×10^{-4}	0.7553615×10^{-3}

$N^{(3)}(Z_1)$	$N^{(5)}(Z_1)$	K_1
$-0.2851005 \times 10^{-2}$	-0.0309004	0.0111385

Area 2	$N(Z_2)$	$N^{(2)}(Z_2)$
0.1179×10^{-5}	0.58136×10^{-5}	0.123671×10^{-3}

$N^{(3)}(Z_2)$	$N^{(5)}(Z_2)$	K_2
-0.52878×10^{-3}	-0.791129×10^{-2}	0.0026413

Hence, the probability of a capacity outage of 40 MW or more is given by

$$\text{Prob[capacity outage} \geq 40 \text{ MW]} = \text{Area 1} + \text{Area 2} + K_1 + K_2$$

$$= 0.0137908$$

Note that if the normal distribution [20] is used to approximate the discrete distribution of system capacity outages, the values are much lower than those obtained by the Fourier transform method. The value, for example, of the probability of a capacity outage of 40 MW or more was found to be 09774×10^{-5} (Area 1). The cumulative probabilities associated with the rest of the capacity outage states can be similarly obtained using the step-by-step procedure. The results obtained by the recursive and Fourier transform methods are shown in Table 2.9.

It can be seen from Table 2.9 that the values obtained using the Fourier transform method are quite different from those obtained using the recursive technique. This is due to the fact that the system under study has a very small number

Table 2.9

Capacity on outage (MW)	Cumulative probability	
	Recursive method	*Fourier transform method*
0.0	1.0	1.0
40.0	0.049009	0.0137883
80.0	0.9800×10^{-3}	0.105844×10^{-12}
120.0	0.900×10^{-5}	$0.2821629 \times 10^{-33}$

of generators. The main intention here is to illustrate the method. The accuracy of the Fourier transform method is to be compared with the recursive technique only when the system is sufficiently large. A comparison has been made for the IEEE–RTS and the results are shown in Appendix 2.

The Fourier transform method is efficient and easy to apply. It provides accurate results when compared with the basic recursive approach in systems with a large number of generating units and particularly when these units have relatively large forced outage rates [22]. It is therefore suited to systems containing a large number of fossil fired units. It can be quite inaccurate at certain outage levels in systems containing hydro units which have relatively low forced outage rates [23].

An alternative approach [24] is to transform the unit capacity tables into the frequency domain using fast Fourier transforms (FFT) and to convolve using a point by point multiplication. An inverse FFT algorithm can then be used to produce the final capacity outage probability table. This method, although not as fast as the previously described Fourier transform method, can be considerably faster than the direct recursive method. On the other hand, because it models the true discrete nature of the generating units, it does not suffer the significant disadvantages of the Fourier transform method and can be applied to both large and small systems alike with no loss of accuracy.

2.3 Loss of load indices

2.3.1 Concepts and evaluation techniques

The generation system model illustrated in the previous section can be convolved with an appropriate load model to produce a system risk index. There are a number of possible load models which can be used and therefore there are a number of risk indices which can be produced. The simplest load model and one that is used quite extensively is one in which each day is represented by its daily peak load. The individual daily peak loads can be arranged in descending order to form a cumulative load model which is known as the daily peak load variation curve. The resultant model is known as the load duration curve when the individual hourly load values are used, and in this case the area under the curve represents the energy required in the given period. This is not the case with the daily peak load variation curve.

In this approach, the applicable system capacity outage probability table is combined with the system load characteristic to give an expected risk of loss of load. The units are in days if the daily peak load variation curve is used and in hours if the load duration curve is used. Prior to combining the outage probability table it should be realized that there is a difference between the terms 'capacity outage' and 'loss of load'. The term 'capacity outage' indicates a loss of generation which may or may not result in a loss of load. This condition depends upon the generating capacity reserve margin and the system load level. A 'loss of load' will occur only when the capability of the generating capacity remaining in service is exceeded by the system load level.

The individual daily peak loads can be used in conjunction with the capacity outage probability table to obtain the expected number of days in the specified period in which the daily peak load will exceed the available capacity. The index in this case is designated as the loss of load expectation (LOLE).

$$\text{LOLE} = \sum_{i=1}^{n} P_i(C_i - L_i) \quad \text{days/period} \tag{2.6}$$

where C_i = available capacity on day i.

L_i = forecast peak load on day i.

$P_i(C_i - L_i)$ = probability of loss of load on day i. This value is obtained directly from the capacity outage cumulative probability table.

This procedure is illustrated using the 100 MW system shown in Table 2.7. The load data for a period of 365 days is shown in Table 2.10.

Using Equation (2.6),

$$\begin{aligned}\text{LOLE} &= 12P(100-57) + 83P(100-52) + 107P(100-46) \\ &\quad + 116P(100-41) + 47P(100-34) \\ &= 12(0.020392) + 83(0.020392) + 107(0.000792) \\ &\quad + 116(0.000792) + 47(0.000792) \\ &= \underline{2.15108 \text{ days/year.}}\end{aligned}$$

Table 2.10 Load data used to evaluate LOLE

Daily peak load (MW)	57	52	46	41	34
No. of occurrences	12	83	107	116	47

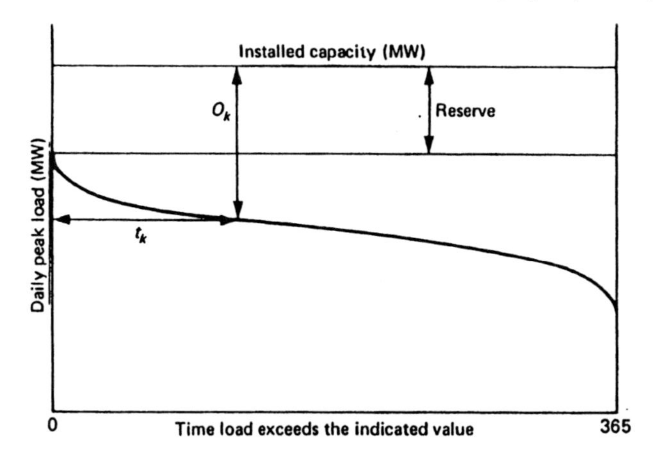

Fig. 2.4 Relationship between load, capacity and reserve
Q_k Magnitude of the kth outage in the system capacity outage probability table
t_k Number of time units in the study interval that an outage magnitude of O_k would result in a loss of load

The same LOLE index can also be obtained using the daily peak load variation curve. Figure 2.4 shows a typical system load–capacity relationship where the load model is shown as a continuous curve for a period of 365 days. A particular capacity outage will contribute to the system LOLE by an amount equal to the product of the probability of existence of the particular outage and the number of time units in the study interval that loss of load would occur if such a capacity outage were to exist. It can be seen from Fig. 2.4 that any capacity outage less than the reserve will not contribute to the system LOLE. Outages of capacity in excess of the reserve will result in varying numbers of time units during which loss of load could occur. Expressed mathematically, the contribution to the system LOLE made by capacity outage O_k is $p_k t_k$ time units where p_k is the individual probability of the capacity outage O_k. The total LOLE for the study interval is

$$\text{LOLE} = \sum_{k=1}^{n} p_k t_k \quad \text{time units} \tag{2.7}$$

The p_k values in Equation (2.7) are the individual probabilities associated with the capacity outage states. The equation can be modified to use the cumulative state probabilities. In this case

$$\text{LOLE} = \sum_{k=1}^{n} (t_k - t_{k-1}) P_k \tag{2.8}$$

Note P_k = cumulative outage probability for capacity state O_k.

If the load characteristic in Fig. 2.4 is the load duration curve, the value of LOLE is in hours. If a daily peak load variation curve is used, the LOLE is in days for the period of study.

The period of study could be a week, a month or a year. The simplest application is the use of the curve on a yearly basis. If no generating unit maintenance were performed, the capacity outage probability table would be valid for the entire period.

The effect of unit maintenance is discussed in Section 2.6. When using a daily peak load variation curve on an annual basis, the LOLE is in days per year. The reciprocal of this value in years per day is often quoted as a reliability index. The use of this reciprocal value has led to considerable confusion, particularly among people who are not aware of the true meaning. The days/year result is simply a mathematical expectation of load loss in time units for the period under study which indicates the average number of days during which a loss of load will be encountered. It must be stressed that it has neither a frequency nor duration connotation.

2.3.2 Numerical examples

(a) *Basic study*

The application of Equations (2.7) and (2.8) can be illustrated by a simple numerical example.

Consider a system containing five 40 MW units each with a forced outage rate of 0.01. The capacity outage probability table for this system is shown in Table 2.11.

Probability values less than 10^{-6} have been neglected. The system load model is represented by the daily peak load variation curve shown in Fig. 2.5. This is assumed to be linear in order to simplify hand calculations, although such a linear representation is not likely to occur in practice.

The study period in this case is assumed to be a year and therefore 100% on the abscissa corresponds to 365 days. In many studies, weekends and holidays are neglected as their contribution to the LOLE is negligible. The time span is then

Table 2.11 Generation model for the five-unit system. System installed capacity = 200 MW

Capacity out of service	*Individual probability*	*Cumulative probability*
0	0.950991	1.000000
40	0.048029	0.049009
80	0.000971	0.000980
120	0.000009	0 000009
	1.000000	

Table 2.12 LOLE using individual probabilities

Capacity out of service (MW)	*Capacity in service (MW)*	*Individual probability*	*Total time t_k (%)*	*LOLE*
0	200	0.950991	0	—
40	160	0.048029	0	—
80	120	0.000971	41.7	0.0404907
120	80	0.000009	83.4	0.0007506
		1.000000		0.0412413

approximately 260 days. The forecast peak load for this system is 160 MW, which corresponds to the 100% condition on the ordinate. The LOLE can be found using either the individual capacity outage probabilities or using the cumulative values. Both methods are illustrated in this example. Table 2.12 shows the calculation using Equation (2.7). The time periods t_k are shown in Fig. 2.6.

The LOLE is 0.0412413% of the time base units. Assuming a 365 day year, this LOLE becomes 0.150410 days or 6.65 years per day. The abscissa and hence the total time t_k could have been in days rather than in percent and identical results obtained.

If the cumulative probability values are used, the time quantities used are the interval or increases in curtailed time represented by T_k in Fig. 2.6. The procedure is shown in Table 2.13.

The LOLE of 0.0412413% is identical to the value obtained previously. Both techniques are shown simply to illustrate that either approach will provide the same result.

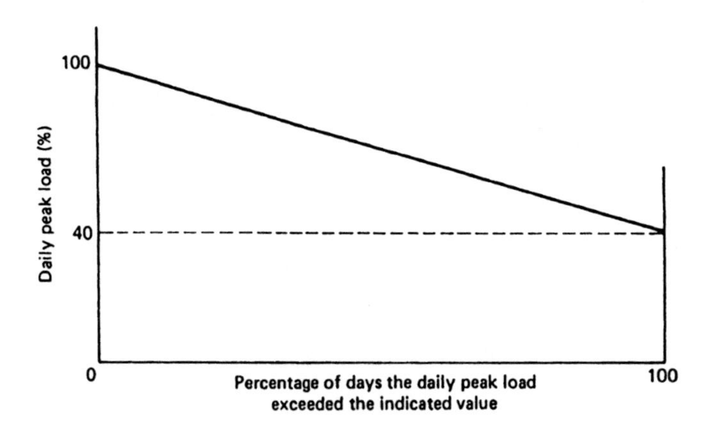

Fig. 2.5 Daily peak load variation curve

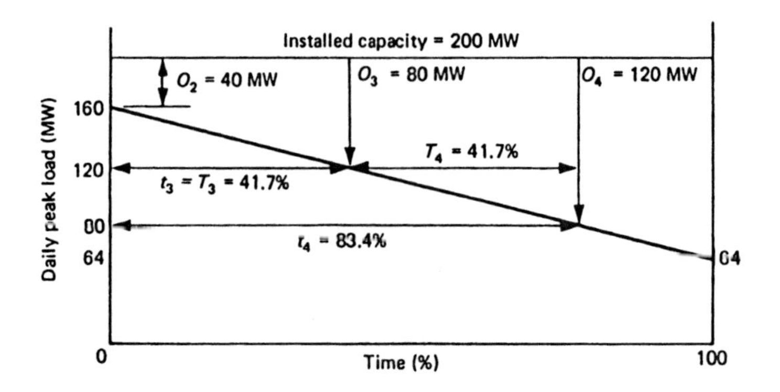

Fig. 2.6 Time periods during which loss of load occurs

(b) *Sensitivity studies*

The system peak load in the above example is 160 MW. Table 2.14 shows the variation in risk as a function of the peak load. The load characteristic for each forecast peak load is that shown in Fig. 2.5. The LOLE is calculated on an annual basis assuming 365 days in the year.

These results can best be displayed in the form of a graph using semi-logarithmic paper as shown in Fig. 2.7.

The system risk for a given capacity composition and forecast peak load is dependent upon the unavailability values for the individual units. This effect is illustrated in Table 2.15. The LOLE values for a range of peak load levels are shown as a function of the unit forced outage rates using the system of Table 2.11.

The system used in this example is very small and therefore the effect of generating unit unreliability is quite pronounced. This effect can also be quite considerable in a big system if the large units have high forced outage rates. This is shown in Fig. 2.8. The system in this case has a total installed capacity of 10100 MW. The largest units have 300 MW and 500 MW capacities and their forced outage rates have been varied as shown. The risk profile as a function of peak load

Table 2.13 LOLE using cumulative probabilities

Capacity out of service (MW)	*Capacity in service (MW)*	*Cumulative probability*	*Time interval T_k (%)*	*LOLE*
0	200	1.000000	0	—
40	160	0.049009	0	—
80	120	0.000980	41.7	0.0408660
120	80	0.000009	41.7	0.0003753
				0.0412413%

Table 2.14 Sensitivity study results

System peak load (MW)	LOLE (days/year)	LOLE (years/day)
200.0	6.083	0.16
190.0	4.837	0.21
180.0	3.447	0.29
170.0	1.895	0.53
160.0	0.1506	6.64
150.0	0.1208	8.28
140.0	0.08687	11.51
130.0	0.04772	20.96
120.0	0.002005	498
110.0	0.001644	608
100.0	0.001210	826

is almost a straight line in Fig. 2.8 as compared to the characteristic shown in Fig. 2.7. A large system with a wide range of unit sizes has a more continuous capacity outage probability table resulting in a smoother risk profile. It can however be perturbed by the addition of a relatively large unit. This point is discussed in Section 2.5.

The system peak load carrying capability (PLCC) can be determined as a function of the risk level. In the system shown in Fig. 2.8 the PLCC at a risk level of 0.1 days/year is 9006 MW for forced outage rates of 0.04. Table 2.16 shows the change in PLCC for FOR values from 0.04 to 0.13. The decrease in PLCC is 815 MW. If the forecast peak load is 9000 MW and the forced outage rates of the large

Table 2.15 Effect of FOR and system peak load

System peak load (MW)	System risk level — Unit FOR 0.01	0.02	0.03	0.04	0.05
200.0	6.083	12.165	18.247	24.330	30.411
190.0	4.834	9.727	14.683	19.696	24.764
180.0	3.446	7.024	10.729	14.556	18.502
170.0	1.895	3.998	6.304	8.804	11.494
160.0	0.150	0.596	1.328	2.337	3.614
150.0	0.121	0.480	1.073	1.894	2.939
140.0	0.087	0.347	0.781	1.388	2.167
130.0	0.048	0.194	0.445	0.805	1.278
120.0	0.002	0.016	0.053	0.124	0.240

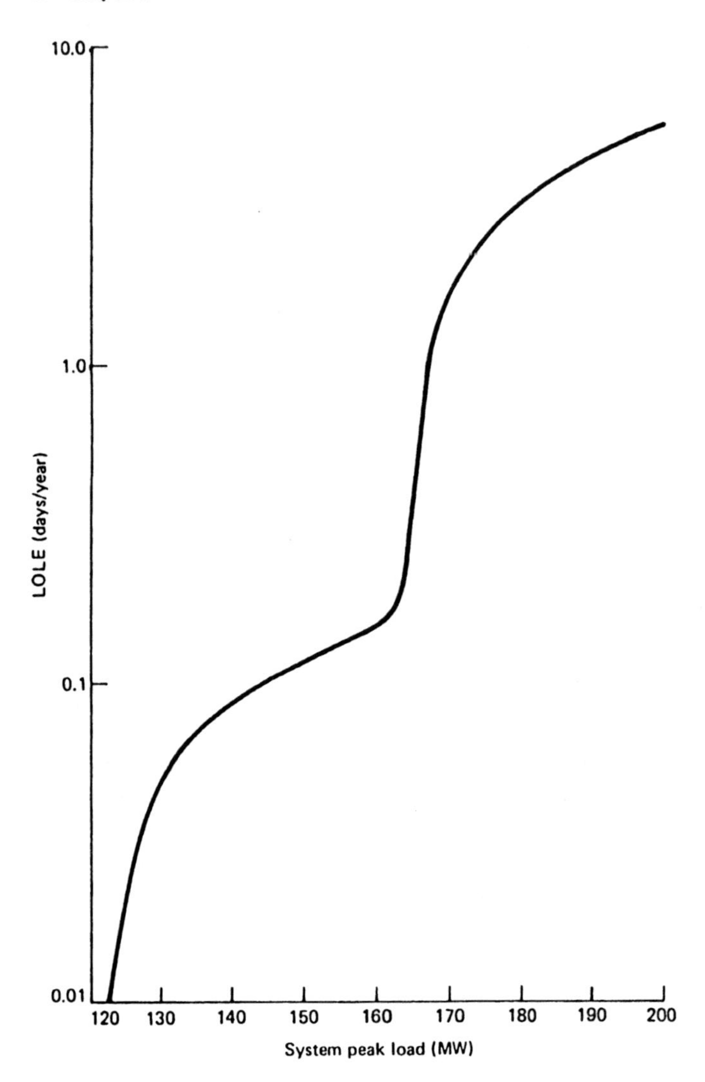

Fig. 2.7 Variation in risk with system peak load

units are 0.13, then this system would have to install approximately 1000 MW additional capacity to satisfy a risk level of 0.1 days/year. At a nominal $1000/kW installed this would cost approximately 10^9 dollars. The consequences of unit unavailability in terms of additional capacity can be seen quite clearly in this example [25]. Additional penalties in the form of expected energy replacement costs are illustrated in Reference [26].

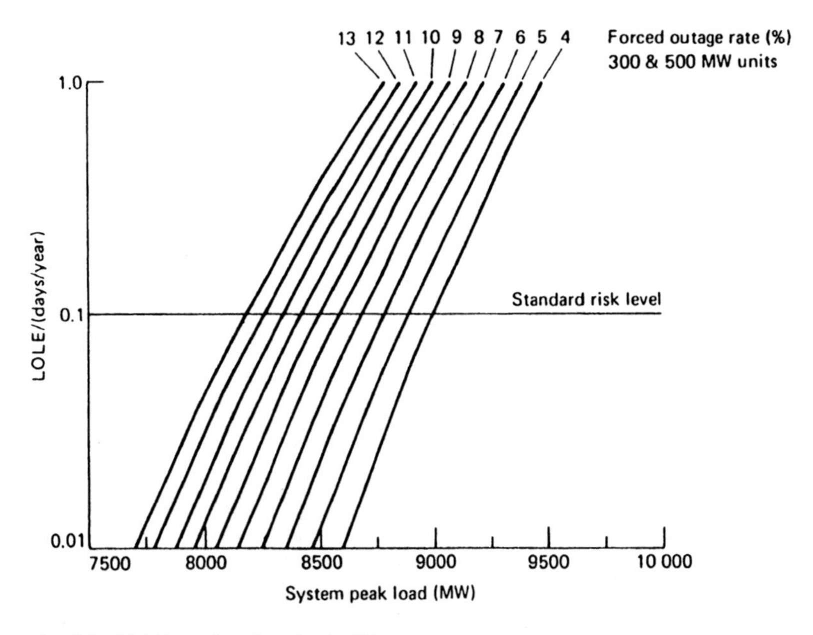

Fig. 2.8 LOLE as a function of unit FOR

Additional investment in terms of design, construction, reliability, maintainability and spare parts provisioning can result in improved generating unit unavailability levels. The worth of the improvement must be appraised on a total system basis and compared with the cost of attaining it.

Table 2.16 Changes in PLCC

Forced outage rate (%)	*Peak load carrying capability (MW)*	*Difference (MW)*	*Cumulative difference (MW)*
4	9006	—	—
5	8895	111	111
6	8793	102	213
7	8693	100	313
8	8602	91	404
9	8513	89	493
10	8427	86	579
11	8345	82	661
12	8267	78	739
13	8191	76	815

2.4 Equivalent forced outage rate (EFOR)

Data collection is an essential constituent of reliability evaluation, and utilities throughout the world have recorded the operational history of their units for many years. These data are then either stored in-house by the utility or processed by a central organization such as the Edison Electric Institute (EEI) who regularly publish data on generating unit reliability. Data in North America are now collected and disseminated by the CEA and NERC. The data collected for generating units usually involves the monitoring of residence times for each of the recorded output levels of the unit. This process may therefore recognize many derated states. It is not necessary or even feasible to accommodate a large number of such states and in practice these can be reduced to a very limited number using a weighted-averaging method using the same concept as rounding, which was discussed in Section 2.2.2. In the limit the number of states can be reduced to two; the up state and the down state and all others are weighted into these two states. This leads to the concept known as 'equivalent forced outage rate' or EFOR, which is sometimes defined as the equivalent probability of finding a unit on forced outage at some distant time

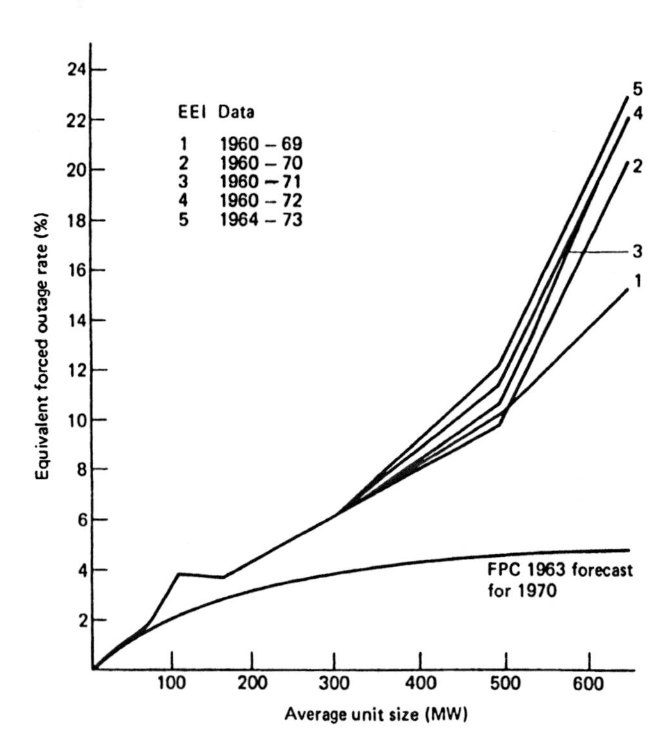

Fig. 2.9 EFOR as a function of unit size

in the future. The evaluation method of EFOR is given in References [16] and [27].

Generating unit unavailability levels have historically increased as unit sizes increase. Figure 2.9 shows the variation in equivalent forced outage rate (EFOR) as a function of fossil fired unit sizes using a series of Edison Electric Institute data. It can be seen that the EFOR increases dramatically with unit size.

The use of the word equivalent tends to imply that the two-state representation has the same impact as the multi-state representation when utilized in capacity evaluation studies. This is not the case, as the EFOR representation gives a pessimistic appraisal of system reliability by grouping weighted derated state residence times into the full forced outage state. The Canadian Electrical Association has chosen to call this statistic the derated adjusted forced outage rate (DAFOR) to avoid the connotation of equivalence. The effect of using a multi-state representation and an EFOR representation in a practical system study is shown in Fig. 2.10.

Figure 2.10 illustrates that the use of a two-state representation for units which do have significant derated states can result in considerable inaccuracy. These units should be modelled with at least three states.

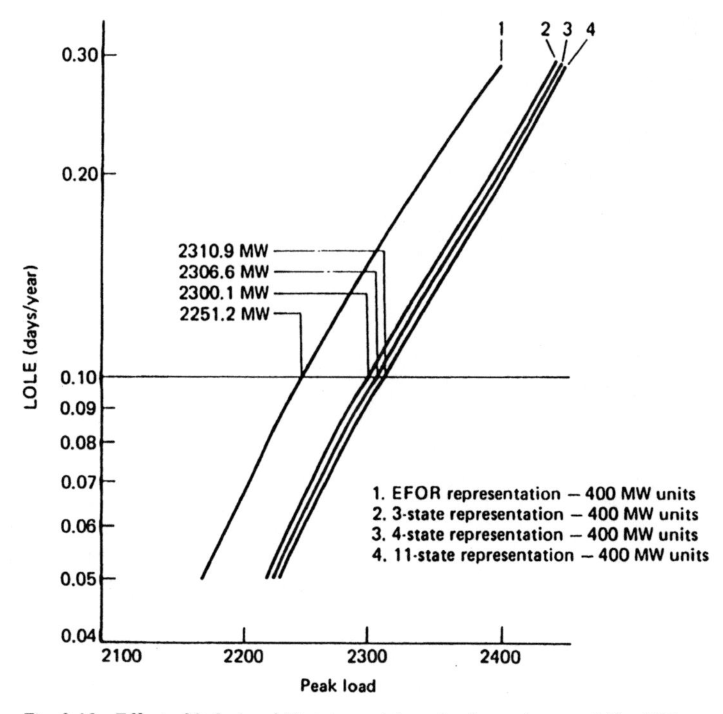

Fig. 2.10 Effect of 2, 3, 4 and 11 state models on load-carrying capability [28]

2.5 Capacity expansion analysis

2.5.1 Evaluation techniques

The time period required to design, construct and commission a large generating station can be quite extensive (5 to 10 years) depending on the environmental and regulatory requirements. It therefore becomes necessary to determine the system requirements considerably in advance of the actual unit in-service date. The actual load at an extended time in the future is also uncertain and should be considered as a random variable. This aspect is discussed in Section 2.8.

The concept of capacity expansion analysis can be illustrated using the system with five 40 MW units, described in Table 2.11. Assume that it has been decided to add additional 50 MW units with forced outage rates of 0.01 to meet a projected future load growth of 10% per year. The question is—in what years must the units be committed in order to meet the accepted system risk level? The change in risk level with the sequential addition of 50 MW units is shown in Table 2.17 for a range of system peak loads. The LOLE is in days for a 365-day year. The load characteristic is the daily peak load variation curve using a straight line from the 100% to 40% points.

The results in Table 2.17 can again be displayed in the form of a graph as shown in Fig. 2.11.

The annual peak load for each of the next eight years is shown in Table 2.18.

Table 2.17 LOLE in generation expansion

System peak load (MW)	*LOLE (days/year)*			
	200 MW capacity	*250 MW capacity*	*300 MW capacity*	*350 MW capacity*
100.0	0.001210	—	—	—
120.0	0.002005	—	—	—
140.0	0.08686	0.001301	—	—
160.0	0.1506	0.002625	—	—
180.0	3.447	0.06858	—	—
200.0	6.083	0.1505	0.002996	—
220.0	—	2.058	0.03615	—
240.0	—	4.853	0.1361	0.002980
250.0	—	6.083	0.1800	0.004034
260.0	—	—	0.6610	0.01175
280.0	—	—	3.566	0.1075
300.0	—	—	6.082	0.2904
320.0	—	—	—	2.248
340.0	—	—	—	4.880
350.0	—	—	—	6.083

If the assumption that an installed capacity of 200 MW is adequate for a system peak load of 160 MW, then the risk criterion is 0.15 days/year. This risk level can be used to measure the adequacy of the system capacity in the successive years. It must be realized that any risk level could have been selected. The actual choice is a management decision. Using the criterion of 0.15 days/year, the timing of unit additions can be obtained using Fig. 2.11. This expansion is shown in Table 2.19.

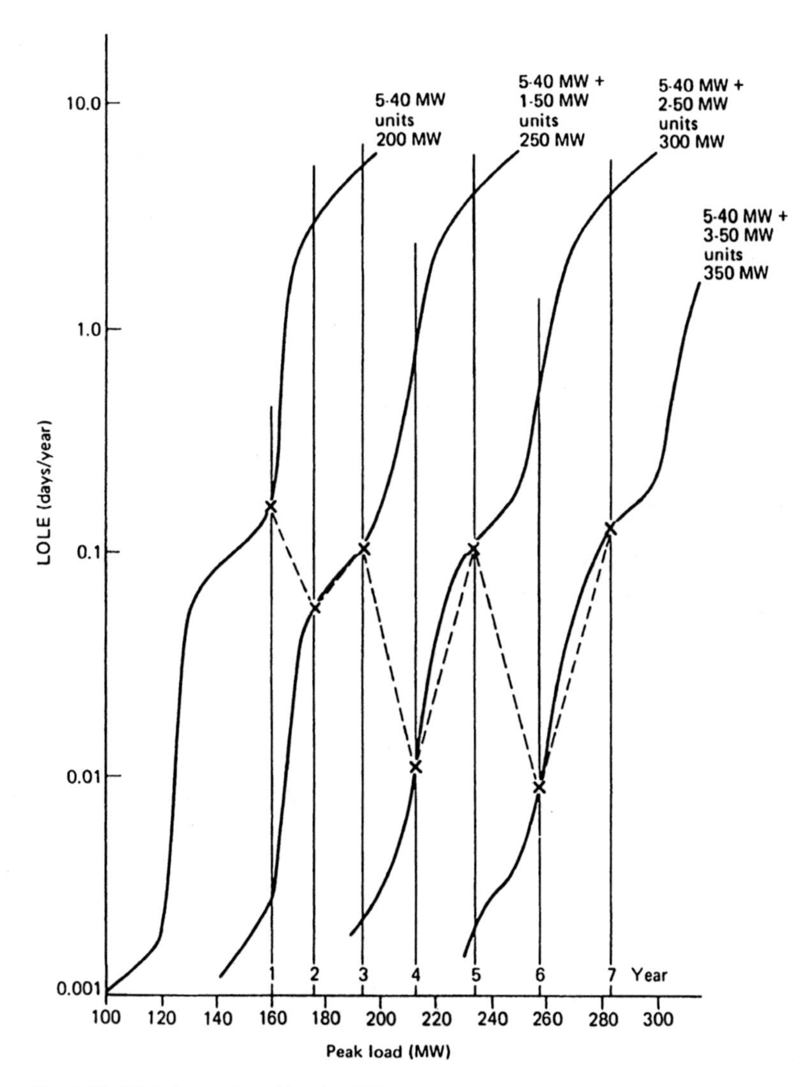

Fig. 2.11 Variation in risk with unit additions

Table 2.18 Load growth at 10% p.a.

Year number	*Forecast peak load (MW)*
1	160
2	176
3	193.6
4	213.0
5	234.3
6	257.5
7	283.1
8	311.4

The 50 MW unit additions would have to be made in years 2, 4 and 6. The variation in annual risk level is shown by the dotted line in Fig. 2.11. This particular expansion study represents a somewhat idealized case. The present worth of this particular scheme would have to be compared with others to determine the optimum expansion pattern for the system. The expansion study should cover a sufficiently long period into the future in order to establish a realistic present worth evaluation and to minimize perturbation effects (see Section 2.5.2). Theoretically this should extend to infinity; however, in practice a period of twenty to thirty years is usually adequate. The generation expansion plan can, and probably will, be varied as real time is advanced.

2.5.2 Perturbation effects

Large capacity unit additions often appear to be economically advantageous due to the so-called 'economy of scale'. Large units generally have relatively low cost per

Table 2.19 Generation expansion results

Year	*Unit added (MW)*	*System capacity (MW)*	*Peak load (MW)*	*LOLE (days/year)*
1	—	200	160.0	0.15
2	—	200	176.0	2.9
	50	250	176.0	0.058
3	—	250	193.6	0.11
4	—	250	213.0	0.73
	50	300	213.0	0.011
5	—	300	234.3	0.11
6	—	300	257.4	0.55
	50	350	257.4	0.009
7	—	350	283.1	0.125
8	—	350	311.4	0.96

Table 2.20 IPLCC for five-unit system

System capacity (MW)	Allowable peak load (MW)	Increase in peak load carrying capability (MW)	
		Individual	Cumulative
(5 × 40) = 200	144	0	0
(5 × 40) + (1 × 50) = 250	186	42	42
(5 × 40) + (2 × 50) = 300	232	46	88
(5 × 40) + (3 × 50) = 350	279	47	135

kW installed and better heat rates than smaller capacity units. Economic evaluation of alternative sizes should, however, include the impact on the system reliability of adding a relatively large unit to the overall system. This effect can be seen in terms of the increased system peak load-carrying capability (IPLCC) due to unit additions. Using Fig. 2.11, the IPLCC can be determined for each 50 MW unit addition at a specified risk level. Table 2.20 shows the individual unit and cumulative IPLCC values for each 50 MW unit at a system risk level of 0.1 days/year.

The 50 MW units added to the system in this case are not much larger than the 40 MW units already in the system and therefore they do not create a large perturbation. The effect of adding relatively large units to a system can be seen by adding the 50 MW units to a system with the same initial 200 MW of capacity but with a different unit composition.

Consider a system composed of 10–20 MW units each with a forced outage rate of 0.01. The total installed capacity in this case is 200 MW and would require the same reserve capacity as the 5 × 40 MW unit system using the percentage reserve criterion. The loss of the largest unit criterion would dictate that the 5 × 40 MW unit system could carry a peak load of 160 MW while the 20 × 10 MW unit system could carry a 190 MW peak load. Note that neither criterion includes any consideration of the actual load shape. Table 2.21 shows the individual unit and cumulative IPLCC values for each 50 MW unit addition to the 20 × 10 MW unit system at a system risk level of 0.1 days/year.

Table 2.21 IPLCC for 20-unit system

System capacity (MW)	Allowable peak load (MW)	Increase in peak load carrying capability (MW)	
		Individual	Cumulative
(20 × 10) = 200	184	0	0
(20 × 10) + (1 × 50) = 250	202	18	18
(20 × 10) + (2 × 50) = 300	250	48	66
(20 × 10) + (3 × 50) = 350	298	48	114

The initial load-carrying capability of the two systems are considerably different as the system with the smaller units can carry a much higher peak load. The first 50 MW unit addition creates a considerable perturbation in this system and results in an IPLCC of only 18 MW. The second unit appears to create an IPLCC of 48 MW. It may be better, however, to think in terms of the cumulative value of 66 MW created due to the addition of the two 50 MW units. Relatively large units cannot be easily added to small systems or to systems composed of relatively small units without a significant initial PLCC penalty. This penalty will diminish as additional units are added and the basic system composition changes. This is one reason why unit additions must be examined in terms of an expansion plan and considered over a reasonable time period rather than on a single year or single unit addition basis.

This effect is further accentuated if the unit forced outage rate is increased in the first few years to accommodate a break-in or infant mortality period. A common utility practice is to double the unit forced outage rate for the first two years, particularly if the unit size or type is significantly different from others in the system and little experience is available. The utilization of probability techniques even in the relatively simple form of LOLE evaluation permits the factors that do influence the system reliability to be included in the analysis and gives proper weight to unit sizes and outage rates and to the system load characteristic.

2.6 Scheduled outages

The system capacity evaluation examples previously considered assumed that the load model applied to the entire period and that the system capacity model was also applicable for the entire period. This will not be the case if units are removed from service for periodic inspection and maintenance in accordance with a planned program. During this period, the capacity available for service is not constant and

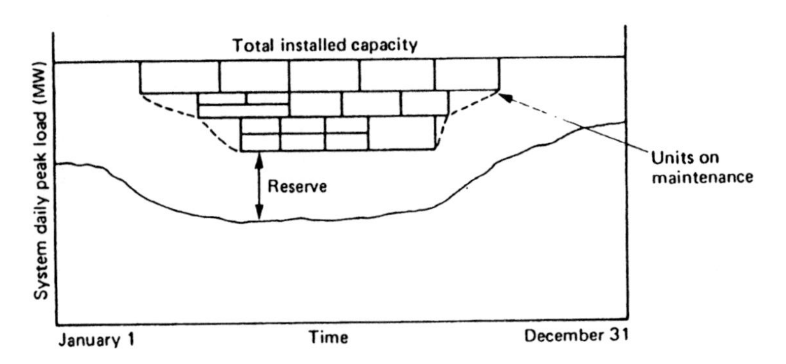

Fig. 2.12 Annual load and capacity model

therefore a single capacity outage probability table is not applicable. Figure 2.12 shows a hypothetical example of a maintenance schedule for a winter peaking system.

The annual $LOLE_a$ can be obtained by dividing the year into periods and calculating the period $LOLE_p$ values using the modified capacity model and the appropriate period load model. The annual risk index is then given by

$$LOLE_a = \sum_{p=1}^{n} LOLE_p \tag{2.9}$$

The modified capacity model can be obtained by creating a new capacity outage probability table for each capacity condition. The unit removal algorithm illustrated by Equations (2.3) and (2.4) can be used in this case. The total installed capacity may also increase during the year due to the commissioning of a new unit. This can also be added to the capacity model in the appropriate periods. If the actual in-service date of the new unit is uncertain, it can be represented by a probability distribution and incorporated on a period basis using the following equation.

$$LOLE_p = (LOLE_{pa})a + (LOLE_{pu})u \tag{2.10}$$

where $LOLE_p$ = period LOLE value
$LOLE_{pa}$ = period LOLE value including the unit
$LOLE_{pu}$ = period LOLE value without the unit
a = probability of the unit coming into service
u = probability of the unit not coming into service.

The unit still has the opportunity to fail given that it comes into service. This is included in the $LOLE_{pa}$ value. The annual risk index is then obtained using Equation (2.9).

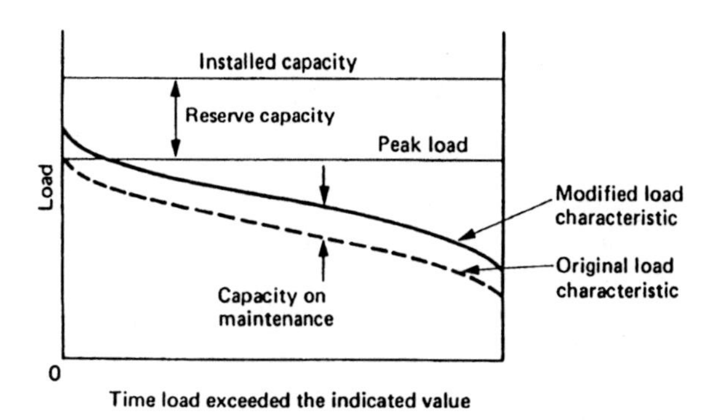

Fig. 2.13 Approximate method of including maintenance

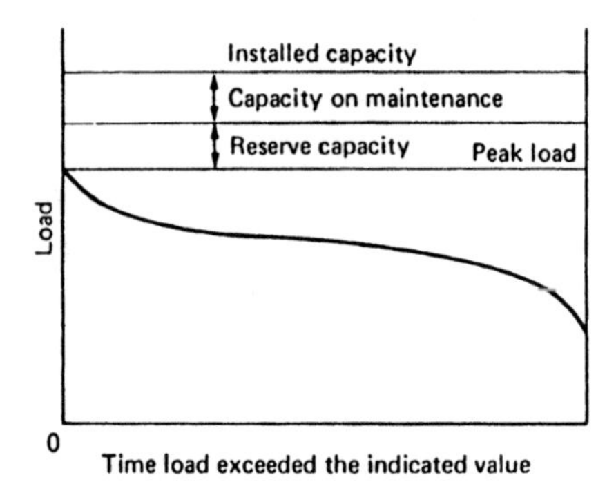

Fig. 2.14 Capacity reduction due to maintenance

Maintenance has been considered by some authors as indicated in Fig. 2.13 by adding the capacity on maintenance to the load and using a single capacity outage probability table.

The approach shown in Fig. 2.13 gives the same results as that of Fig. 2.14, in which the original capacity outage probability table is used, but the total available capacity is reduced by the quantity on outage.

Both of these methods are approximations because the state probabilities in the generation model are unaltered and therefore do not really relate to the system during maintenance.

The most realistic approach is to combine the units actually available to the system into a capacity outage probability table applicable for the period considered as described above. Practical system studies using the approximate methods and the realistic method indicate that adding the capacity on maintenance to the load or subtracting it from the installed capacity without altering the outage probabilities results in higher calculated risk levels and that the error increases with increased maintenance capacity. This error may be negligible in a large system in which the capacity on maintenance is an extremely small percentage of the total installed capacity. Removing units on maintenance from the capacity outage probability table results in negligible error for normal magnitudes of capacity on maintenance for those cases when the units removed are not exact multiples of the rounding increment used in the table.

If the maintenance is scheduled either to minimize [38] risk or in accordance with a constant risk criterion then the reserve shown in Fig. 2.12 may be quite variable. It is important to realize that constant reserve is not the same as constant risk. The system is clearly not as reliable if a unit with a low forced outage rate is removed from service when compared with the situation in which a similar capacity unit with a high forced outage unit is removed from service.

There are a number of approximate techniques for scheduling maintenance. One approach is to reduce the total installed capacity by the expected capacity loss (i.e. the product of the unit capacity and its availability) rather than by the actual unit capacity and then schedule maintenance on a constant reserve basis. A better approach, and one that is often quite accurate if only a few units are on maintenance at any given time, is to determine the decrease in PLCC at the appropriate risk level for each individual unit on maintenance and then use these values in scheduling maintenance on a constant reserve basis. The applicable approach will depend on the capacity composition, the required maintenance level and the system load profile.

2.7 Evaluation methods on period bases

The basic LOLE approach is extremely flexible in regard to the extent to which load models and maintenance considerations can be incorporated. This flexibility also dictates the necessity to thoroughly understand the modelling assumptions used prior to quoting and comparing risk indices for different systems. This important point can be appreciated by considering the following three ways in which the LOLE method can be used to determine an annual risk index:

(a) monthly (or period) basis considering maintenance;
(b) annual basis neglecting maintenance;
(c) worst period basis.

In the monthly approach and assuming constant capacity for the period, the appropriate capacity outage probability table is combined with the corresponding load characteristic. If the capacity on maintenance is not constant during the month, the month can be divided into several intervals during which the capacity is constant. The capacity outage probability table, modified by removing the units on maintenance for each separate interval, can be combined with the monthly peak and a load characteristic using the interval as its time base. This method assumes that the normalized monthly load characteristic holds for any portion of a month and that the monthly peak can occur on any day during the period. The total risk for the month is obtained by summing the interval values. The annual risk is the sum of the twelve monthly risks.

In the annual approach neglecting maintenance, the annual forecast peak and system load characteristic are combined with the system capacity outage probability table to give an annual risk level. The basic assumption in this approach is that a constant capacity level exists for the entire period. The justification for this assumption is dependent upon the time of generating unit additions, the planned maintenance and the monthly load levels relative to the annual peak. If the year can be divided into a peak load season and a light load season, the planned maintenance may be scheduled entirely in this latter period. The contribution of the light load season to the annual risk may be quite low and therefore the assumption of a constant capacity level is justified. The relative period risk contributions for any particular system should be examined before adopting this approach.

In some cases, the load level in a particular season or even month may be so high that this value dominates the annual figure. A reliability criterion for such a system can be obtained using only this 'worst period' value. A study of the Saskatchewan and Manitoba Systems indicated that the month of December generally constitutes the highest monthly risk period. An annual risk figure can be obtained by multiplying the December value by twelve. This approach assumes twelve possible Decembers in a year and is designated the '12 December basis'.

Computing risk levels on a monthly basis considering maintenance can be quite laborious, especially when the maintenance capacity is not constant during a month. This approach can be used to determine if the risk levels for specific maintenance periods exceed a specified amount. This condition can be studied by comparing the risk levels for each of the maintenance intervals converted to a common time base (for example 365 days). If the expectation for a period of ten days is 0.001 hours, then $0.001 \times 36.5 = 0.0365$ hours is the expectation on an annual basis. This technique is necessary to avoid the tendency to assume that for a particular interval, a low expected value indicates little risk. The low value may be due to the interval itself being very small and not due to having a high reserve capacity margin.

In planning unit additions where risk levels for different years are to be compared, the 'annual basis neglecting maintenance' or the '12 worst months basis' are the simplest methods and generally provide satisfactory results. The '12 worst months basis' cannot be used to compare the risk levels in two different systems with different annual load characteristics. This approach is only consistent when applied continually to the same system.

The above approaches are not exhaustive and various alternatives are possible. It should be stressed, however, that the risk index evaluated depends on the approach used and therefore risk indices of different utilities are not necessarily comparable [29]. This is not a point of concern provided the approach used by a given utility is consistent.

2.8 Load forecast uncertainty

(a) *Method 1*

In the previous sections of this chapter it has been assumed that the actual peak load will differ from the forecast value with zero probability. This is extremely unlikely in actual practice as the forecast is normally predicted on past experience. If it is realized that some uncertainty can exist, it can be described by a probability distribution whose parameters can be determined from past experience, future load modelling and possible subjective evaluation.

The uncertainty in load forecasting can be included in the risk computations by dividing the load forecast probability distribution into class intervals, the number of which depends upon the accuracy desired. The area of each class interval represents the probability the load is the class interval mid-value. The LOLE is computed for each load represented by the class interval and multiplied by the probability that that load exists. The sum of these products represents the LOLE for the forecast load. The calculated risk level increases as the forecast load uncertainty increases.

It is extremely difficult to obtain sufficient historical data to determine the distribution describing the load forecast uncertainty. Published data, however, has suggested that the uncertainty can be reasonably described by a normal distribution. The distribution mean is the forecast peak load. The distribution can be divided into a discrete number of class intervals. The load representing the class interval mid-point is assigned the designated probability for that class interval. This is shown in Fig. 2.15, where the distribution is divided into seven steps. A similar approach can be used to represent an unsymmetrical distribution if required. It has been found that there is little difference in the end result between representing the distribution of load forecast uncertainty by seven steps or forty-nine steps. The error is, however, dependent upon the capacity levels for the system.

The computation of the LOLE considering load forecast uncertainty is shown for a small hypothetical system in the following example.

The system consists of twelve 5 MW units, each with a forced outage rate of 0.01. The capacity model is shown in Table 2.22. The forecast peak load is 50 MW, with uncertainty assumed to be normally distributed using a seven-step approximation (Fig. 2.15). The standard deviation is 2% of the forecast peak load. The monthly load–duration curve is represented by a straight line at a load factor of 70%. The LOLE calculation is shown in Table 2.23.

The LOLE increased from 0.025240 with no load forecast uncertainty to 0.07839425 with 2% uncertainty. The index in this case is in hours/month. Load forecast uncertainty is an extremely important parameter and in the light of the financial, societal and environmental uncertainties which electric power utilities

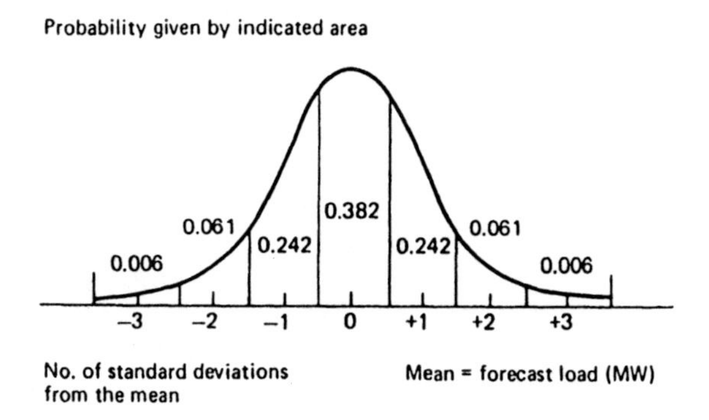

Fig. 2.15 Seven-step approximation of the normal distribution

Table 2.22 Generation model

Capacity on outage (MW)	*Cumulative probability*
0	1.00000000
5	0.11361513
10	0.00617454
15	0.00020562
20	0.00000464
25	0.00000007

(Probability values less than 10^{-8} are neglected.)
Period = 1 month = 30 days = 720 hours
Forecast load = mean = 50 MW
Standard deviation (2%) = 50 × 2/100 = 1 MW

face may be the single most important parameter in operating capacity reliability evaluation. In the example shown in Table 2.23, the risk was evaluated for each peak load level. The seven individual values were then weighted by the probability of existence of that peak load level. The final LOLE is actually the expected value of a group of loss of load expectations.

(b) *Method 2*

The LOLE value including uncertainty can be found using a somewhat different approach. The load characteristic can be modified to produce a load profile which includes uncertainty. This single load characteristic can then be combined with the capacity outage probability table to compute the LOLE index. If the uncertainty is fixed at some specified value and the load shape remains unchanged, then the modified load curve can be used for a range of studies with a considerable saving

Table 2.23 LOLE results

(1) Number of standard deviations from the mean	*(2) Load (MW)*	*(3) Probability of the load in Col. (2)*	*(4) LOLE (hours/month) for the load in Col. (2)*	*(3) × (4)*
–3	47	0.006	0.01110144	0.00006661
–2	48	0.061	0.01601054	0.00097664
–1	49	0.242	0.02071927	0.00501406
0	50	0.382	0.02523965	0.00966679
+1	51	0.242	0.17002797	0.04114677
+2	52	0.061	0.30924753	0.01886410
+3	53	0.006	0.44321350	0.00265928
			Total	0.07839425

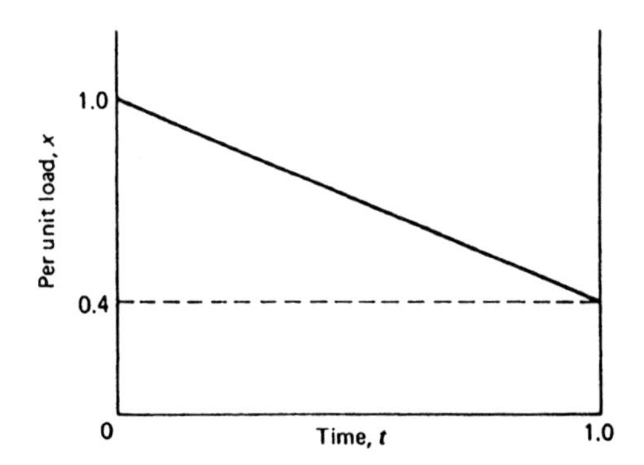

Fig. 2.16 Monthly load-duration curve in per unit

in computer time. This procedure is illustrated using the previous example. The load model used in the example is a monthly load–duration curve represented by a straight line at a load factor of 70% as shown in Fig. 2.16. This is a simplification of a real load–duration curve and in practice the following analysis needs modifying so that either the non-linear equation of the load curve is convolved or the load curve is segmented into a series of straight lines.

The equation for this line is

$$t = \frac{10}{6}(1 - x)$$

if

X = load in MW
L = forecast peak load
$x = X/L$.

The load forecast uncertainty is represented by a seven-step approximation to the normal distribution as shown in Fig. 2.15. The standard deviation of this distribution is equal to 2% of the forecast peak load. In the case of a 50 MW peak this corresponds to 1 MW. There are therefore seven conditional load shapes as shown in Fig. 2.17, each with a probability of existence. Consider two examples of the seven conditional load shapes:

At a peak level of 47 MW,

$$t_1 = \frac{10}{6}\left(1 - \frac{X}{47}\right) \quad \text{for } 0 \le X \le 47$$

which exists with a probability of 0.006.

At a peak load level of 50 MW,

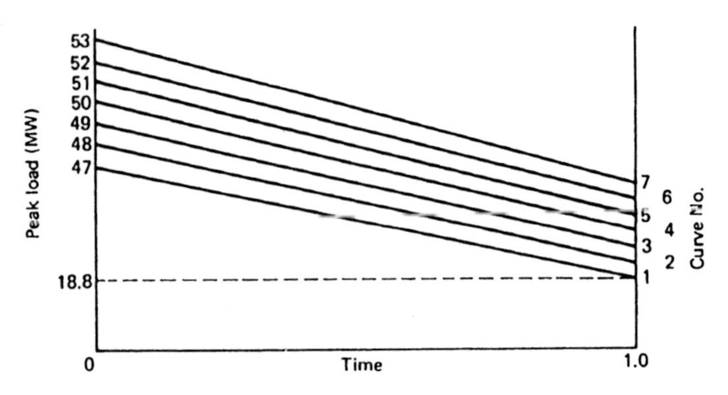

Fig. 2.17 Conditional load–duration curves

$$t_4 = \tfrac{10}{6}\left(1 - \frac{X}{50}\right) \quad \text{for } 0 \le X \le 50$$

which exists with a probability of 0.382.

The modified load–duration curve is now composed of a group of conditional segments as shown in Fig. 2.18. The evaluation of four of these segments is shown below; the remaining segments can be evaluated similarly.

For Segment 1 $t = 1.0$ for $0 < X < 18.8$

For Segment 2
$$t = 0.006t_1 + 0.061t_2 + 0.242t_3$$
$$0.382t_4 + 0.242t_5 + 0.061t_6$$
$$0.006t_7 \qquad \text{for } 18.8 \le X \le 47$$

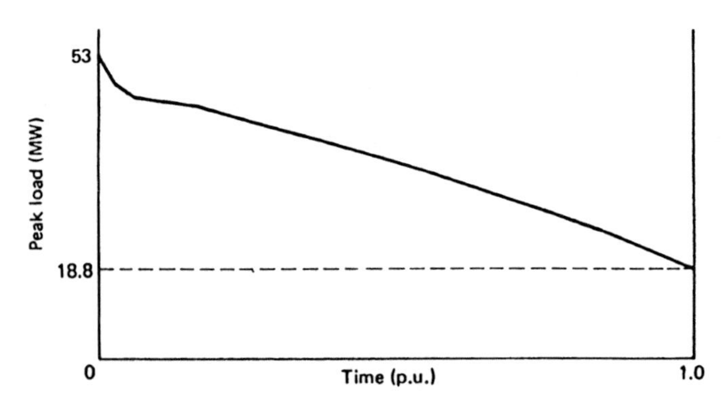

Fig. 2.18 Modified load–duration curve

Table 2.24 LOLE results

Capacity Out	*(MW) In*	*Individual probability*	*Time (pu)*	*Expectation*
0	60	0.8863848717	—	—
5	55	0.1074405905	—	—
10	50	0.0059689217	0.0123847909	0.0000739239
15	45	0.0002009738	0.1661845138	0.0000333987
20	40	0.0000045676	0.3330899382	0.0000015214
25	35	0.0000000738	0.4999453626	0.0000000369
				0.0001088809

For Segment 7 $t = 0.061t_6 + 0.006t_7$ for $51 \le X \le 52$

For Segment 8 $t = 0.006t_7$ for $52 \le X \le 53$

The modified load–duration curve of Fig. 2.18 is shown in terms of MW of peak load. It can also be expressed in percentage or per unit of the forecast peak load and used with any forecast peak assuming the basic characteristic and uncertainty remains constant. The LOLE calculation is shown in Table 2.24.

The LOLE in hours/month = 0.0001088809 × 30 × 24 = 0.07839425. This value is that shown in Table 2.23. In order to illustrate the evaluation of the time values shown in Table 2.24, consider as an example the value at the 50 MW capacity level which corresponds to Segment 6:

$$
\begin{aligned}
t &= 0.242t_5 + 0.061t_6 + 0.006t_7 \\
&= (0.242)(0.0326797) + (0.061)(0.0641026) + (0.006)(0.0943396) \\
&= \underline{0.0123847909}
\end{aligned}
$$

where, for example,

$$t_5 = \tfrac{10}{6}\left(1 - \frac{X}{51}\right) = \underline{0.0326797386}$$

The concept of conditional load curves leading to a modified curve is a useful technique which can be used in certain cases to save computation time in repetitive studies. If applicable, the modified curve can be used as input data in further studies. This idea is used in Section 2.10 as a load modification technique in loss of energy studies and production cost calculations.

2.9 Forced outage rate uncertainty

The loss of load expectation as computed using the techniques illustrated earlier in this chapter assumes that the generating unit unavailability parameters are single point values. In actual fact these parameters are usually single point best estimates based upon the available data and future forecasts. There is therefore considerable uncertainty in these parameters which creates uncertainty in the calculated LOLE parameter. The actual distribution associated with the calculated LOLE can be obtained by Monte Carlo simulation. If the uncertainty associated with the unit unavailabilities is considered to be normally distributed, then the resultant LOLE uncertainty can be considered to be normally distributed. In all other cases, an exact solution is analytically intractable and simulation must be used. The uncertainty associated with unit unavailability was first considered in Reference [30]. The effect of uncertainty in unit unavailability on LOLE and uncertainty in the unit failure rate on spinning reserve requirements were first considered in Reference [31] using an upper bound confidence limit approach. The combination of the actual unit uncertainties was first considered in Reference [32] and the technique was extended in subsequent publications [33, 34]. The basic approach is to calculate the conventional capacity outage probability table using the conventional recursive equations and also compute the variance associated with the cumulative probability at each capacity level. This involves the successive determination of the covariance matrix associated with each unit capacity addition. The final capacity outage probability table and its covariance matrix can be combined with the load model to obtain the expected value for the calculated LOLE and its variance. The calculation and storage of the covariance matrix can become cumbersome in a large system and an approximate technique has been developed [34]. Both the exact and approximate approaches are presented, followed by a numerical example using the simple three-unit system given in Table 2.7.

2.9.1 Exact method [32]

The capacity outage probability table together with its covariance matrix are constructed by adding generating units one at a time to an existing table using the following expressions:

$$P(X) = (1 - r)P'(X) + rP'(X - C)$$

$$\begin{aligned}
\mathrm{Cov}[P(X),P(Y)] = {} & [(1 - r)^2 + V]\mathrm{Cov}[P'(X),P'(Y)] \\
& + [r(1 - r) - v]\{\mathrm{Cov}[P'(X),P'(Y - C)] \\
& + \mathrm{Cov}[P'(X - C),P'(Y)]\} \\
& + [r^2 + v]\mathrm{Cov}[P'(X - C),P'(Y - C)] \\
& + v[P'(X)P'(Y) - P'(X)P'(Y - C) \\
& - P'(X - C)P'(Y) + P'(X - C)P'(Y - C)
\end{aligned}$$

where:

X and Y = capacity on outage levels

$P(X)$ = probability of capacity outage of X MW or more after the unit addition

$P'(X)$ = probability of capacity outage of X MW or more before the unit addition
$\text{Cov}[P(X), P(Y)]$ = covariance of $P(X)$ and $P(Y)$ after the unit addition
$\text{Cov}[P'(X), P'(Y)]$ = covariance of $P'(X)$ and $P'(Y)$ before the unit addition
r = expected value of FOR for the unit being added
C = capacity of unit being added
v = variance of FOR for the unit being added

The initial conditions before the addition of any unit are $P(X \leq 0) = 1.0$, $P(X > 0) = 0$ and $\text{Cov}[P(X), P(Y)] = 0$ for all X and Y.

2.9.2 Approximate method [34]

A method based on the Taylor-series expansion of a function of several variables can be used to compute the elements of the covariance matrix associated with the capacity outage probability table. The required formula is given by

$$\text{Cov}[P(X),P(Y)] \simeq \sum_{i=1}^{m} \left(\frac{\partial P(X)}{\partial r_i}\right)\left(\frac{\partial P(Y)}{\partial r_i}\right) \text{Var}[r_i] + \sum_{i=1}^{m} \sum_{j=i+1}^{m} \left(\frac{\partial^2 P(X)}{\partial r_i \, \partial r_j}\right)\left(\frac{\partial^2 P(Y)}{\partial r_i \, \partial r_j}\right) \text{Var}[r_i]\text{Var}[r_j]$$

where m denotes the number of generating units. The partial derivations used in the above formula are computed using the following equations:

$$\frac{\partial P(X)}{\partial r_i} = P'(X - C_i) - P'(X)$$

$$\frac{\partial^2 P(X)}{\partial r_i \, \partial r_j} = P''(X - C_i - C_j) + P''(X) - P''(X - C_i) - P''(X - C_j)$$

where:

$P'(X)$ = the element in the capacity outage probability table after unit of C_i MW and FOR r_i is removed from the original table.
$P''(X)$ = the element in the capacity outage probability table after two units of capacities C_i and C_j are removed from the original table.

2.9.3 Application

The application of these recursive expressions is illustrated using the simple system shown in Table 2.7, with the variance associated with the unit FOR assumed to be 9×10^{-6} (Table 2.25)

Table 2.25

Unit No.	*Unit capacity C (MW)*	*Unit FOR r*	Var[*FOR*] v
1	25	0.02	9×10^{-6}
2	25	0.02	9×10^{-6}
3	50	0.02	9×10^{-6}

The capacity outage probability table and its covariance matrix are developed as follows.

Step 1 Add the first 25 MW unit. The table becomes Table 2.26, and its covariance matrix is given by

$$\mathrm{Cov}[P(X), P(Y)] = \begin{bmatrix} 0.0 & 0.0 \\ 0.0 & 9 \times 10^{-6} \end{bmatrix}$$

Step 2 Add the second 25 MW unit. The new table is Table 2.27 and its covariance matrix is given by

$$\mathrm{Cov}[P(X), P(Y)] = \begin{bmatrix} 0.0 & 0.0 & 0.0 \\ & 1.7287281 & 0.0352719 \\ \text{Symmetric matrix} & & 0.0007281 \end{bmatrix} \times 10^{-5}$$

Step 3 Add the last unit (50 MW) to the table. The complete table is Table 2.28 and its covariance matrix is given by

$$\mathrm{Cov}[P(X), P(Y)] =$$

$$\begin{bmatrix} 0.0 & 0.0 & 0.0 & 0.0 & 0.0 \\ & 2.4904173 & 0.8979052 & 0.0680962 & 0.0010367 \\ & & 0.8999794 & 0.0363167 & 0.0003742 \\ & & & 0.0021183 & 0.0000287 \\ \text{Symmetric matrix} & & & & 0.0000004 \end{bmatrix} \times 10^{-5}$$

2.9.4 LOLE computation

The mean and variance of the LOLE are given by

Table 2.26

State No.	*Capacity out*	*Cumulative probability*
1	0	1.0
2	25	0.02

Table 2.27

State No.	Capacity out	Cumulative probability
1	0.0	1.0
2	25.0	0.0396
3	50.0	0.0004

$$E[\text{LOLE}] = \sum_{i=1}^{n} E[P_i(C_i - X_i)]$$

$$\text{Var}[\text{LOLE}] = \sum_{i=1}^{n} \sum_{j=1}^{n} \text{Cov}[P_i(C_i - X_i), P_j(C_j - X_j)]$$

where:

n = number of days in the study period
C_i = available capacity on day i
X_i = forecast peak load on day i
$E[P_i]$ = expected value of the loss of load probability on day i
$\text{Cov}[P_i, P_j]$ = covariance of the loss of load probabilities on day i and day j.

Example

$N = 2$ days, Forecast peak loads = 65, 45 MW.

$$E[\text{LOLE}] = \sum_{i=1}^{2} P_i(C_i - X_i) = P_1(100 - 65) + P_2(100 - 45)$$

$$= P_1(35) + P_2(55)$$

$$= 0.020392 + 0.000792 = 0.021184$$

Table 2.28

State No.	Capacity out	Cumulative probability
1	0.0	1.0
2	25.0	0.058808
3	50.0	0.020392
4	75.0	0.000792
5	100.0	0.000008

$$\text{Var}[\text{LOLE}] = \sum_{i=1}^{2} \sum_{j=1}^{2} \text{Cov}[P_i(100 - X_i), P_j(100 - X_j)]$$

$$= \text{Var}[P_1(35)] + \text{Var}[P_2(55)] + 2\,\text{Cov}[P_1(35), P_2(55)]$$

If the exact method is used, the variance of LOLE is given by

$$\text{Var}[\text{LOLE}] = 0.8999794 \times 10^{-5} + 0.0021183 \times 10^{-5} + 2 \times 0.0363167 \times 10^{-5}$$

$$= 0.9747311 \times 10^{-5}$$

If the approximate method is used, the different terms in the variance equation of LOLE are given by

$$\text{Var}[P_1(35)] \simeq \left(\frac{\partial P_1}{\partial r_1}\right)^2 v_1 + \left(\frac{\partial P_1}{\partial r_2}\right)^2 v_2 + \left(\frac{\partial P_1}{\partial r_3}\right)^2 v_3 + \left(\frac{\partial^2 P_1}{\partial r_1 \partial r_2}\right)^2 v_1 v_2$$

$$+ \left(\frac{\partial^2 P_1}{\partial r_1 \partial r_3}\right)^2 v_1 v_3 + \left(\frac{\partial^2 P_1}{\partial r_2 \partial r_3}\right)^2 v_2 v_3$$

$$= 2[0.0396 - 0.02]^2 \times 9 \times 10^{-6} + [1 - 0.0004]^2 \times 9 \times 10^{-6}$$

$$+ [1 + 0.02 - 0.02 - 0.02]^2 \times (9 \times 10^{-6})$$

$$+ 2[1 + 0 - 0.02 - 1]^2 (9 \times 10^{-6})^2$$

$$= 0.0006915 \times 10^{-5} + 0.8992801 \times 10^{-5}$$

$$+ 0.778572 \times 10^{-10}$$

$$= 0.8999793 \times 10^{-5}$$

$$\text{Var}[P_2(55)] \simeq \sum_{i=1}^{3} \left(\frac{\partial P_2}{\partial r_i}\right)^2 v_i + \sum_{i=1}^{3} \sum_{j=i+1}^{3} \left(\frac{\partial^2 P_2}{\partial r_i \partial r_j}\right)^2 v_i v_j$$

$$= 2[0.02 - 0.0004]^2 \times 9 \times 10^{-6} + [0.0396 - 0]^2 \times 9 \times 10^{-5}$$

$$+ [0.02 + 0 - 0.02 - 0.02]^2 (9 \times 10^{-6})^2$$

$$+ 2[1 - 0 - 0 - 0.02]^2 (9 \times 10^{-6})^2$$

$$= 0.0006915 \times 10^{-5} + 0.0014113 \times 10^{-5}$$

$$+ 0.324 \times 10^{-13} + 1.555848 \times 10^{-10}$$

$$= 0.0021183 \times 10^{-5}$$

$$\text{Cov}[P_1(35),P_2(55)] \simeq \sum_{i=1}^{3}\left(\frac{\partial P_1}{\partial r_i}\right)\left(\frac{\partial P_2}{\partial r_i}\right)\nu_1$$

$$+\sum_{i=1}^{3}\sum_{j=i+1}^{3}\left(\frac{\partial^2 P_1}{\partial r_i \partial r_j}\right)\left(\frac{\partial^2 P_2}{\partial r_i \partial r_j}\right)\nu_i\nu_j$$

$$= 2[0.0396 - 0.02]\,[0.02 - 0.0004] \times 9 \times 10^{-6}$$

$$+ [1 - 0.0004]\,[0.0396 - 0] \times 9 \times 10^{-6}$$

$$+ [1 + 0.02 - 0.02 - 0.02]\,[0.02 + 0 - 0.02 - 0.02]$$

$$\times (9 \times 10^{-6})^2$$

$$+ 2[1 + 0 - 0.02 - 1]\,[1 + 0 - 0 - 0.02](9 \times 10^{-6})^2$$

$$= 0.0006915 \times 10^{-5} + 0.0356257 \times 10^{-5}$$

$$- 0.047628 \times 10^{-10}$$

$$= 0.0363168 \times 10^{-5}$$

$$\text{Var[LOLE]} = \text{Var}[P_1(35)] + \text{Var}[P_2(55)] + 2\,\text{Cov}[P_1(35),P_2(55)]$$

$$= 0.8999793 \times 10^{-5} + 0.021183 \times 10^{-5} + 2 \times 0.0363168 \times 10^{-5}$$

$$= 0.9747313 \times 10^{-5}$$

2.9.5 Additional considerations

The expected value associated with the calculated LOLE parameter can be obtained without recognition of the uncertainty associated with the generating unit unavailability. This parameter is affected by load forecast uncertainty. Uncertainties in forced outage rates and load forecasts can be incorporated in the same calculation [33]. The actual distribution associated with the calculated LOLE can only be obtained by Monte Carlo simulation. It has been suggested, however, that in many practical cases the distribution can be approximated by a gamma distribution which can then be used to place approximate confidence bounds on the LOLE for any particular situation.

The 'exact' technique illustrated in Section 2.9.1 becomes difficult to formulate if derated units are added to the capacity model. The 'approximate' method shown in Section 2.9.2 is, however, directly applicable and is not limited in regard to the number of derated states used. This situation is illustrated in Reference [34].

In conclusion, it is important to realize that there is a possible distribution associated with the calculated LOLE parameter. This distribution depends upon the inherent variability in the two basic parameters of load forecast uncertainty and the individual generating unit forced outage rates. The expected value of the LOLE parameter is not influenced by the uncertainty in the unit unavailabilities although the distribution of the LOLE parameter is affected by both uncertainty considerations. The distribution of the LOLE is useful in terms of determining approximate confidence bounds on the LOLE in any given situation. It is unlikely, however, that further use can be made of it at this time in practical system studies. The expected value of the calculated LOLE parameter is used as a conventional criterion for capacity evaluation. The uncertainty associated with the future load to be served by a proposed future capacity configuration is a significant factor which should be considered in long-term system evaluation.

2.10 Loss of energy indices

2.10.1 Evaluation of energy indices

The standard LOLE approach utilizes the daily peak load variation curve or the individual daily peak loads to calculate the expected number of days in the period that the daily peak load exceeds the available installed capacity. A LOLE index can also be calculated using the load duration curve or the individual hourly values. The area under the load duration curve represents the energy utilized during the specified period and can be used to calculate an expected energy not supplied due to insufficient installed capacity. The results of this approach can also be expressed in terms of the probable ratio between the load energy curtailed due to deficiencies in the generating capacity available and the total load energy required to serve the requirements of the system. For a given load duration curve this ratio is independent of the time period considered, which is usually a month or a year. The ratio is generally an extremely small figure less than one and can be defined as the 'energy index of unreliability'. It is more usual, however, to subtract this quantity from unity and thus obtain the probable ratio between the load energy that will be supplied and the total load energy required by the system. This is known as the 'energy index of reliability.'

The probabilities of having varying amounts of capacity unavailable are combined with the system load as shown in Fig. 2.19. Any outage of generating capacity exceeding the reserve will result in a curtailment of system load energy. Let:

$O_k =$ magnitude of the capacity outage
$P_k =$ probability of a capacity outage equal to O_k
$E_k =$ energy curtailed by a capacity outage equal to O_k

This energy curtailment is given by the shaded area in Fig. 2.19.

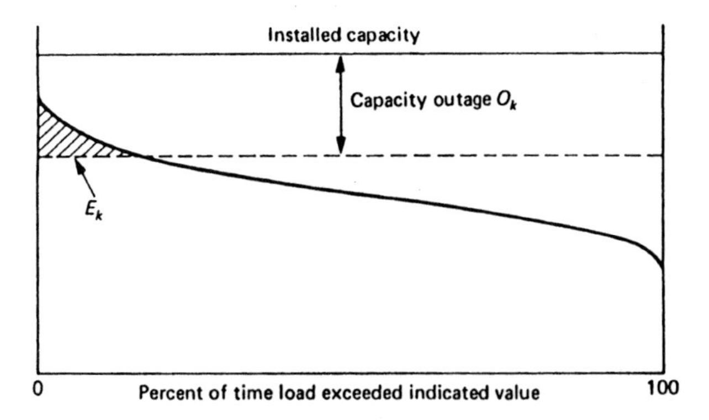

Fig. 2.19 Energy curtailment due to a given capacity outage condition

The probable energy curtailed is E_kP_k. The sum of these products is the total expected energy curtailment or loss of energy expectation LOEE where:

$$\text{LOEE} = \sum_{k=1}^{n} E_k P_k \tag{2.11}$$

This can then be normalized by utilizing the total energy under the load duration curve designated as E.

$$\text{LOEE}_{\text{p.u.}} = \sum_{k=1}^{n} \frac{E_k P_k}{E} \tag{2.12}$$

The per unit LOEE value represents the ratio between the probable load energy curtailed due to deficiencies in available generating capacity and the total load energy required to serve the system demand. The energy index of reliability, EIR, is then

$$\text{EIR} = 1 - \text{LOEE}_{\text{p.u.}} \tag{2.13}$$

This approach has been applied to the 5 × 40 MW unit system previously studied using the LOLE approach (Section 2.3.2). The system load–duration curve was assumed to be represented by a straight line from the 100% to the 40% load level. The risk as a function of the system peak load is given in Table 2.29. These results can be plotted in a similar form to Fig. 2.7. Although the 'loss of energy' approach has perhaps more physical significance than the 'loss of load' approach, it is not as flexible in overall application and has not been used as extensively.

It is important to appreciate, however, that future electric power systems may be energy limited rather than power or capacity limited and therefore future indices may be energy based rather than focused on power or capacity.

Table 2.29 Variation of EIR

System peak load (MW)	*Energy index of reliability*
200	0.997524
190	0.998414
180	0.999162
170	0.999699
160	0.999925
150	0.999951
140	0.999974
130	0.999991
120	0.999998
110	0.999999
100	0.999999

2.10.2 Expected energy not supplied

The basic expected energy curtailed concept can also be used to determine the expected energy produced by each unit in the system and therefore provides a relatively simple approach to production cost modelling. This approach, which is described in detail in Reference [35], is illustrated by the following example. Consider the load duration curve (LDC) shown in Fig. 2.20 for a period of 100 hours and the generating unit capacity data given in Table 2.30.

Assume that the economic loading order is Units 1, 2 and 3. The total required energy in this period is 4575.0 MWh, i.e. the area under the LDC in Fig. 2.20. If there were no units in the system, the expected energy not supplied, EENS, would be 4575.0 MW (=$EENS_0$). If the system contained only Unit 1, the EENS can be calculated as shown in Table 2.31.

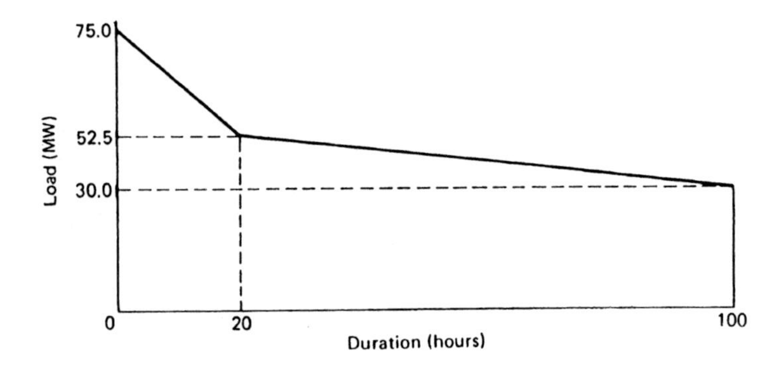

Fig. 2.20 Load model

Table 2.30 Generation data

Unit No.	*Capacity (MW)*	*Probability*
1	0	0.05
	15	0.30
	25	0.65
2	0	0.03
	30	0.97
3	0	0.04
	20	0.96

The contribution from Unit 2 can now be obtained by adding Unit 2 to the capacity model of Table 2.31 and calculating the EENS for Units 1 and 2 combined. This is shown in Table 2.32.

The final capacity outage probability table for all three units is shown in Table 2.33 and the $EENS_3 = 64.08$ MWh. The expected contribution from Unit 3 is 401.7 – 64.08 = 337.6 MWh. The individual unit expected energy outputs are summarized in Table 2.34.

The expected energy not supplied in the above system is 64.08 MWh. This can be expressed in terms of the energy index of reliability, EIR, using Equations (2.12) and (2.13):

$$\text{EIR} = 1 - \frac{64.08}{4575.0} = \underline{0.985993}$$

The situation in which Unit 1 is loaded to an intermediate level in the priority order before loading to full output at a higher priority level is illustrated in Reference [35]. Determination of expected unit energy outputs is a relatively simple matter in a system without energy limitations other than those associated with generating capacity outages. The approach illustrated can consider any number of units, derated capacity levels, load forecast uncertainty, station models and radial transmission limitations. The basic requirement is the ability to develop a sequential capacity outage probability table for the system generating capacity.

Table 2.31 EENS with Unit 1

Capacity out of service (MW)	*Capacity in service (MW)*	*Probability*	*Energy curtailed (MWh)*	*Expectation (MWh)*
0	25	0.65	2075.0	1348.75
10	15	0.30	3075.0	922.50
25	0	0.05	4575.0	228.75
				$EENS_1 = 2500.0$ MWh

The expected energy produced by Unit 1

$= EENS_0 - EENS_1$

$= 4575.0 - 2500.0 = \underline{2075.0 \text{ MWh}}.$

Table 2.32 EENS with Units 1 and 2

Capacity out of service (MW)	*Capacity in service (MW)*	*Probability*	*Energy curtailed (MWh)*	*Expectation (MWh)*
0	55	0.6305	177.8	112.10
10	45	0.2910	475.0	138.23
25	30	0.0485	1575.0	76.39
30	25	0.0195	2075.0	40.46
40	15	0.0090	3075.0	27.68
55	0	0.0015	4575.0	6.86
				$EENS_2 = 401.7$ MWh

The expected energy produced by Unit 2
$= EENS_1 - EENS_2$
$= 2500.0 - 401.7 =$ 2098.3 MWh.

Table 2.33 EENS with Units 1, 2 and 3

Capacity out of service (MW)	*Capacity in service (MW)*	*Probability*	*Energy curtailed (MWh)*	*Expectation (MWh)*
0	75	0.60528	0	—
10	65	0.27936	44.4	12.40
20	55	0.02522	177.8	4.49
25	50	0.04656	286.0	13.32
30	45	0.03036	475.0	14.42
40	35	0.00864	1119.4	9.67
45	30	0.00194	1575.0	3.06
50	25	0.00078	2075.0	1.62
55	20	0.00144	2575.0	3.71
60	15	0.00036	3075.0	1.11
75	0	0.00006	4575.0	0.28
				64.08

Table 2.34 Summary of EENS

Priority level	*Unit capacity (MW)*	*EENS (MWh)*	*Expected energy output (MWh)*
1	25	2500.0	2075.0
2	30	401.7	2098.3
3	20	64.1	337.6

2.10.3 Energy-limited systems

The simplest energy-limited situation to incorporate into the analysis is the condition in which the output capacity of a unit is dictated by the energy available. An example of this energy limitation is a run-of-the-river hydro installation with little or no storage. The flow rate determines the unit output capacity. The unit is then represented as a multi-state unit in which the capacity states correspond to the water flow rates. This representation might also apply to variable flow availabilities of natural gas. The analysis in this case is identical to that used in Section 2.10.2 for a non-energy-limited unit. This is illustrated by adding a 10 MW generating unit with a capacity distribution due to a flow-rate distribution as described in Table 2.35 to the system analyzed in Section 2.10.2.

The unit can be placed in an appropriate place in the priority loading order and the expected energy outputs calculated using the previous techniques. The expected energy not supplied in this case is 35.5 MWh and the EIR = 0.992236.

Generating units which have short-term storage associated with their prime mover can be used to peak shave the load and therefore reduce the requirement from more expensive units. The approach in this case is to modify the load model using the capacity and energy distributions of the limited energy storage unit and then apply the technique described earlier for the non-energy-limited units. If this load modification technique is used in connection with a non-energy-limited unit system analysis, the results are identical to those obtained by the basic method.

The first step is to capacity-modify the load–duration curve using a conditional probability approach. The modified curve is the equivalent load curve for the rest of the units in the system if the unit used to modify it was first in the priority list. The capacity-modified curve is then energy-modified using the energy probability distribution of the unit under consideration. The final modified curve is then used in the normal manner with the rest of the units in the system to determine their expected energy outputs and the resulting expected energy not supplied.

The approach can be illustrated by adding the unit shown in Table 2.36 to the original three-unit system in Table 2.30.

The capacity-modified curve is shown in Fig. 2.21. The curve is obtained by the conditional probability approach used earlier for load forecast uncertainty analysis. The energy-modified load–duration curve is shown in Fig. 2.22.

Table 2.35 Data for 10 MW unit

Capacity (MW)	*Probability*
0	0.040
2.5	0.192
5.0	0.480
10.0	0.288
	1.000

Table 2.36 Energy-limited unit

Capacity model		*Energy model*	
Capacity (MW)	*Probability*	*Energy (MWh)*	*Cumulative probability*
0	0.03	200	1.00
10	0.25	350	0.70
15	0.72	500	0.20

The resulting load–duration curve in Fig. 2.22 becomes the starting curve for subsequent unit analysis. In the example used, an additional unit of 10 MW with a forced outage rate of 0.04 was added to the previous system. A two-state energy distribution was assumed with 70.0 and 150.0 MWh having cumulative probabilities of 1.0 and 0.6 respectively. Under these conditions, the expected energy not supplied is 15.7 MWh and the EIR of the system for the 100 hour period is 0.996562. Reference [35] illustrates the extension of this technique to the situation in which an energy-limited unit is partly base loaded and partly used for peak shaving.

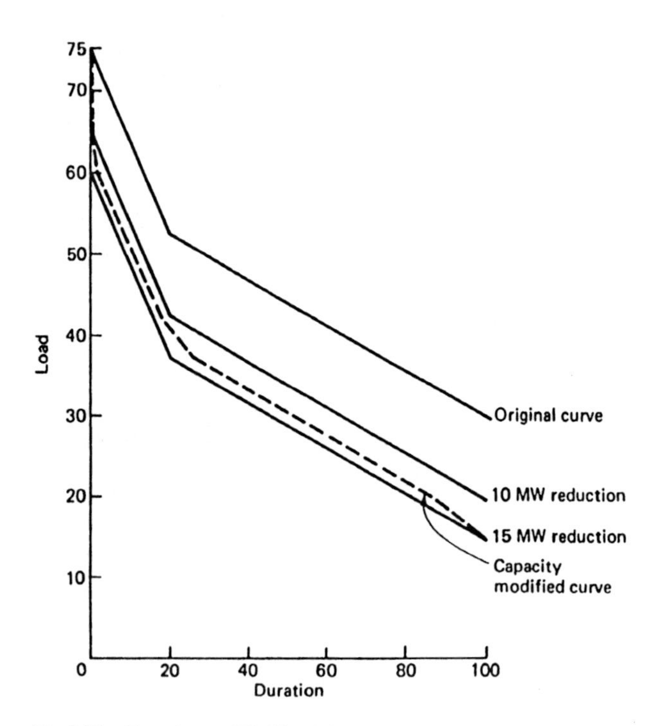

Fig. 2.21 Capacity-modified load-duration curve

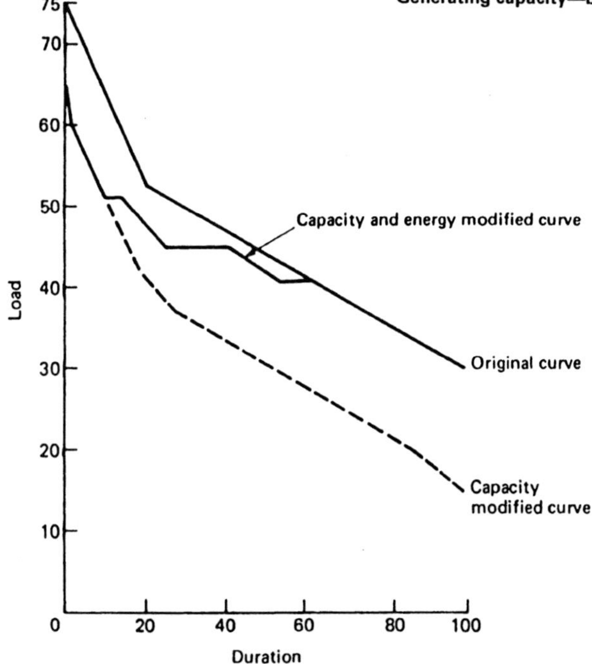

Fig. 2.22 Energy-modified load–duration curve

A further type of storage facility, which is commonly encountered, is one in which the stored energy can be held for some time and used in both a peak shaving and base load manner. In the case of a hydro facility with a large reservoir, the operation would be guided by a rule curve which dictates how much energy should be used during the specified period. The available energy during each period can vary due to in-flow variations and operating policies. The approach in this case is to capacity-modify the load–duration curve using the non-energy-limited units. This leaves an equivalent load curve for the rest of the units. The units with energy limitations can then be used to peak shave the equivalent load–duration curve. The area under the load–duration curve after these units have been dispatched is the expected load energy not supplied. A numerical example for this type of system is shown in Reference [35].

2.11 Practical system studies

The techniques and algorithms presented in this chapter are suitable for the analysis of both small and large systems. Typical practical systems contain a large number of generating units and cannot normally be analyzed by hand calculations. The algorithms presented can be used to create efficient computer programs for the analysis of practical system configurations. The IEEE Subcommittee on the Application of Probability Methods recently published a Reliability Test System containing a generation configuration and an appropriate bulk transmission network [36]. It is expected that this system will become a reference for research in new techniques and in comparing the results obtained using different computer programs. Appendix 2 contains the basic generation model from the IEEE Reliability Test System (IEEE–RTS) and also a range of results from different reliability studies. These results cannot be obtained by hand analysis. The reader is encouraged to develop his digital computer program using the techniques contained in this book and to compare them with those presented in Appendix 2.

2.12 Conclusions

This chapter has illustrated the application of basic probability concepts to generating capacity reliability evaluation. The LOLE technique is the most widely used probabilistic approach at the present time. There are, however, many differences in the resulting indices produced. These differences depend mainly on the factors used in the calculation procedure, i.e. derated representation or EFOR values, uncertainty considerations, maintenance effects, etc. Different indices are created by using different load models. It is not valid to obtain an LOLE index in hours by dividing the days/year value obtained using a daily peak load variation curve, DPLVC, by 24, as the DPLVC has a different shape from the load–duration curve, LDC. If an LOLE in hours/year is required, then the LDC should be used. The LDC is a better representation than the DPLVC as it uses more actual system data. The energy not supplied is an intuitively appealing index as it tends to include some measure of basic inadequacy rather than just the number of days or hours that all the load was not satisfied.

The basic LOLE index has received some criticism in the past on the grounds that it does not recognize the difference between a small capacity shortage and a large one, i.e. it is simply concerned with 'loss of load'. All shortages are therefore treated equally in the basic technique. It is possible, however, to produce many additional indices such as the expected capacity shortage if a shortage occurs, the expected number of days that specified shortages occur, etc. It is mainly a question of deciding what expectation indices are required and then proceeding to calculate them. The derived indices are expected values (i.e. long run average) and should not be expected to occur each year. The indices should also not be considered as absolute measures of capacity adequacy and they do not describe the frequency and duration of inadequacies. They do not include operating considerations such as spinning reserve requirements, dynamic and transient system disturbances, etc. Indices such as LOLE and LOEE are simply indications of static capacity adequacy which respond to the basic elements which influence the adequacy of a given configuration, i.e. unit size and availability, load shape and uncertainty. Inclusion of additional parameters does not change this fundamental concept. Inclusion of elements such as maintenance, etc., make the derived index sensitive to these elements and therefore a more overall index, but still does not make the index an absolute measure of generation system reliability.

2.13 Problems

1 A power system contains the following generating capacity.

3×40 MW hydro units FOR = 0.005
1×50 MW thermal unit FOR = 0.02
1×60 MW thermal unit FOR = 0.02

The annual daily peak load variation curve is given by a straight line from the 100% to the 40% points.

(a) Calculate the loss of load expectation for the following peak load values.
(i) 150 MW (ii) 160 MW (iii) 170 MW
(iv) 180 MW (v) 190 MW (vi) 200 MW

(b) Calculate the loss of load expectation for the following peak load values, given that another 60 MW thermal unit with a FOR of 0.02 is added to the system.
(i) 200 MW (ii) 210 MW (iii) 220 MW (iv) 230 MW
(v) 240 MW (vi) 250 MW (vii) 260 MW

(c) Determine the increase in load carrying capability at the 0.1 day/year risk level due to the addition of the 60 MW thermal unit.

(d) Calculate the loss of load expectation for the load levels in (a) and (b) using the load forecast uncertainty distribution shown in Fig. 2.23.

(e) Determine the increase in load carrying capability at the 0.1 day/year risk level for the conditions in part (d).

2 A generating system contains three 25 MW generating units each with a 4% FOR and one 30 MW unit with a 5% FOR. If the peak load for a 100 day period is 75 MW, what

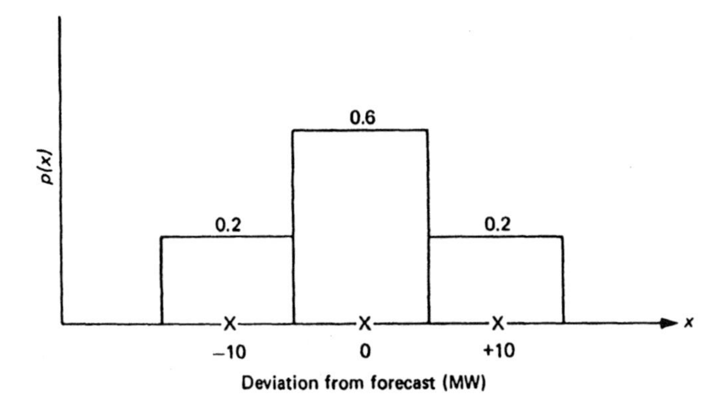

Fig. 2.23

is the LOLE and EIR for this period? Assume that the appropriate load characteristic is a straight line from the 100% to the 60% points.

3 A system contains three non-identical 30 MW generating units each with a 5% FOR and one 50 MW unit with a 6% FOR. The system peak load for a specified 100 day period is 120 MW. The load–duration curve for this period is a straight line from the 100% to the 80% load points.

Calculate the energy index of reliability for this system. The economic loading order for this system is the 50 MW unit first, followed by the 30 MW units A, B and C, in that order. Calculate the expected energy provided to the system by the 50 MW unit and by the 30 MW unit C.

4 A system contains 120 MW of generating capacity in 6 × 20 MW units. These units are connected through step-up station transformers to a high-voltage bus. The station is then connected to a bulk system load point by two identical transmission lines. This configuration is shown in Fig. 2.24.

System data

Generating units	*Transformers*	*Transmission lines*
$\lambda = 3$ f/Year	$\lambda = 0.1$ f/year	$\lambda = 3$ f/year/100 m
$\mu = 97$ r/year	$\mu = 19.9$ r/year	$\mu = 365$ r/year

Assume that the load-carrying capabilities of lines 1 and 2 are 70 MW each. The annual daily peak load variation curve is a straight line from the 100% to the 70% points. The annual load–duration curve is a straight line from the 100% to the 50% point.

(a) Conduct a LOLE study at the generating bus and at the load bus for an annual forecast peak load of 95 MW.

(b) Repeat Question (a) given that each pair of generating units is connected to the high voltage bus by a single transformer.

(c) Calculate the expected energy not supplied and the energy index of reliability at the load bus for a forecast annual peak load of 95 MW.

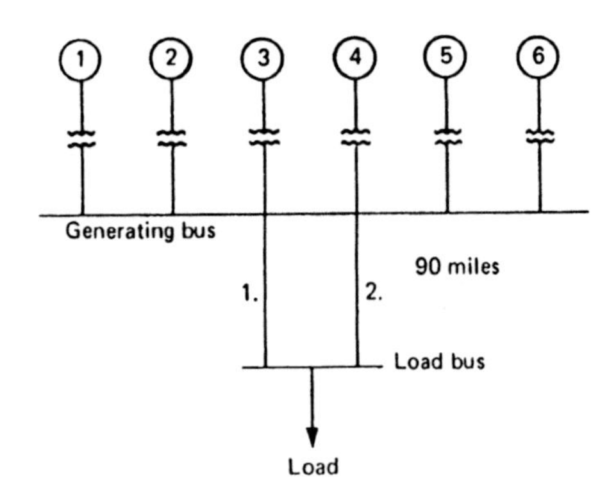

Fig. 2.24

5 A generating system consists of the following units:
(A) 1 × 10 MW unit
(B) 1 × 20 MW unit
(C) 1 × 30 MW unit
(D) 1 × 40 MW unit
The 10, 20 and 30 MW units have forced outage rates of 0.08. The 40 MW unit has a full forced outage rate of 0.08 and a 50% derated state which has a probability of 0.06. (a) Calculate the LOLE for this system for a single daily peak load of 60 MW. (b) What is the LOLE for the same condition if the 40 MW unit is represented as a two-state model using an Equivalent Forced Outage Rate?

6 The generating system given in Question 5 supplies power to an industrial load. The peak load for a specified 100-day period is 70 MW. The load–duration curve for this period is a straight line from the 100% to the 60% load point.
(a) Calculate the energy index of reliability for this system.
(b) Given that the economic loading order for the generating units is (D), (C), (B), (A), calculate the expected energy provided to the system by each unit.

7 A four-unit hydro plant serves a remote load through two transmission lines. The four hydro units are connected to a single step-up transformer which is then connected to the two lines. The remote load has a daily peak load variation curve which is a straight line from the 100% to the 60% point. Calculate the annual loss of load expectation for a forecast peak of 70 MW using the following data.

Hydro units
25 MW. Forced outage rate = 2%.
Transformer
110 MVA. Forced outage rate = 0.2%
Transmission lines
Carrying capability 50 MW each line
Failure rate = 2 f/year
Average repair time = 24 hours

2.14 References

1. Billinton, R., *Bibliography on Application of Probability Methods in the Evaluation of Generating Capacity Requirements*, IEEE Winter Power Meeting (1966), Paper No. 31 CP 66–62.
2. Billinton, R., 'Bibliography on the application of probability methods in power system reliability evaluation', *IEEE Transactions*, **PAS-91** (1972), pp. 649–660.
3. IEEE Subcommittee on the Application of Probability Methods, 'Bibliography on the application of probability methods in power system reliability evaluation 1971–77'. *IEEE Transactions*, **PAS-97** (1978), pp. 2235–2242.
4. Calabrese, G., 'Generating reserve capacity determined by the probability method', *AIEE Transactions*, **66** (1947), pp. 1439–50.

5. Lyman, W. J., 'Calculating probability of generating capacity outages', *AIEE Transactions*, **66** (1947), pp. 1471–7.
6. Seelye, H. P., 'A convenient method for determining generator reserve', *AIEE Transactions*, **68** (Pt. II) (1949), pp. 1317–20.
7. Loane, E. S., Watchorn, C. W., 'Probability methods applied to generating capacity problems of a combined hydro and steam system', *AIEE Transactions*, **66** (1947), pp. 1645–57.
8. Baldwin, C. J., Gaver, D. P., Hoffman, C. H., 'Mathematical models for use in the simulation of power generation outages: I—Fundamental considerations', *AIEE Transactions* (*Power Apparatus and Systems*), **78** (1959), pp. 1251–8.
9. Baldwin, C. J., Billings, J. E., Gaver, D. P., Hoffman, C. H., 'Mathematical models for use in the simulation of power generation outages: II—Power system forced outage distributions', *AIEE Transactions* (*Power Apparatus and Systems*), **78** (1959), pp. 1258–72.
10. Baldwin, C. J., Gaver, D. P., Hoffman, C. H., Rose, J. A., 'Mathematical models for use in the simulation of power generation outages: III—Models for a large interconnection', *AIEE Transactions (Power Apparatus and Systems)*, **78** (1960), pp. 1645–50.
11. Hall, J. D., Ringlee, R. J., Wood, A. J., 'Frequency and duration methods for power system reliability calculations: Part I—Generation system model', *IEEE Transactions*, **PAS-87** (1968), pp. 1787–96.
12. Ringlee, R. J., Wood, A. J., 'Frequency and duration methods for power system reliability calculations: Part II—Demand model and capacity reserve model', *IEEE Transactions*, **PAS-88** (1969), pp. 375–88.
13. Galloway, C. D., Garver, L. L., Ringlee, R. J., Wood, A. J., 'Frequency and duration methods for power system reliability calculations: Part III—Generation system planning', *IEEE Transactions*, **PAS-88** (1969), pp. 1216–23.
14. Cook, V. M., Ringlee, R. J., Wood, A. J., 'Frequency and duration methods for power system reliability calculations: Part IV—Models for multiple boiler-turbines and for partial outage states', *IEEE Transactions*, **PAS-88** (1969), pp. 1224–32.
15. Ringlee, R. J., Wood, A. J., 'Frequency and duration methods for power system reliability calculations: Part V—Models for delays in unit installations and two interconnected systems', *IEEE Transactions*, **PAS-90** (1971), pp.79–88.
16. *IEEE Standard Definitions For Use in Reporting Electric Generations Unit Reliability, Availability and Productivity*. IEEE Standard 762–1980.
17. IEEE Committee, 'A four state model for estimation of outage risk for units in peaking service', IEEE Task Group on Model for Peaking Units of the Application of Probability Methods Subcommittee, *IEEE Transactions*, **PAS-91** (1972), pp. 618–27.

18. Patton, A. D., Singh, C., Sahinoglu, M., 'Operating considerations in generation reliability modelling—an analytical approach', *IEEE Transactions*, **PAS-100** (1981), p. 2656–71.
19. Billinton, R., *Power System Reliability Evaluation*, Gordon and Breach, New York (1970).
20. Bhavaraju, M. P., *An Analysis of Generating Capacity Reserve Requirements*, IEEE (1974), Paper No. C-74-155-0.
21. Schenk, K. F., Rau, N. S., *Application of Fourier Transform Techniques for the Assessment of Reliability of Generating Systems*, IEEE Winter Power Meeting (1979), Paper No. A-79-103-3.
22. Stremel, J. P., 'Sensitivity study of the cumulant method of calculating generation system reliability', *IEEE Transactions*, **PAS-100** (1981), pp.771–78.
23. Billinton, R., Hamoud, G., Discussion of Reference 22.
24. Allan, R. N., Leite da Silva, A. M., Abu-Nasser, A., Burchett, R. C., 'Discrete convolution in power system reliability', *IEEE Transactions on Reliability*, **R-30** (1981), pp. 452–6.
25. Billinton, R., Krasnodebski, J., 'Practical application of reliability and maintainability concepts to generating station design', *IEEE Transactions*, **PAS-92** (1973), pp. 1814–24.
26. Walters, J., 'Realizing reliability', *Proceedings 1979 Reliability Conference for the Electric Power Industry*. Miami, Fla (April 1979).
27. *Instruction Manual*, Generation Equipment Status Canadian Electrical Association (1979).
28. Billinton, R., Kuruganty, P. R. S., 'Unit derating levels in generating capacity evaluation', *Proceedings 1976 Annual Conference for the Electric Power Industry*, Montreal (1976).
29. Billinton, R., 'Reliability criteria used by Canadian utilities in generating capacity planning and operation', *IEEE Transactions*, **PAS-97** (1978), pp. 1097–1103.
30. Billinton, R., *Distributed Representation of Forced Outage Probability Value in Generating Capacity Reliability Studies*, IEEE Winter Power Meeting (1966), Paper No. 31-CP-66-61.
31. Billinton, R., Bhavaraju, M. P., *Outage Rate Confidence Levels In Static and Spinning Reserve Evaluation*, IEEE Summer Power Meeting (1968), Paper No. 68-CP-608-PWR.
32. Patton, A. D., Stasinos, A., 'Variance and approximate confidence levels on LOLP for a single area system', *IEEE Transactions*, **PAS-94** (1975), pp. 1326–33.
33. Wang, L., 'The effects of uncertainties in forced outage rates and load forecast on the loss of load probability (LOLP)', *IEEE Transactions*, **PAS-96** (1977), pp. 1920–27.

34. Hamoud, G., Billinton, R., 'An approximate and practical approach to including uncertainty concepts in generating capacity reliability evaluation', *IEEE Transactions*, **PAS-100** (1981), pp. 1259–65.
35. Billinton, R., Harrington, P. G., 'Reliability evaluation in energy limited generating capacity studies', *IEEE Transactions*, **PAS-97** (1978), pp. 2076–86.
36. Reliability System Task Force of the Application of Probability Methods Subcommittee, 'IEEE reliability test system', *IEEE Transactions*, **PAS-98** (1979), pp. 2047–54.
37. Allan, R. N., Takieddine, F. N., 'Generation modelling in power system reliability evaluation', IEE Conference on Reliability of Power Supply Systems, IEE Conf. Pub. 148 (1977), pp. 47–50.
38. Allan, R. N., Takieddine, F. N., 'Generator maintenance scheduling using simplified frequency and duration reliability criteria', *Proc. IEE*, **124** (1977), pp. 873–80.
39. Allan, R. N., Takieddine, F. N., *Network Limitations on Generating Systems Reliability Evaluation Techniques*, IEEE Winter Power Meeting (1978), Paper No. A 78 070-5.

3 Generating capacity—frequency and duration method

3.1 Introduction

The previous chapter illustrates the application of basic probability methods to the evaluation of static capacity adequacy. The basic indices illustrated in Chapter 2 are the expected number of days (or hours) in a given period that the load exceeded the available capacity and the expected energy not supplied in the period due to insufficient installed capacity. These are useful indices which can be used to compare the adequacy of alternative configurations and expansions. They do not, however, give any indication of the frequency of occurrence of an insufficient capacity condition, nor the duration for which it is likely to exist. The LOLE index of days/year, when inverted to provide years/day, is often misinterpreted as a frequency index. It should, however, be regarded in its basic form as the expected number of days/year that the load exceeds the available installed capacity.

A frequency and duration approach to capacity evaluation was first introduced by Halperin and Adler [1] in 1958. This approach is somewhat cumbersome and the indices were not really utilized until a group of papers by Ringlee, Wood *et al.* [2–5] in 1968–69 presented recursive algorithms for capacity model building and load model combination which facilitated digital computer application [6].

There are increasingly many attempts to incorporate the generation and major transmission elements into an overall or composite system evaluation procedure which can provide both load point and overall system adequacy indices. This is the subject of Chapter 6 of this text. Frequency and duration are the most useful indices for customer or load point evaluation and therefore the creation of similar indices for capacity assessment appears to offer increased compatibility in overall assessment.

The basic approach is portrayed in Fig. 2.1 and therefore this chapter will basically follow the format presented in Chapter 2 by illustrating the development of the capacity models followed by the load models and the subsequent convolution to create the system risk indices. As in the LOLE approach, the basic system representation is that shown in Fig. 2.2. Transmission elements will be introduced in Chapter 6.

The frequency and duration (F&D) method requires additional system data to that used in the basic probabilistic methods. Figure 2.3(a) illustrates the fundamental two-state model for a base load generating unit. The LOLE or LOEE methods utilize the steady-state availability A and the unavailability U parameters for this model. The F&D technique utilizes in addition to A&U, the transition rate parameters λ and μ. The basic concepts associated with frequency and duration analysis are described in detail in *Engineering Systems* and therefore are not repeated in this text. The fundamental relationship however can be obtained from Equation 2.1(b).

$$\text{Availability } A = \frac{\mu}{\lambda + \mu} = \frac{f}{\lambda}$$

$$\therefore \; f = A\lambda \tag{3.1}$$

The frequency of encountering state 0 in Fig. 2.3(a) is the probability of being in the state multiplied by the rate of departure from the state. In the case of the two-state model it is also equal to the probability of not being in the state multiplied by the rate of entry (Equation 2.1(a)). In a more general sense the frequency of a particular condition can be expressed as the mathematical expectation of encountering the boundary wall surrounding that condition. The frequency of entry is equal to the frequency of leaving. This concept of frequency balance was presented in *Engineering Systems* as a means of formulating equations for the solution of state transition model probabilities.

3.2 The generation model

3.2.1 Fundamental development

The concepts can perhaps be most easily seen by using a simple numerical example. The system described in Table 2.7 contains the basic data required for both the LOLE and the F&D methods. This section illustrates the development of a capacity model using the fundamental relationship shown by Equation (3.1). This is not a practical approach for large system analysis using a digital computer; the recursive technique shown in Section 3.2.2 should be used.

If each unit can exist in two states, then there are 2^n states in the total system where n = number of units (i.e. 2^3 in this case). The total number of states in the system of Table 2.7 are enumerated in Table 3.1.

These states can also be represented as a state transition diagram as shown in Fig. 3.1. This diagram enumerates all the possible system states and also shows the transition modes from one state to another. As an example, given that the system is in State 2 in which unit 1 is down and the others are up, the system can transit to States 1, 5 or 6 in the following ways:

Table 3.1 Failure modes and effects

State number	*1*	*2*	*3*	*4*	*5*	*6*	*7*	*8*
Unit No. 1	U	D	U	U	D	D	U	D
Unit No. 2	U	U	D	U	D	U	D	D
Unit No. 3	U	U	U	D	U	D	D	D
Capacity out (MW)	0	25	25	50	50	75	75	100

U = Up D = Down

From State 2 to 1 if unit 1 is repaired.
From State 2 to 5 if unit 2 fails.
From State 2 to 6 if unit 3 fails.

The total rate of departure from State 2 is therefore the sum of the individual rates of departure ($\mu_1 + \lambda_2 + \lambda_3$). The probabilities associated with each state in Table 3.1 can be easily calculated assuming event independence. The frequencies of encountering each state are obtained using Equation (3.1) when the rate of departure or

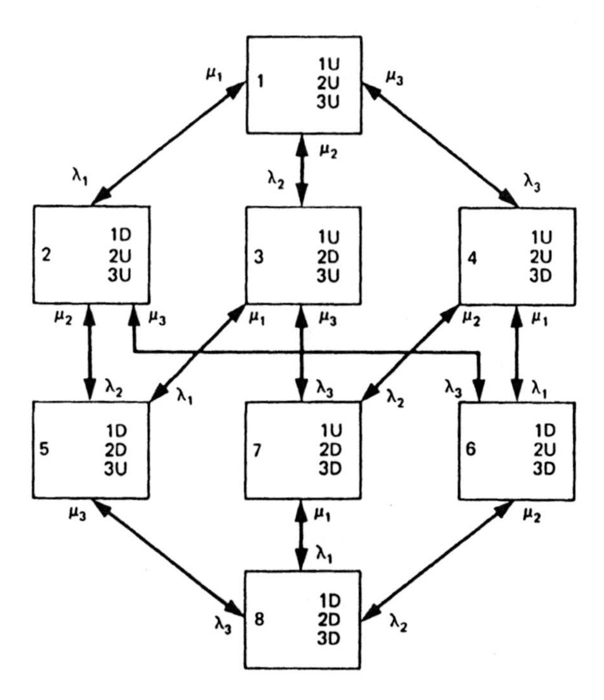

Fig. 3.1 Three-unit state space diagram

Table 3.2 Generation model

State No.	Capacity out C_i (MW)	State probability, p_i	State frequency, f_i (encounters/day
1	0	(0.98)(0.98)(0.98) = 0.941192	(0.941192)(0.03) = 0.02823576
2	25	(0.02)(0.98)(0.98) = 0.019208	(0.019208)(0.51) = 0.00979608
3	25	(0.98)(0.02)(0.98) = 0.019208	(0.019208)(0.51) = 0.00979608
4	50	(0.98)(0.98)(0.02) = 0.019208	(0.019208)(0.51) = 0.00979608
5	50	(0.02)(0.02)(0.98) = 0.000392	(0.000392)(0.99) = 0.00038808
6	75	(0.02)(0.98)(0.02) = 0.000392	(0.000392)(0.99) = 0.00038808
7	75	(0.98)(0.02)(0.02) = 0.000392	(0.000392)(0.99) = 0.00038808
8	100	(0.02)(0.02)(0.02) = 0.000008	(0.000008)(1.47) = 0.00001176
		1.000000	

entry is the sum of the appropriate rates. The basic manipulations are shown in Table 3.2.

Table 3.2 contains a number of identical capacity states which can be combined using the following equations where the subscript *i* refers to the identical states and *k* refers to the new merged state.

The capacity outage of state k $= C_k = C_1 = C_2 = \ldots = C_i$

The probability of state k $= p_k = \sum p_i$ (3.2)

The frequency of state k $= f_k = \sum f_i$ (3.3)

The rates of departure from state $k = \lambda_{\pm k}$

$$\lambda_{\pm k} = \frac{\Sigma p_i \lambda_{\pm i}}{p_k} \tag{3.4}$$

The reduced model is shown in Table 3.3 in which the state numbers refer to the new merged states.

Table 3.3 Reduced generation model

State No.	Capacity out (MW)	Capacity in (MW)	Probability p_k	Frequency (occ/day) f_k
1	0	100	0.941192	0.02823576
2	25	75	0.038416	0.01959216
3	50	50	0.019600	0.01018416
4	75	25	0.000784	0.00077616
5	100	0	0.000008	0.00001176

The generation system model in the form shown in Table 3.3 gives the probability and frequency of having a given level of capacity forced out of service and of the complementary level of capacity in service. This generation system model can be modified to give cumulative probabilities and frequencies rather than values corresponding to a specific capacity level. At any given capacity level the cumulative values give the probability and frequency of having that capacity *or more* forced out of service. The individual state probabilities and frequencies can be combined to form the cumulative state values using the following equations:

$$P_{n-1} = p_k + P_n \tag{3.5}$$

$$F_{n-1} = F_n + p_k\lambda_{+k} - p_k\lambda_{-k} \tag{3.6}$$

where n refers to the cumulative state with known probability and frequency and k is the state which is being combined to form the cumulative state $n - 1$. The process therefore starts from the last state, in which the individual and cumulative values are the same. The transition values λ_{+k} and λ_{-k} are the transition rates to higher and lower available capacity levels respectively.

The process can be started with State 5 in Table 3.3.

State 5

$$C_5 \geq 100 \text{ MW}$$

$$P_5 = p_5 = 0.000008$$

$$F_5 = f_5 = 0.00001176/\text{day}$$

State 4

$$C_4 \geq 75 \text{ MW}$$

$$P_4 = p_4 + P_5 = 0.000792$$

$$F_4 = F_5 + p_4\lambda_{+4} - p_4\lambda_{-4}$$

$$= 0.00001176 + 0.00076832 - 0.00000784$$

$$= 0.00077224/\text{day}$$

The values for $p_4\lambda_{+4}$ and $p_4\lambda_{-4}$ are obtained using the state probabilities in Table 3.2, the given unit transition rates and Equation (3.4), i.e.

$$p_4\lambda_{+4} = 0.000392(0.49 + 0.49) + 0.000392(0.49 + 0.49) = 0.00076832$$

$$p_4\lambda_{-4} = 0.000392(0.01) + 0.000392(0.01) = 0.000784.$$

State 3

$$C_3 \geq 50 \text{ MW}$$

$$P_3 = p_3 + P_4 = 0.020392$$

$$F_3 = F_4 + p_3\lambda_{+3} - p_3\lambda_{-3}$$

$$= 0.00077224 + 0.00979608 - 0.00038808$$

$$= 0.01018024/\text{day}$$

State 2

$$C_2 \geq 25 \text{ MW}$$

$$P_2 = p_2 + P_3 = 0.058808$$

$$F_2 = F_3 + p_2\lambda_{+2} - p_2\lambda_{-2} = 0.02823576/\text{day}$$

State 1

$$C_1 \geq 0 \text{ MW}$$

$$P_1 = p_1 + P_2 = 1.000000$$

$$F_1 = F_2 + p_1\lambda_{+1} - p_1\lambda_{-1} = 0$$

These values are shown in tabular form in Table 3.4.

The generation system model shown in Table 3.4 can be used directly as an indication of system generating capacity adequacy. When used in this form, the approach is known as the 'loss of capacity method'. Table 3.4 contains both cumulative probability and frequency values at each capacity level and can be considered as a complete system capacity model. The conventional capacity outage probability table as used in the LOLE approach does not include any frequency parameters and is given by the values to the left of the dashed line in Table 3.4. Omitting the frequency aspect simplifies the development of the system capacity model but with the attendant loss of some of the physical significance of the model capacity levels.

Table 3.4 Generation system model

State No.	*Capacity out (MW)*	*Cumulative probability P_k*	*Cumulative frequency/day F_k*
1	0	1.000000	0.0
2	25	0.058808	0.02823576
3	50	0.020392	0.01018024
4	75	0.000792	0.00077224
5	100	0.000008	0.00001176

3.2.2 Recursive algorithm for capacity model building [6]

The capacity model can be created using relatively simple algorithms which can also be used to remove a unit from a table. This approach can also be used for a multi-state unit. The technique is illustrated for a two-state unit addition followed by the more general case of a multi-state unit.

Case 1 No derated states

The recursive expressions for a state of 'exactly X MW on forced outage' after a unit of C MW and forced outage rate U is added are shown in Equations (3.7)–(3.9).

$$p(X) = p'(X)(1 - U) + p'(X - C)U \tag{3.7}$$

$$\lambda_+(X) = \frac{p'(X)(1 - U)\lambda_+(X) + p'(X - C)U(\lambda'_+(X - C) + \mu)}{p(X)} \tag{3.8}$$

$$\lambda_-(X) = \frac{p'(X)(1 - U)(\lambda'_-(X) + \lambda) + p'(X - C)(\lambda'_-(X - C))}{p(X)} \tag{3.9}$$

The $p(X)$, $\lambda_+(X)$ and $\lambda_-(X)$ parameters are the individual state probability and the upward and downward capacity departure rates respectively after the unit is added. The primed values represent similar quantities before the unit is added. In Equations (3.7), (3.8) and (3.9), if X is less than C

$$p'(X - C) = 0$$

$$\lambda'_+(X - C) = 0$$

$$\lambda'_-(X - C) = 0$$

The procedure is initiated with the addition of the first unit (C_1). In this case

$$\lambda_+(0) = 0$$

$$\lambda_-(0) = \lambda_1$$

$$\lambda_+(C_1) = \mu_1$$

$$\lambda_-(C_1) = 0$$

$$\lambda_+(X) = \lambda_-(X) = 0 \quad \text{for } X \neq 0, C_1$$

The algorithms are illustrated using the system given in Table 2.7.

Step 1 Add the first unit (Table 3.5)

Table 3.5

State No. i	*Cap. out (MW)*	*Probability p(X)*	$\lambda_+(X)$ *(occ/day)*	$\lambda_-(X)$ *(occ/day)*
1	0	0.98	0	0.01
2	25	0.02	0.49	0

Table 3.6

(1) Cap out X (MW)	(2) $p'(X)(1-U)$	(3) $p'(X-C)U$	(4) Col (4) = Col (2) + Col (3) p (X)
0	0.98×0.98	$0 \quad \times 0.02$	0.9604
25	0.02×0.98	0.98×0.02	0.0392
50	$0 \quad \times 0.98$	0.02×0.02	0.0004

(1) Cap. out X (MW)	(5) Col (2) $(\lambda'_+(X))$	(6) Col (3) $(\lambda'_+(X-C)+\mu)$	(7) Col (5) + Col (6)	(8) Col (7)/Col (4) $\lambda_+(X)$(occ/day)
0	0.9604×0	$0 \quad \times (0+0.49)$	0	0
25	0.0196×0.49	$0.0196 \times (0+0.49)$	0.019208	0.49
50	$0 \quad \times 0$	$0.0004 \times (0.49+0.49)$	0.000392	0.98

(1) Cap. out X (MW)	(9) Col (2) $(\lambda'_-(X)+\lambda)$	(10) Col (3) $(\lambda'_-(X-C))$	(11) Col (9) + Col (10)	(12) Col (11)/Col (4) $\lambda_-(X)$ (occ/day)
0	$0.9604 \times (0.01+0.01)$	$0 \quad \times 0$	0.019208	0.02
25	$0.0196 \times (0+0.01)$	0.0196×0.01	0.000392	0.01
50	$0 \quad \times (0+0.01)$	0.0004×0	0	0

Step 2 Add the second unit (Table 3.6)

Note: The columns in Table 3.6 have been given numbers and are referred to as Col (2) etc. in subsequent manipulations.

Step 3 Add the third unit (Table 3.7)

The individual capacity state probabilities are given in Col (4). They can be combined directly with the values in Col (8) and Col (12) to give the individual state frequencies. These values can also be used to give the cumulative state probabilities and frequencies using the following equations.

$$P(X) = P(Y) + p(X) \tag{3.10}$$

$$F(X) = F(Y) + p(X)(\lambda_+(X) - \lambda_-(X)) \tag{3.11}$$

where Y denotes the capacity outage state just larger than X MW.

The complete capacity model is shown in Table 3.8.

The results shown in Table 3.8 are the same as those shown in Tables 3.3 and 3.4. The approach shown in Section 3.2.1 is ideally suited for very small systems and uses only the basic concepts of frequency analysis. It is not practical to use this approach for large or practical system studies. The recursive algorithms shown in this section, however, are ideally suited to digital computer application and provide a fast technique for building capacity models.

Table 3.7

(1) Cap. out X (MW)	(2) $p'(X)(1-U)$	(3) $P'(X-C)U$	(4) Col (4) = Col (2) + Col (3) p(X)
0	0.9604 × 0.98	0 × 0.02	0.941192
25	0.0392 × 0.98	0 × 0.02	0.038416
50	0.0004 × 0.98	0.9604 × 0.02	0.019600
75	0 × 0.98	0.0392 × 0.02	0.000784
100	0 × 0.98	0.0004 × 0.02	0.000008

(1) Cap. out X (MW)	(5) Col (2) $\lambda_+'(X)$)	(6) Col (3) ($\lambda_+'(X-C)+\mu$)	(7) Col (5) + Col (6)	(8) Col (7)/Col (4) $\lambda_+(X)$ (occ/day)
0	0.941192 × 0	0 × (0 + 0.49)	0	0
25	0.038416 × 0.49	0 × (0 + 0.49)	0.018824	0.4900
50	0.000392 × 0.98	0.019208 × (0 + 0.49)	0.009796	0.4998
75	0 × 0	0.000784 × (0.49 + 0.49)	0.000768	0.9800
100	0 × 0	0.000008 × (0.98 + 0.49)	0.000012	1.4700

(1) Cap. out X (MW)	(9) Col (2) $\lambda'_-(X)+\lambda$)	(10) Col (3) ($\lambda'_-(X-C)$)	(11) Col (9) + Col (10)	(12) Col (11)/Col (4) $\lambda_-(X)$ (occ/day)
0	0.941192 × (0.02 + 0.01)	0 × 0	0.028236	0.0300
25	0.038416 × (0.01 + 0.01)	× 0	0.000768	0.0200
50	0.000392 × (0 + 0.01)	0.019208 × 0.02	0.000388	0.0198
75	0 × (0 + 0.01)	0.000784 × 0.01	0.000008	0.0100
100	0 × (0 + 0.01)	0.000008 × 0	0	0

Additional columns can be added to Table 3.8 to form a more complete capacity model and additional physical indicators of system capacity adequacy. The *cycle time* is the average duration between successive occurrences of the condition under examination:

Table 3.8 Complete generation model

					Cumulative	
Cap. out X (MW)	Probability p(X)	$\lambda_+(X)$ (occ/day)	$\lambda_-(X)$ (occ/day)	Frequency (occ/day) f(X)	Probability P(X)	Frequency (occ/day) F(X)
0	0.941192	0	0.0300	0.028236	1.000000	0
25	0.038416	0.4900	0.0200	0.019592	0.058808	0.028236
50	0.019600	0.4998	0.0198	0.010184	0.020392	0.010180
75	0.000784	0.9800	0.0100	0.000776	0.000792	0.000772
100	0.000008	1.4700	0	0.000012	0.000008	0.000012

$$\text{cycle time} = \frac{1}{\text{cycle frequency}} \tag{3.12}$$

The cycle time could therefore be obtained for either an individual or cumulative capacity condition.

The *average duration* of a particular capacity condition can be obtained as follows:

$$\text{average duration} = \frac{\text{probability of the condition}}{\text{frequency of the condition}} \tag{3.13}$$

The average duration of an individual or cumulative capacity condition can therefore be found using the appropriate values.

Case 2 Derated states included

Generating units with multi-state representations can be included in an F&D analysis using either the basic approach or using a recursive algorithm. Figure 3.2 shows the three-state model for the 50 MW unit given in Table 2.8. The availability values for each state can be obtained using the following equations:

$$P[\text{up}] = \frac{A}{D} = 0.960$$

$$P[\text{derated}] = \frac{B}{D} = 0.033$$

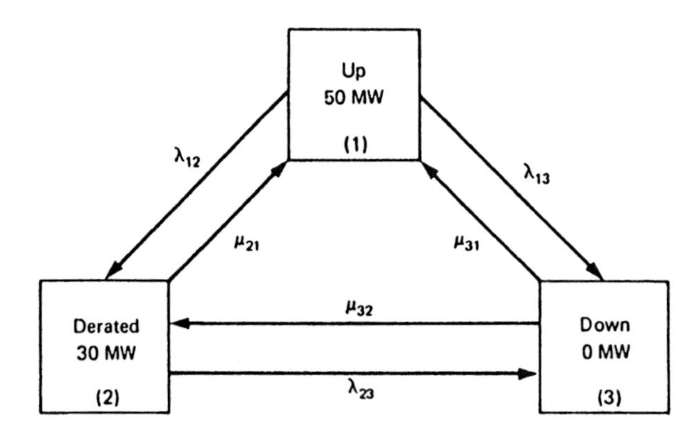

Fig. 3.2 Three-state unit model
$\lambda_{13} = 0.022$ occ/day
$\lambda_{12} = 0.008$
$\lambda_{23} = 0.019$
$\mu_{31} = 2.93571$
$\mu_{21} = 0.25$
$\mu_{32} = 0.171$

Table 3.9 General multi-state model

State i	*Capacity outage C_i*	*Probability p_i*	$\lambda_+(C_i)$	$\lambda_-(C_i)$	$F(C_i)$
1	C_1	p_1	$\lambda_+(C_1)$	$\lambda_-(C_1)$	$F(C_1)$
2	C_2	p_2	$\lambda_+(C_2)$	$\lambda_-(C_2)$	$F(C_2)$
.	.	.	.	.	.
.	.	.	.	.	.
.	.	.	.	.	.
n	C_n	p_n	$\lambda_+(C_n)$	$\lambda_-(C_n)$	$F(C_n)$

$$P[\text{down}] = \frac{C}{D} = 0.007$$

where $A = \mu_{31}\mu_{21} + \mu_{31}\lambda_{23} + \mu_{32}\mu_{21}$

$$B = \lambda_{13}\mu_{32} + \lambda_{12}\mu_{32} + \lambda_{12}\mu_{31}$$

$$C = \lambda_{12}\lambda_{23} + \lambda_{13}\lambda_{12} + \lambda_{13}\lambda_{23}$$

$$D = A + B + C$$

A complete state transition diagram for the three-unit system consisting of the two 25 MW units and the above 50 MW unit has 12 individual states and becomes somewhat complicated. The reader should utilize this approach and compare his results with those shown in Table 3.10.

The capacity model incorporating derated units can be developed using a recursive approach similar to that shown in Equations (3.7)–(3.9). A general multi-state generating unit model is defined in Table 3.9.

Table 3.10 Generation model

Cap. out X (MW)	*Probability p(X)*	$\lambda_+(X)$ *(occ/day)*	$\lambda_-(X)$ *(occ/day)*	*Frequency (occ/day) f(X)*	*Cumulative probability P(X)*	*Cumulative frequency (occ/day) F(X)*
0	0.9219840	0	0.05	0.046099	1.0	0
20	0.0316932	0.25	0.039	0.009159	0.078016	0.046099
25	0.0376320	0.49	0.04	0.019945	0.046323	0.039959
45	0.0012936	0.74	0.029	0.000995	0.008691	0.023025
50	0.0071068	2.991798	0.020540	0.021408	0.007397	0.022128
70	0.0000132	1.23	0.019	0.000016	0.000290	0.001011
75	0.0002744	3.59671	0.01	0.000990	0.000277	0.000996
100	0.0000028	4.08671	0	0.000011	0.000003	0.000011

Equations (3.14)–(3.16) can be used to add the unit to a capacity model:

$$p(X) = \sum_{i=1}^{n} p'(X - C_i)p_i \tag{3.14}$$

$$\lambda_+(X) = \frac{\sum_{i=1}^{n} p'(X - C_i)p_i(\lambda'_+(X - C_i) + \lambda_+(C_i))}{p(X)} \tag{3.15}$$

$$\lambda_-(X) = \frac{\sum_{i=1}^{n} p'(X - C_i)p_i(\lambda'_-(X - C_i) + \lambda_-(C_i))}{p(X)} \tag{3.16}$$

When $n = 2$, these equations reduce to the set used earlier, namely Equations (3.7)–(3.9).

The capacity model for the three-unit system of Table 2.7 with the 50 MW unit representation in Fig. 3.2 is shown in Table 3.10. The reader should use the recursive expressions given in Equations (3.14)–(3.16) to obtain these results. The cumulative frequency can be computed from

$$F_t(X) = \sum_{i=1}^{n} (p_i(F'(X - C_i) + P'(X - C_i)(\lambda_+(C_i) - \lambda_-(C_i))) - P'(X - C_i)\Delta(C_i))$$

$$\Delta(C_i) = \left(\sum_{j=n}^{i} p_j(\lambda_+(C_j) - \lambda_-(C_j)) \right) - F(C_i)$$

$$F(X_1) = 0$$

$$F(X_2) = F_t(X_2) - F_t(X_1)$$

$$F(X_i) = F_t(X_i)$$

$$P'(X - C_i) = 1,\ F'(X - C_i) = 0,\ X < C_i.$$

In the binary model the frequency of crossing the boundary wall between the two states is exactly balanced in both directions. This unique relationship however is not true in the case of a general multi-state generating unit model. The frequency of crossing the imaginary wall between any two states is not balanced in both directions. Additional information ($F(C_i)$) about the multi-state model is therefore required to modify the cumulative frequency expression. The reader should attempt to reduce these recursive expressions to those for adding the binary generators when n becomes 2. Hint: $\Delta(C_i)$ becomes zero for binary models and so $F_t(X) = F(X)$.

The recursive algorithms required to remove a unit from the capacity model can be obtained from Equations (3.14)–(3.16).

$$p'(X) = \frac{p(X) - \sum_{i=2}^{n} p'(X - C_i)p_i}{p_1} \tag{3.17}$$

$$\lambda'_{+}(X) = \frac{p(X)\lambda_{+}(X) - \sum_{i=2}^{n} p'(X - C_i)p_i(\lambda'_{+}(X - C_i) + \lambda_{+}(C_i))}{p'(X)p_1} \tag{3.18}$$

$$\lambda'_{-}(X) = \frac{p(X)\lambda_{-}(X) - \sum_{i=2}^{n} p'(X - C_i)p_i(\lambda'_{-}(X - C_i) + \lambda_{-}(C_i)) - p'(X)p_i\lambda_{-}(0)}{p'(x)p_1} \tag{3.19}$$

The reader should attempt to use Equations (3.17)–(3.19) to obtain the capacity model shown in Step 2 for the two 25 MW units.

The capacity model developed using the frequency and duration approach can be reduced considerably by truncating the table for cumulative probabilities less than a prespecified magnitude. This is the most significant reduction factor in a large system study. The table can also be rounded off [9] to selected increments using a procedure similar to that illustrated in Chapter 2. The frequencies and probabilities associated with the states to be absorbed can be apportioned at the two consecutive states remaining. It should be realized that the new model is now an approximation of the original model. As the rounding increment increases, the accuracy decreases [9]. Under these conditions, the capacity table decreases in size and bears less physical relationship to the original complete table. The rounding increment selected must be larger than the capacity of the smallest unit in the system or the capacity model will expand instead of reducing as intended.

3.3 System risk indices

The generating capacity models illustrated in the previous section can now be combined with the load to obtain system risk indices [7]. As with the basic probability methods discussed in Chapter 2 there are a number of possible load models which can be used. Two possible representations are examined and illustrated in this section. They are designated as the individual state load model and the cumulative state load model. Other possible representations are available in the published literature.

3.3.1 Individual state load model

This load model is the original representation proposed in Reference [3]. The load cycle for the period is a random sequence of N load levels where N is a fixed integer.

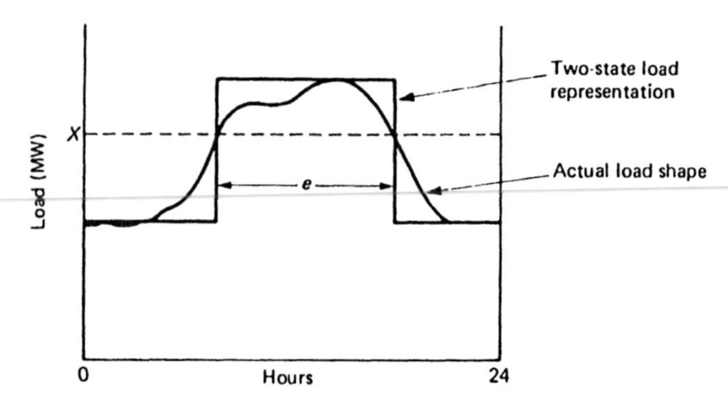

Fig. 3.3 Daily load model

The daily load model contains a peak load level of mean duration e day and a fixed low load of $1 - e$ day. This situation is illustrated in Fig. 3.3.

The load cycle for a specified period is shown in Fig. 3.4. Each peak load returns to the low load level each day before transferring to another peak on the next day. It is also assumed that each peak load has the same constant 'mean' duration e day separated by a low load of mean duration $1 - e$ day. The element e is called the exposure factor and is obtained from the daily load curve as shown in Fig. 3.3. There is no precise method for selecting the value X at which to determine the magnitude of the exposure factor. As X decreases, e increases, resulting in a more severe load model. When $e = 1$, the load is represented by its daily peak value as in the conventional LOLE approach. A value for X of 85% of the daily peak load is sometimes used. The assumption is also made that the load state transitions occur independently of the capacity state transitions. F&D analysis is normally done on a period basis as the assumed constant low load level and random peak load sequence do not usually apply over a long period of time. Maintenance of generating capacity also results in a modified capacity model.

The parameters required to completely define the period load model are as follows:

Number of load levels	N
Peak loads	$L_i, i = 1, \ldots, N$ such that $L_1 > L_2 > \ldots > L_N$
Low load	L_0
Number of occurrences of L_i	$n(L_i), i = 1, \ldots, N$

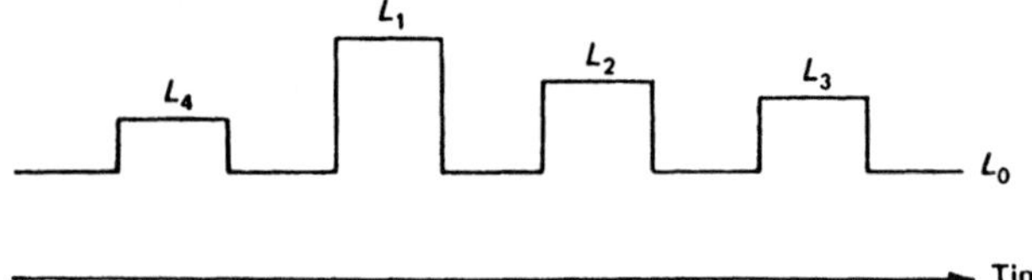

Fig. 3.4 Period load model

Period	$D = \sum_{i=1}^{N} n(L_i)$	
	Peak load L_i	*Low load* L_0
Mean duration	e	$1 - e$
Probability	$p(L_i) = \frac{n(L_i)}{D} e$	$p(L_0) = 1 - e$
Upward load departure rate	$\lambda_+(L_i) = 0$	$\lambda_+(L_0) = \frac{1}{1 - e}$
Downward load departure rate	$\lambda_-(L_i) = \frac{1}{e}$	$\lambda_-(L_0) = 0$
Frequency	$f(L_i) = \frac{n(L_i)}{D}$	$f(L_0) = 1$

The combination of discrete levels of available capacity and discrete levels of system demand or load creates a set of discrete capacity margins m_k. A margin is defined as the difference between the available capacity and the system load. A negative margin therefore represents a state in which the system load exceeds the available capacity and depicts a system failure condition. A cumulative margin state contains all states with a margin less than or equal to the specified margin. A margin state m_k is the combination of the load state L_i and the capacity state C_n, where

$$m_k = C_n - L_i \tag{3.20}$$

The rates of departure associated with m_k are as follows:

$$\lambda_{+m} = \lambda_{+c} + \lambda_{-L} \tag{3.21a}$$

$$\lambda_{-m} = \lambda_{-c} + \lambda_{+L} \tag{3.21b}$$

The probability of occurrence of two or more events in a single small increment of time is assumed to be negligible. The transition from one margin state to another can be made by a change in load or a change in capacity but not by both simultaneously. The upward margin transition rate λ_{+m} is, therefore, the summation of the upward capacity transition rate and the downward load transition rate. The opposite is true for λ_{-m}, the downward margin transition rate.

The probability of the margin state is the product of the capacity state and load state probabilities.

$$\text{Probability } p_k = p_n p_i \tag{3.22}$$

The frequency of encountering the margin state m_k is the product of the steady-state probability of the margin state and the summation of the rates of departure from the state:

$$\text{Frequency } f_k = p_k(\lambda_{+k} + \lambda_{-k}) \tag{3.23}$$

Having obtained the individual probabilities and frequencies of the margin states, the cumulative values can be obtained in a similar manner to that used in the capacity model. Different combinations of capacity and load states can result in identical margin states. These identical states are independent of each other and can only transit from one to another by going through the low load state L_0 and by a change in capacity.

The identical margin states can be combined as follows. For a given margin state m_k made up of s identical margin states:

$$p_k = \sum_{i=1}^{s} p_i \tag{3.24}$$

$$f_k = \sum_{i=1}^{s} f_i \tag{3.25}$$

$$\lambda_{\pm k} = \frac{\sum_{i=1}^{s} p_i \lambda_{\pm i}}{p_k} \tag{3.26}$$

The upward and downward departure rates for each margin are required in the calculation of cumulative margin frequencies. Equation (3.4) can be used for this purpose. Equation (3.10) is used to calculate the cumulative margin probabilities.

The calculation of the margin values can be accomplished using a fundamental approach which is ideally suited for developing an appreciation of the concept or by a series of algorithms ideally suited to digital computer application. Both techniques are presented in this section. Consider the load data shown in Table 3.11.

Assume that the exposure factor e for this load condition is 0.5, i.e. one half of a day. The capacity model for the three-unit 100 MW system is shown in Table

Table 3.11 Load data

Peak load level	*No. of occurrences*
65	8
55	4
50	4
46	4
	20 days

Table 3.12 Loss of load expectation

Capacity outage C_k *(MW)*	*Time units* t_k *(days)*	*Probability* p_k	$p_k t_k$
0	0	0.941192	0.0
25	0	0.038416	0.0
50	12	0.019600	0.235200
75	20	0.000784	0.015680
100	20	0.000008	0.000160
			$\Sigma\, p_k t_k = 0.251040$

3.8. A conventional LOLE calculation using daily peak loads is shown in Table 3.12 for this system.

The LOLE is 0.25104 days for the 20-day period. If the year is assumed to be composed of a series of such periods, the annual LOLE is 0.25104 × (365/20) = 4.58148 days/year.

The low load level does not normally contribute substantially to the negative margins and is sometimes omitted from the calculation. This can easily be done by assuming that the low load level is zero. Table 3.13 shows the system margin array for this capacity and load condition. The probability of any given margin is obtained using Equation (3.22) and the upward and downward departure rates using Equation (3.21). The frequency of a given margin state is obtained by using Equation (3.23). Identical margin states can be combined using Equations (3.24) and (3.25) and the equivalent rates of departure from Equation (3.26) used to obtain the cumulative margin values. These results are shown in Table 3.14.

The 20-day period in this case has been assumed to be the entire period of study and therefore the $\Sigma\, p(L_i)$ 1.0 in Table 3.13.

The margin state values shown in Table 3.13 contain the full set of information regarding capacity adequacy for the period considered. The positive margins are clearly influenced by the assumption of a zero low load level. If positive margins are of interest, then the low load level should be represented. The most useful single index is the cumulative probability and frequency of the first negative margin. These are as follows.

Cumulative probability of the first negative margin = 0.006276

Cumulative frequency of the first negative margin = 0.015761 occ/day

The cumulative probability can be used to obtain the LOLE value calculated in Table 3.12.

$$\text{LOLE} = \begin{bmatrix}\text{cumulative probability of the} \\ \text{first negative margin}\end{bmatrix} \times \frac{365}{e}$$

$$= (0.006276)\left(\frac{365}{0.5}\right) = \underline{4.58148\ \text{days/year.}} \tag{3.27}$$

Table 3.13 Margin state array

Load									
			i		1	2	3	4	0
			L_i (MW)		65	55	50	46	0
			P_i		0.2	0.1	0.1	0.1	0.5
			$\lambda_-(L_i)$		2	2	2	2	0
			$\lambda_+(L_i)$		0	0	0	0	2
Generation									
n	C_n	λ_+	λ_-	m	35	45	50	54	100
1	0	0	0.03	P	0.1882384	0.0941192	0.0941192	0.0941192	0.470596
($P_1 = 0.941192$)				λ_+	2	2	2	2	0
				λ_-	0.03	0.03	0.03	0.03	2.03
2	25	0.49	0.02	m	10	20	25	29	75
($P_2 = 0.038416$)				P	0.0076832	0.0038416	0.0038416	0.0038416	0.019208
				λ_+	2.49	2.49	2.49	2.49	0.49
				λ_-	0.02	0.02	0.02	0.02	2.02
3	50	0.4998	0.0198	m	–15	–5	0	4	50
($P_3 = 0.019600$)				P	0.00392	0.00196	0.00196	0.00196	0.0098
				λ_+	2.4998	2.4998	2.4998	2.4998	0.4998
				λ_-	0.0198	0.0198	0.0198	0.0198	2.0198
4	75	0.98	0.01	m	–40	–30	-25	–21	+25
($P_4 = 0.000784$)				P	0.0001568	0.0000784	0.0000784	0.0000784	0.000392
				λ_+	2.98	2.98	2.98	2.98	0.98
				λ_-	0.01	0.01	0.01	0.01	2.01
5	100	1.47	0	m	–65	–55	–50	–46	0
($P_5 = 0.000008$)				P	0.0000016	0.0000008	0.0000008	0.0000008	0.000004
				λ_+	3.47	3.47	3.47	3.47	1.47
				λ_-	0	0	0	0	2

Table 3.14 Margin state probabilities and frequencies

Margin state (MW)	*Individual probability*	*Individual frequency*	*Cumulative probability*	*Cumulative frequency*
100	0.470596	0.955310	1.000000	—
75	0.019208	0.048212	0.529404	0.955310
54	0.094119	0.191062	0.511096	0.984698
50	0.103919	0.215754	0.416077	0.799283
45	0.094119	0.191062	0.312158	0.628764
35	0.188238	0.382124	0.218038	0.443349
29	0.003842	0.009642	0.029800	0.072520
25	0.004234	0.010814	0.025958	0.063031
20	0.003842	0.009642	0.021725	0.053946
10	0.007683	0.019285	0.017883	0.044457
4	0.001960	0.004938	0.010200	0.025480
0	0.001964	0.004952	0.008240	0.020619
−5	0.001960	0.004938	0.006276	0.015761
−15	0.003920	0.009877	0.004316	0.010900
−21	0.000078	0.000234	0.000396	0.001178
−25	0.000078	0.000234	0.000318	0.000945
−30	0.000078	0.000234	0.000239	0.000712
−40	0.000157	0.000469	0.000161	0.000480
−46	0.000001	0.000003	0.000004	0.000014
−50	0.000001	0.000003	0.000003	0.000011
−55	0.000001	0.000003	0.000002	0.000008
−65	0.000002	0.000006	0.000002	0.000006

The cumulative probabilities and frequencies associated with the margin states can be obtained using a set of algorithms which are ideally suited to digital computer applications.

Let $P(m)$ and $F(m)$ be the cumulative probability and frequency respectively associated with the specified margin m.

$$P(m) = \sum_{j=1}^{N} p(L_j)P(X_j) \tag{3.28}$$

$$F(m) = \sum_{j=1}^{N} p(L_j)(F(X_j) + P(X_j)(\lambda_-(L_j) - \lambda_+(L_j))) \tag{3.29}$$

$$D(m) = \frac{P(m)}{F(m)} \tag{3.30}$$

$$T(m) = \frac{1}{F(m)} \tag{3.31}$$

Table 3.15 Calculation of *P(m)* and *F(m)*

(1) j	(2) L_j	(3) $L_j + m$	(4) X_j	(5) $p(L_j)$	(6) $P(X_j)$	(7) Col (5) × Col (6)
1	65	60	50	0.2	0.020392	0.004078
2	55	50	50	0.1	0.020392	0.002039
3	50	45	75	0.1	0.000792	0.000079
4	46	41	75	0.1	0.000792	0.000079
5	0	−5	—	0.5	—	—
						0.006275

(1) j	(8) $\lambda_-(L_j) - \lambda_+(L_j)$	(9) Col (6) × Col (8)	(10) $F(X_j)$	(11) Col (9) + Col (10)	(12) Col (5) × Col (11)
1	2	0.040784	0.010180	0.050964	0.010193
2	2	0.040784	0.010180	0.050964	0.005096
3	2	0.001584	0.000772	0.002356	0.000236
4	2	0.001584	0.000772	0.002356	0.000236
5	−2	—	—	—	—
					0.015761

$P(-5) = 0.006275$
$F(-5) = 0.015761$ occ/day
$D(-5) = 0.398135$ days
$T(-5) = 63.4477$ days.

The application of these equations is illustrated in Table 3.15. The first negative margin provides the basic reliability index. In the previous example $m = -5$ MW and $N = 5$ (including the zero low load level).

The low load level in the calculation shown in Table 3.15 has been taken as zero. If a low load level is included, then $P(-5)$ increases slightly depending on the value of the low load level. $F(-5)$ decreases slightly as the load level transitions do not add to the frequency when the available capacity level is less than the low load level.

The IEEE-RTS load data shown in Appendix 2 has been analyzed to produce a suitable load model for the 100 MW system previously analyzed [6]. This data is shown in Table 3.16.

The reader is encouraged to determine the frequency and duration of the individual and cumulative margin states for this system using the fundamental approach shown in Table 3.13, and to check these results using Equations (3.28) and (3.29). In this case the first negative margin is −2 MW.

$$P(-2) = 0.003343$$

$$F(-2) = 0.007098 \text{ occ/day}$$

Table 3.16 Modified RTS data

Level No. j	*Load level* L_j *(MW)*	*No. of occurrences*	*Probability* $P(L_j)$	$\lambda_+(L_j)$	$\lambda_-(L_j)$
1	57	12	0.016438	0	2
2	52	83	0.113699	0	2
3	46	107	0.146575	0	2
4	41	116	0.158904	0	2
5	34	47	0.064384	0	2
6	31	365	0.500000	2	0

3.3.2 Cumulative state load model

The system load can be described in a different form which does not require the exposure factor or low load level considerations used in the two-state representation [8]. There are two conditions for any arbitrary demand level L.

State 1: Load $\geq L$

State 2: Load $< L$

The probability that the load is equal to or greater than the arbitrary level L is obtained from the hourly load data or directly from the load–duration curve. The frequency associated with the two states is obtained by counting the transitions from one state to another and dividing by the load period. A load–frequency characteristic can be obtained by varying the value of L, the arbitrary load level. The general shape of the load–frequency characteristic is shown in Fig. 3.5.

The frequency is zero for loads lower than the minimum value and greater than the maximum value in the load period. The individual state capacity model and the cumulative state load model expressed by the load–duration curve and the load–frequency characteristic can be combined to produce cumulative probabilities and frequencies for selected margin states. The first negative margin cannot always be defined uniquely as in the case of the individual state load model, and it is usual in this case to consider the zero margin as the load loss situation.

Given a margin m and a capacity outage X for a system of installed capacity C, a loss of load situation arises when

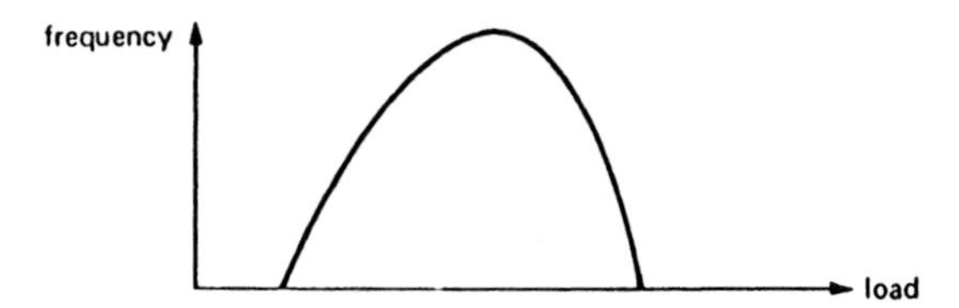

Fig. 3.5 Load–frequency characteristic

Table 3.17 Load data

Level No. i	*Load level L (MW)*	*Probability*	*Frequency (occ/day)*
1	>60	0	0
2	>55	0.0384	0.0055
3	>50	0.2027	0.0301
4	>46	0.4192	0.1041
5	>41	0.7041	0.0986
6	>36	0.8795	0.0877
7	>31	0.9973	0.0658
8	≤31	1.0000	0

$L > C - X - m.$

The appropriate indices can be obtained from [6]

$$P(m) = \sum_X p(X)P(C - X - m) \tag{3.32}$$

$$F(m) = \sum_X p(X)((\lambda_+(X) - \lambda_-(X))P(C - X - m) + F(C - X - m)) \tag{3.33}$$

The application of Equations (3.32) and (3.33) is illustrated in Table 3.18. The load data for the IEEE–RTS given in Appendix 2 has been analyzed and scaled down to the 100 MW generation system model. The load–duration and load-frequency data are shown in Table 3.17.

The installed capacity C is 100 MW in this case and selecting a value of m = 0 then $C - X - m = 100 - X$.

$$P(0) = \sum_X p(X)P(100 - X)$$

$$F(0) = \sum_X p(X)((\lambda_+(X) - \lambda_-(X))P(100 - X) + F(100 - X))$$

The calculation is shown in Table 3.18.

The final indices are

$P(0) = 0.009008$

$F(0) = 0.006755$ occ/day

The reader is encouraged to use Equations (3.32) and (3.33) to calculate the cumulative probability and frequency of the zero margin for the 100 MW system in which the 50 MW unit has the three-state representation given in Table 3.10. The final indices for the load model of Table 3.17 are

$P(0) = 0.003532$

$F(0) = 0.010828$ occ/day.

Table 3.18 Cumulative probability and frequency analysis

(1) Capacity state *X* (*MW*)	*(2)* $\lambda_+(X) - \lambda_-(X)$	*(3)* $(100 - X)$	*(4)* $P(100 - X)$	*(5)* *Col (2) × Col (4)*
0	–0.0300	100	0	0
25	0.4700	75	0	0
50	0.4800	50	0.4192	0.201216
75	0.9700	25	1.0000	0.970000
100	1.4700	0	1.0000	1.470000

(6) $F(100 - X)$	*(7)* *(5) + (6)*	*(8)* $p(X)$	*(9)* *(4) × (8)*	*(10)* *(7) × (8)*
0	0	0.941192	0	0
0	0	0.038416	0	0
0.1041	0.305316	0.019600	0.008216	0.005984
0	0.970000	0.000784	0.000784	0.000760
0	1.470000	0.000008	0.000008	0.000011
			0.009008	0.006755

The load model data shown in Tables 3.16 and 3.17 were obtained from the IEEE–RTS data given in Appendix 2. In each case the data is presented on an annual basis. In the actual study, the data should be utilized on a period basis, i.e. one in which the low load level is constant for the individual state load model and the load–frequency characteristic is valid for the cumulative load model. The appropriate capacity model for each period should then be used. This will include any reductions due to maintenance or unit additions or retirements in the annual period.

3.4 Practical system studies

3.4.1 Base case study

It is not practical to analyze a large system containing many units using transition diagrams and complete state enumeration. The algorithms presented in this chapter can be utilized to develop efficient digital computer programs for capacity model construction and convolution with the appropriate load models to determine the system risk indices.

Table 3.19 presents a 22-unit generating system containing a total of 1725 MW of capacity. The load model for a 20-day period is shown in Table 3.20. The generation model for the system of Table 3.19 is shown in Table 3.21.

An exposure factor *e* of 0.5 was used and the low load level assumed to be negligible. As noted earlier, the assumption of a negligible low load level adds very little to the negative margin but will influence the positive margins. The cumulative

Table 3.19 Generation system

No. of identical units	*Unit size (MW)*	*Mean down time (r) (years)*	*Mean up time (m) (years)*
1	250	0.06	2.94
3	150	0.06	2.94
2	100	0.06	2.94
4	75	0.06	2.94
9	50	0.06	2.94
3	25	0.06	2.94
	Total installed capacity = 1725 MW		

probability of the first negative margin for the 20-day period is 0.8988137×10^{-4} which corresponds to a LOLE of 0.065613 days/period. This can be expressed on an annual basis by assuming that the year consists of a series of identical 20-day periods with the load model shown in Table 3.20. In actual practice, the year should be divided into periods during which the generating capacity on scheduled outage is specified and the non-stationary effects of seasonal load changes are incorporated by using a valid load model for each interval. The annual LOLE and outage frequency are obtained by summing the period values.

The cumulative frequency of the first negative margin is 0.1879385×10^{-3} encounters/day which, when expressed in the reciprocal form as cycle time, becomes 0.5320887×10^{4} days. If the assumption is made that the annual period consists of a group of identical 20-day periods, then the annual indices are as follows:

$$\text{LOLE} = 0.065613 \times \frac{365}{20} = \underline{1.197 \text{ days/year}}$$

$$\text{Frequency}\, F = \underline{0.003430 \text{ occurrences/day}}$$

$$= \underline{1.252 \text{ occurrences/year}}$$

$$\text{Cycle time}\, T = \underline{291.6 \text{ days}}$$

$$= \underline{0.7988 \text{ years}}$$

These LOLE and F indices are somewhat higher than normally encountered due to the assumption that the 20-day peak load period is repeated over the entire

Table 3.20 Load model

Load level (MW)	*L_i (%)*	*No. of occurrences*
1450	100	8
1255	85.6	4
1155	79.6	4
1080	74.5	4
		20 days

Table 3.21 Generation system model of the 22-unit system

Capacity outage equal to or greater than (MW)	*Cumulative probability P*	*Cumulative frequency (F) (per day)*	*Cycle time (1/F) (days)*	*Mean duration (P/F) (days)*
0.0	1.000000	—	—	—
25.0	0.358829	0.131448×10^{-1}	0.760757×10^{2}	0.272981×10^{2}
50.0	0.319573	0.121205×10^{-1}	0.825046×10^{2}	0.263663×10^{2}
75.0	0.201006	0.898946×10^{-2}	0.111241×10^{3}	0.223602×10^{2}
100.0	0.141450	0.709902×10^{-2}	0.140864×10^{3}	0.199252×10^{2}
125.0	0.102314	0.546692×10^{-2}	0.182918×10^{3}	0.187151×10^{2}
150.0	0.904434×10^{-1}	0.457351×10^{-2}	0.218650×10^{3}	0.197755×10^{2}
175.0	0.436883×10^{-1}	0.295266×10^{-2}	0.338678×10^{3}	0.147963×10^{2}
200.0	0.379311×10^{-1}	0.247558×10^{-2}	0.403946×10^{3}	0.153221×10^{2}
225.0	0.295167×10^{-1}	0.181697×10^{-2}	0.550368×10^{3}	0.162450×10^{2}
250.0	0.253559×10^{-1}	0.146596×10^{-2}	0.682147×10^{3}	0.172965×10^{2}
275.0	0.968847×10^{-2}	0.887232×10^{-3}	0.112710×10^{4}	0.109199×10^{2}
300.0	0.809136×10^{-2}	0.729771×10^{-3}	0.137029×10^{4}	0.110875×10^{2}
325.0	0.438965×10^{-2}	0.433177×10^{-3}	0.230853×10^{4}	0.101336×10^{2}
350.0	0.291192×10^{-2}	0.302569×10^{-3}	0.330503×10^{4}	0.962397×10
375.0	0.189202×10^{-2}	0.202682×10^{-3}	0.493384×10^{4}	0.933492×10
400.0	0.154317×10^{-2}	0.158243×10^{-3}	0.631938×10^{4}	0.975185×10
425.0	0.528440×10^{-3}	0.719494×10^{-4}	0.138986×10^{5}	0.734460×10
450.0	0.391825×10^{-3}	0.533764×10^{-4}	0.187349×10^{5}	0.734079×10
475.0	0.203941×10^{-3}	0.293670×10^{-4}	0.340519×10^{5}	0.694457×10
500.0	0.114013×10^{-3}	0.173100×10^{-4}	0.577700×10^{5}	0.658653×10
525.0	0.587161×10^{-4}	0.951282×10^{-5}	0.105121×10^{6}	0.617232×10
550.0	0.412809×10^{-4}	0.649118×10^{-5}	0.154055×10^{6}	0.635955×10
575.0	0.145389×10^{-4}	0.270617×10^{-5}	0.369526×10^{6}	0.537251×10
600.0	0.899488×10^{-5}	0.169028×10^{-5}	0.591619×10^{6}	0.532154×10
625.0	0.438848×10^{-5}	0.859062×10^{-6}	0.116406×10^{7}	0.510845×10
650.0	0.217446×10^{-5}	0.443870×10^{-6}	0.225291×10^{7}	0.489887×10
675.0	0.926480×10^{-6}	0.204309×10^{-6}	0.489455×10^{7}	0.453470×10
700.0	0.531049×10^{-6}	0.116074×10^{-6}	0.861516×10^{7}	0.457507×10
725.0	0.199338×10^{-6}	0.477871×10^{-7}	0.209261×10^{8}	0.417138×10
750.0	0.963639×10^{-7}	0.236983×10^{-7}	0.421971×10^{8}	0.406627×10
775.0	0.431916×10^{-7}	0.109852×10^{-7}	0.910313×10^{8}	0.393179×10
800.0	0.188172×10^{-7}	0.493791×10^{-8}	0.202515×10^{9}	0.381076×10

year. This is basically similar to the 'multiple worst period' concept discussed in Section 2.7.

The cycle time T is also a useful index and measures the average duration between encountering the specified negative margin (i.e. in this case the first negative margin). The LOLE and T indices are shown as a function of the peak load

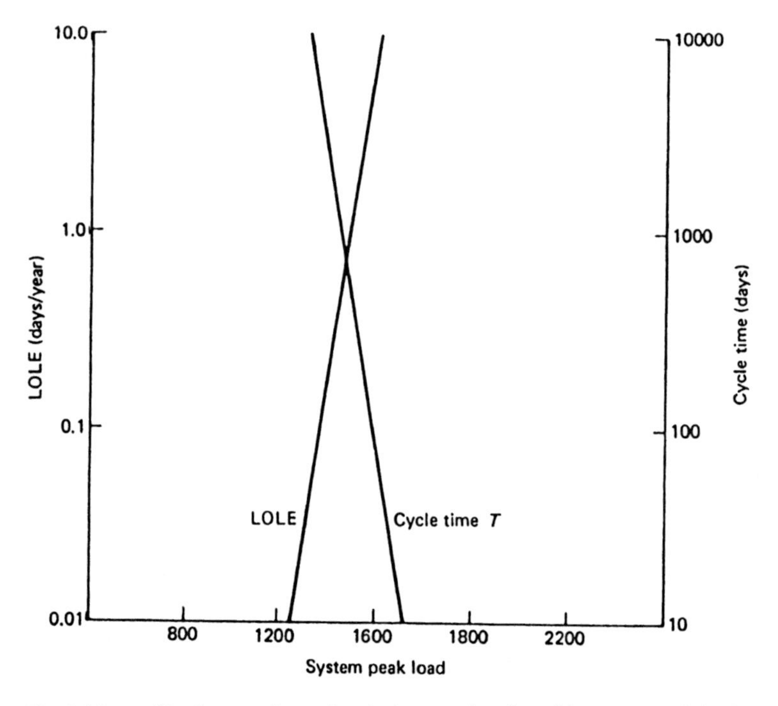

Fig. 3.6 Loss of load expectation and cycle time as a function of the system peak load

in Fig. 3.6. The load model for the 20-day period is assumed to have the relative shape shown in Table 3.20.

3.4.2 System expansion studies

Figures 3.7 and 3.8 show the effects of adding a group of 250 MW units to the 22-unit system. The risk index in Fig. 3.7 is the annual LOLE value. The cycle time T is used in Fig. 3.8.

The peak load in Year 1 is 1450 MW and it has been assumed that the installed capacity of 1725 MW is adequate for this condition. The two previous risk indices of LOLE = 1.197 days/year and $T = 291.6$ days can therefore be considered as the system standard indices and used to schedule capacity additions. The dotted lines in Figs 3.7 and 3.8 show the changes in annual risk as units are added in a given year. The effective load-carrying capability of a unit is the increase in system load-carrying capability at a given risk level due to the unit addition. It is difficult to determine these values precisely from Figs 3.7 and 3.8 but they can be obtained from the results used to plot these figures. In this case, there is virtually no difference between the results obtained using LOLE or cycle time T. In general, there may be

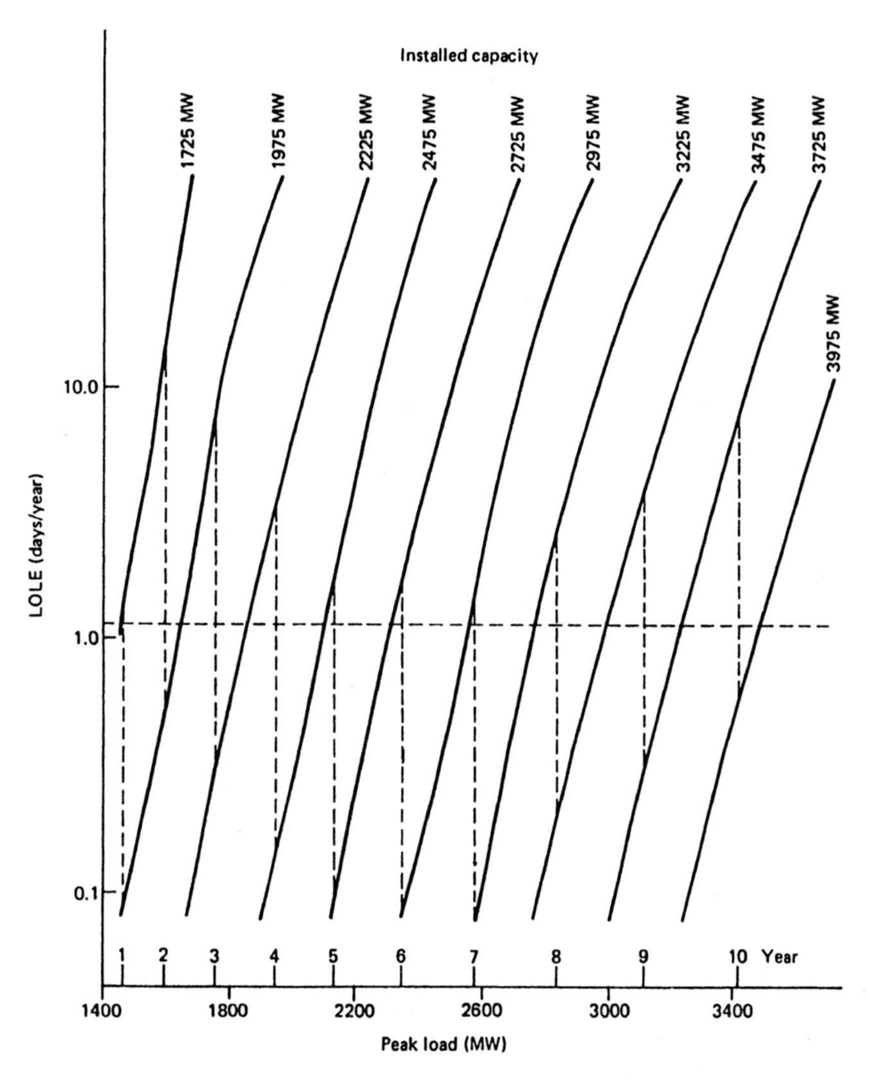

Fig. 3.7 System expansion using loss of load expectation as the criterion

slight differences. Generation expansion sequences based upon the probability of the first cumulative negative margin which is directly related to LOLE or cycle time of the same margin is basically the same when the two-state load model is used. This assumes that equivalent risk indices are used to initiate the expansion sequence. This may not be the case if other load models such as the cumulative state model described earlier or multi-level exposure factor models are used.

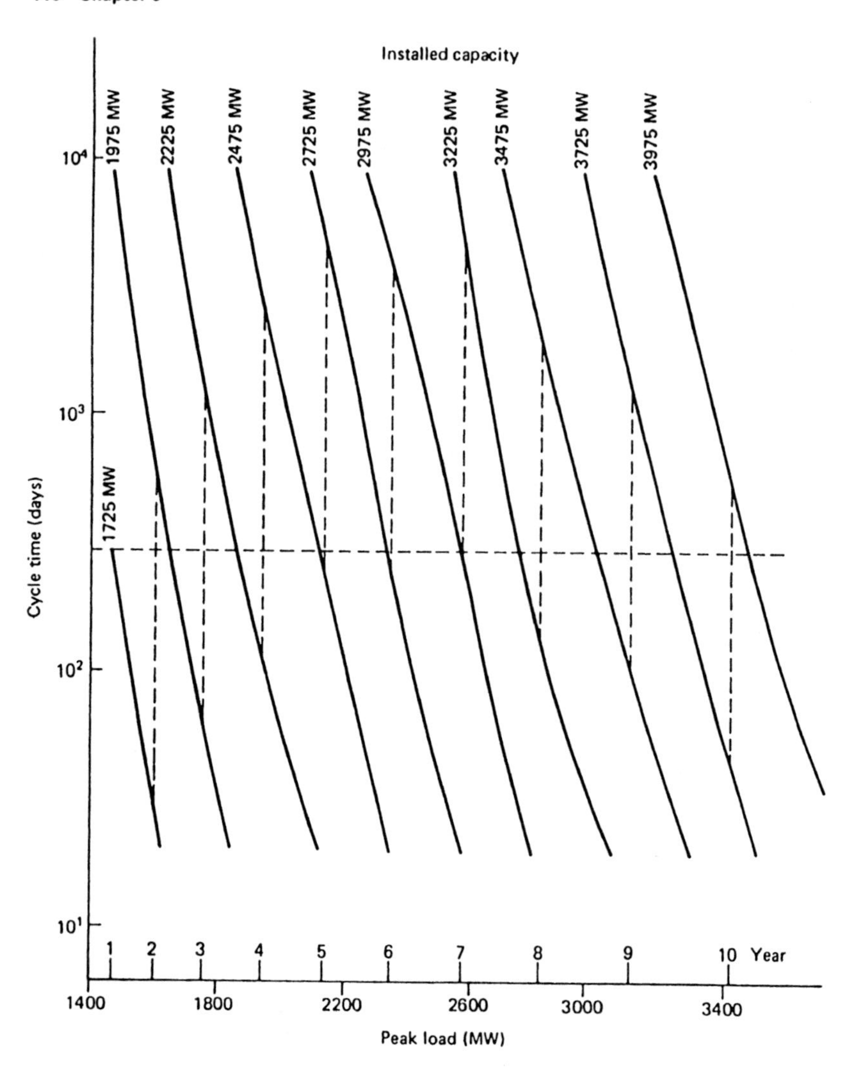

Fig. 3.8 System expansion using cycle time as the criterion

The variation in risk with unit additions is clearly shown in Figs 3.7 and 3.8. It can be seen that the actual risk in any given year is not the actual specified system index but a somewhat lower value. The object is to arrive at the most economical expansion plan while staying on the acceptable side of the standard risk index. The unit expansion plan in Figs 3.7 and 3.8 is composed of 250 MW units, which results in certain years having a much lower risk than the specified value. A better sequence might involve a mixture of unit sizes resulting in a much lower variation in overall risk. Risk evaluation is only one part of capacity planning. The system energy

Table 3.22 Expanding with 200 MW units

Year of study	*Peak load (MW)*	*Total capacity (MW)*	*Risk level (days)*	*Whether unit addition required*
1	1450.0000	1725.00	291.56	No
2	1595.0000	1725.00	26.48	Yes
Capacity of unit added (MW) = 200.000		$\lambda = 0.0009318795$		$\mu = 0.04566210$
2	1595.0000	1925.00	504.76	No
3	1754.0000	1925.00	38.26	Yes
Capacity of unit added (MW) = 200.000		$\lambda = 0.0009318795$		$\mu = 0.04566210$
3	1754.0000	2125.00	600.74	No
4	1930.0000	2125.00	31.86	Yes
Capacity of unit added (MW) = 200.000		$\lambda = 0.0009318795$		$\mu = 0.04566210$
4	1930.0000	2325.00	491.83	No
5	2123.0000	2325.00	48.28	Yes
Capacity of unit added (MW) = 200.000		$\lambda = 0.0009318795$		$\mu = 0.04566210$
5	2123.0000	2525.00	659.64	No
6	2335.0000	2525.00	21.59	Yes
Capacity of unit added (MW) = 200.000		$\lambda = 0.0009318795$		$\mu = 0.04566210$
6	2335.0000	2725.00	272.28	Yes
Capacity of unit added (MW) = 200.000		$\lambda = 0.0009318795$		$\mu = 0.04566210$
6	2335.0000	2925.00	4325.02	No
7	2569.0000	2925.00	196.46	Yes
Capacity of unit added (MW) = 200.000		$\lambda = 0.0009318795$		m = 0.04566210
7	2569.0000	3125.00	2760.77	No
8	2826.0000	3125.00	73.70	Yes
Capacity of unit added (MW) = 200.000		$\lambda = 0.0009318795$		$\mu = 0.04566210$
8	2826.0000	3325.00	770.17	No
9	3108.0000	3325.00	28.23	Yes
Capacity of unit added (MW) = 200.000		$\lambda = 0.0009318795$		$\mu = 0.04566210$
9	3108.0000	3525.00	234.92	Yes
Capacity of unit added (MW) = 200.000		$\lambda = 0.0009318795$		$\mu = 0.04566210$
9	3108.0000	3725.00	2652.26	No
10	3419.0000	3725.00	64.91	Yes
Capacity of unit added (MW) = 200.000		$\lambda = 0.0009318795$		$\mu = 0.04566210$
10	3419.0000	3925.00	602.71	No

requirements in addition to power or capacity needs must also be examined. The representation used in the two-state load model is not suitable for energy evaluation and a multi-level or continuous model is required. The most suitable model in this case is the conventional load–duration curve utilized in the loss of energy expectation method.

Figures 3.7 and 3.8 provide a pictorial representation of the variation in risk with peak load and unit additions. They are not required, however, if only the unit

Table 3.23 Expanding with 250 MW units

Year of study	*Peak load (MW)*	*Total capacity (MW)*	*Risk level (days)*	*Whether unit addition required*
1	1450.0	1725.0	291.56	No
2	1595.0	1725.0	26.48	Yes
Capacity of unit added (MW) = 250.0		$\lambda = 0.0009318795$		$\mu = 0.04566210$
2	1595.0	1975.0	709.43	No
3	1754.0	1975.0	49.29	Yes
Capacity of unit added (MW) = 250.0		$\lambda = 0.0009318795$		$\mu = 0.04566210$
3	1754.0	2225.0	1345.24	No
4	1930.0	2225.0	115.00	Yes
Capacity of unit added (MW) = 250.0		$\lambda = 0.0009318795$		$\mu = 0.04566210$
4	1930.0	2475.0	2624.68	No
5	2123.0	2475.0	265.59	Yes
Capacity of unit added (MW) = 250.0		$\lambda = 0.0009318795$		$\mu = 0.04566210$
5	2123.0	2725.0	5582.01	No
6	2335.0	2725.0	224.61	Yes
Capacity of unit added (MW) = 250.0		$\lambda = 0.0009318795$		$\mu = 0.04566210$
6	2335.0	2975.0	4138.32	No
7	2569.0	2975.0	283.60	Yes
Capacity of unit added (MW) = 250.0		$\lambda = 0.0009318795$		$\mu = 0.04566210$
7	2569.0	3225.0	5116.20	No
8	2826.0	3225.0	141.31	Yes
Capacity of unit added (MW) = 250.0		$\lambda = 0.0009318795$		$\mu = 0.04566210$
8	2826.0	3475.0	2041.30	No
9	3108.0	3475.0	107.29	Yes
Capacity of unit added (MW) = 250.0		$\lambda = 0.0009318795$		$\mu = 0.04566210$
9	3108.0	3725.0	1374.17	No
10	3419.0	3725.0	56.91	Yes
Capacity of unit added (MW) = 250.0		$\lambda = 0.0009318795$		$\mu = 0.04566210$
10	3419.0	3975.0	614.36	No

addition sequence in a particular expansion plan is required. Tables 3.22–24 show the unit in-service sequence for three expansion sequences involving 200, 250 and 300 MW units respectively. In each case the standard risk level chosen for the system is 291.6 days. The expansion plan involving the addition of 200 MW units requires two units in years 6 and 9 and one unit in each of the other years. In the case of 300 MW unit additions, the system does not require a unit in year 4 but units must be added in each of the other years. The unit sizes could probably be mixed in the expansion plan to achieve the lowest present worth.

Table 3.24 Expanding with 300 MW units

Year of study	*Peak load (MW)*	*Total capacity (MW)*	*Risk level (days)*	*Whether unit additional required*
1	1450.0	1725.0	291.56	No
2	1595.0	1725.0	26.48	Yes
Capacity of unit added (MW) = 300.0		$\lambda = 0.0009318795$		$\mu = 0.04566210$
2	1595.0	2025.0	1067.90	No
3	1754.0	2025.0	81.27	Yes
Capacity of unit added (MW) = 300.0		$\lambda = 0.0009318795$		$\mu = 0.04566210$
3	1754.0	2325.0	2949.44	No
4	1930.0	2325.0	317.31	No
5	2123.0	2325.0	35.26	Yes
Capacity of unit added (MW) = 300.0		$\lambda = 0.0009318795$		$\mu = 0.04566210$
5	2123.0	2625.0	800.32	No
6	2335.0	2625.0	36.50	Yes
Capacity of unit added (MW) = 300.0		$\lambda = 0.0009318795$		$\mu = 0.04566210$
6	2335.0	2925.0	802.87	No
7	2569.0	2925.0	123.34	Yes
Capacity of unit added (MW) = 300.0		$\lambda = 0.0009318795$		$\mu = 0.04566210$
7	2569.0	3225.0	2303.73	No
8	2826.0	3225.0	129.78	Yes
Capacity of unit added (MW) = 300.0		$\lambda = 0.0009318795$		$\mu = 0.04566210$
8	2826.0	3525.0	2109.46	No
9	3108.0	3525.0	137.13	Yes
Capacity of unit added (MW) = 300.0		$\lambda = 0.0009318795$		$\mu = 0.04566210$
9	3108.0	3825.0	2098.85	No
10	3419.0	3825.0	112.80	Yes
Capacity of unit added (MW) = 300.0		$\lambda = 0.0009318795$		$\mu = 0.04566210$
10	3419.0	4125.0	1545.85	No

3.4.3 Load forecast uncertainty

Load forecast uncertainty can be incorporated in the F&D approach in a similar manner to that used in the LOLE method (Section 2.8). Risk indices can be established for each conditional load level and weighted by the probability of that level to produce overall indices. Unit additions incorporating load forecast uncertainty are at a different rate than that determined without recognizing uncertainty. In general, the reserve required to satisfy a future uncertain load is always higher than that required for an equivalent known load. The expansion plan cost is therefore higher in this case. The most important uncertainty in any expansion plan is that uncertainty which still exists at the time the actual decision has to be made regarding unit additions. This factor is becoming more important as the lead time for unit conception, construction and commissioning increases.

3.5 Conclusions

This chapter presents the fundamental concepts associated with the frequency and duration approach to generating capacity evaluation. It also provides a group of recursive algorithms which can be used to calculate the required indices.

It is not possible to state that one method of capacity reliability evaluation is better than another under all conditions. The LOLE method is certainly simpler to appreciate than the F&D approach and in general results in a virtually identical expansion plan for a given reliability index. Expectation does lack a certain physical significance which is provided by the frequency or cycle time indices of the F&D approach. The LOLE method uses the unit forced outage rates and is therefore not as sensitive to variations in the two individual elements, unit failure rate and unit repair rate, as the F&D technique. Both methods, however, are decidedly superior to rule-of-thumb techniques such as percentage reserve and largest unit criteria, as they do permit the factors which influence the reliability of the generation system to be included in the analysis.

3.6 Problems

1. A generating plant containing three identical 40 MW generating units is connected to a constant 82 MW load. The unit failure and average repair times are 3 failures/year and 8 days respectively. Develop frequency, duration and probability risk indices for this system.
2. A generating plant containing three 25 MW generating units is connected to a constant 50 MW load by two transmission lines each rated at 40 MW. Develop a frequency and duration risk index for this system using the following data:

   Generation $\lambda = 0.02$ f/day $r = 4$ days

   Transmission $\lambda = 1.0$ f/yr $r = 4$ hours

3 Consider the system given in Problem 4 of Chapter 2. Develop a generation–transmission capacity model including frequency and rate of departure indices at the load bus. Do not attempt to combine this with the load model.

4 A system contains five 40 MW generating units with the following parameters:

Failure rate = 0.01 failures/day

Repair rate = 0.49 repairs/day

This system serves a load which on a specified 20-day period has the following model:

Peak load level (MW)	160	100	80	60	0
Number of occurrences	2	5	8	5	20

(a) Develop a generation model for this system including both individual and cumulative state probabilities and frequencies.

(b) Develop a complete margin array containing both individual and cumulative state probabilities and frequencies for this system. Assume that the exposure factor is 0.5 and that the 20 days represents a one-period contribution to an annual study.

3.7 References

1. Halperin, H., Adler, H. A., 'Determination of reserve generating capacity', *AIEE Transactions*, **PAS-77** (1958), pp. 530–44.
2. Hall, J. D., Ringlee, R. J., Wood, A. J., 'Frequency and duration methods for power system reliability calculations: Part I—Generation system model', *IEEE Transactions*, **PAS-87** (1968), pp. 1787–96.
3. Ringlee, R. J., Wood, A. J., 'Frequency and duration methods for power system reliability calculations: Part II—Demand model and capacity reserve model', *IEEE Transactions*, **PAS-88** (1969), pp. 375–88.
4. Galloway, C. D., Garver, L. L., Ringlee, R. J., Wood, A. J., 'Frequency and duration methods for power system reliability calculations: Part III—Generation system planning', *IEEE Transactions*, **PAS-88** (1969), pp. 1216–23.
5. Cook, V. M., Ringlee, R. J., Wood, A. J., 'Frequency and duration methods for power system reliability calculations: Part IV—Models for multiple boiler-turbines and for partial outage states', *IEEE Transactions*, **PAS-88** (1969), pp. 1224–32.
6. Billinton, R., Wee, C. L., Hamoud, G., 'Digital computer algorithms for the calculation of generating capacity reliability indices', *Proceedings of the PICA Conference* (May 1981), pp. 46–54.
7. Billinton, R., *Power System Reliability Evaluation*, Gordon and Breach, New York (1970).

8. Ayoub, A. K., Patton, A. V., 'Frequency and duration method for generating system reliability evaluation', *IEEE Transactions*, **PAS-95** (1976), pp. 1929–33.
9. Allan, R. N., Takieddine, F. N., 'Generator maintenance scheduling using simplified frequency and duration reliability criteria', *Proc. IEE*, **124** (1977), pp. 873–80.

4 Interconnected systems

4.1 Introduction

The adequacy of the generating capacity in a power system is normally improved by interconnecting the system to another power system [1]. Each interconnected system can then operate at a given risk level with a lower reserve than would be required without the interconnection [2]. This condition is brought about by the diversity in the probabilistic occurrence of load and capacity outages in the different systems [3]. The actual interconnection benefits depend on the installed capacity in each system, the total tie capacity, the forced outage rates of the tie lines, the load levels and their residual uncertainties in each system and the type of agreement in existence between the systems [4].

There are several probabilistic methods available at the present time which provide a quantitative reliability assessment of interconnected system generation facilities. The loss of load expectation (LOLE) approach is the most widely used technique due to its flexibility and simplicity of application. Two different approaches are presented in this chapter for calculating the LOLE indices in interconnected systems. They are the probability array method [2] and the equivalent assisting unit method [4]. In the first approach a capacity model is developed for each system and an array of simultaneous capacity outage existence probabilities is then obtained from the individual models. The second method models the assisting system as an equivalent assisting unit which can be moved through the tie lines and added into the existing capacity model of the assisted system. The computation of the risk in the assisted system proceeds as in the case of a normal single system study.

The array method and the equivalent assistance unit method can also be utilized in a frequency and duration (F&D) approach to interconnected system capacity evaluation. The basic technique is illustrated in this chapter.

The effects of tie capacity, tie line parameters, peak load and the type of agreement between systems are illustrated by a relatively simple hypothetical example. The basic concepts involved are then extended to adequacy assessment in three interconnected systems. Detailed examples are provided in each case to illustrate the principles behind the probability array and equivalent assisting unit methods. Results of selected studies using the IEEE Reliability Test System (IEEE–RTS) are presented in Appendix 2 to illustrate the application of these concepts to a practical system and to provide the reader with the opportunity to develop a computer program and test his results.

4.2 Probability array method in two interconnected systems

4.2.1 Concepts

A loss of load situation is considered to arise in a system when the available assistance through the interconnections cannot offset a capacity deficiency arising due to capacity outages and load demands in that system. The calculated LOLE index therefore depends on the simultaneous occurrences of the capacity outages in both systems. The generating facilities in each system can be represented by a two-dimensional probability array covering all possible combinations of capacity outages in the two systems. This amalgamated array represents the overall interconnected system capacity model with ideal interconnections. This representation can then be modified by including the load levels in each system and the tie line constraints. This concept is shown diagrammatically in Fig. 4.1 which illustrates the boundaries between good and bad states.

The first step in this approach is to develop the capacity model of each system. This is normally expressed in the form of a capacity outage probability table. It can be obtained recursively by successively adding generating units to the table using the algorithms given in Chapter 2. The probability array of the various simultaneous outage probabilities in the two systems is then obtained from the individual system tables. The loss of load situations in each system are determined from the simultaneous capacity outage conditions, the emergency assistance available over the tie lines, and the respective system daily peak loads. When the period of study is a single day, the sum of the simultaneous outage probabilities associated with the loss of load situations in each system is the system risk expressed as a loss of load expectation in days per day. Reliability evaluation for a longer study period can be

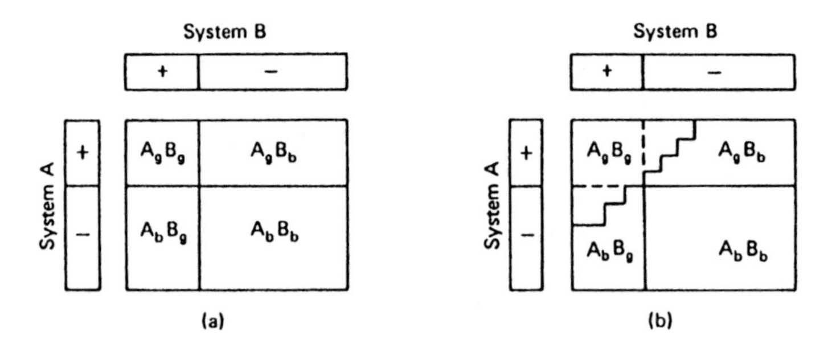

Fig. 4.1 Concept of probability array approach: (a) without tie line; (b) with tie line
Subscripts: g = good states, b = bad states
Symbols: plus = positive reserve states, minus = negative reserve states

Table 4.1 System data

System	*Number of units*	*Unit cap (MW)*	*FOR per unit*	*Installed cap. (MW)*	*Daily peak load (MW)*
A	5	10	0.02	75	50
	1	25	0.02		
B	4	10	0.02	60	40
	1	20	0.02		

Tie line connecting		*Number of tie lines*	*Tie cap (MW)*	*FOR per tie line*
System A	*System* B	1	10	0

accomplished by repeating the calculation for each of the subsequent days and summing the LOLE expectation values or by using the approximate technique discussed later in this chapter.

4.2.2 Evaluation techniques

The method can be illustrated by application to the hypothetical example shown in Fig. 4.2. Assume that two systems A and B are interconnected by a finite capacity tie line and have the data shown in Table 4.1. The operating agreement between the two systems is that each system will provide assistance only up to the point at which it begins to share the neighboring system difficulty. It is required to evaluate the risk in System A. A similar procedure can be followed to calculate the risk in System B.

The capacity outage probability table for each system is shown in Table 4.2. Probabilities less than 10^{-8} have been neglected in this table. The peak load in System A is assumed to be 50 MW and therefore the system will fail to meet its load demand when the capacity on outage is greater than 25 MW. The cumulative probability for a capacity outage of 30 MW becomes the LOLE index in System A (LOLE_A). System B has a reserve of 60 – 40 = 20 MW and will encounter a loss of load situation for any capacity outage greater than 20 MW to give LOLE_B.

$$\text{LOLE}_\text{A} = \underline{0.00199767 \text{ days}}$$

$$\text{LOLE}_\text{B} = \underline{0.00158353 \text{ days}}.$$

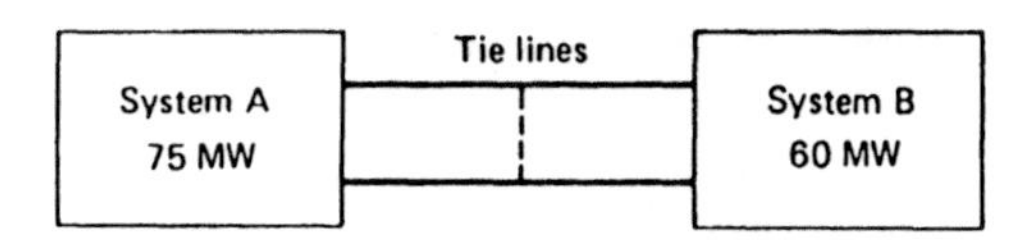

Fig. 4.2 Two systems interconnected by repairable tie lines

Table 4.2 Capacity outage probability tables

System A				System B			
State i	*Cap. out* $C_i(MW)$	*Individual prob.* $p(C_i)$	*Cum. prob.* $P(C_i)$	*State i*	*Cap. out* $C_i(MW)$	*Individual prob.* $p(C_i)$	*Cum. prob.* $P(C_i)$
1	0	0.88584238	1.00000000	1	0	0.90392080	1.00000000
2	10	0.09039207	0.11415762	2	10	0.07378945	0.09607920
3	20	0.00368947	0.02376555	3	20	0.02070622	0.02228975
4	25	0.01807841	0.02007608	4	30	0.00153664	0.00158353
5	30	0.00007530	0.00199767	5	40	0.00004626	0.00004689
6	35	0.00184474	0.00192237	6	50	0.00000063	0.00000063
7	40	0.00000077	0.00007763	7	60	0.00000000	0.00000000
8	45	0.00007530	0.00007686				
9	50	0.00000000	0.00000156				
10	55	0.00000154	0.00000156				
11	65	0.00000002	0.00000002				
12	75	0.00000000	0.00000000				

The various simultaneous capacity outage probabilities in the two systems can be calculated from the individual system tables. Table 4.3 presents this two-dimensional array, which is the overall interconnected system capacity outage probability table with ideal interconnections. Each array cell in Table 4.3 has two numbers which represent the existence probability of a particular simultaneous capacity outage condition and the load loss in MW in System A considering the emergency assistance from System B for the tie-line constraint given in Table 4.1. The sum of the simultaneous outage probabilities associated with non-zero load losses is the system risk in System A ($LOLE_{AB}$). This result is shown in Table 4.4. Also shown in Table 4.4 are the expected load losses ELL_A and ELL_{AB}.

The probability array as shown in Table 4.3 may require a significant amount of computer memory and the processing may also use a considerable amount of execution time when the interconnected systems are large. These requirements can be reduced by rounding the table using a suitable increment. There is, however, an alternative approach which can produce the same results more directly and with appreciable savings in both computer memory and execution time. This approach is described in Section 4.3.

4.3 Equivalent assisting unit approach to two interconnected systems [5–6]

The equivalent unit approach represents the benefits of interconnection between the two systems in terms of an equivalent multi-state unit which describes the

Table 4.3 Probability of simultaneous outages in Systems A and B and load loss in System A

System A MW out \ System B MW out	0		10		20		30		40		50		60	
0	0.80073135	0	0.06536582	0	0.01834245	0	0.00136122	0	0.00004098	0	0.00000056	0	0.00000000	0
10	0.08170728	0	0.00666998	0	0.00187168	0	0.00013890	0	0.00000418	0	0.00000006	0	0.00000000	0
20	0.00333499	0	0.00027224	0	0.00007640	0	0.00000567	0	0.00000017	0	0.00000000	0	0.00000000	0
25	0.01634146	0	0.00133400	0	0.00037434	0	0.00002778	0	0.00000084	0	0.00000001	0	0.00000000	0
30	0.00006806	0	0.00000556	0	0.00000156	5	0.00000012	5	0.00000000	5	0.00000000	5	0.00000000	5
35	0.00166750	0	0.00013612	0	0.00003820	10	0.00000283	10	0.00000009	10	0.00000000	10	0.00000000	10
40	0.00000070	5	0.00000006	5	0.00000002	15	0.00000000	15	0.00000000	15	0.00000000	15	0.00000000	15
45	0.00006806	10	0.00000556	10	0.00000156	20	0.00000012	20	0.00000000	20	0.00000000	20	0.00000000	20
50	0.00000000	15	0.00000000	15	0.00000000	25	0.00000000	25	0.00000000	25	0.00000000	25	0.00000000	25
55	0.00000139	20	0.00000011	20	0.00000003	30	0.00000000	30	0.00000000	30	0.00000000	30	0.00000000	30
65	0.00000001	30	0.00000000	30	0.00000000	40	0.00000000	40	0.00000000	40	0.00000000	40	0.00000000	35
75	0.00000000	40	0.00000000	40	0.00000000	50	0.00000000	50	0.00000000	50	0.00000000	50	0.00000000	40

Table 4.4 Results for System A

	Individual system	*Interconnected system*
LOLE (days)	0.00199767	0.00012042
ELL (MW)	0.02038785	0.00122465

potential ability of one system to accommodate capacity deficiencies in the other. This is described considering System A as the assisted system and System B as the assisting system. The capacity assistance level for a particular outage state in System B is given by the minimum of the tie capacity and available system reserve at that outage state. All capacity assistance levels greater than or equal to the tie capacity are replaced by one assistance level which is equal to the tie capacity. This capacity assistance table can then be converted into a capacity model of an equivalent multi-state unit which is added to the existing capacity model of System A. Using the algorithms given in Chapter 2, the risk level is then evaluated as if the assisted system is a single area system.

The equivalent assisting unit approach has been applied to the example system shown in Table 4.1.

System B has a reserve of 20 MW and this is therefore the maximum assistance it can provide without sharing potential difficulties in System A. The capacity assistance levels of System B are obtained from the capacity outage probability table given in Table 4.2. These levels are shown in Table 4.5. This table can be converted to a capacity outage probability table (Table 4.6) for an equivalent assisting unit model of System B.

The 10 MW tie line described in Table 4.1 constrains the capacity assistance from System B and therefore the equivalent assisting unit is also tie-line constrained as shown in Table 4.7.

This equivalent multi-state unit is now added to the existing capacity model of System A, giving a new installed capacity of 75 + 10 = 85 MW. The new System A capacity outage probability table is shown in Table 4.8.

The daily peak load in System A is 50 MW and therefore the loss-of-load situation occurs when the capacity outage in System A is greater than the reserve

Table 4.5 Assistance probability table from System B

Assistance (MW)	*Individual prob.*
20	0.90392080
10	0.07378945
0	0.02228975

Table 4.6 Equivalent assisting unit model of System B

Cap. out (MW)	*Individual prob.*
0	0.90392080
10	0.07378945
20	0.02228975

Table 4.7 Tie-line constrained equivalent unit model of System B

Cap. out (MW)	*Individual prob.*
0	0.97771025
10	0.02228975

85 – 50 = 35 MW. The cumulative probability for a capacity outage of 40 MW is therefore the risk with interconnection:

$$\text{LOLE}_{\text{AB}} = P(C_7 = 40) = \underline{0.00012042 \text{ days}}$$

This result is identical to that given in Table 4.4 and illustrates the fact that both approaches provide the same numerical index. This chapter utilizes the equivalent assisting unit approach in later studies due to its particular advantage. The value of LOLE_{BA} can be found using System A as the assisting system and System B as the assisted system.

Table 4.8 Modified capacity outage probability table of System A

State i	*Cap. out C_i(MW)*	*Individual prob. $p(C_i)$*	*Cum. prob. $P(C_i)$*
1	0	0.86609717	1.00000000
2	10	0.10812248	0.13390283
3	20	0.00562205	0.02578035
4	25	0.01767545	0.02015830
5	30	0.00015585	0.00248285
6	35	0.00220658	0.00232700
7	40	0.00000243	0.00012042
8	45	0.00011474	0.00011799
9	50	0.00000002	0.00000325
10	55	0.00000318	0.00000323
11	60	0.00000000	0.00000005
12	65	0.00000005	0.00000005
13	75	0.00000000	0.00000000

4.4 Factors affecting the emergency assistance available through the interconnections

4.4.1 Introduction

A loss of load in a single system occurs when the available capacity cannot meet the load demand. In the case of interconnected systems, the capacity deficiency may be accommodated by available assistance from other systems. This assistance depends on the available capacity and the operating reserve in the other systems, the interconnection limitations and the type of agreement between systems. These factors are all interlinked in regard to their impact on the reliability levels of an interconnected group of systems. Their individual impacts are illustrated in this chapter using the systems shown in Table 4.1 by varying one factor at a time and considering the change in system risk level.

4.4.2 Effect of tie capacity

This effect on the risk level is illustrated by varying the tie capacity from 0 to 30 MW in steps of 5 MW. The maximum assistance available from System B using the load condition given in Table 4.1 is 20 MW, if System B is not to share the difficulties of System A. Either of the techniques illustrated in the preceding sections can be used to obtain the results shown in Table 4.9.

The interconnections considered in Table 4.9 are assumed to be fully reliable. Table 4.9 shows the rapid reduction in risk which occurs with increase in the interconnection capacity between the two systems. The risk converges to a value which represents the minimum risk that System A can have under these conditions. The risk at this limit can be designated as the infinite tie capacity risk as there will be no further decrease in risk with the addition of further tie capacity. The infinite tie capacity value is directly related to the maximum assistance available from System B. This is easily seen in the case of a single load level example but becomes less specific when a group of load levels is considered.

Table 4.9 Effect of tie capacity

Tie cap. (MW)	*System A* $LOLE_{AB}$ *(days/day)*
0	0.00199767
5	0.00192403
10	0.00012042
15	0.00011972
20	0.00005166
25	0.00005166
30	0.00005166

4.4.3 Effect of tie line reliability

Systems may be interconnected by several tie lines, each of which has an availability that is less than unity. The various tie-line capacity states impose capacity limits on the emergency assistance available through the interconnections. This effect can be evaluated using one of two approaches.

(a) *Approach 1*

The first approach is to convolve the capacity states of the tie lines with those of the equivalent assisting unit obtained from the assisting system. In this case, the output model of the combination then effectively represents a tie-line constrained multi-state generating unit in the assisting system.

Consider the case in which there is only one tie line of 10 MW interconnecting systems A and B. Assume the failure and repair rates of the tie line are $\lambda = 3$ f/yr and $\mu = 1$ r/day respectively, giving an unavailability of 0.00815217.

There are two states in the capacity outage probability table for the tie line as shown in Table 4.10. The equivalent assisting unit model of System B for a 100% reliable tie line is given in Table 4.7.

System A may receive the emergency assistance of 10 MW if the tie line is up and therefore the probability of receiving 10 MW under this condition is $0.99184783 \times 0.97771025 = 0.96973979$. On the other hand, the assistance is not available to System A when the tie line is down or System B also encounters difficulty even if the tie line is up. In this case, the probability is $0.00815217 + 0.99184783 \times 0.02228975 = 0.03026021$, or it can be obtained by subtracting from unity the probability of receiving 10 MW $(1-0.96973979)$. The tie-line constrained equivalent assisting unit model of System B is shown in Table 4.11.

This can be added to the existing capacity model of System A to give a new installed capacity of $75 + 10 = 85$ MW. Table 4.12 shows the modified capacity outage probability table for System A.

System A with interconnection has a reserve of $85 - 50 = 35$ MW and fails to meet its load demand when the capacity outage is greater than 35 MW. The cumulative probability for a capacity outage of 40 MW is therefore the system risk.

$$\text{LOLE}_{\text{AB}} = P(C_7 = 40) = \underline{0.00013572 \text{ days}}$$

Table 4.10 Tie capacity outage probability table

Tie cap. out (MW)	*Individual prob.*
0	0.99184783
10	0.00815217

Table 4.11 Tie-line constrained equivalent assisting unit model of System B

Cap. out (MW)	*Individual prob.*
0	0.96973979
10	0.03026021

(b) *Approach 2*

The second approach evaluates the same risk using the probability array method and the conditional probability rule. The overall system risk is given by the sum of the products of conditional LOLE_{AB} and the corresponding tie capacity probability:

$$\begin{aligned}\text{LOLE}_{\text{AB}} = {} & \text{LOLE}_{\text{AB}}\,(0 \text{ MW tie cap out}) \times p(0 \text{ MW tie cap out}) \\ & + \text{LOLE}_{\text{AB}}\,(10 \text{ MW tie cap out}) \times p(10 \text{ MW tie cap out})\end{aligned}$$

The conditional LOLE_{AB} values in column (2) of Table 4.13 are obtained from Table 4.4.

4.4.4 Effect of number of tie lines

Consider the case in which the interconnection between Systems A and B consists of two identical tie lines rated 10 MW each having $\lambda = 3$ f/yr and $\mu = 1$ r/day. The capacity outage probability table for the two tie lines is shown in Table 4.14.

The effect of tie line constraints is included by combining the capacity out states of both the equivalent assisting unit and the tie lines as shown in Table 4.15.

Table 4.12 Modified capacity outage probability table of System A

State i	*Cap. out C_i (MW)*	*Individual prob. $p(C_i)$*	*Cum. prob. $P(C_i)$*
1	0	0.85903660	1.00000000
2	10	0.11446258	0.14096340
3	20	0.00631311	0.02650082
4	25	0.01753136	0.02018771
5	30	0.00018466	0.00265635
6	35	0.00233597	0.00247169
7	40	0.00000302	0.00013572
8	45	0.00012884	0.00013270
9	50	0.00000003	0.00000386
10	55	0.00000377	0.00000383
11	60	0.00000000	0.00000006
12	65	0.00000006	0.00000006
13	75	0.00000000	0.00000000

Table 4.13 System risk using the conditional probability rule

Tie cap. out (MW)	*Tie cap. in (MW)*	*(1) Individual prob.*	*(2) Conditional $LOLE_{AB}$*	*(1) × (2)*
0	10	0.99184783	0.00012042	0.00011943
10	0	0.00815217	0.00199767	0.00001629
			$LOLE_{AB}$ =	0.00013572

Table 4.14 Tie capacity outage probability table

Tie cap. out (MW)	*Individual prob.*
0	0.98376211
10	0.01617143
20	0.00006646

The equivalent unit in Table 4.15 can be added to the existing capacity model of System A to give a new installed capacity of 75 + 20 = 95 MW. Table 4.16 shows the modified capacity outage probability table for System A.

System A with interconnection has a reserve of 95 – 50 = 45 MW and will fail to meet its load demand when the capacity outage is greater than 45 MW. The cumulative probability for a capacity outage of 50 MW is therefore the system risk:

$$LOLE_{AB} = P(C_9 = 50) = \underline{0.00005290 \text{ days}}$$

Table 4.15 Tie-line and assisting unit capacity states

		Tie cap. out states (MW)		
		0	*10*	*20*
Cap. out states of equivalent assisting unit (MW)	0	0 0.88924303	10 0.01461768	20 0.00006008
	10	10 0.07259127	10 0.00119328	20 0.00000491
	20	20 0.02192781	20 0.00036046	20 0.00000148

Tie-line constrained equivalent assisting unit model of System B

Cap. out (MW)	*Individual prob.*
0	0.88924303
10	0.08840223
20	0.02235474

Table 4.16 Modified capacity outage probability table of System A

State i	*Cap. out (MW)* C_i	*Individual prob.* $p(C_i)$	*Cum. prob.* $P(C_i)$
1	0	0.78772918	1.00000000
2	10	0.15869097	0.21227082
3	20	0.03107446	0.05357985
4	25	0.01607611	0.02250539
5	30	0.00241380	0.00642928
6	35	0.00323859	0.00401548
7	40	0.00008982	0.00077689
8	45	0.00063417	0.00068707
9	50	0.00000175	0.00005290
10	55	0.00004926	0.00005115
11	60	0.00000002	0.00000189
12	65	0.00000183	0.00000187
13	70	0.00000000	0.00000004
14	75	0.00000004	0.00000004
15	85	0.00000000	0.00000000
16	95	0.00000000	0.00000000

This value can be compared with the index of 0.00005166 days obtained for 20 MW of fully reliable tie capacity shown in Table 4.9. Inclusion of the tie-line availability results in an increase in system risk level. This increase is usually only slight because the unavailability of the tie line is usually much less than that of generating units.

Alternatively the same risk can be obtained using the probability array method and the conditional probability rule. The tie lines can reside in only one of the tie-capacity outage states at any point in time. The states are therefore mutually exclusive events. The overall system risk is given by the sum of the products of conditional LOLE_{AB} and the respective tie capacity probability.

$$\begin{aligned}\text{LOLE}_{\text{AB}} = {} & \text{LOLE}_{\text{AB}}(0 \text{ MW tie cap. out}) \times p(0 \text{ MW tie cap. out}) \\ & + \text{LOLE}_{\text{AB}}(10 \text{ MW tie cap. out}) \times p(10 \text{ MW tie cap. out}) \\ & + \text{LOLE}_{\text{AB}}(20 \text{ MW tie cap. out}) \times p(20 \text{ MW tie cap. out})\end{aligned}$$

Table 4.17 System risk using conditional probability rule

Tie cap. out	*Tie cap. in*	(1) *Individual prob.*	(2) *Conditional* $LOLE_{AB}$	(1) × (2)
0	20	0.98376211	0.00005166	0.00005082
10	10	0.01617143	0.00012042	0.00000195
20	0	0.00006646	0.00199767	0.00000013
			LOLE_{AB} =	0.00005290

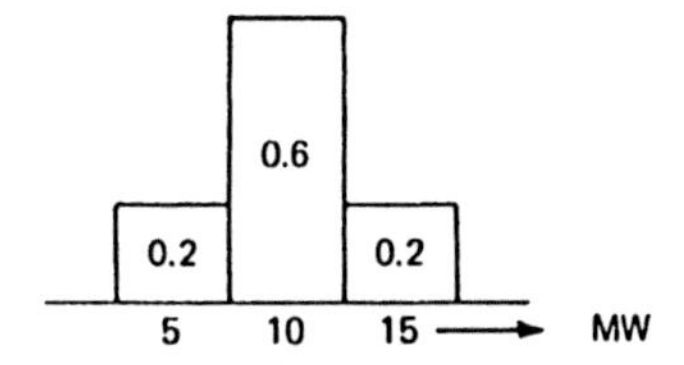

Fig. 4.3 A probability distribution of the tie capacity

The conditional $LOLE_{AB}$ in column (2) of Table 4.17 can be obtained from the two-dimensional array shown in Table 4.3 and its values are equal to those shown in Table 4.9.

4.4.5 Effect of tie-capacity uncertainty

The previous sections have assumed that the tie capacity is a fixed value. This may not be the case due to changing transmission conditions in the two systems. The random variation in tie capacity can be represented by a discrete or continuous probability distribution. The conditional probability rule can be applied using an approximate discrete probability distribution to compute the overall system risk. This is similar in concept to the assessment of load forecast uncertainty described in Chapter 2.

Consider the case in which the tie lines between the systems have a probability of 0.6 of attaining the forecast tie capacity of 10 MW and that the tie capacity can be 5 or 15 MW with a probability of 0.2 respectively. The area under the histogram shown in Fig. 4.3 represents the probability of the tie capacity being the designated values.

The LOLE is then computed for each tie capacity and multiplied by the probability of existence of that tie capacity. The sum of these products represents the overall system risk for the forecast tie capacity.

These results are shown in Table 4.18.

Table 4.18 Effect of tie capacity uncertainty

Tie cap.	(1) *Individual prob.*	(2) *Conditional $LOLE_{AB}$*	(1) × (2)
5	0.2	0.00192403	0.00038481
10	0.6	0.00012042	0.00007225
15	0.2	0.00011972	0.00002394
		$LOLE_{AB}$ =	0.00048100

4.4.6 Effect of interconnection agreements

The analyses in the previous sections are based on one particular interconnection agreement. Many other agreements between different utilities exist and it is not possible to consider all of these exhaustively. Some basic agreements are illustrated in this section however by application to the previous hypothetical example.

(a) *Firm purchase*

Consider the case in which a firm purchase by System A is backed up by the complete System B and determine the risk in System A for different values of firm purchase and tie capacity. The tie line is assumed to be 100% reliable in these cases.

(i)	*Firm purchase*	*Tie capacity*
	10 MW	10 MW

The availability of 10 MW through the interconnection is guaranteed by the entire System B when System A is in need of emergency assistance. The firm purchase can be modelled as a 10 MW generating unit with an effective zero forced outage rate added to System A. The increased installed capacity of System A results in the following risk level:

$$\text{LOLE}_{AB} = \underline{0.00007763 \text{ days}}$$

This value can be obtained from Table 4.2 by either adding 10 MW to the installed capacity of System A or subtracting 10 MW from its load level.

(ii)	*Firm purchase*	*Tie capacity*
	10 MW	15 MW

The firm purchase of 10 MW from System B is added to the capacity model of System A as a perfectly reliable 10 MW generating unit. The installed capacity of System B is reduced by 10 MW. Additional emergency assistance from System B is then possible over the remaining tie capacity of 15 – 10 = 5 MW.

The risk level in System A is obtained using the modified capacity models of Systems A and B and a 5 MW tie capacity. The risk in System A becomes

$$\text{LOLE}_{AB} = \underline{0.00007691 \text{ days}}$$

(b) *Firm purchase tied to a specific 10 MW unit in System B*

Consider the case in which the purchase of 10 MW of power is under the contractual condition that if the 10 MW unit in System B is out of service then the purchase capacity is not guaranteed.

(i)	*Firm purchase*	*Tie capacity*
	10 MW	10 MW

Table 4.19 Risk with 10 MW tie line

Firm purchase (MW)	*Tie cap. (MW)*	(1) *Individual prob.*	(2) *Conditional $LOLE_{AB}$*	(1) × (2)
10	10	0.98	0.00007763	0.00007608
		0.02	0.00022667	0.00000453
				$LOLE_{AB}$ = 0.00008061

The 10 MW unit can exist in one of two states—up or down. When it is up, the firm purchase of 10 MW is added to the capacity model of System A as a 10 MW unit of zero forced outage rate. If it is down, no firm capacity is available and System A has to take the chance that some random assistance may be available. The overall expected risk value is obtained by weighting the corresponding conditional LOLE with the availability and forced outage rate of the 10 MW unit to which the purchase is tied.

(ii) *Firm purchase* *Tie capacity*

10 MW 15 MW

When the 10 MW unit in System B is up, the firm purchase of 10 MW is realized and added directly to the capacity model of System A as a perfectly reliable 10 MW unit. This installed capacity of System B is reduced by 10 MW. Additional emergency assistance from System B is still possible over the remaining tie capacity 15 – 10 = 5 MW. When the 10 MW unit is down, the firm purchase of 10 MW is not available. System A can still receive assistance over the 15 MW tie line from System B whose capacity model does not include the unavailable 10 MW unit. Both conditional $LOLE_{AB}$ values are then weighted by the respective probability of availability and forced outage rate of the 10 MW unit to give the overall expected risk as shown in Table 4.20. A summary of these results is shown in Table 4.21.

The contractual arrangements shown are relatively simple. The results show, however, that they have a definite impact on the risk indices and their consideration should be an integral part of any interconnected system reliability evaluation.

Table 4.20 Risk with 15 MW tie line

Firm purchase (MW)	*Tie cap. (MW)*	(1) *Individual prob.*	(2) *Conditional $LOLE_{AB}$*	(1) × (2)
10	15	0.98	0.00007691	0.00007537
		0.02	0.00022667	0.00000453
				$LOLE_{AB}$ = 0.00007991

Table 4.21 Summary of results

Firm purchase (MW)	*Tie cap. (MW)*	*Contractual condition*	$LOLE_{AB}$
10	10	Firm purchase is backed up by the complete system.	0.00007763
10	15	Firm purchase is backed up by the complete system and additional assistance may be possible over remaining 5 MW tie.	0.00007691
10	10	Firm purchase is tied to the 10 MW unit.	0.00008061
10	15	Firm purchase is tied to the 10 MW unit and additional assistance may be possible over remaining 5 MW tie.	0.00007991

4.4.7 Effect of load forecast uncertainty

The load forecast uncertainty existing in either the assisting system or the assisted system or both can be included in the analysis using either the array or equivalent unit methods. In the case of the array approach there are $n \times m$ risk levels for System A, where n and m are the discrete load steps in the uncertainty models for Systems A and B respectively. The risk in each case is weighted by the probability of the simultaneous load conditions and the summation of these risks is the expected risk for System A.

In the case of the equivalent unit approach, a conditional equivalent unit can be obtained for System B. This is then added directly to System A and the analysis continues with a single system study for System A using either a discrete uncertainty model or a modified load curve as shown in Chapter 2.

4.5 Variable reserve versus maximum peak load reserve

There is basically no conceptual difficulty in evaluating the total risk for a longer study period than for the one-day example used. A multi-day period can be broken down into a sequence of one-day periods and the overall risk can be computed as the sum of the daily risks. The variation of the daily loads in the assisting system gives rise to a variable daily reserve and therefore governs the extent to which the assisting system can provide emergency assistance. This type of reliability analysis on a daily basis can be designated as the variable reserve approach, in which case it becomes necessary to forecast the simultaneous daily peak loads in each system for each day over the whole study period. This can be difficult to accomplish, particularly for periods well into the future.

The problem can be simplified by replacing the daily peak loads in the assisting system by the maximum daily peak load for the multi-day study period. The reserve

Table 4.22 Daily peak load

Day	*System A (MW)*	*System B (MW)*
1	46.5	37.2
2	50.0	40.0
3	49.0	39.2
4	48.0	38.4
5	47.0	37.6

available in the assisting system is thereby confined to the minimum value of the daily reserves in the period. This concept of maximum peak load reserve or minimum reserve therefore constitutes a pessimistic assessment of the interconnected system reliability.

The load data for a 5-day period curve was extracted from the load parameters shown in Appendix 2 for the IEEE–RTS. The load levels were then scaled down for the hypothetical example to give the levels shown in Table 4.22. The tie line is assumed to have a tie capacity of 10 MW and to be fully reliable. The results are shown in Table 4.23.

Although the individual values of $LOLE_{AB}$ for the case of the maximum peak load reserve are all identical in this example, this is only a reflection of the data used and the discrete levels of the generation model. In other examples, the results may be different.

A similar computation was carried out for different tie capacities to measure the differences between the overall risks for the two reserve concepts. The results are shown in Table 4.24. The difference increases as the tie capacity becomes larger. The degree of pessimism embedded in the minimum reserve approach may be

Table 4.23 Comparison of variable and maximum peak load reserves

			Variable reserve		*Maximum peak load reserve*	
Tie cap. (MW)	*Day*	*System A Peak load (MW)*	*System B Peak load (MW)*	*Daily $LOLE_{AB}$*	*System B Peak load (MW)*	*Daily $LOLE_{AB}$*
10	1	46.5	37.2	0.00011886	40.0	0.00012042
	2	50.0	40.0	0.00012042	40.0	0.00012042
	3	49.0	39.2	0.00012042	40.0	0.00012042
	4	48.0	38.4	0.00012042	40.0	0.00012042
	5	47.0	37.6	0.00011886	40.0	0.00012042
			5-day $LOLE_{AB}$ =	0.00059898	5-day $LOLE_{AB}$ =	0.00060210

Table 4.24 Effect of tie-line capacity

	5-day $LOLE_{AB}$		
Tie cap. (MW)	*Variable reserve*	*Maximum peak load reserve*	*Error (%)*
0	0.00998825	0.00998825	0
5	0.00961705	0.00962017	0.03
10	0.00059898	0.00060210	0.52
15	0.00059538	0.00059861	0.54
20	0.00025507	0.00025830	1.27
25	0.00025507	0.00025830	1.27

significant when the interconnection capacity is much greater than the reserve. This effect is illustrated in Appendix 2 using the IEEE–RTS.

4.6 Reliability evaluation in three interconnected systems

4.6.1 Direct assistance from two systems

Consider the case in which System A has an additional interconnection with a third system as shown in Fig. 4.4. The system data is shown in Table 4.25.

The three systems A, B and C are interconnected by fully reliable tie lines and it is required to evaluate the reliability level in System A. The approach is to develop equivalent assisting units for Systems B and C and to add them to the capacity model of System A. These models are shown in Tables 4.26 and 4.27.

The modified installed capacity of System A is 75 + 15 + 30 = 120 MW when both tie-line constrained equivalent assisting units of Systems B and C are added. This gives the modified capacity outage probability table of System A shown in Table 4.28.

The daily peak load in System A is 50 MW and therefore a loss-of-load situation occurs when the capacity outage in System A is greater than the reserve 120 – 50 = 70 MW. The cumulative probability for a capacity outage of 75 MW therefore becomes the risk.

$$\mathrm{LOLE}_{\mathrm{ABC}} = P(C_{16}) = \underline{0.00000074 \text{ days}}$$

The risk level in System A as computed for a range of peak loads in System A with the peak loads in Systems B and C assumed to be fixed at 40 and 90 MW

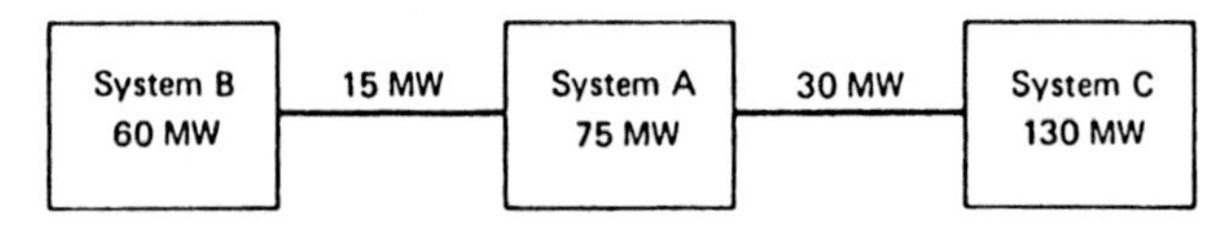

Fig. 4.4 Three interconnected systems

Table 4.25 System data

System	*Number of units*	*Unit cap. (MW)*	*FOR per unit*	*Installed cap. (MW)*	*Daily peak load (MW)*
A	5	10	0.02	75	50
	1	25	0.02		
B	4	10	0.02	60	40
	1	20	0.02		
C	5	20	0.02	130	90
	1	30	0.02		

Tie-line system	*Connecting system*	*Number of tie lines*	*Tie cap. (MW)*	*FOR per tie line*
A	B	1	15	0
A	C	1	30	0

respectively are shown in Table 4.29. These values are obtained directly from Table 4.28.

4.6.2 Indirect assistance from two systems

This case can be described by considering System B in Fig. 4.4 as the assisted systems and Systems A and C as the assisting systems. The peak loads in Systems A and C in this case are fixed at 50 and 90 MW respectively. The procedure is similar to that of Section 4.6.1. The first step is to evaluate the benefit derived by System A being connected to System C. From Table 4.27 this gives the assistance model shown in Table 4.30.

Table 4.26 Assistance by System B

Equivalent assisting unit model of System B

Cap. out (MW)	*Individual prob.*
0	0.90392080
10	0.07378945
20	0.02228975

Tie line constraint: 1 tie line of 15 MW, 100% reliable.

Tie-line constrained equivalent assisting unit model of System B

Cap. out (MW)	*Individual prob.*
0	0.90392080
5	0.07378945
15	0.02228975

Table 4.27 Assistance by System C

Capacity outage probability table of System C	
Cap. out	*Individual prob.*
0	0.88584238
20	0.09039208
30	0.01807842
40	0.00368947
50	0.00184474
60	0.00007530
70	0.00007530
80	0.00000077
90	0.00000154
100	0.00000000
110	0.00000002
130	0.00000000

Assistance probability table from system C	
Asistance	*Individual prob.*
40	0.88584238
20	0.09039208
10	0.01807842
0	0.00568712

Equivalent assisting unit model of System C	
Cap. out	*Individual prob.*
0	0.88584238
10	0.09039208
20	0.01807842
40	0.00568712

Tie-line contrained equivalent assisting unit model of System C	
Cap. out (MW)	*Individual prob.*
0	0.88584238
10	0.09039208
20	0.01807842
30	0.00568712

Tie-line contraint: 1 tie line of 30 MW, 100% reliable.

This model is added to the capacity model of System A to give the modified capacity model shown in Table 4.31, which is then combined with the load model to determine an equivalent assisting unit appearing at System B as shown in Table 4.32.

The assisting unit shown in Table 4.32 is, however, tie-line constrained as the tie capacity is 15 MW. The final unit is shown in Table 4.33.

This model is added to the capacity model of System B and the computation of the risk in System B follows as if there remained only one single system. This modified model is shown in Table 4.34.

The reserve in System B is 75 – 40 = 35 MW when the daily peak load is 40 MW. A loss-of-load situation occurs for any capacity outage greater than 35 MW. The $LOLE_{BAC}$ is therefore the cumulative probability for a capacity outage of 40 MW:

$$LOLE_{BAC} = P(C_9 = 40) = \underline{0.00004717 \text{ days}}$$

The bottleneck in the capacity assistance can clearly be seen in this example. The assistance from the modified Systems A and C is constrained by the finite tie

Table 4.28 Modified capacity outage probability table of System A

State i	*Cap. out (MW)* C_i	*Individual prob.* $p(C_i)$	*Cum. prob.* $P(C_i)$
1	0	0.70932182	1.00000000
2	5	0.05790381	0.29067818
3	10	0.14475951	0.23277437
4	15	0.02930824	0.08801486
5	20	0.02481592	0.05870662
6	25	0.02007136	0.03389070
7	30	0.00757446	0.01381934
8	35	0.00408807	0.00624488
9	40	0.00112987	0.00215681
10	45	0.00070749	0.00102694
11	50	0.00013445	0.00031945
12	55	0.00014523	0.00018500
13	60	0.00002354	0.00003977
14	65	0.00001138	0.00001623
15	70	0.00000411	0.00000485
16	75	0.00000042	0.00000074
17	80	0.00000030	0.00000032
18	85	0.00000001	0.00000002
19	90	0.00000001	0.00000001

Table 4.29 Effect of peak load in System A

Peak load (MW)	$LOLE_{ABC}$
50	0.00000074
55	0.00000485
60	0.00001623
65	0.00003977
70	0.00018500

Table 4.30 Tie-line constrained equivalent assisting unit model of System C appearing to System A

Cap. out (MW)	*Individual prob.*
0	0.88584238
10	0.09039208
20	0.01807842
30	0.00568712

Table 4.31 Modified capacity outage probability table of System A

Cap. out (MW)	*Individual prob.*
0	0.78471669
10	0.16014629
20	0.02745365
25	0.01601463
30	0.00707224
35	0.00326829
40	0.00058826
45	0.00056028
50	0.00002242
55	0.00014433
60	0.00000044
65	0.00001201
75	0.00000046
85	0.00000001

Table 4.32 Equivalent assisting unit model of modified System A appearing to System B

Assistance (MW)	*Cap. out (MW)*	*Individual prob.*
55	0	0.78471669
45	10	0.16014629
35	20	0.02745365
30	25	0.01601463
25	30	0.00707224
20	35	0.00326829
15	40	0.00058826
10	45	0.00056028
5	50	0.00002242
0	55	0.00015725

Table 4.33 Tie-line constrained equivalent assisting unit model of modified System A appearing to System B

Cap. out (MW)	*Individual prob.*
0	0.99926005
5	0.00056028
10	0.00002242
15	0.00015725

Table 4.34 Modified capacity outage probability table of System B

State i	*Cap. out (MW)* C_i	*Individual prob.* $p(C_i)$	*Cum. prob.* $P(C_i)$
1	0	0.90325194	1.00000000
2	5	0.00050645	0.09674806
3	10	0.07375511	0.09624161
4	15	0.00018348	0.02248650
5	20	0.02069256	0.02230302
6	25	0.00002320	0.00161046
7	30	0.00153597	0.00158726
8	35	0.00000412	0.00005129
9	40	0.00004626	0.00004717
10	45	0.00000027	0.00000091
11	50	0.00000063	0.00000064
12	55	0.00000001	0.00000001
13	60	0.00000000	0.00000000

capacity between Systems A and B. System B therefore does not benefit as much as System A from this interconnection configuration.

4.7 Multi-connected systems

The methods described in Section 4.6 for evaluating risk levels in three interconnected systems can be extended to find the risk levels in multi-interconnected systems including those systems that are networked or meshed. The techniques can be based on either the probability array approach or the equivalent unit approach. The most important factor to define before commencing this evaluation is the interconnection agreement that exists between the interconnected utilities. Consider as an example the case of three systems A, B and C connected as shown in Fig. 4.5, with A as the system of interest. First consider the following two system conditions (others are also possible):

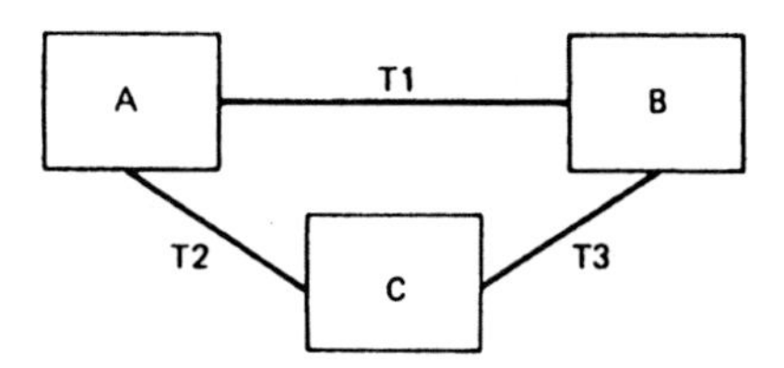

Fig. 4.5 Three interconnected systems

(a) *A deficient, B and C in surplus*

In this case System A can receive assistance from System B either directly through tie line T1 or indirectly via System C via tie lines T2 and T3. The limitation to the assistance will be dependent on the reserve in System B and the tie-line capacities. Similarly System A can receive assistance from System C. In this case, there is no difficulty over which system has priority of available reserves.

(b) *A and C deficient, B in surplus*

In this case, both System A and System C require assistance from System B and a clear appreciation of priority is essential before the risk can be evaluated. There are many possible interconnection agreements including the following: System A (or C) has total priority over the other and its deficiencies are made up before the other system has recourse to the reserves of System B if any remain; System A and System C share the reserves of System B in one of many ways.

Once the interconnection agreement has been established, the risk in any of the systems can be evaluated using either the probability array approach or the equivalent unit approach. These are described briefly below.

(a) *Probability array approach*

In this approach, a probability array is created which has as many dimensions as there are systems. Clearly this cannot be achieved manually but can be created on a digital computer. A multi-dimensional boundary wall is then constructed through this array which partitions the good states of a system from its bad states taking into account the agreement between the systems, the reserve in each system and the tie-line capacities. This is identical in concept to the two-dimensional array shown in Fig. 4.1. The risk in System A (also Systems B and C) can then be evaluated using the techniques described previously. Although this is computationally possible, it has the major disadvantage of excessive storage requirements.

(b) *Equivalent unit approach*

In this case equivalent assisting units can be developed for both of the assisting systems (in the equivalent unit approach only one of the systems is considered as the assisted system) that take into account the agreement between the systems, the reserve in each system and the tie-line capacities. These equivalent units are then combined with the generation model of the assisted system, which is then analyzed as a single system using the previous techniques. As an example, consider that System A has first priority on the reserves in System B, limited only by the capacity of tie line T1 in Fig. 4.5, plus any additional reserve in System B that can be transported via tie line T2 after System C has made up any of its own deficiencies via tie line T3. An equivalent unit can therefore be developed for System B considering tie line T1. Any additional potential assistance can be moved through tie line T3 and become part of the equivalent assistance unit that can be moved through tie line T2. System A can then be analyzed as a single system. There are many alternatives to this, depending on the agreement between the systems, but clearly these are too numerous to be discussed in this book. The concepts for other alternatives however are based on those described above.

4.8 Frequency and duration approach

4.8.1 Concepts

The technique outlined in Chapter 3 can also be used to obtain risk indices for interconnected systems. As in the case of the LOLE technique described in Section 4.2, either the array method or the equivalent assistance unit method can be used. The conventional approach [7–9] developed in 1971 is to create a margin array for the two interconnected systems and to impose an appropriate boundary wall dividing the good and bad states. This approach assumes that the generation and

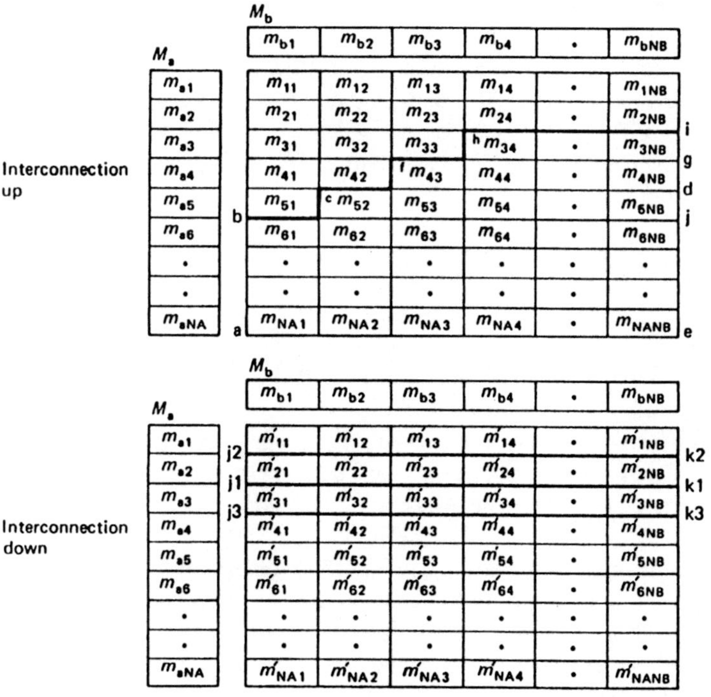

Interconnection up

M_a \ M_b	m_{b1}	m_{b2}	m_{b3}	m_{b4}	•	m_{bNB}
m_{a1}	m_{11}	m_{12}	m_{13}	m_{14}	•	m_{1NB}
m_{a2}	m_{21}	m_{22}	m_{23}	m_{24}	•	m_{2NB}
m_{a3}	m_{31}	m_{32}	m_{33}	h m_{34}	•	m_{3NB}
m_{a4}	m_{41}	m_{42}	f m_{43}	m_{44}	•	m_{4NB}
m_{a5}	m_{51}	c m_{52}	m_{53}	m_{54}	•	m_{5NB}
m_{a6}	m_{61}	m_{62}	m_{63}	m_{64}	•	m_{6NB}
•	•	•	•	•	•	•
•	•	•	•	•	•	•
m_{aNA}	m_{NA1}	m_{NA2}	m_{NA3}	m_{NA4}	•	m_{NANB}

(boundary labels: a, b, i, g, d, j, e)

Interconnection down

M_a \ M_b	m_{b1}	m_{b2}	m_{b3}	m_{b4}	•	m_{bNB}
m_{a1}	m'_{11}	m'_{12}	m'_{13}	m'_{14}	•	m'_{1NB}
m_{a2}	m'_{21}	m'_{22}	m'_{23}	m'_{24}	•	m'_{2NB}
m_{a3}	m'_{31}	m'_{32}	m'_{33}	m'_{34}	•	m'_{3NB}
m_{a4}	m'_{41}	m'_{42}	m'_{43}	m'_{44}	•	m'_{4NB}
m_{a5}	m'_{51}	m'_{52}	m'_{53}	m'_{54}	•	m'_{5NB}
m_{a6}	m'_{61}	m'_{62}	m'_{63}	m'_{64}	•	m'_{6NB}
•	•	•	•	•	•	•
•	•	•	•	•	•	•
m_{aNA}	m'_{NA1}	m'_{NA2}	m'_{NA3}	m'_{NA4}	•	m'_{NANB}

(boundary labels: j2, j1, j3, k2, k1, k3)

Fig. 4.6 Effective margin state matrices for System A connected to System B

load models in each system are stochastically independent. This situation is shown in Fig. 4.6, where M_a and M_b represent the margin states in Systems A and B respectively. The load model in each system is the two-state representation shown in Fig. 3.4. The margin states are therefore discrete and mutually exclusive.

Figure 4.6 contains two mutually exclusive margin arrays as the interconnection is either available or unavailable. The transitions between the two arrays will depend upon the interconnection failure and repair rates. The margin vector M_a contains all the required information for System A when the interconnection is unavailable. The solid line obtained by joining points i, h, f, c, b denotes the boundary wall dividing positive or zero and negative margins when the interconnection is available. This approach is discussed in detail in References [7–9].

The equivalent assistance unit approach [10,11] can be used to obtain identical results to those determined by the margin array method. In this case, the equivalent assistance unit is created from the assisting system margin vector. As in the array method this assumes stochastic independence between the generation and load models in the two systems. This is a very flexible approach which can be extended to multi-interconnected systems with tie constraints and capacity purchase agreements. This technique is discussed in detail in Reference [11]. The equivalent assistance from the interconnected facility appears as a multi-level derated unit [10] which is then added into the margin model using the algorithms given in Chapter 2.

The assistance from the interconnected facility can also be considered on a one-day basis rather than on a period basis in a similar manner to that shown in Section 4.3. Correlation between the loads in the two systems can be recognized using this approach. This method can be extended to using a fixed assistance model for a period, which is the same as the maximum peak load reserve approach described in Section 4.5.

4.8.2 Applications

The basic technique is illustrated using the systems given in Table 4.1. The units in Systems A and B have forced outage rates of 0.02. Assume that each unit has a failure rate of 0.01 f/day and a repair rate of 0.49 r/day. The complete capacity models for the two systems are shown in Table 4.35.

The equivalent assistance unit from System B is shown in Table 4.36.

The 10 MW tie line between A and B constrains the capacity assistance from System B and therefore the equivalent assisting unit is constrained as shown in Table 4.37.

This equivalent multi-state unit is now added to the existing capacity model of System A giving a new capacity of 85 MW. The new model for System A is shown in Table 4.38.

The probability and frequency of a load loss situation in System A without interconnection are 0.00199767 and 0.00195542 occurrences/day respectively from Table 4.35. With the interconnection, these values change to 0.00012042 and

Table 4.35 System capacity models

State i	Cap. out $C_i(MW)$	Individual prob. $p(C_i)$	Departure rate/day $\lambda_+(C_i)$	$\lambda_-(C_i)$	Cum. prob. $P(C_i)$	Cum. freq./day $F(C_i)$
			SYSTEM A			
1	0	0.88384238	0.00000000	0.06000000	1.00000000	0.00000000
2	10	0.09039207	0.49000000	0.05000000	0.11415762	0.05415054
3	20	0.00368947	0.98000000	0.04000000	0.02376555	0.01337803
4	25	0.01807841	0.49000000	0.05000000	0.02007608	0.00990992
5	30	0.00007530	1.47000000	0.03000000	0.00199767	0.00195542
6	35	0.00184474	0.98000000	0.04000000	0.00192237	0.00184700
7	40	0.00000077	1.96000000	0.02000000	0.00007763	0.00011294
8	45	0.00007530	1.47000000	0.03000000	0.00007686	0.00011145
9	50	0.00000000	2.45000000	0.01000000	0.00000156	0.00000303
10	55	0.00000154	1.96000000	0.02000000	0.00000156	0.00000303
11	65	0.00000002	2.45000000	0.01000000	0.00000002	0.00000005
12	75	0.00000000	2.94000000	0.00000000	0.00000000	0.00000000
			SYSTEM B			
1	0	0.90392080	0.00000000	0.05000000	1.00000000	0.00000000
2	10	0.07378945	0.48000000	0.04000000	0.09607920	0.04519604
3	20	0.02070622	0.54345454	0.03890909	0.02228975	0.01199079
4	30	0.00153664	0.98980001	0.02980000	0.00158353	0.00154355
5	40	0.00004626	1.47166100	0.01996610	0.00004689	0.00006838
6	50	0.00000063	1.96000000	0.01000000	0.00000063	0.00000123
7	60	0.00000000	2.45000000	0.00000000	0.00000000	0.00000000

Table 4.36 Equivalent assisting unit model of System B

Cap. out (MW)	Individual prob.	$\lambda_+(occ/day)$	$\lambda_-(occ/day)$	Cum. freq./day
0	0.90392080	0.00000000	0.05000000	0.00000000
10	0.07378945	0.49000000	0.04000000	0.04519604
20	0.02228975	0.53795067	0.00000000	0.01199079

Table 4.37 Tie-line constrained equivalent unit model of System B

Cap. out (MW)	Individual prob.	Departure rate/day λ_+	λ_-	Cum. freq./day
0	0.97771025	0.00000000	0.01226415	0.00000000
10	0.02228975	0.53795067	0.00000000	0.01199079

Table 4.38 Modified capacity outage table of System A

State i	Cap. out C_i (MW)	Individual prob. $p(C_i)$	Departure rate/day $\lambda_+(C_i)$	$\lambda_-(C_i)$	Cum. prob. $P(C_i)$	Cum. freq./day $F(C_i)$
1	0	0.86609717	0.00000000	0.07226415	1.00000000	0.00000000
2	10	0.10812248	0.49875670	0.06185068	0.13390283	0.06258778
3	20	0.00562205	0.99718445	0.05145273	0.02578035	0.01534842
4	25	0.01767545	0.49000000	0.06226415	0.02015830	0.01003147
5	30	0.00015585	1.49530140	0.04106946	0.00248285	0.00247104
6	35	0.00220658	0.98875671	0.05185068	0.00232700	0.00224439
7	40	0.00000243	1.99312450	0.03070007	0.00012042	0.00017703
8	45	0.00011474	1.48718450	0.04145273	0.00011799	0.00017227
9	50	0.00000002	2.49066940	0.02034381	0.00000325	0.00000639
10	55	0.00000318	1.98530140	0.03106946	0.00000323	0.00000634
11	60	0.00000000	2.98795070	0.01000000	0.00000005	0.00000012
12	65	0.00000005	2.48312450	0.02070007	0.00000005	0.00000012
13	75	0.00000000	2.98066940	0.01034381	0.00000000	0.00000000

0.00017703 occurrences/day from Table 4.38. The load model in System A has not been included in these calculations and a constant daily peak load has been assumed. The effect on these indices of varying the tie capacity is shown in Table 4.39.

Section 4.4 illustrated a series of factors which affect the emergency assistance through an interconnection. These factors included the effect of tie capacity and reliability, tie capacity uncertainty and interconnection agreements. All these aspects can be easily incorporated into a frequency and duration appraisal of the capacity adequacy using the equivalent assistance approach. The studies in connection with Systems A and B of Table 4.1 are given as problems in Section 4.10.

Table 4.39 Effect of tie capacity

Tie cap. (MW)	$LOLE_{AB}$ (days/day)	F_{AB} (occ/day)
0	0.00199767	0.00195542
5	0.00192403	0.00185031
10	0.00012042	0.00017703
15	0.00011972	0.00017575
20	0.00005166	0.00008112
25	0.00005166	0.00008112
30	0.00005166	0.00008112

4.8.3 Period analysis

The analysis for a period can be accomplished as shown in Section 4.5. The assistance can be considered to be a variable or held constant for the period. Table 4.40 gives the frequency indices for System A using the load data given in Table 4.22. The results from Table 4.40 can be added to those shown in Table 4.23 to give a complete picture of capacity adequacy in System A.

The results shown in Table 4.40 do not include any load model considerations as the load is a constant daily peak value. The calculated frequency indices do not therefore include any load model transition values. If the assistance from System B is held constant at the maximum peak reserve value then the modified generation model for System A can be convolved with the load model for the period using either the discrete or the continuous load models described in Chapter 3.

The assistance available from an interconnected system can be obtained from the margin vector of that system if the assumption is made that the generation and load models in each system are stochastically independent. This assumption is implicit in the array approach shown by Fig. 4.6. A set of load data for a 5-day period in small Systems A and B is shown in Table 4.41.

The results using the variable reserve, maximum peak load reserve and margin array reserve methods are shown in Table 4.42 for variable tie capacities.

The frequency values for the margin array reserve approach include the load model transitions in both systems.

The equivalent assistance unit approach can be applied to multi-interconnected systems using the concepts outlined in Sections 4.6 and 4.7. The frequency component can be included by using one of the methods proposed in the analysis shown in Table 4.42. The operating agreement in regard to emergency assistance must be clearly understood prior to commencing the analysis as noted in Section 4.7.

Table 4.40 Comparison of variable and maximum peak load reserves

	Variable reserve		*Maximum peak load reserve*	
System A peak load (MW)	*System B peak load (MW)*	*Daily F_{AB}*	*System B peak load (MW)*	*Daily F_{AB}*
46.5	37.2	0.00013678	40.0	0.00017703
50.0	40.0	0.00017703	40.0	0.00017703
49.0	39.2	0.00017398	40.0	0.00017703
48.0	38.4	0.00013827	40.0	0.00017703
47.0	37.6	0.00013678	40.0	0.00017703
		5-day F_{AB} = 0.00076284		5-day F_{AB} = 0.00088515

Table 4.41 Load occurrence tables for the margin array reserve approach

System A

No. of Load States = 3
Exposure Factor = 0.5
Period of Study = 5 days

	Load level (MW)	*No. of occurrences*	*Individual prob.*	*Departure λ_+*	*Rate/day λ_-*
	50	2	0.2	0	2
	47	3	0.3	0	2
low load	46	5	0.5	2	0

System B

No. of Load States = 3
Exposure Factor = 0.5
Study Period = 5 days

	Load level (MW)	*No. of occurrences*	*Individual prob.*	*Departure λ_+*	*Rate/day λ_-*
	40	2	0.2	0	2
	38	3	0.3	0	2
low load	37	5	0.5	2	0

Table 4.42 Variable reserve, maximum peak load reserve, margin array reserve

	5-day $LOLE_{AB}$		
Tie cap. (MW)	*Variable reserve*	*Maximum peak load reserve*	*Margin array reserve*
0	0.00998825	0.00998825	0.00998825
5	0.00961705	0.00962017	0.00961518
10	0.00059898	0.00060210	0.00059709
15	0.00059538	0.00059861	0.00059344
20	0.00025507	0.00025830	0.00025313
25	0.00025507	0.00025830	0.00025312

	5-day F_{AB}		
Tie cap. (MW)	*Variable reserve*	*Maximum peak load reserve*	*Margin array reserve*
0	0.00977710	0.00977710	0.01870319
5	0.00924016	0.00925157	0.01777780
10	0.00076283	0.00088517	0.00169412
15	0.00085305	0.00087861	0.00168270
20	0.00039162	0.00040558	0.00089843
25	0.00039384	0.00040558	0.00089838

4.9 Conclusions

The determination of the benefits associated with interconnection is an important aspect of generating system planning and operating.

Chapters 2 and 3 illustrated basic techniques for evaluating the adequacy of the planned and installed generating capacity in single systems. This chapter has presented extensions of these techniques for the evaluation of two or more interconnected systems. Two basic techniques leading to LOLE and F&D indices have been examined in detail. Other indices such as expected loss of load in MW, or expected energy not supplied in MWh, etc., can be evaluated for the assisted system using the modified capacity model. The concept of using an equivalent assistance unit which can be added to the single system model as a multi-step derated unit addition is a powerful tool for interconnected studies. The equivalent can easily include the interconnecting transmission between the two systems and any operating constraints or agreements. This approach can also be used for operating reserve studies in interconnected systems using the unit representations shown in Chapter 5.

The numerical value of the reliability associated with a particular interconnected system study will depend on the assumptions used in the analysis in addition to the actual factors which influence the reliability of the system. It is important to clearly understand these assumptions before arriving at specific conclusions regarding the actual benefits associated with a given interconnected configuration.

4.10 Problems

1. Two power systems are interconnected by a 20 MW tie line. System A has three 20 MW generating units with forced outage rates of 10%. System B has two 30 MW units with forced outage rates of 20%. Calculate the LOLE in System A for a one-day period, given that the peak load in both System A and System B is 30 MW.
2. Consider the following two systems:
   System A

   6×50 MW units—FOR = 4%

   Peak load 240 MW

   System B

   6×100 MW units—FOR = 6%

   Peak load 480 MW

   The two systems are interconnected by a 50 MW tie line. Calculate the loss of load expectation in each system on a one-day basis for the above data.
3. Two systems are interconnected by two 16 MW tie lines. System A has four 30 MW generating units with forced outage rates of 10%. System B has eight 15 MW generating units with forced outage rates of 8%. Calculate the expected loss of load in each system in days and in MW for a one-day period, given that the peak load in both System A and System B is 100 MW, (a) if the tie lines are 100% reliable, (b) if the tie lines have failure rates of 5 failures/year and average repair times of 24 hours.

4 A generating system designated as System A contains three 25 MW generating units each with a forced outage rate of 4% and one 30 MW unit with a 5% forced outage rate. If the peak load for a 100-day period is 75 MW, what is the LOLE for this period? Assume a straight-line load characteristic from the 100% to the 60% points.

This system is connected to a system containing 10–20 MW hydraulic generating units each with a forced outage rate of 1%. The tie line is rated at 15 MW capacity. If the peak load in the hydraulic system is 175 MW, what is the LOLE for System A assuming that the maximum assistance from the hydraulic system is fixed at 25 MW, i.e. the peak load reserve margin?

5 The system given in Problem 4 of Chapter 2 (System A) is interconnected at the load bus by a 40 MW transmission line to System X which has 4 × 30 MW units, each having a failure rate and repair rate of 5 f/yr and 95 repairs/yr respectively. System X has a peak load of 85 MW and the same load characteristics as System A. (a) Given that the assistance is limited to the peak load reserve margin, calculate the LOLE in each system. Assume that the interconnection terminates at the load bus in System A and that the interconnection is 100% reliable. (b) What are the LOLE indices if the interconnection has an availability of 95%?

4.11 References

1. Watchorn, C. W., 'The determination and allocation of the capacity benefits resulting from interconnecting two or more generating systems', *AIEE Transactions*, **69**, Part II (1950), pp. 1180–6.
2. Cook, V. M., Galloway, C. D., Steinberg, M. J., Wood, A. J., 'Determination of reserve requirements of two interconnected systems, *AIEE Transactions*, **PAS-82**, Part III (1963), pp. 110–16.
3. Billinton, R., Bhavaraju, M. P., 'Loss of load approach to the evaluation of generating capacity reliability of two interconnected systems', *CEA Transactions*, Spring Meeting, March (1967).
4. Billinton, R., Bhavaraju, M. P., Thompson, M. P., 'Power system interconnection benefits', *CEA Transactions*, Spring Meeting, March (1969).
5. Billinton, R., Wee, C. L., Kuruganty, P. R. S., Thompson, P. R., 'Interconnected system reliability evaluation concepts and philosophy', *CEA Transactions*, **20**, Part 3 (1981), Paper 81-SP-144.
6. Kuruganty, P. R. S., Thompson, P. R., Billinton, R., Wee, C. L., 'Interconnected system reliability evaluation—applications', *CEA Transactions*, **20**, Part 3 (1981), Paper 81-SP-145.
7. Billinton, R., Singh, C., 'Generating capacity reliability evaluation in interconnected systems using a frequency and duration approach: Part I—Mathematical analysis', *IEEE Transactions*, **PAS-90** (1971), pp. 1646–54.
8. Billinton, R., Singh, C., 'Generating capacity reliability evaluation in interconnected systems using a frequency and duration approach: Part II—System applications', *IEEE Transactions*, **PAS-90** (1971), pp. 1654–64.

9. Billinton, R., Singh, C., 'Generating capacity reliability evaluation in interconnected systems using a frequency and duration approach: Part III—Correlated load models', *IEEE Transactions*, **PAS-91** (1972), pp. 2143–53.
10. Billinton, R., Wee, C. L., Hamoud, G., 'Digital computer algorithms for the calculation of generating capacity reliability indices', *Transactions PICA Conference* (1981), pp. 46–54.
11. Billinton, R., Wee, C. L., *A Frequency and Duration Approach for Interconnected System Reliability Evaluation* ', IEEE Summer Power Meeting (1981), Paper No. 81 SM 446–4.

5 Operating reserve

5.1 General concepts

As discussed in Section 2.1, the time span for a power system is divided into two sectors: the planning phase, which was the subject of Chapters 2–4, and the operating phase. In power system operation, the expected load must be predicted (short-term load forecasting) and sufficient generation must be scheduled accordingly. Reserve generation must also be scheduled in order to account for load forecast uncertainties and possible outages of generation plant. Once this capacity is scheduled and spinning, the operator is committed for the period of time it takes to achieve output from other generating plant; this time may be several hours in the case of thermal units but only a few minutes in the case of gas turbines and hydroelectric plant.

The reserve capacity that is spinning, synchronized and ready to take up load is generally known as spinning reserve. Some utilities include only this spinning reserve in their assessment of system adequacy, whereas others also include one or more of the following factors: rapid start units such as gas turbines and hydro-plant, interruptable loads, assistance from interconnected systems, voltage and/or frequency reductions. These additional factors add to the effective spinning reserve and the total entity is known as operating reserve.

Historically, operating reserve requirements have been done by *ad hoc* or rule-of-thumb methods, the most frequently used method being a reserve equal to one or more largest units. This method was discussed in Section 2.2.3 in which it was shown that it could not account for all system parameters. In the operational phase, it could lead to overscheduling which, although more reliable, is uneconomic, or to underscheduling which, although less costly to operate, can be very unreliable.

A more consistent and realistic method would be one based on probabilistic methods. A risk index based on such methods would enable a consistent comparison to be made between various operating strategies and the economics of such strategies. Several methods [1, 2] have been proposed for evaluating a probabilistic risk index and these will be described in the following sections of this chapter. Generally two values of risk can be evaluated: unit commitment risk and response risk. Unit commitment risk is associated with the assessment of which units to commit in any given period of time whilst the response risk is associated with the dispatch decisions of those units that have been committed.

The acceptable risk level is and must remain a management decision based on economic and social requirements. An estimate of a reasonable level can be made by evaluating the probabilistic risk index associated with existing operational reserve assessment methods. Once a risk level has been defined, sufficient generation can be scheduled to satisfy this risk level.

5.2 PJM method

5.2.1 Concepts

The PJM method [3] was proposed in 1963 as a means of evaluating the spinning requirements of the Pennsylvania–New Jersey–Maryland (USA) interconnected system. It has been considerably refined and enhanced since then but still remains a basic method for evaluating unit commitment risk. In its more enhanced form, it is probably the most versatile and readily implementable method for evaluating operational reserve requirements.

The basis of the PJM method is to evaluate the probability of the committed generation just satisfying or failing to satisfy the expected demand during the period of time that generation cannot be replaced. This time period is known as the lead time. The operator must commit himself at the beginning of this lead time ($t = 0$) knowing that he cannot replace any units which fail or start other units, if the load grows unexpectedly, until the lead time has elapsed. The risk index therefore represents the risk of just supplying or not supplying the demand during the lead time and can be re-evaluated continuously through real time as the load and the status of generating units change.

The method in its basic and original form [3] simplifies the system representation. Each unit is represented by a two-state model (operating and failed) and the possibility of repair during the lead time is neglected.

5.2.2 Outage replacement rate (ORR)

It was shown in *Engineering Systems* (Section 9.2.2) that, if failures and repairs are exponentially distributed, the probability of finding a two-state unit on outage at a time T, given that it was operating successfully at $t = 0$, is

$$P(\text{down}) = \frac{\lambda}{\lambda + \mu} - \frac{\lambda}{\lambda + \mu} e^{-(\lambda + \mu)T} \tag{5.1}$$

If the repair process is neglected during time T, i.e., $\mu = 0$, then Equation (5.1) becomes

$$P(\text{down}) = 1 - e^{-\lambda T} \tag{5.2}$$

which, as should be expected, is the exponential equation for the probability of failure of a two-state, non-repairable component. Finally, if $\lambda T \ll 1$, which is generally true for short lead times of up to several hours,

$$P(\text{down}) \simeq \lambda T \tag{5.3}$$

Equation (5.3) is known as the outage replacement rate (ORR) and represents the probability that a unit fails and is not replaced during the lead time T. It should be noted that this value of ORR assumes exponentially distributed times to failure. If this distribution is inappropriate, the ORR can be evaluated using other more relevant distributions and the same concepts.

The ORR is directly analogous to the forced outage rate (FOR) used in planning studies. The only difference is that the ORR is not simply a fixed characteristic of a unit but is a time-dependent quantity affected by the value of lead time being considered.

5.2.3 Generation model

The required generation model for the PJM method is a capacity outage probability table which can be constructed using identical techniques to those described in Chapter 2. The only difference in the evaluation is that the ORR of each unit is used instead of the FOR.

Consider a committed generating system (System A) consisting of 2 × 10 MW units, 3 × 20 MW units and 2 × 60 MW units. Let each be a thermal unit having the failure rates shown in Table 5.1. The ORR of each unit for lead times of 1, 2 and 4 hours are also shown in Table 5.1.

The units of this system can be combined using the techniques described in Chapter 2 and the values of ORR shown in Table 5.1 to give the capacity outage probability tables shown for System A in Table 5.2.

The remaining two columns in Table 5.2 relate to:

System B—basically the same as System A but with one of the 60 MW thermal units replaced by a 60 MW hydro unit having a failure rate of 1 f/yr (equivalent to an ORR of 0.000228 for a lead time of 2 hours). Generally hydro units have much smaller failure rates than thermal units.

Table 5.1 Failure rates and ORR

		ORR for lead times of		
Unit (MW)	*λ (f/yr)*	*1 hour*	*2 hours*	*4 hours*
10	3	0.000342	0.000685	0.001370
20	3	0.000342	0.000685	0.001370
60	4	0.000457	0.000913	0.001826

Table 5.2 Capacity outage probability tables

Capacity		Cumulative probability				
		System A and lead times of			*System B*	*System C*
Out (MW)	*In (MW)*	*1 hour*	*2 hours*	*4 hours*	*2 hours*	*2 hours*
0	200	1.000000	1.000000	1.000000	1.000000	1.000000
10	190	0.002620	0.005238	0.010455	0.004556	0.006829
20	180	0.001938	0.003874	0.007740	0.003192	0.000021
30	170	0.000915	0.001829	0.003665	0.001145	
40	160	0.000914	0.001826	0.003654	0.001142	
50	150	0.000914	0.001825	0.003648	0.001141	
60	140	0.000914	0.001825	0.003648	0.001141	
70	130	0.000002	0.000007	0.000028	0.000004	
80	120	0.000001	0.000005	0.000018	0.000002	
120	80		0.000001	0.000003		

System C—a scheduled system of 20 × 10 MW units each having an ORR equal to that of the 10 MW units of System A.

5.2.4 Unit commitment risk

The PJM method assumes that the load will remain constant for the period being considered. The value of unit commitment risk can therefore be deduced directly from the generation model since this model does not need to be convolved with a load model.

In order to illustrate the deduction of unit commitment risk, consider System A of Section 5.2.3 and an expected demand of 180 MW. From Table 5.2, the risk is 0.001938, 0.003874 and 0.007740 for lead times of 1, 2 and 4 hours respectively. The risk in Systems B and C for a lead time of 2 hours and the same load level are 0.003192 and 0.000021 respectively.

It is necessary in a practical system to first define an acceptable risk level in order to determine the maximum demand that a particular committed system can meet. Consider, for example, that a risk of 0.001 is acceptable. If additional generation can be made available in System A within 1 hour, the required spinning reserve is only 30 MW and a demand of 170 MW can be supplied. If the lead time is 4 hours, however, the required spinning reserve increases to 70 MW and a demand of only 130 MW can be supplied. It is therefore necessary to make an economic comparison between spinning a large reserve and reducing the lead time by maintaining thermal units on hot reserve or investing in rapid start units such as hydro plant and gas turbines.

The results shown in Table 5.2 indicate that the risk, for a given level of spinning reserve and lead time, is less for System B than for System A, although the systems are identical in size and capacity. This is due solely to the smaller failure rate of the hydro plant. It follows therefore that it is not only beneficial to use hydro plants because of their reduced operational costs but also because of their better reliability. It will be seen in Section 5.7.4 that it may be preferable, however, not to fully dispatch these hydro units because of their beneficial effect on the response risk.

In practice an operator would use the PJM risk assessment method by adding, and therefore committing, one unit at a time from a merit order table until the unit commitment risk given by the generation model became equal to or less than the acceptable value for the demand level expected.

5.3 Extensions to PJM method

5.3.1 Load forecast uncertainty

In Section 5.2.4, it was assumed that the load demand was known exactly. In practice, however, this demand must be predicted using short-term load forecasting methods. This prediction will exhibit uncertainties which can be taken into account in operational risk evaluation. These load forecast uncertainties can be included in the same way that was described for planning studies in Section 2.7. The uncertainty distribution, generally assumed to be normal, is divided into discrete intervals (see Fig. 2.15). The operational risk associated with the load level of each interval is weighted by the probability of that interval. The total operational risk is the sum of the interval risks.

In order to illustrate this assessment, reconsider System A in Section 5.2.3 and a lead time of 2 hours. Assume the load forecast uncertainties are normally distributed, the expected load is 160 MW and the forecast uncertainty has a standard deviation of 5% (≡8 MW). Using the information shown in Table 5.2, the risk assessment shown in Table 5.3 can be evaluated.

Table 5.3 Unit commitment risk including load forecast uncertainty

No. of standard deviations	*Load level (MW)*	*Prob. of load level*	*Risk at load level (from Table 5.2)*	*Expected unit commitment risk (col. 3 × col. 4)*
−3	136	0.006	0.001825	0.000011
−2	144	0.061	0.001825	0.000111
−1	152	0.242	0.001826	0.000442
0	160	0.382	0.001826	0.000698
+1	168	0.242	0.001829	0.000443
+2	176	0.061	0.003874	0.000236
+3	184	0.006	0.005238	0.000031
				0.001972

The results shown in Table 5.3 indicate that, as generally found, the unit commitment risk increases when load forecast uncertainty is included, the difference in risk increasing as the degree of uncertainty increases. This means that more units must be selected from a merit order table and committed in order to meet the acceptable risk level.

5.3.2 Derated (partial output) states

When large units are being considered, it can become important to model [4] the units by more than two states in order to include the effect of one or more derated or partial output states. This concept was previously discussed in Section 2.4. For planning studies and the reason for derated states is described in Section 11.2.

Consider a unit having three states as shown in Fig. 5.1(a): operating (O), failed (F) and derated (D).

If repair during the lead time can be neglected, the complete set of transitions shown in Fig. 5.1(a) can be reduced to those shown in Fig. 5.1(b). Furthermore, if the probability of more than one failure of each unit is negligible during the lead time, this state space diagram reduces to that of Fig. 5.1(c). These simplifications are not inherently essential, however, and the state probabilities of the most appropriate model can be evaluated and used. Considering the model of Fig. 5.1(c) and assuming $\lambda T << 1$, it follows from Equation (5.3) that

$$P(\text{down}) \simeq \lambda_2 T \tag{5.4a}$$

$$P(\text{derated}) \simeq \lambda_1 T \tag{5.4b}$$

$$P(\text{operating}) \simeq 1 - (\lambda_1 + \lambda_2)T \tag{5.4c}$$

Reconsider System A of Section 5.2.3 and let each 60 MW unit have a derated output capacity of 40 MW with $\lambda_1 = \lambda_2 = 2$ f/yr. Then for a lead time of 2 hours, $P(\text{down}) = P(\text{derated}) = 0.000457$ and these units can be combined with the remaining units to give the generation model shown in Table 5.4.

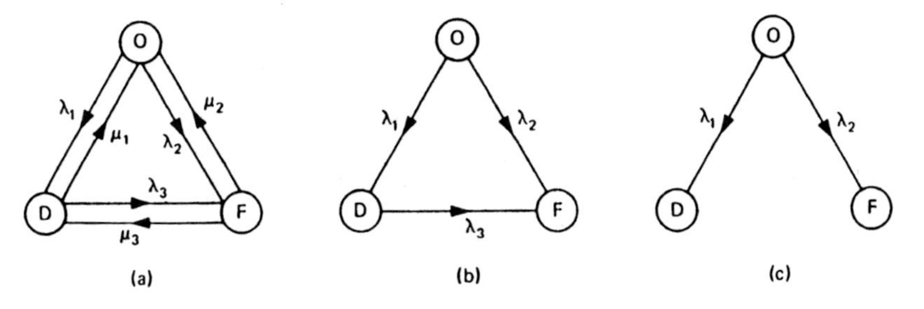

Fig. 5.1 Three-state model of a generating unit

Table 5.4 Generation model including derated states

Capacity out (MW)	*Cumulative probability for a lead time of 2 hours*
0	1.000000
10	0.005239
20	0.003875
30	0.000920
40	0.000916
50	0.000913
60	0 000913
70	0.000003
80	0.000002

5.4 Modified PJM method

5.4.1 Concepts

The modified PJM method [5] is essentially the same in concept to the original PJM method. Its advantage is that it extends the basic concepts and allows the inclusion of rapid start units and other additional generating plant having different individual lead times. These units are in a standby mode at the decision time of $t = 0$ and must be treated differently from those that are presently spinning and synchronized because, not only must the effect of running failures be considered, but also the effect of start-up failures must be included. These effects can be assessed by creating and analyzing a model of the standby generating units that recognizes the standby or ready-for-service state as well as the failure and in-service states.

5.4.2 Area risk curves

The unit commitment risk is defined as the probability of the generation just satisfying or failing to satisfy the system load. At the decision time of $t = 0$, the condition of the system is deterministically known; the risk is either unity or zero, depending on whether the load is greater or less than the available generation at that time. The problem facing the operator is therefore to evaluate the risk and change in risk for a certain time into the future. One very convenient way of representing this risk pictorially is the risk function or area risk concept [5].

Consider first a single unit represented by a two-state model as defined by Equation (5.2). The risk (or density) function $f(R)$ for this model is

$$f(R) = \frac{\mathrm{d}p}{\mathrm{d}t} = \lambda \mathrm{e}^{-\lambda t} \tag{5.5}$$

and probability of the unit failing in the time period (0 to T) is given by

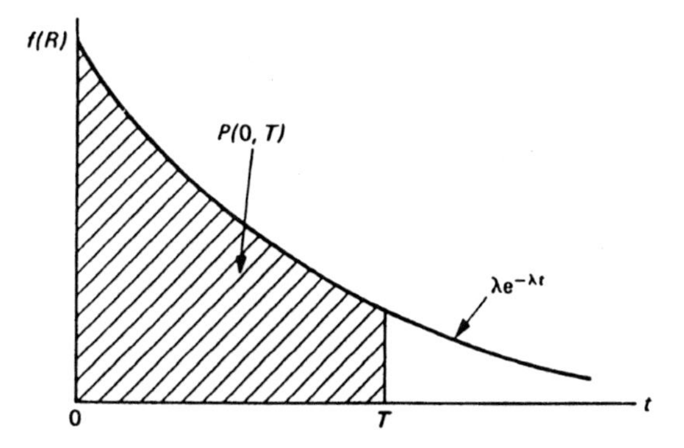

Fig. 5.2 Concept of area risk curves

$$P(0, T) = \int_0^T \lambda e^{-\lambda t}\, dt \tag{5.6}$$

Equation (5.5) is shown pictorially in Fig. 5.2. The probability of the unit failing in time $(0, T)$ is the area under the curve between 0 and T. This representation is known as an area risk curve. It should be noted, however, that these curves are only pictorial representations used to illustrate system behaviour, and it is not normally necessary to explicitly evaluate $f(R)$.

Evaluation of the risk of a complete system for a certain time into the future can be depicted using area risk curves, two examples of which are shown in Fig. 5.3. The area risk curve in Fig. 5.3(a) represents the behavior of the system when the only reserve units are those that are actually spinning and synchronized. This is therefore equivalent to the basic PJM method. The curve in Fig. 5.3(b), however, represents the modified PJM method. After lead times of T_1 and T_2 respectively, rapid start units and hot reserve units become available. The total risk in the period $(0, T_3)$ is therefore less when these standby units are taken into account, the reduction in risk being indicated by the shaded area of Fig. 5.3(b); this increases as more standby units are considered. It should be noted that this reduction in risk is achieved, not simply by the presence of standby units in the system, but by an operational decision to start them at the decision point of $t = 0$.

The assessment of the risk using the modified PJM method therefore requires the evaluation of the risks in the individual intervals $(0, T_1)$, (T_1, T_2), (T_2, T_3), etc, the total risk being the summation of the interval risks. This evaluation process requires suitable models for the standby units that realistically account for the fact that a decision to start them is made at $t = 0$, that they may or may not come into service successfully after their respective lead times, and that they may suffer running failures after their lead times.

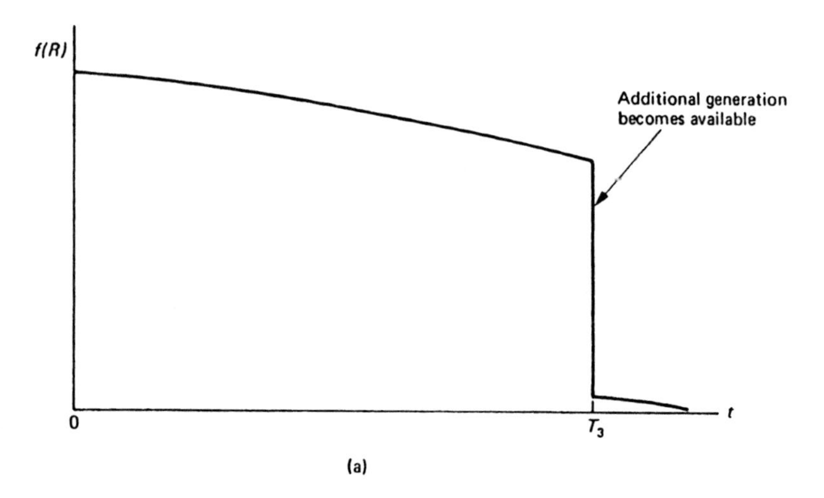

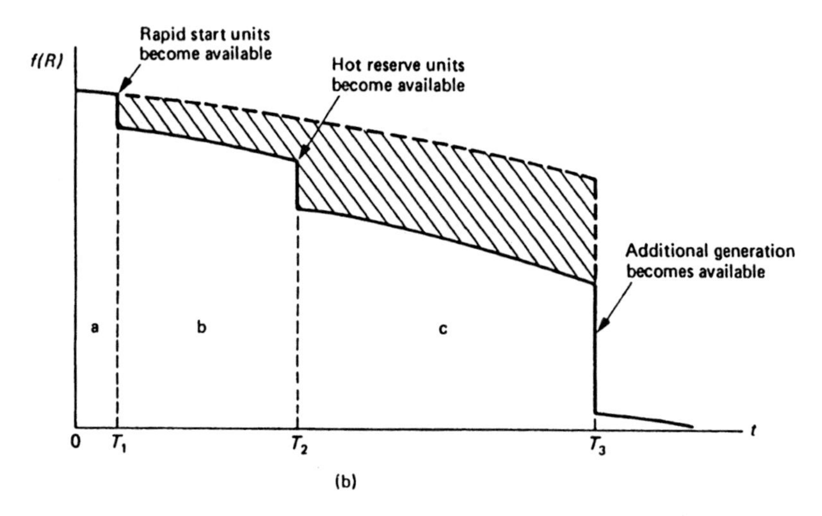

Fig. 5.3 Area risk curves for complete systems

5.4.3 Modelling rapid start units

(a) *Unit model*

Rapid start units such as gas turbines and hydro plant can be represented by the four-state model [5] shown in Fig. 5.4 in which

$$\lambda_{ij} = N_{ij} / T_i, \tag{5.7}$$

where

λ_{ij} = transition rate from state i to state j
N_{ij} = number of transitions from state i to state j during the period of observation
T_i = total time spent in state i during the same period of observation.

(b) *Evaluating state probabilities*

The model shown in Fig. 5.4 cannot be simplified as readily as that shown in Fig. 5.1(a). The probability of residing in any of the states however can be evaluated using Markov techniques for any time into the future. As described in Section 9.6.2 of *Engineering Systems*, these time-dependent probabilities are most easily evaluated using matrix multiplication techniques

$$[P(t)] = [P(0)][P]^n \tag{5.8}$$

where

$[P(t)]$ = vector of state probabilities at time t
$[P(0)]$ = vector of initial probabilities
$[P]$ = stochastic transitional probability matrix
n = number of time steps used in the discretization process.

The stochastic transitional probability matrix for the model of Fig. 5.4 is

$$[P] = \begin{array}{c} \\ 1 \\ 2 \\ 3 \\ 4 \end{array} \begin{array}{c} \begin{array}{cccc} 1 & 2 & 3 & 4 \end{array} \\ \begin{bmatrix} 1-(\lambda_{12}+\lambda_{14})dt & \lambda_{12}dt & \text{—} & \lambda_{14}dt \\ \lambda_{21}dt & 1-(\lambda_{21}+\lambda_{23})dt & \lambda_{23}dt & \text{—} \\ \text{—} & \lambda_{32}dt & 1-(\lambda_{32}+\lambda_{34})dt & \lambda_{34}dt \\ \lambda_{41}dt & \lambda_{42}dt & \text{—} & 1-(\lambda_{41}+\lambda_{42})dt \end{bmatrix} \end{array} \tag{5.9}$$

The value of dt in $[P]$ must be chosen judiciously; it must not be so small that the number of matrix multiplications, i.e. n, becomes too large, but it must not be so large that the error introduced in the values of probabilities becomes too large. A value of 10 minutes is usually satisfactory for most systems.

During the start-up time (lead time), a rapid start unit does not contribute to system generation and resides in the ready-for-service state (2 in Fig. 5.4) with a probability of unity. If a positive decision was made to enter the unit into service at the decision time of $t = 0$, the unit either starts successfully after the lead time and resides in the in-service state (1) or fails to start and enters the failure state (4). If such a decision was not made at $t = 0$, the unit is not considered in the risk evaluation and is totally ignored. Therefore the vector of initial probabilities at the time when the unit may contribute to system generation is [6]

$$[P(0)] = \begin{array}{c} \begin{array}{cccc} 1 & 2 & 3 & 4 \end{array} \\ [P_{10} \quad 0 \quad 0 \quad P_{40}] \end{array} \tag{5.10}$$

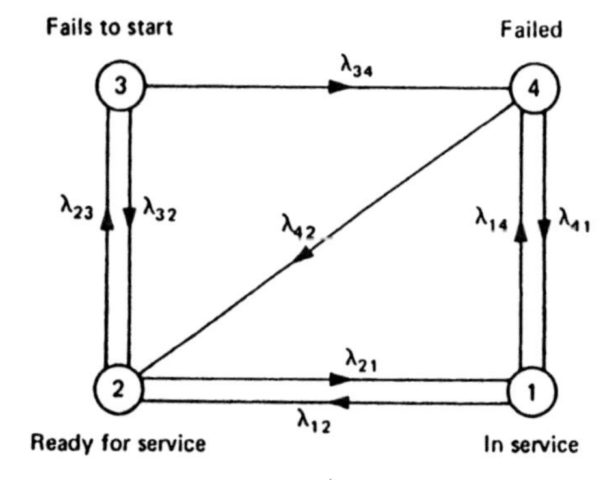

Fig. 5.4 Four-state model for rapid start units

where

P_{40} = probability of failing to start (P_{fs}), i.e. probability of being in state 4 given that it was instructed to start at $t = 0$.

$$P_{fs} = \frac{\text{total number of times units failed to take up load}}{\text{total number of starts}}$$

$$= N_{23}/(N_{21} + N_{23})$$

$$= \lambda_{23}/(\lambda_{21} + \lambda_{23}) \tag{5.11}$$

$$P_{10} = 1 - P_{fs}$$

(c) *Evaluating unavailability statistics*

After evaluating the individual state probabilities of a rapid start unit at times greater than the lead time using Equations (5.8)–(5.11), it is necessary to combine these to give the probability of finding the unit in the failed state. The required index [7] is 'the probability of finding the unit on outage given that a demand has occurred'. Using the concept of conditional probability, this is given [6] by

$$P(\text{down}) = \frac{P_3(t) + P_4(t)}{P_1(t) + P_3(t) + P_4(t)} \tag{5.12}$$

since the numerator of Equation (5.12) represents the probability that the unit is in the failed state and the denominator represents the probability that a demand occurs.

Similarly [6]

$$P(\text{up}) = 1 - P(\text{down})$$

$$= \frac{P_1(t)}{P_1(t) + P_3(t) + P_4(t)} \tag{5.13}$$

5.4.4 Modelling hot reserve units

The daily load cycle necessitates units being brought into service and taken out of service. When taken out of service, the status of the unit can be left in one of two states; hot reserve or cold reserve. Cold reserve means that the unit, including its boiler, is completely shut down. Hot reserve means that the turbo-alternator is shut down but the boiler is left in a hot state. Consequently the time taken for hot reserve units to be brought back into service is very much shorter than cold reserve units. There is clearly a cost penalty involved in maintaining units in hot reserve and the necessity to do so should be assessed using a consistent risk evaluation technique such as the modified PJM method.

The concepts associated with hot reserve units are the same as for rapid start units. The only basic difference between the two modelling processes is that the hot reserve units require a five-state model [5] as shown in Fig. 5.5.

The state probabilities at any future time greater than the lead time are evaluated using Equation (5.8) with the stochastic transitional probability matrix replaced by that derived from the state space diagram of Fig. 5.5 and the vector of initial probabilities replaced [6] by

$$\begin{array}{ccccc} 1 & 2 & 3 & 4 & 5 \\ \end{array}$$
$$[P(0)] = [P_{10} \quad 0 \quad 0 \quad P_{40} \quad 0] \tag{5.14}$$

in which P_{40} and P_{10} are given by Equation (5.11) as before.

The unavailability statistics are evaluated using similar concepts [6] to those for rapid start units with the modification that state 5 should also be included. This gives

$$P(\text{down}) = \frac{P_3(t) + P_4(t) + P_5(t)}{P_1(t) + P_3(t) + P_4(t) + P_5(t)} \tag{5.15}$$

$$P(\text{up}) = \frac{P_1(t)}{P_1(t) + P_3(t) + P_4(t) + P_5(t)} \tag{5.16}$$

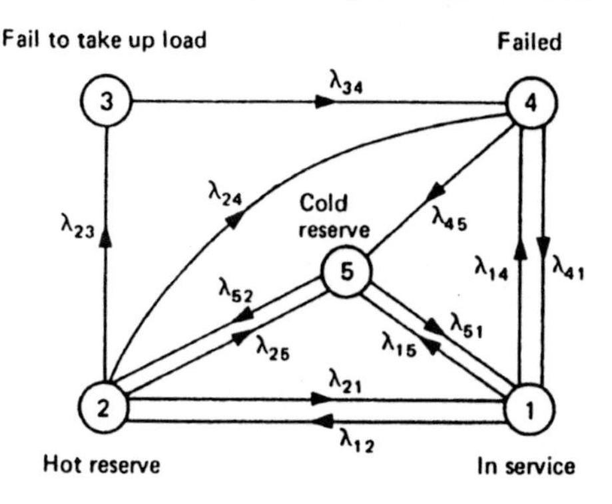

Fig. 5.5 Five-state model for hot reserve units

5.4.5 Unit commitment risk

The unit commitment risk is evaluated in a similar manner to that used in the basic PJM method. In the modified version, however, a set of partial risks must be evaluated for each of the time intervals of the area risk curves typically represented by Fig. 5.3(b). There is no conceptual limit to the number of intervals which can be used in practice and each unit on standby mode can be associated with its own lead time. Too many intervals, however, leads to excessive computation and it is generally reasonable to group similar units and specify the same lead time for each group, which typically may be 10 minutes and 1 hour for rapid start and hot reserve units respectively. This is the concept used in Fig. 5.3(b) in which one group represents the rapid start units and another group represents the hot reserve units. The following discussion will limit the assessment only to these two groups, i.e. rapid start units become available at T_1 and hot reserve units at T_2.

(a) *Risk in the first period* $(0, T_1)$

A generation model is formed using only the on-line generation at $t = 0$ and the appropriate values of ORR evaluated for a lead time of T_1. This is essentially equivalent to the basic PJM method. Combining [5] this model with the system load gives

$$R_a = R_{T1-} \tag{5.17}$$

(b) *Risk in the second period* (T_1, T_2)

Two generation models are formed and hence two partial risks are evaluated [5, 6] for this period: one for the start of the period (T_1) and one for the end of the period (T_2). At T_1 the generation model formed in step (a) at T_1 is combined with the rapid start units for which the state probabilities are P_{fs} and $(1 - P_{fs})$ as defined by Equation (5.11). This gives a partial risk of R_{T1+}. At T_2, a generation model is formed using the initial generation with values of ORR evaluated for a time T_2 and the rapid start units with state probabilities evaluated for a time $(T_2 - T_1)$ as defined by Equations (5.8), (5.12) and (5.13). This gives a partial risk of R_{T2-}. The risk for the second period is then

$$R_b = R_{T2-} - R_{T1+} \tag{5.18}$$

(c) *Risk in the third period* (T_2, T_3)

The risk in the third period is evaluated similarly to that in the second period, i.e. two partial risks [5, 6] are evaluated: one at T_2 and the other at T_3. At T_2, the generation model formed in step (b) at T_2 is combined with the hot reserve units for which the state probabilities are again P_{fs} and $(1 - P_{fs})$. This gives a partial risk of R_{T2+}. At T_3, the generation model includes the initial generation with a lead time of T_3, the rapid start units with a lead time of $(T_3 - T_1)$ and the hot reserve units with state probabilities evaluated for a time $(T_3 - T_2)$ as defined by Equations (5.8), (5.15) and (5.16). This gives a partial risk of R_{T3-} and the risk in the third period becomes

$$R_c = R_{T3-} - R_{T2+} \tag{5.19}$$

This process can be continued for any number of intervals and groups of units. In the present case of three intervals, the total risk for the period of interest $(0, T_3)$ is

$$R = R_a + R_b + R_c \tag{5.20}$$

5.4.6 Numerical examples

(a) *Modified PJM method*

In order to illustrate the application of the modified PJM method, reconsider the example of System A in Section 5.2.3. In addition to the previously scheduled on-line generation, let a 20 MW gas turbine be available at $t = 0$, having a start-up time of 10 minutes and state transitions per hour (Fig. 5.4) of $\lambda_{12} = 0.0050$, $\lambda_{21} = 0.0033$, $\lambda_{14} = 0.0300$, $\lambda_{41} = 0.0150$, $\lambda_{23} = 0.0008$, $\lambda_{32} = 0.0000$, $\lambda_{34} = 0.0250$, $\lambda_{42} = 0.0250$. Consider a total lead time of 1 hour and an expected demand of 180 MW.

There are two periods to consider: before the gas turbine becomes available (0, 10 minutes) and after it becomes available (10 minutes, 1 hour).

(i) *Period (0, 10 minutes)*

The ORR at a lead time of 10 minutes for the 10 and 20 MW units is 0.000057 and for the 60 MW units is 0.000076. Combining these units as in Section 5.2.3 gives the generation model shown in Table 5.5. The risk in the first period is

$$R_a = 0.000323$$

(ii) *Period (10 minutes, 1 hour)*

From Equation (5.11), the gas turbine has values of

$$P_{fs} = 0.0008/(0.0033 + 0.0008) = 0.195122 \quad \text{and} \quad 1 - P_{fs} = 0.804878$$

Combining this unit with the generation model shown in Table 5.5 gives the generation model for all units at 10 minutes shown in Table 5.6.

From the above values of P_{fs}, the vector of initial probabilities (Equation (5.10)) of the gas turbine is

$$[P(0)] = [0.804878 \quad 0 \quad 0 \quad 0.195122]$$

Also the stochastic transitional probability matrix (Equation (5.9)) using the specified transition rates and discretizing the period into 10 minute intervals is

Table 5.5 Generation model for on-line units at 10 minutes

Capacity out (MW)	*Capacity in (MW)*	*Cumulative probability*
0	200	1.000000
10	190	0.000437
20	180	0.000323
30	170	0.000152
40	160	0.000152
50	150	0.000152
60	140	0.000152

$$[P] = \begin{array}{c} \\ 1 \\ 2 \\ 3 \\ 4 \end{array} \begin{array}{c} \begin{array}{cccc} \quad 1 \quad & \quad 2 \quad & \quad 3 \quad & \quad 4 \quad \end{array} \\ \begin{bmatrix} 0.994167 & 0.000833 & — & 0.005000 \\ 0.000550 & 0.999317 & 0.000133 & — \\ — & 0.000000 & 0.995833 & 0.004167 \\ 0.002500 & 0.004167 & — & 0.993333 \end{bmatrix} \end{array}$$

Using the vector $[P(0)]$, the matrix $[P]$ and the matrix multiplication concept of Equation (5.8) gives the following sequential state probability vectors for the gas turbine

$$[P(10 \text{ minutes})] = [0.800670 \quad 0.001484 \quad 0.000000 \quad 0.197846]$$

$$[P(20 \text{ minutes})] = [0.796496 \quad 0.002974 \quad 0.000000 \quad 0.200530]$$

$$[P(30 \text{ minutes})] = [0.792353 \quad 0.004471 \quad 0.000000 \quad 0.203176]$$

$$[P(40 \text{ minutes})] = [0.788241 \quad 0.005975 \quad 0.000001 \quad 0.205783]$$

$$[P(50 \text{ minutes})] = [0.784161 \quad 0.007485 \quad 0.000002 \quad 0.208352]$$

From $[P(50 \text{ minutes})]$ and Equations (5.12) and (5.13)

$$P(\text{down}) = 0.209925$$

$$P(\text{up}) = 0.790075$$

Table 5.6 Generation model for all units at 10 minutes

Capacity out (MW)	*Capacity in (MW)*	*Cumulative probability*
0	220	1.000000
10	210	0.195474
20	200	0.195382
30	190	0.000208
40	180	0.000185
50	170	0.000152
60	160	0.000152
70	150	0.000030
80	140	0.000030

Table 5.7 Generation model for all units at 1 hour

Capacity out (MW)	*Capacity in (MW)*	*Cumulative probability*
0	220	1.000000
10	210	0.211995
20	200	0.211456
30	190	0.001273
40	180	0.001129
50	170	0.000914
60	160	0.000914
70	150	0.000193
80	140	0.000192

Combining the gas turbine having these values of state probabilities with the generation model of System A for a lead time of 1 hour as shown in Table 5.2 give the generation model at a time of 1 hour. This is shown in Table 5.7.

From Tables 5.6 and 5.7, the risk in the second period is

$$R_b = 0.001129 - 0.000185 = 0.000944$$

and the total risk for a period of 1 hour is

$$R = 0.000323 + 0.000944 = 0.001267$$

This value compares with a risk of 0.001938 (Table 5.2) if the gas turbine is not brought into service.

The concepts outlined in this example can clearly be extended to cover any number of periods and any required total time of interest.

(b) *Sensitivity study*

Consider a system in which the load is expected to remain constant at 1900 MW for the next 4 hour period. The committed on-line generation is 2100 MW as detailed in Table 5.8. The transition rates (λ_1 and λ_2) for these on-line units relate to the three-state model shown in Fig. 5.1(c).

Table 5.8 Committed on-line generation

Number of committed units	*Derated output capacity (MW)*	*Full output capacity (MW)*	*Transition rates/h*	
			λ_1	λ_2
7	80	100	0.0010	0.0003
8	120	150	0.0050	0.0006
1	160	200	0.0002	0.0005

In addition the operator has a number of 30 MW gas turbines and 100 MW hot reserve units which he can call into operation after lead times of 10 minutes and 1 hour respectively. The transition rates for each gas turbine are the same as those for the previous example (Section 5.4.6(a)). The transition rates per hour (Fig. 5.5) for each hot reserve unit are

$$\lambda_{12} = 0.0240, \quad \lambda_{14} = 0.0080, \ \lambda_{15} = 0.0000, \ \lambda_{21} = 0.0200,$$

$$\lambda_{23} = 0.00002, \ \lambda_{24} = 0.0000, \ \lambda_{25} = 0.0000, \ \lambda_{34} = 0.0300,$$

$$\lambda_{41} = 0.0350, \quad \lambda_{45} = 0.0250, \ \lambda_{51} = 0.0030, \ \lambda_{52} = 0.0025.$$

The effect of increasing the number of rapid start and hot reserve units on the unit commitment risk for a total lead time of 4 hours is shown in Fig. 5.6. Several interesting points can be observed from Fig. 5.6.

(a) The risk, for a given number of gas turbines, decreases significantly when one hot reserve unit is added, decreases further when a second unit is added but changes insignificantly when further units are added. For this system it is evident that no benefit is derived by starting more than two hot reserve units. The reason is that the risk in the period (1 hour to 4 hours) when the hot reserve

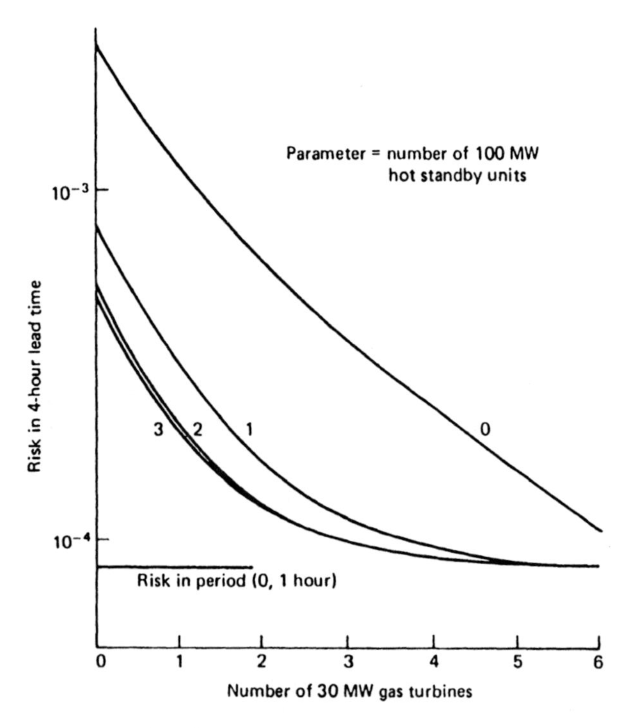

Fig. 5.6 Effect of standby units on risk level

Table 5.9 Results of sensitivity study

Hot reserve units (MW)	*Period 1*	*Period 2*	*Period 3*	*Total*
(a) No rapid start units				
0, 0	0.00008363	0.00044196	0.00203528	0.00256087
1, 100	0.00008363	0.00044196	0.00025880	0.00078439
2, 200	0.00008363	0.00044196	0.00001435	0.00053994
3, 300	0.00008363	0.00044196	0.00000078	0.00052637
(b) One rapid start unit, 30 MW				
0, 0	0.00008363	0.00012713	0.00101340	0.00122416
1, 100	0.00008363	0.00012713	0.00011866	0.00032942
2, 200	0.00008363	0.00012713	0.00000642	0.00021718
3, 300	0.00008363	0.00012713	0.00000018	0.00021094
(c) Two rapid start units, 60 MW				
0, 0	0.00008363	0.00004232	0.00049955	0.00062550
1, 100	0.00008363	0.00004232	0.00004864	0.00017459
2, 200	0.00008363	0.00004232	0.00000150	0.00012745
3, 300	0.00008363	0.00004232	—	0.00012595
(d) Three rapid start units, 90 MW				
0, 0	0.00008363	0.00001991	0.00028851	0.00039205
1, 100	0.00008363	0.00001991	0.00001272	0.00011626
2, 200	0.00008363	0.00001991	0.00000031	0.00010385
3, 300	0.00008363	0.00001991	—	0.00010354
(e) Four rapid start units, 120 MW				
0, 0	0.00008363	0.00000844	0.00014009	0.00023216
1, 100	0.00008363	0.00000844	0.00000373	0.00009580
2, 200	0.00008363	0.00000844	0.00000014	0.00009221
3, 300	0.00008363	0.00000844	—	0.00009207
(f) Five rapid start units, 150 MW				
0, 0	0.00008363	0.00000456	0.00007848	0.00016667
1, 100	0.00008363	0.00000456	0.00000126	0.00008945
2, 200	0.00008363	0.00000456	0.00000002	0.00008821
3, 300	0.00008363	0.00000456	—	0.00008819
(g) Six rapid start units, 180 MW				
0, 0	0.00008363	0.00000148	0.00002319	0.00010830
1, 100	0.00008363	0.00000148	0.00000076	0.00008587
2, 200	0.00008363	0.00000148	—	0.00008511
3, 300	0.00008363	0.00000148	—	0.00008511

units are able to contribute has been reduced to negligible levels by the addition of two hot reserve units.

(b) The risk is reduced most significantly by the addition of the gas turbines and these have a greater effect than the hot reserve units. The reason is that these units are able to affect the risk level over a much longer time period than the hot reserve units.

(c) The risk is almost unaffected by adding more than about six gas turbines. The reason for this is that the risk level is dominated, as shown in Fig. 5.6, by the partial risk in the period up to 1 hour. The standby units do not contribute during this time and therefore only marginally affect the total risk.

5.5 Postponable outages

5.5.1 Concepts

The techniques described in Sections 5.2–4 assume that a unit which fails during operation must be removed immediately from service. This is usually an acceptable concept in planning studies but becomes less acceptable in operational reserve evaluation. Many of the failures that occur in practice can be tolerated for a certain period of time and the removal of the unit for repair can be delayed or postponed for up to 24 hours or more. In such cases, the unit continues to supply energy and need not be removed until the load decreases to a level at which the unit is no longer required. If all unit failures could be postponed in this way, capacity outages would not contribute to the operational risk and spinning reserve would only be required to compensate for load forecast uncertainties. In practice, many unit failures do necessitate immediate removal and the effect must therefore be included. Models and evaluation techniques have been published [8] that permit consideration of postponable outages and these are described in the following sections.

5.5.2 Modelling postponable outages

Consider a generating unit that is represented by a two-state model, operating (up) and failed (down). Equations (5.1)–(5.3) showed that

$$P(\text{down}) = \frac{\lambda}{\lambda + \mu} - \frac{\lambda}{\lambda + \mu}\,\mathrm{e}^{-(\lambda + \mu)T}$$

or if repair is neglected

$$P(\text{down}) = \text{ORR} = 1 - \mathrm{e}^{-\lambda T} \simeq \lambda T$$

Consider now the same generating unit but with an additional state to represent the postponable outage as shown in Fig. 5.7, in which

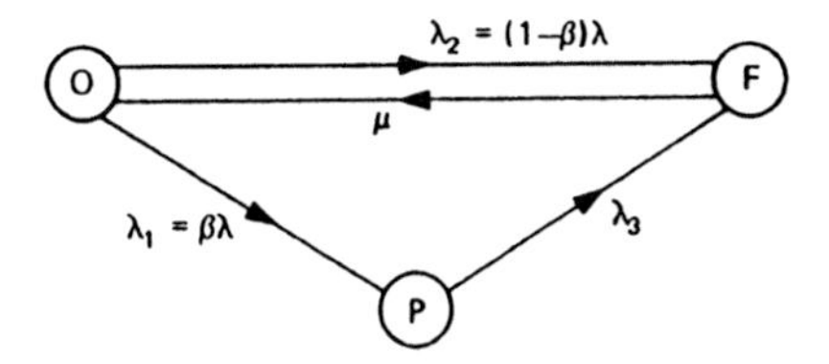

Fig. 5.7 Postponable outage model

O, F, P = operating, failed and removed from service, postponable outage states respectively
λ_1 = rate of outages that can be postponed
λ_2 = rate of outages that cannot be postponed
λ = total failure rate = $\lambda_1 + \lambda_2$
λ_3 = rate at which postponed outages are forced into the 'failed and removed from service' state
β = proportion of λ that can be postponed = λ_1/λ.

It can be shown [8] that

$$P(\text{down}) = \frac{\lambda_2\lambda_3 + \lambda_1\lambda_3}{\alpha\gamma} + \frac{(\lambda_2\lambda_3 + \lambda_1\lambda_3 - \alpha\lambda_2)}{\alpha(\alpha-\gamma)}\,e^{-\alpha T}$$

$$-\frac{(\lambda_2\lambda_3 + \lambda_1\lambda_3 - \gamma\lambda_2)}{\gamma(\alpha-\gamma)}\,e^{-\gamma T} \qquad (5.21)$$

where

$$\alpha = \frac{1}{2}\left[(\lambda_2 + \lambda_1 + \mu + \lambda_3) + \{(\lambda_2 + \lambda_1 + \mu + \lambda_3)^2 - 4(\mu\lambda_3 + \lambda_2\lambda_3 + \lambda_1\lambda_3 + \lambda_2\mu + \lambda_1\mu)\}^{1/2}\right]$$

$$\gamma = \frac{1}{2}\left[(\lambda_2 + \lambda_1 + \mu + \lambda_3) - \{(\lambda_2 + \lambda_1 + \mu + \lambda_3)^2 - 4(\mu\lambda_3 + \lambda_2\lambda_3 + \lambda_1\lambda_3 + \lambda_2\mu + \lambda_1\mu)\}^{1/2}\right]$$

If $\alpha t \ll 1$ and $\gamma t \ll 1$, Equation (5.21) reduces to

$$P(\text{down}) \simeq \lambda_2 T = (1-\beta)\lambda T \qquad (5.22)$$

Since λT is the ORR if no outages can be postponed, Equation (5.22) can be designated as the 'modified outage replacement rate' MORR and

$$P(\text{down}) = \text{MORR} = (1-\beta)\text{ORR} \qquad (5.23)$$

Table 5.10 Generation models including postponable outages

Capacity out (MW)	Cumulative probability for β of			
	0.0	*0.1*	*0.3*	*0.5*
0	1.000000	1.000000	1.000000	1.000000
10	0.005238	0.005056	0.004692	0.004330
20	0.003874	0.003692	0.003327	0.002965
30	0.001829	0.001646	0.001281	0.000918
40	0.001826	0.001643	0.001278	0.000915
50	0.001825	0.001642	0.001277	0.000914
60	0.001825	0.001642	0.001277	0.000914
70	0.000007	0.000005	0.000004	0.000003
80	0.000005	0.000003	0.000003	0.000002
120	0.000001			

5.5.3 Unit commitment risk

In order to illustrate the evaluation of the unit commitment risk when some outages are postponable and to demonstrate their effect, reconsider System A of Section 5.2.3. In the present case assume that the 60 MW units can have some of their outages postponed with values of β of 0.1, 0.3 and 0.5. The respective values of MORR for a lead time of 2 hours are 0.000822, 0.000639 and 0.000457. These values of MORR can be used to form appropriate generation models using the same concepts as those in Section 5.2.3. This evaluation gives the capacity outage probability tables shown in Table 5.10.

The values shown in Table 5.10 indicate, as expected, that the risk decreases as the number of failures which can be postponed increases. Therefore, less spinning reserve needs to be carried in order to achieve a given level of risk.

5.6 Security function approach

5.6.1 Concepts

One approach that has been considered in terms of its application to the operational phase is known as the security function approach [9–11]. This defines events as breaches of security which have been defined as 'inadequacies of spinning generation capacity, unacceptable low voltage somewhere in the system, transmission line or equipment overload, loss of system stability or other intolerable operating conditions'. These concepts relate to a composite reliability assessment of generation and transmission (see Chapter 6). In this complete form of analysis, the evaluated security index is a global system risk comprising generation and transmission violations.

One of the problems facing an operator is to make rapid on-line decisions based on pertinent information which is displayed in front of him. Consequently, this information should not only inform the operator of the risk in his system but, of equal importance, should inform him why a high risk period may exist and what actions can be performed to reduce the risk. For this reason, a complete global security index may not be fruitful since it does not contain the relevant information necessary to perform remedial actions.

It is also a very time-consuming process to include all features of system insecurity and this is again not conducive to on-line operation requiring fast responses to system dynamical changes.

This chapter will therefore be limited to a discussion of the security function approach in terms of its application to the risk assessments of operational reserve. In this application, it is little different from the PJM method of risk assessment.

5.6.2 Security function model

The mathematical model [9–11] for the security function $S(t)$ is

$$S(t) = \sum_i P_i(t) Q_i(t) \tag{5.24}$$

where

$P_i(t)$ = probability that the system is in state i at time t in the future

$Q_i(t)$ = probability that state i constitutes a breach of security at time t in the future

For a rigorous application of Equation (5.24), the summation must be performed over the exhaustive set of all possible system states. It is this exhaustive evaluation that restricts its general application to operational risk assessment.

The security function model defined by Equation (5.24) is an application of conditional probability (see Section 2.5 of *Engineering Systems*) because $Q_i(t)$ is the probability of system failure *given* a particular system state and $P_i(t)$ is the probability of that system state. Furthermore, the exhaustive list of all system states are mutually exclusive.

If the analysis is restricted only to a risk evaluation of the operational reserve, $P_i(t)$ is the individual probability of residing in the ith capacity state and $Q_i(t)$ is the probability that the load is greater than or equal to the available output capacity of state i. If the load is assumed to be constant during the period of interest, then $Q_i(t)$ is either equal to unity when the load is greater than or equal to the state capacity or equal to zero when the load is less than the state capacity. An interpretation of this shows that the evaluation is identical to the PJM method since Equation (5.24) becomes

$$S(t)=\sum_{i} P_i(t) \tag{5.25}$$

where i represents all capacity states in which the output capacity is less than or equal to the load demand and $\Sigma_i\ P_i(t)$ is the cumulative probability of these capacity states.

If load forecast uncertainty is included, the application of Equation (5.24) becomes identical to the extension of PJM method that included load forecast uncertainty (Section 5.3.1).

5.7 Response risk

5.7.1 Concepts

The concepts described in Sections 5.2–6 can be used to evaluate unit commitment risk, i.e. they enable an operator to decide which units and how many should be committed at the decision time of $t = 0$ in order that the system risk is equal to or less than a required or acceptable value. These previous techniques, however, do not indicate how these committed units should be dispatched, i.e. which units or how much of each unit should generate power or be held as spinning reserve. Furthermore, these previous techniques do not consider or account for the pick-up rate of those units that constitute the spinning reserve. This is an important feature of spinning reserve studies because, if a dispatched unit suddenly fails or the load suddenly increases, the units acting as spinning reserve must respond within a certain period of time in order that the system does not undergo undesirable or catastrophic consequences.

Generally, two time periods are of interest in connection with the response of spinning reserve units; a short time of about 1 minute and a longer time of about 5 minutes. The short response time represents the period in which sufficient capacity must respond in order to protect system frequency and tie-line regulation. The longer response time represents the period in which the units must respond in order to eliminate the need to take other emergency measures such as disconnecting load.

The ability to respond to system changes and to pick-up load on demand depends very much on the type of units being used as spinning reserve. Typically the response rate may vary from about 30% of full capacity per minute for hydro-electric plant to only 1% of full capacity per minute for some types of thermal plant. Furthermore, in practice, units usually have a non-linear response rate. Some types of rapid start units such as gas turbines can usually reach full output within 5 minutes from standstill.

The essence of allocating spinning reserve between units is therefore to decide which of the committed units should be dispatched and which should be held as reserve. These decisions can be assisted by evaluating the probability of achieving

Table 5.11 Failure probability and response rate data

Unit (MW)	*λ (f/yr)*	*Prob. of failure in 5 minutes*	*Response rate (MW/minute)*
10	3	0.0000285	1
20	3	0.0000285	1
60	4	0.0000381	1

a certain response or regulating margin within the required response time. This assessment is known as response risk evaluation [12].

5.7.2 Evaluation techniques

The evaluation [12] of response risk is similar to the evaluation of the risk in any generating system. Essentially the problem is to evaluate the probability of achieving within the required response time, the various possible output states of the units held in reserve. This evaluation should include the effect of the response rate of a unit and the probability that the responding unit fails during the response period and therefore cannot contribute to the required response. This evaluation method and the effect of the evaluation is best described using a numerical example.

Consider System A in Section 5.2.3 and a required response time of 5 minutes. The probability that a unit fails in this time period is shown in Table 5.11. This table also includes the response rate of each unit which, for this example, is assumed to be linear.

From the response rate shown in Table 5.11, only 5 MW can be obtained from each unit within the 5 minute response period, given that a response is required and the unit expected to respond does not fail during this period.

Let the system load be 140 MW, i.e. 60 MW of spinning reserve has been committed. Initially, let all this reserve be allocated to one of the 60 MW thermal units. Under these operating circumstances (Dispatch A), the available generation G, the load L dispatched on each unit and the available 5 minute response R of each unit is shown in the schedule of Table 5.12.

The unit commitment risk for this system was shown in Table 5.2 to be 0.001825 for a lead time of 2 hours. The 5 minute response risk for the system is shown in Table 5.13.

Table 5.12 Dispatch A (all values in MW)

G	10	10	20	20	20	60	60
L	10	10	20	20	20	60	0
R	0	0	0	0	0	0	5

Table 5.13 Response risk for Dispatch A

Response (MW)	*Individual probability*	*Cumulative probability (risk)*
5	0.9999619	1.0000000
0	0.0000381	0.0000381

The results shown in Table 5.13 indicate the basic problem of allocating all the spinning reserve to one thermal unit since, in this case, only 5 MW of the 60 MW of spinning reserve can respond in the time period of interest. Furthermore, if the response risk is defined as achieving a given capacity response or less, it is seen that the risk associated with this dispatch is very high.

5.7.3 Effect of distributing spinning reserve

The limited response of Dispatch A of Section 5.7.2 was due to all the reserve being allocated to one unit only. Consider now the effect of distributing this reserve between two or more units. Two examples will be considered; the first has the 60 MW reserve equally distributed between the two 60 MW units as shown in Dispatch B (Table 5.14), and the second has the 60 MW reserve distributed between all units in such a way that the maximum 35 MW response is achieved, as shown in Dispatch C (Table 5.15).

These two schedules give the risk tables shown in Table 5.16.

A comparison of the results in Table 5.16 with those in Table 5.13 shows the great advantage that is gained by distributing the spinning reserve between the available thermal units that have been committed. This distribution not only increases the available response capacity but also decreases the risk associated with a given response requirement. It should be noted that the unit commitment risk is the same for all three dispatches (A, B and C).

Table 5.14 Dispatch B (all values in MW)

G	10	10	20	20	20	60	60
L	10	10	20	20	20	30	30
R	0	0	0	0	0	5	5

Table 5.15 Dispatch C (all values in MW)

G	10	10	20	20	20	60	60
L	0	0	0	15	15	55	55
R	5	5	5	5	5	5	5

Table 5.16 Response risks for Dispatches B and C

Dispatch B		*Dispatch C*	
Response (MW)	*Cumulative risk*	*Response (MW)*	*Cumulative risk*
10	1.0000000	35	1.0000000
5	0.0000762	30	0.0002187
0	0.0000000	25	0.0000000

Table 5.17 Dispatch D (all values in MW)

G	10	10	20	20	20	60 (Th)	60 (Hyd)
L	10	10	20	20	20	60	0
R	0	0	0	0	0	0	60

5.7.4 Effect of hydro-electric units

Hydro units can usually respond to changes much more rapidly than conventional thermal units. They are therefore very useful as spinning reserve units. They are also, however, cheaper to operate than thermal units, and this implies that they should be considered for use as base load units. These two conflicting uses lead to a dilemma because no unit can be used to generate fully under dispatch and simultaneously be used as spinning reserve. Frequently a compromise is used in which part of its available output is dispatched and the remaining part is held in reserve.

In order to illustrate this effect, reconsider System B of Section 5.2.3 and let the response rate of the hydro unit be 20 MW/minute. The probability of the hydro unit failing in 5 minutes is 0.0000095. Consider two possible dispatches for this system as shown in Tables 5.17 and 5.18. These two dispatches give the risk tables shown in Table 5.19.

The results shown in Table 5.19 indicate that, as expected, a greater response can be achieved using Dispatch D than using Dispatch E or any of the previously considered dispatches. The hydro unit cannot be used as on-line generation using Dispatch D because it has been totally dedicated as spinning reserve. In addition, all the reserve is carried by one unit in the case of Dispatch D and therefore the risk of failing to respond is greater than for Dispatch E; e.g., the risk associated with a

Table 5.18 Dispatch E (all values in MW)

G	10	10	20	20	20	60 (Th)	60 (Hyd)
L	10	10	20	20	20	30	30
R	0	0	0	0	0	5	30

Table 5.19 Response risks for Dispatches D and E

Dispatch D		*Dispatch E*	
Response (MW)	*Cumulative risk*	*Response (MW)*	*Cumulative risk*
60	1.0000000	35	1.0000000
0	0.0000095	30	0.0000476
		5	0.0000095
		0	0.0000000

30 MW response is unity in the case of Dispatch D but only 0.0000476 in the case of Dispatch E. It follows from this discussion therefore that it is preferable to distribute the reserve between several units in order to minimize the response risk and also to allocate some reserve to hydro units (if any exist) in order to obtain more rapid and greater values of response.

5.7.5 Effect of rapid start units

An alternative to using conventional hydro units, such as dam systems and run-of-the-river systems, as spinning reserve is to use other rapid start units such as gas turbines and pumped storage systems. These can respond extremely quickly from standstill and hence significantly decrease the response risk and increase the response magnitude.

(a) *Rapid start units do not fail to start*

If the rapid start units do not fail to start, their inclusion in response risk assessment is very simple. In this case, it is only necessary to deduce the amount of response that they can contribute and add this value to the response values that have been previously deduced. This value of response is usually equal to their capacity because their response rate is rapid.

Consider that the operator responsible for System A of Section 5.2.3 has two 30 MW gas turbines at his disposal. If these units can respond fully in the 5 minute

Table 5.20 Modified response risk for Dispatches B and C

Dispatch B		*Dispatch C*	
Response (MW)	*Cumulative risk*	*Response (MW)*	*Cumulative risk*
70	1.0000000	95	1.0000000
65	0.0000762	90	0.0002187
60	0.0000000	85	0.0000000

period, the risk associated with Dispatch B and Dispatch C in Section 5.7.3 are modified to the values shown in Table 5.20.

The response risks shown in Table 5.20 are very much better than those obtained previously and shown in Table 5.16. Although the numerical values of risk are identical to those shown in Table 5.16, the associated level of response is much greater. The present results show that, when the gas turbines are included, both Dispatch B and Dispatch C permit a response to the loss of a 60 MW unit within 5 minutes with a probability of less than 10^{-7}.

(b) *Rapid start units may fail to start*

As discussed in Section 5.4.3, rapid start units may fail to start in practice and in such cases will not be able to contribute to the required response. This is particularly the case with gas turbine units which generally have a relatively high probability of failing to start. Pump storage systems on the other hand generally have a very low probability of failing to start. The effect of failing to start can be included for both types of units using the concept of conditional probability (Section 2.5 of *Engineering Systems*). This concept gives

$$\begin{aligned}\text{risk} = {} & \text{risk (given all rapid start units do not fail to start)}\\ & \times \text{prob. (all rapid start units not failing to start)}\\ & + \text{risk (given one rapid start unit fails to start)}\\ & \times \text{prob. (one unit failing to start)} + \cdots\\ & + \text{risk (given all rapid start units fail to start)}\\ & \times \text{prob. (all units failing to start)}\end{aligned} \tag{5.26}$$

Consider the application of this technique to Dispatch B of Section 5.7.3. Again assume that two 30 MW gas turbines are available to the operator as in (a) above, each having a probability of failing to start of 20%. The risk tables associated with each condition, 'both units start,' 'one unit starts' and 'no units start,' are shown in Table 5.21. The overall risk table is shown in Table 5.22.

Table 5.21 Response risks for each condition

Risk table for condition of:					
No units start		*One unit starts*		*Both units starts*	
MW	*Probability*	*MW*	*Probability*	*MW*	*Probability*
10	0.9999238	40	0.9999238	70	0.9999238
5	0.0000762	35	0.0000762	65	0.0000762
0	0.0000000	30	0.0000000	60	0.0000000
The conditional probabilities are:					
	0.04		0.32		0.64

Table 5.22 Weighted response risk for Dispatch B

Response (MW)	*Cumulative risk*
70	1.0000000
65	0.3600488
40	0.3600000
35	0.0400244
10	0.0400000
5	0.0000030
0	0.0000000

The response risks and magnitudes shown in Table 5.22 are evidently better than those shown in Table 5.16 where it was assumed that no gas turbines were available, but as would be expected they are not as good as those shown in Table 5.20, for which the gas turbines were assumed not to fail.

5.8 Interconnected systems

Although many systems are operated completely separately from all other systems, there are also many systems that have limited capacity tie lines between them. These systems are operated as interconnected systems and can assist each other when operational deficiencies arise.

This concept has been discussed in Chapter 4 in terms of planning of systems. There are no essential differences between the concepts used in planning and the required concepts in operation. Consequently the techniques [13, 14] are similar to those described in Chapter 4 and can be equally applied to the operational phase. The only significant difference that arises is in the values of probability used in the two phases. In Chapter 4, the required value of probability was the forced outage rate (FOR). In operational studies, the required value of probability is the outage replacement rate (ORR) or a similar concept of probability in the case when derated states of on-line generation are considered.

5.9 Conclusions

This chapter has described the various concepts and evaluation techniques that can be used to assess operational risk. This area of reliability evaluation has probably received the least attention of all areas of power systems, yet the techniques are sufficiently well developed for the on-line assessment of operational risk and for assisting operators in their day-today and minute-to-minute decision-making.

An operator is continually faced with the problem of making good decisions rapidly. This imposes many burdens in order to ensure the system is operated economically but with minimum risk. The techniques described in this chapter assist in this decision-making and permit considerations to be taken and a balance to be made between dispatching increased on-line generation, committing more standby plant such as gas turbines and leaving de-synchronized plant on hot standby. These considerations are not easy and cannot be taken lightly. Since rule-of-thumb or deterministic methods cannot compare these alternatives using consistent criteria, the need for probabilistic assessment methods and criteria become apparent.

Any information displayed to an operator must be pertinent but also it must not only inform him of operational difficulties but also indicate why these difficulties have arisen and what can be done to overcome them. If the displayed information conforms to less than these requirements, it can lead to confusion, panic and erroneous decisions. The techniques used therefore must not be too complex or sophisticated and must not attempt to convolve too many disparate effects within one piece of information which the operator cannot disentangle. The techniques described in this chapter are relatively simple, easy to code and employ and can be tailored to suit individual utilities' requirements for information to be displayed to their operators.

5.10 Problems

1 A system consists of 10 × 60 MW units. Evaluate the unit commitment risk for a lead time of 2 hours and loads of 540 MW and 480 MW if
   (a) each unit has a mean up time of 1750 hours;
   (b) each unit has a mean up time of 1750 hours and the loads are forecast with an uncertainty represented by a standard deviation of 5%;
   (c) each unit has a 50 MW derated state, a derated state transition rate of 2 f/yr and a down state transition rate of 3 f/yr;
   (d) each unit has a mean up time of 1750 hours and 20% of the failures of each unit can be postponed until the following weekend;
   (e) the system is connected to another identical system through a tie line of 30 MW capacity and each unit of both systems has a mean up time of 1750 hours.

2 Evaluate the response risk for the system of Problem 1(a) if 50 MW must respond within 5 minutes, during which time the output of each of the 60 MW units can be increased by 6 MW
   (a) when no rapid start units are available;
   (b) when an additional 5 MW gas turbine unit that does not fail to start is available to the operator;
   (c) when the 5 MW gas turbine unit has a starting failure probability of 0.1.

3 Evaluate the unit commitment risk of Problem 1(a) for a lead time of 1 hour if a gas turbine unit of 30 MW capacity can be started in 10 minutes. Referring to the model of Fig. 5.4, the gas turbine has the following transition rates/hour: $\lambda_{12} = 0.0050$, $\lambda_{21} = 0.0033$, $\lambda_{14} = 0.0300$, $\lambda_{41} = 0.0150$, $\lambda_{23} = 0.0008$, $\lambda_{32} = 0.0000$, $\lambda_{34} = 0.0250$, $\lambda_{42} = 0.0250$.

4 An operator expects the system load to be constant for the next few hours at 360 MW.
(a) How many identical 60 MW thermal units must he commit and spin if the failure rate of each is 5f/yr, the lead time is 2 hours and the unit commitment risk must be less than 0.005?
(b) How should these units be dispatched in order to minimize the 5 minutes response risk if the response rate of each is linear at 1 MW/minute?
(c) Evaluate the response risk if the system requires a minimum of 35 MW to respond within 5 minutes.

5 Two systems A and B are interconnected with a 100% reliable tie line. The capacity of the tie line is 10 MW. System A commits five 30 MW units and System B commits five 20 MW units. Each unit has an expected failure rate of 3 f/yr. Evaluate the unit commitment risk in System A for a lead time of 3 hours assuming
(a) the loads in System A and System B remain constant at 120 MW and 60 MW respectively;
(b) System B is willing to assist System A up to the point at which it itself runs into deficiencies.

Compare this risk with that which would exist in System A if the tie line did not exist.

5.11 References

1. IEEE Committee Report, 'Bibliography on the application of probability methods in power system reliability evaluation', *IEEE Trans. on Power Apparatus and Systems*, **PAS-91** (1972), pp. 649–60.
2. IEEE Committee Report, 'Bibliography on the application of probability methods in power system reliability evaluation 1971–1977', *IEEE Trans. on Power Apparatus and Systems*, **PAS-97** (1978), pp. 2235–42.
3. Anstine, L. T., Burke, R. E., Casey, J. E., Holgate, R., John, R. S., Stewart, H. G., 'Application of probability methods to the determination of spinning reserve requirements for the Pennsylvania–New Jersey–Maryland interconnection', *IEEE Trans. on Power Apparatus and Systems*, **PAS-82** (1963), pp. 720–35.
4. Billinton, R., Jain, A. V., 'Unit derating levels in spinning reserve studies', *IEEE Trans. on Power Apparatus and Systems*, **PAS-90** (1971), pp. 1677–87.
5. Billinton, R., Jain, A. V., 'The effect of rapid start and hot reserve units in spinning reserve studies', *IEEE Trans. on Power Apparatus and Systems*, **PAS-91** (1972), pp. 511–16.
6. Allan, R. N., Nunes, R. A. F., *Modelling of Standby Generating Units in Short-term Reliability Evaluation*, IEEE Winter Power Meeting, New York, 1979, paper A79 006–8.
7. IEEE Task Group on Models for Peaking Service Units, *A Four State Model for Estimation of Outage Risk for Units in Peaking Service*, IEEE Winter Power Meeting, New York, 1971, paper TP 90 PWR.

8. Billinton, R., Alam, M., *Outage Postponability Effects in Operating Capacity Reliability Studies*, IEEE Winter Power Meeting, New York, 1978, paper A78 064–8.
9. Patton, A. D., 'A probability method for bulk power system security assessment: I—Basic concepts', *IEEE Trans. on Power Apparatus and Systems*, **PAS-91** (1972), pp. 54–61.
10. Patton, A. D., 'A probability method for bulk power system security assessment: II—Development of probability methods for normally operating components', *IEEE Trans. on Power Apparatus and Systems*, **PAS-91** (1972), pp. 2480–5.
11. Patton, A. D., 'A probability method for bulk power system security assessment: III—Models for standby generators and field data collection and analysis', *IEEE Trans. on Power Apparatus and Systems*, **PAS-91** (1972), pp. 2486–93.
12. Jain, A. V., Billinton, R., *Spinning Reserve Allocation in a Complex Power System*, IEEE Winter Power Meeting, New York, 1973, paper C73 097–3.
13. Billinton, R., Jain, A. V., 'Interconnected system spinning reserve requirements', *IEEE Trans. on Power Apparatus and Systems*, **PAS-91** (1972), pp. 517–26.
14. Billinton, R., Jain, A. V., 'Power system spinning reserve determination in a multi-system configuration', *IEEE Trans. on Power Apparatus and Systems*, **PAS-92** (1973), pp. 433–41.

6 Composite generation and transmission systems

6.1 Introduction

One of the most basic elements in power system planning is the determination of how much generation capacity is required to give a reasonable assurance of satisfying the load requirements. This evaluation is normally done using the system representation shown in Fig. 2.2. The concern in this case is to determine whether there is sufficient capacity in the system to generate the required energy to meet the system load.

A second but equally important element in the planning process is the development of a suitable transmission network to convey the energy generated to the customer load points [1]. The transmission network can be divided into the two general areas of bulk transmission and distribution facilities. The distinction between these two areas cannot be made strictly on a voltage basis but must include the function of the facility within the system [2–4]. Bulk transmission facilities must be carefully matched with the generation to permit energy movement from these sources to points at which the distribution or sub-transmission facilities can provide a direct and often radial path to the customer. Distribution design, in many systems, is almost entirely decoupled from the transmission system development process. Given the location and size of the terminal station emanating from the bulk transmission system, distribution system design becomes a separate and independent process. Coupling between these two systems in reliability evaluation can be accommodated by using the load point indices evaluated for the bulk transmission system as the input reliability indices of the distribution system.

In addition to providing the means to move the generated energy to the terminal stations, the bulk transmission facilities must be capable of maintaining adequate voltage levels and loadings within the thermal limits of individual circuits and also maintaining system stability limits. The models used to represent the bulk facilities should be capable of including both static and dynamic considerations. The static evaluation of the system's ability to satisfy the system load requirements can be designated as *adequacy* evaluation and is the subject of this chapter. Concern regarding the ability of the system to respond to a given contingency can be designated as *security* evaluation. This is an extremely important area which has not yet received much attention in regard to the development of probabilistic indices. One aspect is the determination of the required operating or spinning reserve and this is discussed in Chapter 5. Work has also been done on probabilistic evaluation of transient stability.

The total problem of assessing the adequacy of the generation and bulk power transmission systems in regard to providing a dependable and suitable supply at the terminal stations can be designated as composite system reliability evaluation [5].

6.2 Radial configurations

One of the first major applications of composite system evaluation was the consideration of transmission elements in interconnected system generating capacity evaluation. This aspect has been discussed in detail in Chapter 4 using the array method and the equivalent unit approach. The latter method includes the development of an equivalent generating capacity model and then moving this model through the interconnection facility to the assisted system. This approach can be readily applied to systems such as that shown in Fig. 6.1.

The analysis at the load point L can be done using the LOLE, LOEE or F&D techniques described in Chapter 4. The linking configuration between the generation source and the load point may not be of the simple series–parallel type shown in Fig. 6.1 but could be a relatively complicated d.c. transmission configuration where the transmission capability is dependent upon the availability of the rectifier and inverter bridges, the filters at each end and the associated pole equipment. These concepts are described in Chapter 11. The development of the transmission model may be relatively complex but once obtained can be combined with the generation model to produce a composite model at the load point.

The progressive development of an equivalent model is relatively straightforward for a radial configuration such as that shown in Fig. 6.1. This approach, however, is not suitable for networked configurations including dispersed generation and load points. A more general approach is required which can include the ability of the system to maintain adequate voltage levels, line loadings and steady state stability limits.

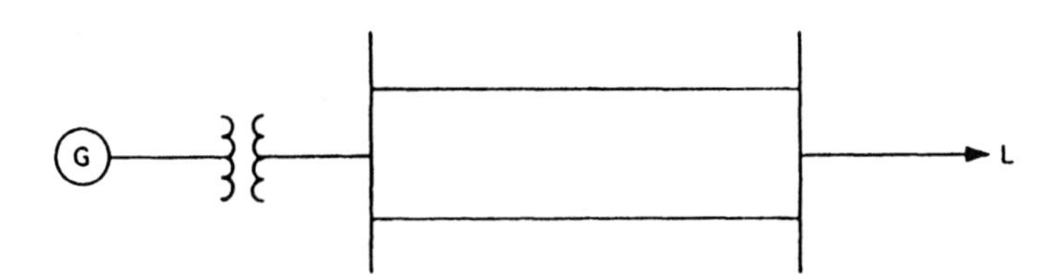

Fig. 6.1 Simple radial generation transmission system

6.3 Conditional probability approach

Many probability applications in reliability evaluation assume that component failures within a fixed environment are independent events. This may or may not be true. It is, however, entirely possible that component failure can result in system failure in a conditional sense. This can occur in parallel facilities that are not completely redundant. If the load can be considered as a random variable and described by a probability distribution, then failure at any terminal station due to component failure is conditional upon the load exceeding the defined carrying capability of the remaining facilities. Load point failure in this case may be defined as inadequate voltage or available energy at the customer bus [5].

If the occurrence of an event A is dependent upon a number of events B_i which are mutually exclusive, then (see Chapter 2 of *Engineering Systems*)

$$P(\mathrm{A}) = \sum_{i=1}^{n} P(\mathrm{A} \mid \mathrm{B}_i)P(\mathrm{B}_i) \tag{6.1}$$

If the occurrence of A is dependent upon only two mutually exclusive events for component B, success and failure, designated B_S and B_f respectively, then:

$$P(\mathrm{A}) = P(\mathrm{A} \mid \mathrm{B_S})P(\mathrm{B_S}) + P(\mathrm{A} \mid \mathrm{B_f})P(\mathrm{B_f}) \tag{6.2}$$

If the event A is system failure, then

$$P(\text{system failure}) = P(\text{system failure} \mid \text{B is good})P(\mathrm{B_S}) + P(\text{system failure} \mid \text{B is bad})P(\mathrm{B_f}) \tag{6.3}$$

This approach can be applied to the simple radial configuration shown in Fig. 6.1.

Define

P_g = probability of generation inadequacy,
P_c = probability of transmission inadequacy
e.g. $P_c(1) = P$(load exceeds the capability of line 1)
Q_s = probability of system failure
A_1, U_1 = line 1 availability and unavailability respectively.
A_2, U_2 = line 2 availability and unavailability respectively.

From Equation (6.3),

$$Q_S = Q_S(\text{L1 in})\, A_1 + Q_S(\text{L1out})\, U_1$$

Given L1 in,

$$Q_S = Q_S(\text{L2in})A_2 + Q_S(\text{L2out})\, U_2$$

Given L1 in and L2 in,

$$Q_S = P_g + P_c(1, 2) - P_g\, P_c(1, 2)$$

It has been assumed that the probabilities of capacity deficiencies and transmission inadequacies are independent.

Given L1 in and L2 out,

$$Q_S = P_g + P_c(1) - P_g P_c(1)$$

Therefore given L1 in,

$$Q_S = A_2(P_g + P_c(1,2) - P_g P_c(1,2)) + U_2(P_g + P_c(1) - P_g P_c(1))$$

Given L1 out,

$$Q_S = Q_S(\text{L2 in})A_2 + Q_S(\text{L2 out})\ U_2$$

$$= A_2(P_g + P_c(2) - P_g P_c(2)) + U_2$$

For the complete system,

$$Q_S = A_1[A_2(P_g + P_c(1,2) - P_g P_c(1,2)) + U_2(P_g + P_c(1) - P_g P_c(1))] + U_1[A_2(P_g + P_c(2) - P_g P_c(2)) + U_2] \tag{6.4}$$

If the two lines are identical this reduces to

$$Q_S = A^2[P_g + P_c(1,2) - P_g P_c(1,2)] + 2AU[P_g + P_c(1) - P_g P_c(1)] + U^2 \tag{6.5}$$

The solution of this simple system could have been obtained directly by using the terms of the binomial expansion of $(A + U)^2$, each term being weighted by the relevant probability of generation and line inadequacy. A set of general equations can be written to give the same result as that shown in Equation (6.5).

The probability of failure Q_K at bus K in a network can be expressed as

$$Q_K = \sum_j [P(\text{B}_j)(P_{gj} + P_{lj} - P_{gj}P_{lj})] \tag{6.6}$$

where

B_j = an outage condition in the transmission network (including zero outages)
P_{gj} = probability of the generating capacity outage exceeding the reserve capacity
P_{lj} = probability of load at bus K exceeding the maximum load that can be supplied at that bus without failure.

If the generating unit outages and the load variation are considered in terms of probability only and not in terms of frequency of occurrence, then an estimate of the expected frequency of failure F_K at bus K is given by

$$F_K = \sum_j [F(\text{B}_j)(P_{gj} + P_{lj} - P_{gj}P_{lj})] \tag{6.7}$$

where $F(\text{B}_j)$ is the frequency of occurrence of outage B_j.

Equations (6.6) and (6.7) consider the generating facility as a single entity. This may be acceptable in a radial configuration but may not be in cases where the generation is dispersed throughout the system. A more general set of equations [6] can be obtained directly from Equation (6.1).

$$Q_K = \sum_j [P(B_j)P_{1j}] \tag{6.8}$$

$$F_K = \sum_j [F(B_j)P_{1j}] \tag{6.9}$$

In this case, the generation outages are treated individually, as are the transmission outage events, and the generation schedule and resulting load flow are modified accordingly. It should be noted, however, that Equation (6.9) does not include a frequency component due to load model transitions. This could be included but it would require the assumption that all system loads transit from high to low load levels at the same time. Equations (6.7) and (6.9) also include possible frequency components due to transitions between states each of which represent a failure condition.

Equations (6.8) and (6.9) are applied to the system shown in Fig. 6.1 using the following data.

Generating units

6 × 40 MW units	$\lambda = 0.01$ f/day	= 3.65 f/yr
	$\mu = 0.49$ r/day	= 178.85 r/yr
	$U = 0.02$	

Transmission elements

2 lines	$\lambda = 0.5$ f/yr
	$r = 7.5$ hours/repair
	$U = 0.0004279$

Load

Peak load = 180 MW

The load is represented by a straight-line load–duration curve from the 100% to the 70% load points.

The generating capacity model (capacity outage probability table) is shown in Table 6.1, and the transmission capability model in Table 6.2.

The capability of each line is designated as X in Table 6.2. The actual carrying capability will depend on the criterion of success at the load point. If it is assumed that an adequate voltage level is required in addition to the required load demand, then the characteristics of the line, associated VAR support and the sending end

Table 6.1 Generation system model

State	Number of generators on outage	Cap. available (MW)	Probability	Dep. rate (occ/yr)	Frequency (occ/yr)
1	0	240	0.88584238	21.9	19.399948
2	1	200	0.10847049	197.1	21.379534
3	2	160	0.00553421	372.3	2.060386
4	3	120	0.00015059	547.5	0.082448
5	4	80	0.00000230	722.7	0.001666
6	5	40	0.00000002	897.9	0.000017
7	6	0	0.00000000	1073.1	0.000000

voltage constraints must be considered. If a line rating can be nominally assigned, the problem becomes one of transport rather than service quality and it becomes somewhat simpler [7].

Table 6.3 shows the composite state probabilities and frequencies assuming that the individual line-carrying capability X is 160 MW.

Equation (6.9) includes possible transitions between failure states and will therefore give an expected failure frequency at the load point which is slightly higher than that determined by creating the complete 21-state Markov model and evaluating the frequency of transitions across a specified capacity boundary wall. In this case transitions between failure states would not be included. The probability and frequency component for each state is weighted by the probability that the load will exceed the capability of that state to give the failure probability and frequency. Table 6.4 shows the load point failure probability and frequency for a peak load of 180 MW at different assumed line-carrying capability levels.

The peak load level in this case is 180 MW and it can be seen from Table 6.4 that the indices are constant for line capacities equal to or greater than this value. Under these conditions, failure would occur only for the loss of both lines or for

Table 6.2 Transmission system model

State	Number of lines on outage	Cap. available (MW)	Probability	Dep. rate (occ/yr)	Frequency (occ/yr)
1	0	2X	0.99914438	1.0	0.999144
2	1	1X	0.00085543	1168.5	0.999574
3	2	0X	0.00000018	2336.0	0.000428

X = rating of each line in MW

Table 6.3 State probabilities and frequencies

			State			*Failure*	
State	*Condition*	*Cap avail. (MW)*	*Probability*	*Frequency (occ/yr)*	P_{1j}	*Probability*	*Frequency (occ/yr)*
1	0*G* 0*L*	240.00	0.88508444	20.268433	0.00000000	0.00000000	0.000000
2	0*G* 1*L*	160.00	0.00075778	0.902061	0.37037038	0.00028066	0.334097
3	0*G* 2*L*	0.00	0.00000016	0.000382	1.00000000	0.00000016	0.000382
4	1*G* 0*L*	200.00	0.10837768	21.469619	0.00000000	0.00000000	0.000000
5	1*G* 1*L*	160.00	0.00009279	0.126713	0.37037038	0.00003437	0.046931
6	1*G* 2*L*	0.00	0.00000002	0.000050	1.00000000	0.00000002	0.000050
7	2*G* 0*L*	160.00	0.00552947	2.064152	0.37037038	0.00204795	0.764501
8	2*G* 1*L*	160.00	0.00000473	0.007294	0.37037038	0.00000175	0.002702
9	2*G* 2*L*	0.00	0.00000000	0.000003	1.00000000	0.00000000	0.000003
10	3*G* 0*L*	120.00	0.00015046	0.082528	1.00000000	0.00015046	0.082528
11	3*G* 1*L*	120.00	0.00000013	0.000221	1.00000000	0.00000013	0.000221
12	3*G* 2*L*	0.00	0.00000000	0.000000	1.00000000	0.00000000	0.000000
13	4*G* 0*L*	80.00	0.00000230	0.001667	1.00000000	0.00000230	0.001667
14	4*G* 1*L*	80.00	0.00000000	0.000004	1.00000000	0.00000000	0.000004
15	4*G* 2*L*	0.00	0.00000000	0.000000	1.00000000	0.00000000	0.000000
16	5*G* 0*L*	40.00	0.00000002	0.000017	1.00000000	0.00000002	0.000017
17	5*G* 1*L*	40.00	0.00000000	0.000000	1.00000000	0.00000000	0.000000
18	5*G* 2*L*	0.00	0.00000000	0.000000	1.00000000	0.00000000	0.000000
19	6*G* 0*L*	0.00	0.00000000	0.000000	1.00000000	0.00000000	0.000000
20	6*G* 1*L*	0.00	0.00000000	0.000000	1.00000000	0.00000000	0.000000
21	6*G* 2*L*	0.00	0.00000000	0.000000	1.00000000	0.00000000	0.000000
						0.00251783	1.233102

Line capacity = 160 MW
G = number of generators on outage
L = number of lines on outage

Table 6.4 Load point indices

	Failure	
Line capacity (MW)	*Probability*	*Frequency (occ/yr)*
100	0.00305635	1.885441
110	0.00305635	1.885441
120	0.00305635	1.885441
130	0.00299300	1.808646
140	0.00283461	1.616831
150	0.00267622	1.424967
160	0.00251783	1.233102
170	0.00236032	1.042589
180	0.00220280	0.852075
190	0.00220280	0.852075
200	0.00220280	0.852075

the loss of two or more generating units. The indices in Table 6.4 are also constant for line capacities of 120 MW or less. The low load level for a peak load of 180 MW is 126 MW. The system is therefore in the failed state for all load levels when the transmission capacity is less than 126 MW. Tables 6.5 and 6.6 show the load point indices for a range of line capacities and peak loads.

The overstatement of failure frequency due to the inclusion of transitions between failure states is more evident in a simple radial configuration than in a networked or meshed system. In the latter case, the loss of an element or a group of elements will only affect a relatively small number of load points in the immediate vicinity of the failure and the major contribution to the frequency will come from transmission outages involving relatively high frequencies and short durations.

Table 6.5 Load point indices—probability

Peak load (MW)	*Line capacity (MW)*				
	120	*140*	*160*	*180*	*200*
200	0.00469472	0.00469472	0.00440962	0.00412609	0.00384257
180	0.00305635	0.00283461	0.00251783	0.00220280	0.00220280
160	0.00084075	0.00048438	0.00012800	0.00012800	0.00012800
140	0.00048150	0.00007422	0.00007422	0.00007422	0.00007422
120	0.00000251	0.00000251	0.00000251	0.00000251	0.00000251

Table 6.6 Load point indices—frequency (occ/yr)

Peak load (MW)	*Line capacity (MW)*				
	120	*140*	*160*	*180*	*200*
200	2.497042	2.497042	2.151686	1.808762	1.465837
180	1.885441	1.616831	1.233101	0.852075	0.852075
160	0.934471	0.502776	0.071081	0.071081	0.071081
140	0.534893	0.041527	0.041527	0.041527	0.041527
120	0.002123	0.002123	0.002123	0.002123	0.002123

6.4 Network configurations

The concepts illustrated in Section 6.3 can be applied to networked or meshed configurations [8, 9]. This technique is illustrated using the system shown in Fig. 6.2.

Assume that the daily peak load curve for the period under study is a straight line from the 100% to the 60% point and that the load–duration curve is a straight line from the 100% to the 40% point. The peak load for the period is 110 MW. A basic generating capacity reliability study for this system can be done using the model of Fig. 2.2 in which the transmission elements are assumed to have no capacity restrictions and are fully reliable. Under these conditions, the basic system indices for a period of 365 days are as follows:

$$\left.\begin{array}{l}\text{LOLE} = 1.3089\text{days/year}\\ \text{LOLP} = 0.003586\end{array}\right\}\text{ using the daily peak load curve,}$$

$$\left.\begin{array}{l}\text{LOEE} = 267.6\text{MWh}\\ \text{LOLP} = 0.002400\end{array}\right\}\text{ using the load duration curve.}$$

The conditional probability approach can be used to develop the following expression for the probability of load point failure.

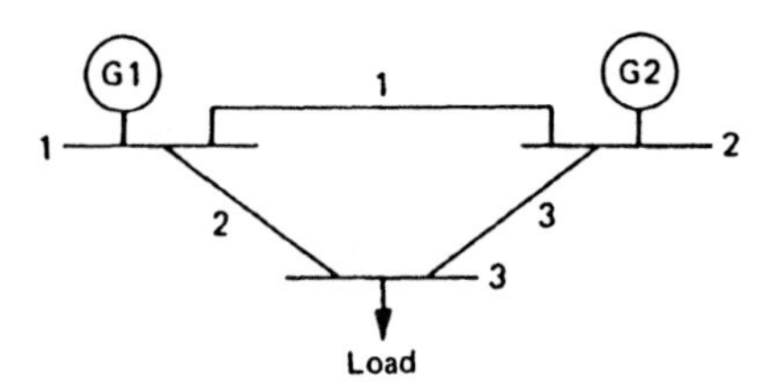

Fig. 6.2 Simple network configuration

Table 6.7 Generation data

Plant	*No. of units*	*Capacity (MW)*	*Unavailability*	*λ (f/yr)*	*μ(r/yr)*
1	4	20	0.01	1	99
2	2	30	0.05	3	57
Total	6	140			

$$
\begin{aligned}
Q_S = A_2[A_3[[P_g(1,2) + P_c(1) - P_g(1,2)P_c(1)]A_1 \\
+ [P_g(1,2) + P_c(2) - P_g(1,2)P_c(2)]U_1] \\
+ U_3[A_1[P_g(1,2) + P_c(3) - P_g(1,2)P_c(3)] \\
+ U_1[P_g(1) + P_c(4) - P_g(1)P_c(4)]]] \\
+ U_2[A_3[A_1[P_g(1,2) + P_c(5) - P_g(1,2)P_c(5)] \\
+ U_1[P_g(2) + P_c(6) - P_g(2)P_c(6)]] + U_3]
\end{aligned}
\quad (6.10)
$$

The term $P_c(j)$ is the probability associated with load curtailment in configurations j shown in Fig. 6.3.

The probability of inadequate transmission capability in each of these configurations can be found after performing a load flow study on each configuration using the appropriate load model. There is a range of possible solution techniques which can be used in this case. It should be fully appreciated that each approach involves different modeling techniques and therefore gives different load point reliability indices. The simplest approach is to assume that there are no transmission curtailment constraints and that continuity is the sole criterion. The next level is to use a transportation approach in which the line capability is prespecified at some maximum value. If line overload is to be considered, then a d.c. load flow may be sufficient, but if voltage is also to be included as a load point criterion, then an a.c. load flow must be used.

Table 6.8 Transmission line data

	Connected to							*Rating on 100 MVA base*	
Line	*Bus*	*Bus*	*λ (f/yr)*	*r (hours)*	*R (ohms)*	*X (ohms)*	*B/2 (mhos)*	*(MVA)*	*(p.u.)*
1	1	2	4	8	0.0912	0.4800	0.0282	80	0.8
2	1	3	5	8	0.0800	0.5000	0.0212	100	1.0
3	2	3	3	10	0.0798	0.4200	0.0275	90	0.9

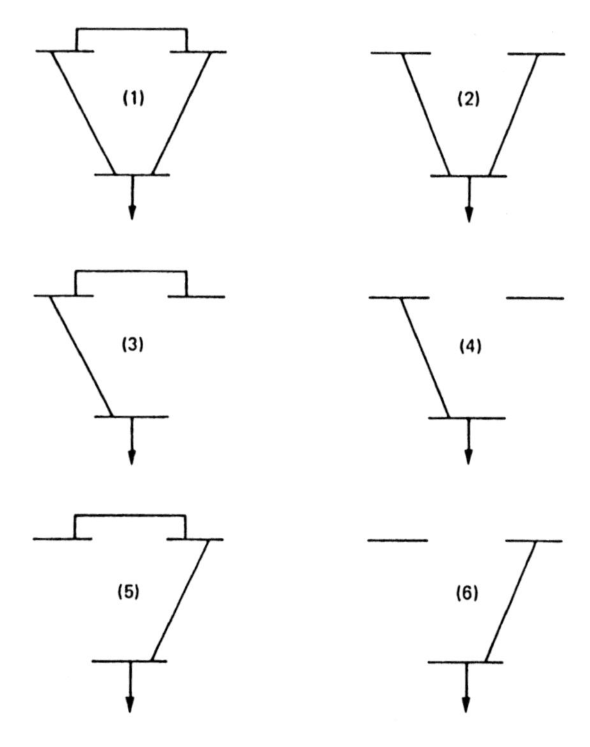

Fig. 6.3 Conditional configurations

The transmission line availabilities (A) and unavailabilities (U) for the system in Fig. 6.2 are given in Table 6.9 using the data from Table 6.8.

The probability of system failure at the load point can be found using Equation (6.10). If the assumption is made that there are no transmission line constraints and that connection to sufficient generating capacity is the sole criterion, then:

$$Q_S = \underline{0.09807433}$$

This value was calculated assuming that the load remains constant at the 110 MW level for the entire year. This index can be designated as an annualized value.

Table 6.9 Transmission line statistics

Line	*Availability*	*Unavailability*
1	0.99636033	0.00363967
2	0.99545455	0.00454545
3	0.99658703	0.00341297

Table 6.10 State probabilities

State j	*Lines out*	$P(B_j)$	P_g	P_{1j}	*P(system failure) individual*
1	0	0.98844633	0.09803430	0	0.09690164
2	1	0.00361076	0.09803430	0	0.00035398
3	2	0.00451345	0.09803430	0	0.00044247
4	3	0.00339509	0.09803430	0	0.00033185
5	1,2	0.00001649	1.0	0	0.00001649
6	1,3	0.00001237	1.0	0	0.00001237
7	2,3	0.00001546	1.0	1	0.00001546
8	1,2,3	0.00000006	1.0	1	0.00000006
					annualized Q_s = 0.09807433

i.e. expressed on an annual base. It can be compared with the value of 0.09803430 which is the probability of having 30 MW or more out of service in the generation model. The 30 MW outage state is considered to represent system failure as there would be some additional transmission loss in addition to the 110 MW load. This annualized index is clearly not a true value of the system reliability as it does not account for the load variation. It is a simple and very useful index, however, for relating and comparing weaknesses in alternative system proposals.

Equations (6.6) and (6.7) can also be used to find the probability and frequency of load point failure. Table 6.10 shows the required transmission and generation state probabilities for the no transmission constraint case.

The load model can be included in the calculation, rather than assuming the load will remain at the 110 MW peak value. Under these conditions the P_g and

Table 6.11 State frequencies

State	*Lines out*	*Departure rate (occ/yr)*	$F(B_j)$	*Failure frequency (occ/yr)*
1	0	12	11.86135596	1.16281973
2	1	1103	3.98266828	0.39043810
3	2	1102	4.97382190	0.48760515
4	3	885	2.99580465	0.29369161
5	1,2	2193	0.03616257	0.03616257
6	1,3	1976	0.02444312	0.02444312
7	2,3	1975	0.03053350	0.03053350
8	1,2,3	3066	0.00018396	0.00018396
				Annualized F_S = 2.42587774 f/yr

P_{1j} values in Table 6.10 reduce because the contribution to Q_S by lower load levels is less. This can be included using conditional probability.

The calculation of the expected frequency of failure using Equation (6.7) requires, in addition to the data shown in Table 6.10, the departure rates for each state. These values together with the state frequencies are shown in Table 6.11.

If transmission line overload conditions result in transmission lines being removed from service, then the load point indices increase. This can be illustrated by assuming that overload occurs whenever line 2 or 3 is unavailable. Under these conditions, load must be curtailed, causing increased load point failures. In this case

$$Q_S = 0.10520855$$

$$F_S = \underline{9.61420753 \text{ f/yr}}$$

6.5 State selection

6.5.1 Concepts

Equations (6.6) and (6.7) consider the generating facilities as one equivalent model and therefore reduce the total number of individual states which must be considered, i.e. 8 transmission states. Equations (6.8) and (6.9) consider each generating unit and transmission line as a separate element, thereby increasing the flexibility of the approach but simultaneously increasing the number of states which must be considered. In this system there are 9 elements which represent a total of 512 states. It becomes necessary therefore to limit the number of states by selecting the contingencies which will be included. This can be done in several basic ways.

The most direct is to simply specify the contingency level [10], i.e. first order, second order etc. This can be modified by neglecting those contingencies which have a probability of occurrence less than a certain minimum value. An alternative method is to consider those outages which create severe conditions within the system [11]. The intention in all methods is to curtail the list of events that can occur in a practical system. A useful approach is to consider those outage conditions which result from independent events and have a probability exceeding some minimum value and, in addition, to consider those outage conditions resulting from outage dependence such as common mode or station related events again having the same probability constraint. At this stage only independent overlapping outages are considered, the problem of outage dependence being discussed later in this chapter.

6.5.2 Application

The state selection process is illustrated by considering first and second order generating unit and transmission line outages in the system shown in Fig. 6.2 and using Equations (6.8) and (6.9). The unavailability associated with a transmission

Table 6.12 State values of generating units

Generating station 1			*Rate*	*Generating station 2*			*Rate*
State	*Units down*	*Probability*	λ_+ λ_- *(occ/yr)*	*State*	*Units down*	*Probability*	λ_+ λ_- *(occ/yr)*
1	0	0.96059601	0 4	1	0	0.9025	0 6
2	1	0.03881196	99 3	2	1	0.0950	57 3
3	2	0.00058806	198 2	3	2	0.0025	114 0
4	3	0.00000396	297 1				
5	4	0.00000001	396 0				

line is normally much lower than that for a generating unit, and therefore a higher order contingency level should be used when generating units are considered. The state information for the generating units is shown in Table 6.12 and for the transmission lines in Tables 6.10 and 6.11. The combined generation and transmission states are shown in Table 6.13.

As in Table 6.10 it has been assumed that a loss of 30 MW will result in a load point failure due to the transmission loss added to the 110 MW load level. It can be seen from Table 6.13 that if the load level is less than the point at which there is a load loss when one unit at Generating Station 2 is unavailable, then the values of Q_S and F_S will change considerably. Under these conditions

$$Q_S = 0.00658129$$

$$F_S = \underline{1.1364886 \text{ f/yr}}$$

The values in Table 6.13 are again for a constant load level of 110 MW and therefore are annualized values. The load model can be incorporated in the analysis, however, by considering the probability that the load will exceed the capability of each state. The P_{lj} values in Table 6.13 will then be modified accordingly and the Q_S and F_S indices will be on a periodic or annual base. The difference between F_S in Table 6.11 and Table 6.13 is due to the fact that the frequency contribution due to generating unit transitions is omitted in Equation (6.7) but included in Equation (6.9). The difference would be much smaller if the generation reserve margin were increased.

The effect of transmission line overloading can be illustrated by assuming, as in Section 6.4, that overload occurs whenever lines 2 or 3 are unavailable. Under these conditions loads must be curtailed, causing increased load point failures. In this case

$$Q_S = 0.10495807$$

$$F_S = \underline{16.44407264 \text{ f/yr}}$$

Table 6.13 System state values

State B_i	Elements out	Probability $P(B_j)$	Frequency $F(B_j)$ (occ/yr)	P_{1j}	Failure Probability P_j	Failure Frequency F_j (occ/yr)
1	—	0.85692158	18.85227476	0		
2	G1	0.03462309	4.15477080	0		
3	G1, G1	0.00052449	0.11436062	1.0	0.00052449	0.11436062
4	G1, G2	0.00364454	0.63414996	1.0	0.00364454	0.63414996
5	G1, L1	0.00012648	0.15329376	0		
6	G1, L2	0.00015810	0.19145910	0		
7	G1, L3	0.00011857	0.11774001	0		
8	G2	0.09020227	6.85537252	1.0	0.09020227	6.85537252
9	G2, G2	0.00237374	0.30858620	1.0	0.00237374	0.30858620
10	G2, L1	0.00032951	0.38783327	1.0	0.00032951	0.38783327
11	G2, L2	0.00041188	0.48438029	1.0	0.00041188	0.48438029
12	G2, L3	0.00030891	0.29315559	1.0	0.00030891	0.29315559
13	L1	0.00313030	3.48402390	0		
14	L1, L2	0.00001430	0.03150290	1.0	0.00001430	0.03150290
15	L1, L3	0.00001072	0.02128992	1.0	0.00001072	0.02128992
16	L2	0.00391288	4.35112256	0		
17	L2, L3	0.00001340	0.02659900	1.0	0.00001340	0.02659900
18	L3	0.00293466	2.62652070	0		
				$Q_S =$	0.09783386	$F_S = 9.15723027$ f/yr

Overloading can be eliminated by curtailing or dropping some load to alleviate the situation. Use of this technique therefore requires a load flow technique which can accommodate it. Load reduction can also be used in the case of an outage condition in the generation configuration provided that the busbars at which load will be curtailed are prespecified. This is clearly not a problem in a single load example.

6.6 System and load point indices

6.6.1 Concepts

The system shown in Fig. 6.2 is a very simple configuration. In a more practical network there are a number of load points and each point has a distinct set of reliability indices [12]. The basic parameters are the probability and frequency of

Table 6.14 Annualized load point indices

Basic values
Probability of failure
Expected frequency of failure
Expected number of voltage violations
Expected number of load curtailments
Expected load curtailed
Expected energy not supplied
Expected duration of load curtailment
Maximum values
Maximum load curtailed
Maximum energy curtailed
Maximum duration of load curtailment
Average values
Average load curtailed
Average energy not supplied
Average duration of curtailment
Bus isolation values
Expected number of curtailments
Expected load curtailed
Expected energy not supplied
Expected duration of load curtailment

failure at the individual load points, but additional indices can be created from these generic values. The individual load point indices can also be aggregated to produce system indices which include, in addition to consideration of generation adequacy, recognition of the need to move the generated energy through the transmission network to the customer load points [12, 13]. Table 6.14 lists a selection of load point indices which can be used.

It is important to appreciate that, if these indices are calculated for a single load level and expressed on a base of one year, they should be designated as *annualized* values. Annualized indices calculated at the system peak load level are usually much higher than the actual annual indices. The indices listed in Table 6.14 can be calculated using the following equations:

$$\text{Probability of failure } Q_K = \sum P_j P_{Kj} \tag{6.11}$$

$$\text{Frequency of failure } F_K = \sum F_j P_{Kj} \tag{6.12}$$

where j is an outage condition in the network
P_j is the probability of existence of outage j
F_j is the frequency of occurrence of outage j
P_{Kj} is the probability of the load at bus K exceeding the maximum load that can be supplied at that bus during the outage j.

Equations (6.11) and (6.12) are the same as Equations (6.8) and (6.9) the notation has been modified slightly to facilitate the development of further equations.

$$\text{Expected number of voltage violations} = \sum_{j \in V} F_j \tag{6.13}$$

where $j \in V$ includes all contingencies which cause voltage violation at bus K.

$$\text{Expected number of load curtailments} = \sum_{j \in x,y} F_j \tag{6.14}$$

where $j \in x$ includes all contingencies resulting in line overloads which are alleviated by load curtailment at bus K.

$j \in y$ includes all contingencies which result in an isolation of bus K.

$$\text{Expected load curtailed} = \sum_{j \in x,y} L_{Kj} F_j \text{ MW} \tag{6.15}$$

where L_{Kj} is the load curtailment at bus K to alleviate line overloads arising due to the contingency j; or the load not supplied at an isolated bus K due to the contingency j.

$$\text{Expected energy not supplied} = \sum_{j \in x,y} L_{Kj} D_{Kj} F_j \text{ MWh} \tag{6.16}$$

$$= \sum_{j \in x,y} L_{Kj} P_j \times 8760 \text{ MWh} \tag{6.17}$$

where D_{Kj} is the duration in hours of the load curtailment arising due to the outage j; or the duration in hours of the load curtailment at an isolated bus K due to the outage j.

$$\text{Expected duration of load curtailment} = \sum_{j \in x,y} D_{Kj} F_j \text{ hours} \tag{6.18}$$

$$= \sum_{j \in x,y} P_j \times 8760 \text{ hours} \tag{6.19}$$

$$\text{Maximum load curtailed} = \text{Max } \{L_{K1}, L_{K2}, \ldots, L_{Kj}, \ldots\} \tag{6.20}$$

$$\text{Maximum energy curtailed} = \text{Max } \{L_{K1} D_{K1}, L_{K2} D_{K2}, \ldots, L_{Kj} D_{Kj}, \ldots\} \tag{6.21}$$

Maximum duration of load curtailment

$$= \text{Max } \{D_{K1}, D_{K2}, \ldots, D_{Kj}, \ldots\} \tag{6.22}$$

Additional information on the contingencies which cause the above maxima is also desirable in order to appreciate their severity.

$$\text{Average load curtailed} = \frac{\sum_{j \in x,y} L_{Kj} F_j}{\sum_{j \in x,y} F_j} \text{ MW/curtailment} \tag{6.23}$$

$$\text{Average energy not supplied} = \frac{\sum_{j \in x,y} L_{Kj} D_{Kj} F_j}{\sum_{j \in x,y} F_j} \text{ MWh/curtailment} \tag{6.24}$$

$$\text{Average duration of curtailment} = \frac{\sum_{j \in x,y} D_{Kj} F_j}{\sum_{j \in x,y} F_j} \text{ h/curtailment} \tag{6.25}$$

Indices due to the isolation of bus K

$$\text{Expected number of curtailments} = \sum_{j \in y} F_j \tag{6.26}$$

$$\text{Expected load curtailed} = \sum_{j \in y} L_{Kj} F_j \text{ MW} \tag{6.27}$$

$$\text{Expected energy not supplied} = \sum_{j \in y} L_{Kj} D_{Kj} F_j \tag{6.28}$$

$$= \sum_{j \in y} L_{Kj} P_j \times 8760 \text{ MWh} \tag{6.29}$$

$$\text{Expected duration of load curtailment} = \sum_{j \in y} D_{Kj} F_j \tag{6.30}$$

$$= \sum P_j \times 8760 \text{ hours} \tag{6.31}$$

6.6.2 Numerical evaluation

The load point indices can be calculated for the system shown in Fig. 6.2 by extending the results shown in Table 6.13. This calculation is shown in Table 6.15. A complete study would require an actual load flow to determine the line loadings and the line loss under each contingency. It has therefore been assumed that the power factor associated with flow on a line is 0.95 and that an arbitrary line loss of 5 MW is added to the actual demand at the load bus. The load point indices shown in Table 6.14 are listed below.

Table 6.15 Load point indices

State	Elements out	Probability p_j	Frequency F_j (occ/yr)	Capacity available (MW)	P_{Kj}	D_{Kj} (hours)	L_{Kj} (MW)	ELC (MW)	NLC	EENS (MWh)	EDLC (hours)
1	—	0.85692158	18.85227476	140	0	398.18	0	0	0	0	0
2	$G1$	0.03462309	4.15477080	120	0	73.00	0	0	0	0	0
3	$G1, G1$	0.00052449	0.11436062	100	1	40.18	15	1.7154	0.11436	68.92	4.5454
4	$G1, G2$	0.00364454	0.63414996	90	1	50.34	25	15.8537	0.63415	798.08	31.9262
5	$G1, L1$	0.00012648	0.15329376	120	0	7.23	0	0	0	0	0
6	$G1, L2$	0.00015810	0.19145910	86	1	7.23	29	5.5500	0.19142	40.13	1.3850
7	$G1, L3$	0.00011857	0.11774001	95	1	8.82	20	2.3548	0.11774	20.77	1.0387
8	$G2$	0.09020227	6.85537252	110	1	115.26	5	34.2769	6.85537	3980.76	790.1719
9	$G2, G2$	0.00237374	0.30858620	80	1	67.38	35	10.8005	0.30859	727.74	20.7940
10	$G2, L1$	0.00032951	0.38783327	110	1	7.44	5	1.9392	0.38783	14.43	2.8864
11	$G2, L2$	0.00041188	0.48438029	86	1	7.45	29	14.0470	0.48438	104.65	3.6081
12	$G2, L3$	0.00030891	0.29115559	95	1	9.23	20	5.8631	0.29316	54.12	2.7061
13	$L1$	0.00313030	3.48402390	140	0	7.87	0	0	0	0	0
14	$L1, L2$	0.00001430	0.03150290	60	1	3.84	55	1.7327	0.03150	6.90	0.1253
15	$L1, L3$	0.00001072	0.02128992	95	1	4.41	35	0.7451	0.02129	3.29	0.0939
16	$L2$	0.00391288	4.35112256	86	1	7.88	29	126.1025	4.35112	993.69	34.2768
17	$L2, L3$	0.00001340	0.02659900	0	1	4.41	110	2.9254	0.02660	12.90	0.1174
18	$L3$	0.00293466	2.62652070	95	1	9.79	20	52.5304	2.62652	514.27	25.7076
								274.441	16.444	7310.65	919.4327

$$D_{Kj} = \frac{P_j}{F_j} \times 8760$$

ELC = expected load curtailed
NLC = expected number of load curtailments
EENS = expected energy not supplied
EDLC = expected duration of load curtailment

Annualized load point indices at Bus 3

Basic values

Probability of failure = 0.10495807

Frequency of failure = 16.444 f/yr

Expected number of curtailments

Total = 16.444

Isolated = 0.0266

Expected load curtailed

Total = 274.44 MW

Isolated = 2.93 MW

Expected energy not supplied

Total = 7310.65 MWh

Isolated = 12.90 MWh

Expected duration of load curtailment

Total = 919.43 hours

Isolated = 0.12 hour

Maximum values

Bus No. 3

Maximum load curtailed = 110 MW

Condition = $L2$ and $L3$ out

Probability = 0.00001340

Maximum energy curtailed = 3980.76 MWh

Condition = G2 out

Probability = 0.09020227

Maximum duration of load curtailment = 790.17 hours

Condition = G2 out

Probability = 0.09020227

Average values

Average load curtailed $= \frac{274.44}{16.444}$

$= \underline{16.69}$ MW/curtailment

Average energy not supplied $= \frac{7310.65}{16.444}$

$= \underline{444.58}$ MWh/curtailment

Average duration of curtailment $= \frac{919.43}{16.444}$

$= \underline{55.91}$ hours/curtailment

The individual load point indices can be aggregated to produce a set of system indices which can provide an overall assessment of the system adequacy. The list of system indices is given in Table 6.16.

Table 6.16 Annualized system indices

Basic values
Bulk power interruption index
Bulk power supply average MW curtailment/disturbance
Bulk power energy curtailment index
Modified bulk power energy curtailment index
Average values
Average number of curtailments/load point
Average load curtailed/load point
Average energy curtailed/load point
Average duration of load curtailed/load point
Average number of voltage violations/load point
Maximum values
Maximum system load curtailed under any contingency condition

These indices can be calculated for the three bus system as follows:

$$\text{Bulk power interruption index} = \frac{\Sigma_K \Sigma_{j\in x,y}\, {}_{Kj} F_j}{L_S} \tag{6.32}$$

$$= \frac{274.44}{110} = \underline{2.4949\ \text{MW/MW-yr}}$$

where L_S is the total system load.

$$\begin{array}{l}\text{Bulk power supply average}\\ \text{MW curtailment/disturbance}\end{array} = \frac{\Sigma_K \Sigma_{j\in x,y} L_{Kj} F_j}{\Sigma_{j\in x,y} F_j} \tag{6.33}$$

$$= \frac{274.44}{16.444} = \underline{16.6894\ \text{MW/disturbance}}$$

$$\text{Bulk power energy curtailment index} = \frac{\Sigma_K \Sigma_{j\in x,y} 60\, L_{Kj} D_{Kj} F_j}{L_S} \tag{6.34}$$

$$= \frac{60 \times 7310.65}{110}$$

$$= \underline{3732.33\ \text{MW-min/MW-yr}}$$

$$= \underline{62.21\ \text{MWh/MW-yr}}$$

$$\begin{array}{l}\text{Modified bulk power}\\ \text{energy curtailment index}\end{array} = \frac{\Sigma_K \Sigma_{j\in x,y} L_{Kj} D_{Kj} F_j}{8760\, L_S} \tag{6.35}$$

$$= \frac{7310.65}{8760 \times 110} = \underline{0.00758681}$$

The bulk power energy curtailment index has also been designated as the severity index. The total unsupplied energy expressed in MW-minutes is divided by the peak system load in MW. Severity is therefore expressed in system minutes. One system minute is equivalent to one interruption of the total system load for one minute at the time of system peak. It does not represent a real system outage time because the interruption need not occur at the time of system peak load.

$$\text{Average number of curtailments/load point} = \sum_{K}\sum_{j \in x,y} F_j / C \tag{6.36}$$

$$= \frac{16.444}{1} = \underline{16.444}$$

where C is the number of load points, i.e. 1 in the present example.

$$\text{Average load curtailed/load point} = \sum_{K}\sum_{j \in x,y} L_{Kj} F_j / C \tag{6.37}$$

$$= \frac{274.44}{1} = \underline{274.44 \text{ MW/yr}}$$

$$\text{Average energy curtailed/load point} = \sum_{K}\sum_{j \in x,y} L_{Kj} D_{Kj} F_j / C \tag{6.38}$$

$$= \frac{7310.65}{1} = \underline{7310.65 \text{ MWh/yr}}$$

$$\text{Average duration of load curtailed/load point} = \sum_{K}\sum_{j \in x,y} D_{Kj} / C \tag{6.39}$$

$$= \frac{919.43}{1} = \underline{919.43 \text{ h/yr}}$$

$$\frac{\text{Average number of voltage violations}}{\text{load point per yr}} = \sum_{K}\sum_{j \in V} F_j / C \tag{6.40}$$

The average number of voltage violations/load point is determined in those studies in which a load flow is performed for each contingency and acceptable voltage levels are defined for each load bus.

Maximum system load curtailed under any contingency condition

$$= \text{Max}\left\{\sum L_{K1}, \sum L_{K2}, \ldots, L_{Kj}, \ldots\right\} \tag{6.41}$$

$= \underline{110 \text{ MW}}$ for $L2$, $L3$ out, Probability = 0.00001340

Maximum system energy not supplied under any contingency condition

$$= \text{Max}\left\{\sum_K L_{K1}D_{K1}, \sum_K L_{K2}D_{K2}, \ldots, \sum L_{Kj}\,D_{Kj}, \ldots\right\} \tag{6.42}$$

$= \underline{3980.76\text{ MWh}}$ for $G2$ out, Probability = 0.09020227

6.7 Application to practical systems

At the present time there is no consensus within the electric power industry on what constitutes a complete set of adequacy indices for composite system reliability evaluation. The selection depends on the use to be made of the indices. Tables 6.14 and 6.16 provide a comprehensive list which can be used in a wide range of applications. The two sets of load point and system indices do not replace each other, they complement each other. The load point values are extremely valuable in system design and in comparing alternative configurations and system additions. They are also useful as input indices in the reliability evaluation of the distribution system which is fed by the relevant bulk supply point. The overall system indices are useful to management and to the system planner as overall adequacy indices which indicate the ability of the system to satisfy its load and energy requirements. The system shown in Fig. 6.2 is a relatively simple system in which all the calculations can be done by hand without too much difficulty. This is not the case in larger systems. In these cases, it is necessary to develop a digital computer program to perform the necessary computations. The application of these concepts and Equations (6.11) to (6.42) to a multi-load point system is illustrated using the system shown in Fig. 6.4. The transmission line data are shown in Table 6.17.

The load variation curve is approximated by a straight line joining the 1.0 peak level at zero probability value and the 0.4 peak level at unit probability value.

Load probability steps,

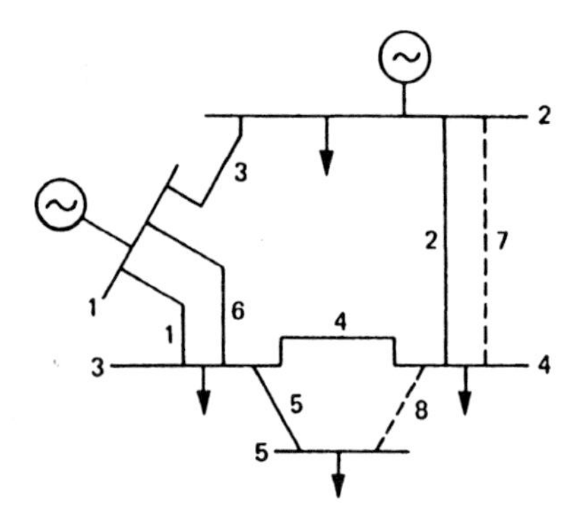

Fig. 6.4 Five-bus composition system: — Base case; - - - Added branches

Table 6.17 Transmission line data

Line	*Length (miles)*	*Impedance (p.u.)*	*Susceptance (b/2)*	*Failure rate (f/yr)*	*Probability of failure*
1, 6	30	$0.0342 + j0.1800$	0.0106	1.5	0.001713
2, 7	100	$0.1140 + j0.6000$	0.0352	5.0	0.005710
3	80	$0.0912 + j0.4800$	0.0282	4.0	0.004568
4, 5, 8	20	$0.0228 + j0.1200$	0.0071	1.0	0.001142

Line data

Lines are assumed to be 795 ACSR 54/7.
Current carrying capability = 374 A = 0.71 p.u.
Failure rate = 0.05 failures/year/mile.
Expected repair duration = 10 hours.

1.00 0.90 0.80 0.70 0.60 0.50
0.40 0.30 0.20 0.10 0.0 (10 steps)

The generation and load data are shown in Table 6.18.

The probability and frequency of failure at each bus for a variety of cases are shown in Table 6.19. These values are annualized results for a peak load of 155 MW. Case 1 is the base case and all further cases are variations from this base. It

Table 6.18 Generation and load data

Bus No.	*No. of units*	*Capacity of each unit*	*Total bus capacity (MW)*	*Type of units*	*Failure rate/unit (f/yr)*	*Repair rate/unit (r/yr)*	*Probability of outage*
1	4	20	80	Thermal	1.1	73	0.015
2	7	5	130	Hydro	0.5	100	0.005
	1	15		Hydro	0.5	100	0.005
	4	20		Hydro	0.5	100	0.005

Swing bus: 1
(If bus 1 is isolated from the network due to an outage condition, bus 2 is selected as the swing bus.)

Bus	*Peak load (MW)*	*Power factor*	*Generation allotted under peak load (MW)*	*VAR limits (MVAR)*	*Voltage limits (p.u.) Max.*	*Voltage limits (p.u.) Min.*
1	0	—	Swing bus	−20 to +20	1.05	0.97
2	20	1.0	110	−10 to 40	1.05	0.97
3	85	1.0	—	—	1.05	0.97
4	40	1.0	—	—	1.05	0.97
5	10	1.0	—	—	1.05	0.97
	155					

Base MVA = 100, Base kV= 110

Table 6.19 Load point failure probability and frequency

Case	Description	Bus 2		Bus 3		Bus 4		Bus 5	
		Probability	*Frequency (f/yr)*	*Probability*	*Frequency (f/yr)*	*Probability*	*Frequency (f/yr)*	*Probability*	*Frequency (f/yr)*
1	Base case	0.00000255	0.00453407	0.00898056	8.19758380	0.00562033	5.12946780	0.00671280	6.10382120
2	V_{min} = 1.0 p.u.	0.00000255	0.00453407	0.00899418	8.22173140	0.00674003	6.15209210	0.00672642	6.12796880
3	V_{min} = 0.95 p.u.	0.00000255	0.00453407	0.00898056	8.19758380	0.00003092	0.05481900	0.00113700	1.05332000
4	Single outages	0.00000000	0.00000000	0.00796343	7.15040730	0.00497714	4.46247200	0.00597257	5.35894810
5	Line 7 added	0.00000254	0.00452107	0.00007250	0.12877826	0.00003075	0.05455766	0.00116441	1.11306940
6	Lines 7&8 added	0.00000000	0.00000000	0.00006452	0.11465395	0.00003071	0.05452625	0.00003184	0.05653558
7	Load model	0.00000070	0.00125053	0.00142480	1.31727910	0.00165222	1.52729250	0.00255386	2.35650470
8	Common cause on Lines 1&6	0.00005887	0.03943445	0.00916632	8.35201140	0.00567706	5.16676540	0.00676832	6.13931290

can be seen that Bus 2 is both a generation and a load bus and this proves to be the most reliable load bus in the network. Bus 3 is connected directly to Bus 1 by two transmission lines on separate circuits. Bus 4 is supplied by two lines while Bus 5 is supplied by a single circuit. The indices at Bus 2 are insensitive to the ability to relax the acceptable voltage at a load bus, as Bus 2 is also a generating bus. Similarly those at Bus 3, which is strongly connected to a generating bus, are virtually insensitive. On the other hand, Bus 4 and Bus 5 are extremely sensitive to the value of acceptable voltage at the load point. As in the previous example, up to two independent outages are considered in the base case and in cases 2 and 3. Case 4 illustrates the indices when only single outages are considered. The results in this case are not very different from the base case due to the configuration and voltage conditions in the system. This may not be true in other cases and even for this system at lower load levels. The addition to the system of line 7 (case 5) makes a considerable difference to the ability in maintaining acceptable voltage and supplying the load at Bus 3, 4 and 5. When both lines 7 and 8 are added to the base case (case 6), the load point indices are improved at each bus and particularly those at Bus 5 which now has a two-line supply and acceptable voltage support.

Case 7 gives the load point indices when the loads are represented by a straight line for the annual period extending between the 100% to the 40% load levels. The load-carrying capability was examined using 10 equal steps as defined in the above data. The indices in this case should be referred to as *annual* indices and not annualized indices as they represent the entire annual period.

Case 8 shows the effect on the system of having lines 1 and 6 on a common tower configuration where they are susceptible to a common cause outage. It was assumed that the common cause outage rate is $\lambda_c = 0.15$ f/yr. It can be seen that this change in system design has a significant impact on the bus indices. This is, of course, a very small system and the effects will vary from system to system. Table 6.20 shows a complete set of load point results for the base case which includes all the indices listed in Table 6.14. Table 6.21 shows a complete set of system results for the base case which includes all the indices listed in Table 6.16.

The impact on the system indices for three of the cases shown in Table 6.19 is shown in Table 6.22. The inclusion of a common-cause outage condition [14] on two of the major circuits is clearly seen to have an incremental effect on all of the overall indices. On the other hand the addition of two extra lines (7 and 8) to the system improves considerably the overall system indices. This effect is quite dramatic in such a small system.

The results given in this section were obtained using a specific computer program. They have been included in order to demonstrate the application of these techniques to a practical system and to illustrate the type of results that can be achieved. The cases considered illustrate the need for sensitivity studies during system design. The use of other programs with the same system data may produce somewhat different results depending upon the assumptions and approximations inherent in the program.

Table 6.20 Annualized load point indices—base case

Bus	Failure probability	Failure frequency (f/yr)	No. of curtailments Total	No. of curtailments Isolated	Load curtailed (MW) Total	Load curtailed (MW) Isolated	Energy curtailed (MWh) Total	Energy curtailed (MWh) Isolated	Duration (hours) Total	Duration (hours) Isolated	Voltage violations
2	0.00000255	0.00453407	0.00	0.00	0.01	0.00	0.07	0.00	0.02	0.00	0.00
3	0.00898056	8.19758380	3.14	0.00	49.14	0.00	448.72	0.00	29.78	0.00	5.07
4	0.00562033	5.12946780	0.04	0.01	0.60	0.40	2.95	1.99	0.22	0.05	5.10
5	0.00671280	6.10382120	1.05	1.04	10.43	10.41	99.08	98.98	9.96	9.90	6.10

(A) Maximum load curtailed (MLC)

Bus	MLC (MW)	Outage condition	Probability
2	3.2258	L1, L6, out	0.00000260
3	85.0000	L1, L2, out	0.00000850
4	40.0000	L2, L4, out	0.00000570
5	10.0000	L5, out	0.00111520

(B) Maximum energy curtailed (MEC)

Bus	MEC (MWh)	Outage condition	Probability
2	15.9130	L1, L6, out	0.00000260
3	420.1354	L1, L2, out	0.00000850
4	197.6550	L2, L4, out	0.00000570
5	96.2859	L5, out	0.0011520

(C) Maximum duration of load curtailment (MDLC)

Bus	MDLC (hours)	Outage condition	Probability
2	4.9330	L1, L6, out	0.00000260
3	9.6339	L1, out	0.00167270
4	4.9428	L1, L2, out	0.00000850
5	9.6286	L5, out	0.00111520

(D) Average bus indices

Bus	Load curtailed (MW)	Energy not supplied (MWh)	Duration of curtailment (hours)
2	3.226	15.913	4.933
3	15.664	143.039	9.132
4	13.331	65.874	4.941
5	9.900	94.064	9.502

Table 6.21 Annualized system indices—base case

Basic values	
Bulk power interruption index	= 0.38824 MW/MW-yr
Bulk power energy curtailment index	= 3.55366 MWh/yr
Bulk power supply average MW curtailment index	= 14.39039 MW/disturbance
Modified bulk power energy curtailment index	= 0.00040567
Severity index	= 213.219 system minutes
Average values	
Av. no. of load curtailment/load pt./year	= 1.05991
Av. no. of voltage violations/load pt./year	= 4.06504
Av. load curtailed/load pt./year	= 15.04432 MW
Av. hours of load curtailment/load pt./year	= 9.99494 hours
Av. energy curtailed/load pt./year	= 137.70422 MWh

Maximum values

Max. load curtailed	*Outage condition*	*Probability*
90.48 MW	*L*2, *L*6 out	0.00000851

Max. energy curtailed	*Outage condition*	*Probability*
447.21 MWh	*L*2, *L*6 out	0.00000851

Table 6.22 Comparison of basic annualized system indices

	Base case *Case 1—Table 6.19*	*Base case with common cause outages* *Case 8—Table 6.19*	*Base case with lines 7&8 added* *Case 6—Table 6.19*
Bulk power interruption index (MW/MW-yr)	0.38824	0.40596	0.00627
Bulk power energy curtailment index (MWh/yr)	3.55366	3.74458	0.03085
Bulk power supply average MW curtailment index (MW/disturbance)	14.39039	14.51102	14.5687
Modified bulk power energy curtailment index	0.00040567	0.00042746	0.00000352
Severity index (system-minutes)	213.219	224.675	1.851

6.8 Data requirements for composite system reliability evaluation

6.8.1 Concepts

The evaluation of a composite system including both generation and bulk transmission is a very complex problem. The data required to analyze this problem can be divided into two basic parts as shown in Fig. 6.5. In a simplistic sense, these two requirements can be considered as deterministic data and stochastic data.

6.8.2 Deterministic data

This data is required at both the system and at the actual component level. The component data includes known parameters such as line impedances and susceptances, current-carrying capacities, generating unit parameters and other similar factors normally utilized in conventional load flow studies. This is not normally difficult to determine as this data is used in a range of studies. The system data, however, is more difficult to appreciate and to include and should take into account the response of the system under certain outage conditions. An example of this would occur if one of the lines between Buses 1 and 3 in Fig. 6.4 suffered an outage; would the loading on the remaining line be such that it would be removed from service, would it carry the overload, or would some remedial action be taken in the system in order to maintain overall system integrity? This data is extremely important in a composite reliability study. The computer model must behave in the same way as the actual system or the results are not appropriate. This is an important aspect in composite system reliability evaluation and is a problem that has not been properly recognized up to this time.

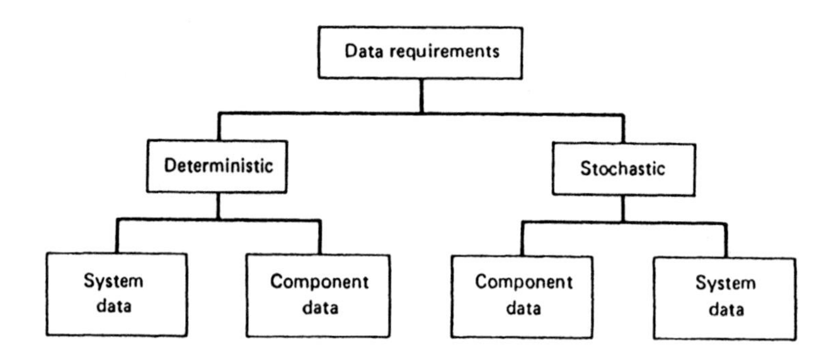

Fig. 6.5 Data requirements

6.8.3 Stochastic data

The stochastic data can again be divided into two parts: component and system data. The component requirements pertain to the failure and repair parameters of the individual elements within the system. This data is generally available. There is also a need to consider and to include system events which involve two or more components. This type of data is system specific and will usually have to be inserted as a second and third level of data input in an overall composite system reliability analysis. System data includes relevant multiple failures resulting from common transmission line configurations or station-induced effects.

The different types of outages can be categorized as follows

(a) independent outages;
(b) dependent outages;
(c) common cause or common mode outages;
(d) station originated outages.

6.8.4 Independent outages

Independent outages are the easiest to deal with and involve two or more elements. They are referred to as overlapping or simultaneous independent outages. The probability of such an outage is the product of the failure probabilities for each of the elements. The basic component model used in these applications is the simple two-state representation in which the component is either up or down.

The rate of departure from a component up-state to its down-state is designated as the failure rate λ. The restoration process from the down-state to the up-state is somewhat more complicated and is normally designated by the repair rate μ. The restoration of a forced outage can take place in several distinct ways which can result in a considerable difference in the probability of finding the component in the down-state (usually designated as the unavailability). Some of these restoration processes are

(a) high speed automatically re-closed;
(b) slow speed automatically re-closed;
(c) without repair;
(d) with repair.

These processes involve different values of outage times and therefore different repair rates. The state space diagram for a two-element configuration considering independent failures is shown in Fig. 6.6. In addition to forced outages, the component may also be removed from service for a scheduled outage. The scheduled outage rate, however, must not be added directly to the failure rate as scheduled outages are not random events. For instance, the component is not normally removed for maintenance if the actual removal results in customer interruption. These aspects and features are discussed in more detail in Chapter 8 in connection with the reliability evaluation of transmission and distribution networks.

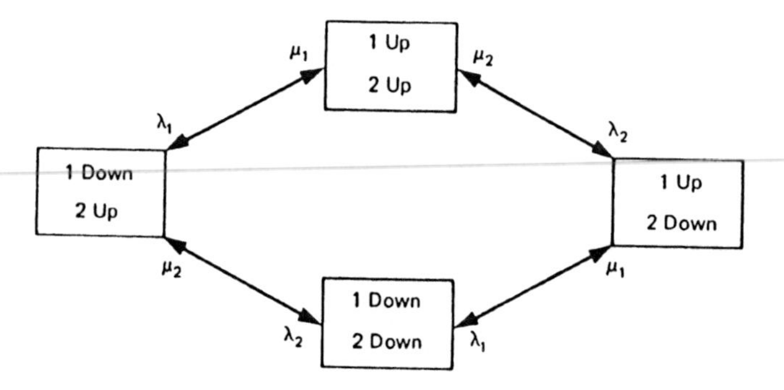

Fig. 6.6 Basic simultaneous independent failure model

Most of the presently published techniques for composite system reliability evaluation assume that the outages constituting a contingency situation are independent.

6.8.5 Dependent outages

As the name implies, these outages are dependent on the occurrence of one or more other outages. An example is an independent outage of one line of a double circuit followed by the removal of the second line due to overload. These outages are not normally included in the reliability evaluation of composite systems and require appreciation of system data in addition to individual component data.

6.8.6 Common mode outages

As stated in Section 6.8.4, the probability of occurrence of an event consisting of two or more simultaneous independent outages is the product of the individual outage probabilities. If the probabilities of the individual outages are low, the product can become extremely small. The probability of a common cause outage resulting in a similar contingency event can however be many times larger. The effect of these outages on bus reliability indices can therefore be significant compared with the effect of second and higher-order independent outages.

A common cause outage is an event having an external cause with multiple failure effects where the effects are not consequences of each other. The most obvious example of a common cause outage is the failure of a transmission tower supporting two or more transmission circuits [14]. This event can be compared with the outcome for a similar configuration but one in which the two circuits are on separate tower structures and possibly separated physically by quite a large margin.

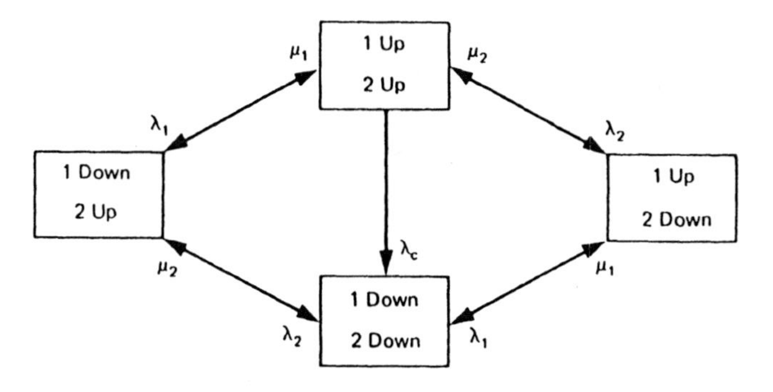

Fig. 6.7 IEEE model—a common mode outage model

The Task Force on Common Mode Outages of Bulk Power Supply Facilities in the IEEE Subcommittee on the Application of Probability Methods suggested a common mode outage model for two transmission lines on the same right of way or on the same transmission tower. This model, shown in Fig. 6.7, is basically similar to that shown in Fig. 6.6, except for the direct transition rate λ_c from state 1 to state 4. This model assumes that the same repair process applies for all failures including common cause failures. Various other possible common cause outage models have been described and analyzed [14] and these are described in more detail in Chapter 8.

6.8.7 Station originated outages [15]

The outage of two or more transmission elements not necessarily on the same right of way and/or generating units can arise due to station originated causes. Station originated outages can occur due to a ground fault on a breaker, a stuck breaker condition, a bus fault, etc., or a combination of these outages. Such outages have been previously accounted for in the line and/or generator outage rates by combining these outages with independent outage rates. Such a treatment cannot, however, recognize a situation in which more than one element of the system is simultaneously out because of a single event in the terminal station. It is, therefore, necessary to consider [15, 17] these outages as separate events. The effect of station originated outages in composite system reliability has not been extensively analyzed and can have an appreciable effect on load point reliability indices. The impact of these can be clearly seen in the station shown in Fig. 6.8.

It can be seen from the configuration shown in Fig. 6.8 that a ground fault on breaker 901 would open breakers 902 and 907 and hence isolate four generating units from the system. This type of event is not normally included in either generating capacity or composite system reliability studies. The duration of the outage in this case, however, might not be associated with the repair of breaker 901

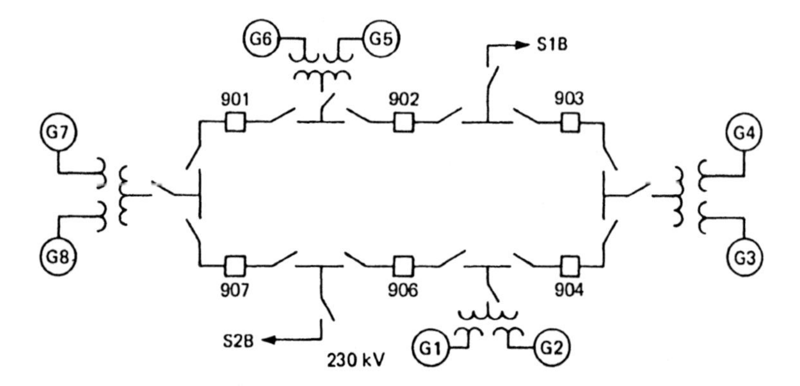

Fig. 6.8 Squaw Rapids Generating Station configuration in the Saskatchewan Power Corporation System

but simply with the switching action required to remove the breaker from the system and restore the four units to system service. It becomes important therefore to recognize that restoration in the case of terminal station faults may not involve repair directly but may be by switching action, and therefore a different model is required. Figure 6.9 shows a possible model which includes both common mode and station related events.

The state transition diagram shown in Fig. 6.9 illustrates two possible common mode failure events characterized by λ_{c1} and λ_{c2}. These events are physically different. In the first case, repair follows the same process as the independent event, while in the second case repair is by a common mode repair process. Either one or both may exist in a particular situation.

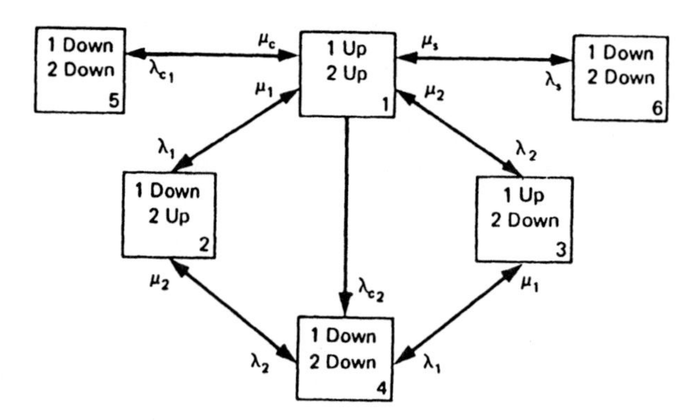

Fig. 6.9 General model for common cause, independent and station associated events

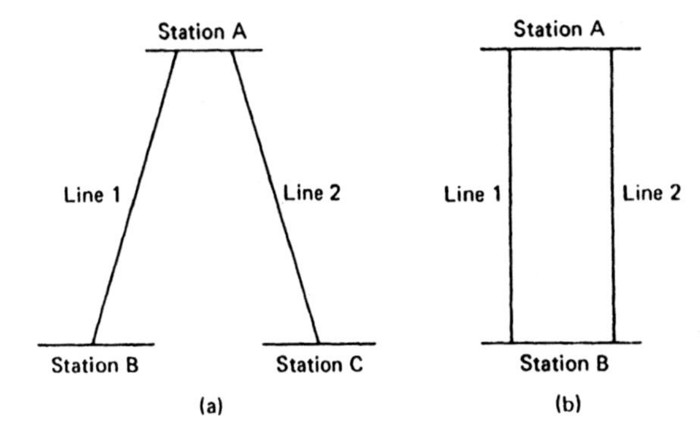

Fig. 6.10 Two-line configurations

It is important to appreciate the different impact that common mode and station originated events can have on the system transmission components. Figure 6.10 shows two line configurations. Lines 1 and 2 in Fig. 6.10(a) start at station A and terminate at two different stations B and C. These two lines may be removed from service by two overlapping independent failures or by a single element failure in station A. The two lines in Fig. 6.10(b) both terminate at station B. In this case one additional outage cause can result; the two lines may be removed from service by a common mode failure if the lines are on a common tower structure or a common right of way. All these factors must be included for a comprehensive analysis of a composite system and it appears that the most suitable way is to input them as separate levels of component and system data.

The stochastic data requirements for composite system reliability therefore include both individual component parameters and higher levels of data which involve more than one component and may be system specific.

It appears that there is a relatively large amount of data related to individual component outages. Many companies have set up or are in the process of setting up comprehensive outage data collection procedures which should provide individual component outage data with an acceptable level of confidence. This is not the case in regard to common mode, dependent and station associated failures. Increased awareness, however, of the necessity for such data by virtue of the need to conduct overall system adequacy studies should lead to better and more enhanced data collection.

6.9 Conclusions

Composite system reliability evaluation in a practical configuration is a complex problem [16]. It involves a physical appreciation of how the actual system would perform under outage conditions and also a realistic awareness of what the calculated indices represent and do not include. The domain within which the application and therefore indices lie is very clearly adequacy evaluation. It is important therefore that the evaluated indices be interpreted strictly from that point of view and not be given a physical significance which does not exist. The individual load point indices and the system indices are both valuable. They do not replace each other; they complement each other and serve two entirely different functions.

This area of power system reliability evaluation is probably the least developed and also one of the most complicated. In view of the environmental, ecological, societal and economic constraints faced by most electric power utilities it is expected that this area will receive considerable attention in the next decade.

6.10 Problems

1 Consider Problem 7 of Chapter 2. Calculate the annual loss of load expectation for this problem using the conditional probability approach. Use Equation (6.6).

2 Consider the system shown in Fig. 6.11.

System data

Generating stations

1. 4×25 MW units $\lambda = 2.0$ f/yr
   $\mu = 98.0$ r/yr
2. 2×40 MW units $\lambda = 3.0$ f/yr
   $\mu = 57.0$ r/yr

Loads

A – 80 MW
B – 60 MW

Transmission lines

1. $\lambda = 4$f/yr, r = 8 hrs
   Load Carrying Capability (LCC) = 80 MW
2. $\lambda = 5$ f/yr, r = 8 hrs, LCC = 60 MW
3. $\lambda = 3$ f/yr, r = 12 hrs, LCC = 50 MW

Calculate an appropriate set of indices at load points A and B and for the system. Assume that the loads are constant at the values shown, for a one-year period and that the transmission loss is zero. In calculating the composite system adequacy indices, consider up to two simultaneous outages and assume that all load deficiencies are shared equally whenever possible. The reader should investigate the effect of including higher order simultaneous outages in the generating system.

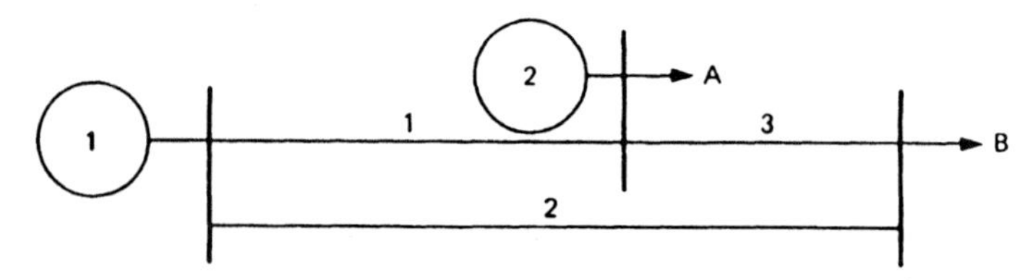

Fig. 6.11

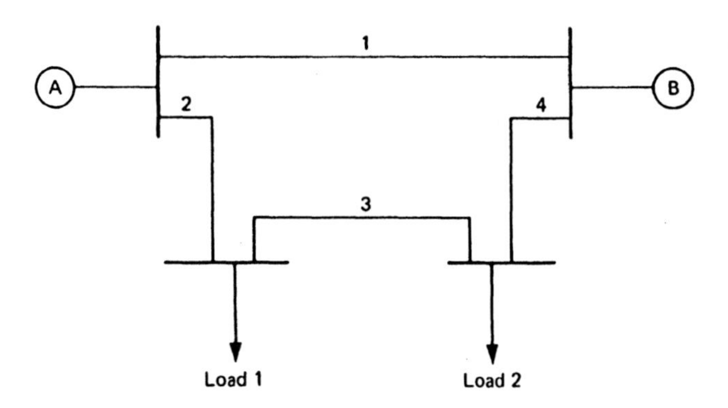

Fig. 6.12

3 Consider the system shown in Fig. 6.12.

Generation data

Plant	*No. of units*	*Unit capacity (MW)*	λ *(f/yr)*	μ *(r/yr)*
A	3	25	2	98
B	2	30	4	46

Transmission line data

Line	λ *(f/yr)*	*r (hours)*	*Load carrying capability (MW)*
1	2	12	50
2	5	15	100
3	4	12	50
4	3	15	90

Load data

Load point	*Load (MW)*
1	60
2	40

Assume the load to be constant over the year.

(a) Calculate an appropriate set of indices for the system considering only the generation and load data.

(b) Calculate the probability and frequency of load point failure using the approach shown in Tables 6.10 and 6.11 respectively.

(c) Calculate the probability and frequency of load point failure as shown in Table 6.13.

(d) Calculate a complete set of annualized load point and system indices for this configuration using the results obtained in part (c).

(e) Repeat parts (c) and (d) for the case in which an identical line is placed in parallel with line 2.

(f) Repeat part (e) if the two parallel lines have a common mode failure rate of 0.5 f/yr.

In calculating the composite system adequacy indices in parts (c), (d) and (e), consider up to two simultaneous outages and assume that all load deficiencies are shared equally whenever possible.

The reader should investigate the effect in parts (c), (d), (e) and (f) of including higher order simultaneous outage probabilities in the generating system.

6.11 References

1. Billinton, R., *Power System Reliability Evaluation*, Gordon and Breach, New York (1970).
2. Billinton, R., 'Bibliography on the application of probability methods in power system reliability evaluation', *IEEE Trans. on Power Apparatus and Systems*, **PAS-91**, No. 2 (March/April 1972), pp. 649–60.
3. IEEE Subcommittee on the Application of Probability Methods, Power System Engineering Committee, *Bibliography on the Application of Probability Methods in Power System Reliability Evaluation: 1971–1977*, Paper No. F78 073–9 presented at the IEEE PES Winter Power Meeting, New York, January 1978.
4. Billinton, R., Ringlee, R. J., Wood, A. J., *Power System Reliability Calculations*, M.I.T. Press, Mass., USA (1973).
5. Billinton, R. 'Composite system reliability evaluation', *IEEE Trans. on Power Apparatus and Systems*, **PAS-88** (1969), pp. 276–81.
6. Billinton, R., Bhavaraju, M. P., 'Transmission planning using a reliability criterion—Pt. I—A reliability criterion', *IEEE Trans. on Power Apparatus and Systems*, **PAS-89** (1970), pp. 28–34.
7. Allan, R. N., Takieddine, F. N., *Network Limitations on Generating Systems Reliability Evaluation Techniques*, Paper No. A 78 070–5 presented at the IEEE PES Winter Power Meeting, New York, January 1978.
8. Billinton, R., Medicherla, T. K. P., Sachdev, M. S., *Composite Generation and Transmission System Reliability Evaluation*, Paper No. A 78 237-0 presented at the IEEE PES Winter Power Meeting, New York, January 1978.
9. Billinton, R., Medicherla, T. K. P., 'Overall approach to the reliability evaluation of composite generation and transmission systems', *IEE Proc.*, **127**, Pt. C, No. 2 (March 1980), pp. 72–81.
10. Marks, G. E., *A Method of Combining High-Speed Contingency Load Flow Analysis with Stochastic Probability Methods to Calculate a Quantitative

Measure of Overall Power System Reliability, Paper No. A 78 053–1 presented at the IEEE PES Winter Power Meeting, New York, January 1978.

11. Dandeno, P. L., Jorgensen, G. E., Puntel, W. R., Ringlee, R. J., *A Program for Composite Bulk Power Electric System Adequacy Evaluation*, A paper presented at the IEE Conference on Reliability of Power Supply Systems, February 1977, IEE Conference Publication No. 148.
12. Billinton, R., Medicherla, T. K. P., Sachdev, M. S., *Adequacy Indices for Composite Generation and Transmission System Reliability Evaluation*, Paper No. A 79 024–1, presented at the IEEE PES Winter Power Meeting, New York, February 1979.
13. Working Group on Performance Records for Optimizing System Design of the Power System Engineering Committee, IEEE Power Engineering Society, 'Reliability indices for use in bulk power supply adequacy evaluation', *IEEE Trans. on Power Apparatus and Systems*, **PAS-97** (July/August 1978), pp. 1097–1103.
14. Billinton, R., Medicherla, T. K. P., Sachdev, M. S., *Application of Common-cause Outage Models in Composite Systems Reliability Evaluation*, Paper No A 79 461–5 presented at the IEEE PES Summer Power Meeting, Vancouver, July 1979.
15. Billinton, R., Medicherla, T. K. P., 'Station originated multiple outages in the reliability analysis of a composite generation and transmission system', *IEEE Trans. on Power Apparatus and Systems*, **PAS-100** (1981), pp. 3869–79.
16. Endrenyi, J., Albrecht, P. F., Billinton, R., Marks, G. E., Reppen, N. P., Salvaderi, L., *Bulk Power System Reliability Assessment—Why and How?* (Part I Why, Part II How), IEEE Winter Power Meeting, 1982.
17. Allan, R. N., Adraktas, A. N., *Terminal Effects and Protection System Failures in Composite System Reliability Evaluation*, Paper No 82 SM 428–1, presented at the IEEE Summer Power Meeting, 1982.

7 Distribution systems—basic techniques and radial networks

7.1 Introduction

Over the past few decades distribution systems have received considerably less of the attention devoted to reliability modelling and evaluation than have generating systems. The main reasons for this are that generating stations are individually very capital intensive and that generation inadequacy can have widespread catastrophic consequences for both society and its environment. Consequently great emphasis has been placed on ensuring the adequacy and meeting the needs of this part of a power system.

A distribution system, however, is relatively cheap and outages have a very localized effect. Therefore less effort has been devoted to quantitative assessment of the adequacy of various alternative designs and reinforcements. On the other hand, analysis of the customer failure statistics of most utilities shows that the distribution system makes the greatest individual contribution to the unavailability of supply to a customer. This is illustrated by the statistics [1] shown in Table 7.1, which relate to a particular distribution utility in the UK. Statistics such as these reinforce the need to be concerned with the reliability evaluation of distribution systems, to evaluate quantitatively the merits of various reinforcement schemes available to the planner and to ensure that the limited capital resources are used to achieve the greatest possible incremental reliability and improvement in the system.

Several other aspects must also be considered in the need to evaluate the reliability of distribution systems. Firstly, although a given reinforcement scheme may be relatively inexpensive, large sums of money are expended collectively on such systems. Secondly, it is necessary to ensure a reasonable balance in the reliability of the various constituent parts of a power system, i.e. generation, transmission, distribution. Thirdly, a number of alternatives are available to the distribution engineer in order to achieve acceptable customer reliability, including alternative reinforcement schemes, allocation of spares, improvements in maintenance policy, alternative operating policies. It is not possible to compare quantitatively the merits of such alternatives nor to compare their effect per monetary unit expended without utilizing quantitative reliability evaluation.

These problems are now fully recognized and an increasing number of utilities [2, 3] throughout the world are introducing and routinely using quantitative

Table 7.1 Typical customer unavailability statistics [1]

Contributor	*Average unavailability per customer year* (minutes)	(%)
Generation/transmission	0.5	0.5
132 kV	2.3	2.4
66 kV and 33 kV	8.0	8.3
11 kV and 6.6 kV	58.8	60.7
Low voltage	11.5	11.9
Arranged shutdowns	15.7	16.2
Total	96.8 minutes	100.0

reliability techniques. Simultaneously, additional evaluation techniques are continuously being developed and enhanced, as shown by the rapidly growing number of papers being published [4, 5] in this area.

It is not easy to identify the year in which interest developed in quantitative reliability evaluation of distribution systems because the techniques used initially were based with little or no modification on the classical methods of series and parallel systems. The greatest impetus, however, was made in 1964–65, when a set of papers [6, 7] was published which proposed a technique based on approximate equations for evaluating the rate and duration of outages. This technique has formed the basis and starting point of most of the later and more modern developments.

Since these initial developments, many papers have been published which have considerably enhanced the basic techniques and which permit very realistic and detailed modelling of power system networks. The available papers are too numerous to identify individually and the two bibliographies [4, 5] should be studied to ascertain this information. together with the references given in Chapters 8–10.

The techniques required to analyze a distribution system depend on the type of system being considered and the depth of analysis needed. This chapter is concerned with the basic evaluation techniques. These are completely satisfactory for the analysis of simple radial systems. Chapters 8 and 9 extend these basic techniques to the evaluation of parallel and meshed systems and to the inclusion of more refined modelling aspects.

7.2 Evaluation techniques

A radial distribution system consists of a set of series components, including lines, cables, disconnects (or isolators), busbars, etc. A customer connected to any load point of such a system requires all components between himself and the supply

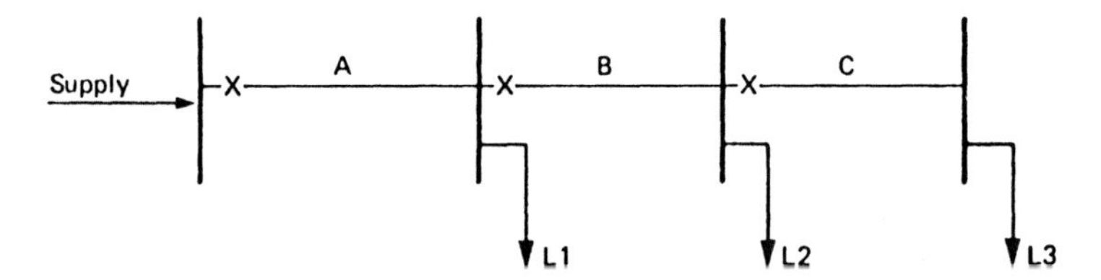

Fig. 7.1 Simple 3-load point radial system

point to be operating. Consequently the principle of series systems discussed in Section 11.2 of *Engineering Systems* can be applied directly to these systems: it was shown that the three basic reliability parameters of average failure rate, λ_s, average outage time, r_s, and average annual outage time, U_s, are given by

$$\lambda_s = \sum_i \lambda_i \tag{7.1}$$

$$U_s = \sum_i \lambda_i r_i \tag{7.2}$$

$$r_s = \frac{U_s}{\lambda_s} = \frac{\Sigma_i \lambda_i r_i}{\Sigma_i \lambda_i} \tag{7.3}$$

Consider the simple radial system shown in Fig. 7.1. The assumed failure rates and repair times of each line A, B and C are shown in Table 7.2 and the load-point reliability indices are shown in Table 7.3.

This numerical example illustrates the typical and generally accepted feature of a radial system—that the customers connected to the system farthest from the supply point tend to suffer the greatest number of outages and the greatest unavailability. This is not a universal feature, however, as will be demonstrated in later sections of this chapter.

The results for this example were evaluated using the basic concepts of network reliability described in Chapter 11 of *Engineering Systems* and Equations (7.1)–(7.3). This assumes that the failure of line elements A, B and C are simple

Table 7.2 Component data for the system of Fig. 7.1

Line	*λ (f/yr)*	*r (hours)*
A	0.20	6.0
B	0.10	5.0
C	0.15	8.0

Table 7.3 Load-point reliability indices for the system of Fig. 7.1

Load point	λ_L *(f/yr)*	r_L *(hours)*	U_L *(hours/yr)*
L1	0.20	6.0	1.2
L2	0.30	5.7	1.7
L3	0.45	6.4	2.9

open circuits with no compound effects, i.e. the failure of line element C does not effect L1 or L2. This is the same as assuming perfect isolation of faults on line elements A, B and C by the breakers shown in Fig. 7.1. These aspects are discussed in depth in Section 7.4.

7.3 Additional interruption indices

7.3.1 Concepts

The reliability indices that have been evaluated using classical concepts are the three primary ones of average failure rate, average outage duration and average annual unavailability or average annual outage time. These indices will be generally referred to in this book only as failure rate, outage duration and annual outage time. It should be noted, however, that they are not deterministic values but are the expected or average values of an underlying probability distribution and hence only represent the long-run average values. Similarly the word 'average' or 'expected' will be generally omitted from all other indices to be described, but again it should be noted that this adjective is always implicit in the use of these terms.

Although the three primary indices are fundamentally important, they do not always give a complete representation of the system behavior and response. For instance, the same indices would be evaluated irrespective of whether one customer or 100 customers were connected to the load point or whether the average load at a load point was 10 kW or 100 MW. In order to reflect the severity or significance of a system outage, additional reliability indices can be and frequently are evaluated. The additional indices that are most commonly used are defined in the following sections.

7.3.2 Customer-orientated indices

(i) *System average interruption frequency index, SAIFI*

$$\text{SAIFI} = \frac{\text{total number of customer interruptions}}{\text{total number of customers served}} = \frac{\Sigma\, \lambda_i N_i}{\Sigma\, N_i} \tag{7.4}$$

where λ_i is the failure rate and N_i is the number of customers of load point i.

(ii) *Customer average interruption frequency index, CAIFI*

$$\text{CAIFI} = \frac{\text{total number of customer interruptions}}{\text{total number of customers affected}} \tag{7.5}$$

This index differs from SAIFI only in the value of the denominator. It is particularly useful when a given calendar year is compared with other calendar years since, in any given calendar year, not all customers will be affected and many will experience complete continuity of supply. The value of CAIFI therefore is very useful in recognizing chronological trends in the reliability of a particular distribution system.

In the application of this index, the customers affected should be counted only once, regardless of the number of interruptions they may have experienced in the year.

(iii) *System average interruption duration index, SAIDI*

$$\text{SAIDI} = \frac{\text{sum of customer interruption durations}}{\text{total number of customers}} = \frac{\Sigma\, U_i N_i}{\Sigma\, N_i} \tag{7.6}$$

where U_i is the annual outage time and N_i is the number of customers of load point i.

(iv) *Customer average interruption duration index, CAIDI*

$$\text{CAIDI} = \frac{\text{sum of customer interruption durations}}{\text{total number of customer interruptions}} = \frac{\Sigma\, U_i N_i}{\Sigma\, \lambda_i N_i} \tag{7.7}$$

where λ_i is the failure rate, U_i is the annual outage time and N_i is the number of customers of load point i.

(v) *Average service availability (unavailability) index, ASAI (ASUI)*

$$\text{ASAI} = \frac{\text{customer hours of available service}}{\text{customer hours demanded}}$$

$$= \frac{\Sigma\, N_i \times 8760 - \Sigma\, U_i N_i}{\Sigma\, N_i \times 8760} \tag{7.8}$$

$$\text{ASUI} = 1 - \text{ASAI} = \frac{\text{customer hours of unavailable service}}{\text{customer hours demanded}}$$

$$= \frac{\Sigma\, U_i N_i}{\Sigma\, N_i \times 8760} \tag{7.9}$$

where 8760 is the number of hours in a calendar year.

7.3.3 Load- and energy-orientated indices

One of the important parameters required in the evaluation of load- and energy-orientated indices is the average load at each load-point busbar.

The average load L_a is given by

(a) $L_a = L_p f$ (7.10)

where L_p = peak load demand
f = load factor

(b) $L_a = \dfrac{\text{total energy demanded in period of interest}}{\text{period of interest}} = \dfrac{E_d}{t}$ (7.11)

where E_d and t are shown on the load–duration curve (see Section 2.3) of Fig. 7.2 and t is normally one calendar year.

(i) *Energy not supplied index, ENS*

$$\text{ENS} = \text{total energy not supplied by the system} = \Sigma\, L_{a(i)} U_i \quad (7.12)$$

where $L_{a(i)}$ is the average load connected to load point i.

(ii) *Average energy not supplied, AENS or average system curtailment index, ASCI*

$$\text{AENS} = \frac{\text{total energy not supplied}}{\text{total number of customers served}} = \frac{\Sigma\, L_{a(i)} U_i}{\Sigma\, N_i} \quad (7.13)$$

(iii) *Average customer curtailment index, ACCI*

$$\text{ACCI} = \frac{\text{total energy not supplied}}{\text{total number of customers affected}} \quad (7.14)$$

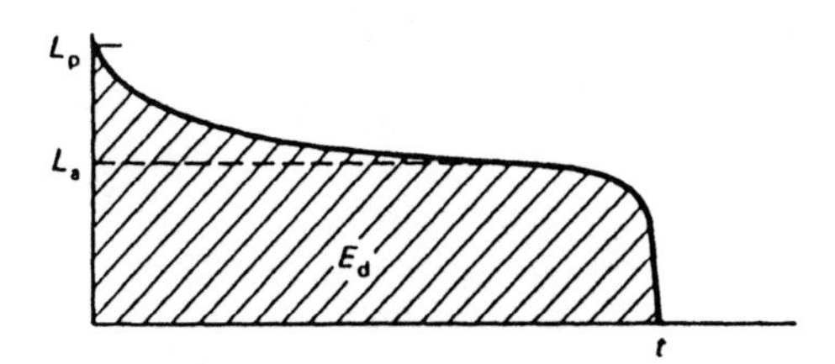

Fig. 7.2 Illustration of L_p, L_a, E_d and t

This index differs from AENS in the same way that CAIFI differs from SAIFI. It is therefore a useful index for monitoring the changes of average energy not supplied between one calendar year and another.

7.3.4 System performance

The customer- and load-orientated indices described in Sections 7.3.2 and 7.3.3 are very useful for assessing the severity of system failures in future reliability prediction analysis. They can also be used, however, as a means of assessing the past performance of a system. In fact, at the present time, they are probably more widely used in this respect than as measures of future performance. Assessment of system performance is a valuable procedure for three important reasons:

(a) It establishes the chronological changes in system performance and therefore helps to identify weak areas and the need for reinforcement.
(b) It establishes existing indices which serve as a guide for acceptable values in future reliability assessments.
(c) It enables previous predictions to be compared with actual operating experience.

The evaluation of system performance indices can be illustrated by considering a portion of a distribution system having six load-point busbars. The number of customers and average load connected to these busbars are shown in Table 7.4.

Assume that four system failures occur in one given calendar year of interest, having the interruption effects shown in Table 7.5.

The information given in Tables 7.4 and 7.5 permits all the customer and load-orientated indices to be evaluated as shown below:

$$\text{SAIFI} = \frac{\Sigma N_C}{\Sigma N} = \frac{3100}{4000} = 0.775 \text{ interruptions/customer}$$

$$\text{CAIFI} = \frac{\Sigma N_C}{N_a} = \frac{3100}{2200} = 1.409 \text{ interruptions/customer affected}$$

Table 7.4 Details of the distribution system

Load point	*Number of customers, N*	*Average load connected, L_a (kW)*
1	1000	5000
2	800	3600
3	600	2800
4	800	3400
5	500	2400
6	300	1800
Total	4000	19000

Table 7.5 Interruption effects in a given calendar year

Interruption case	*Load point affected*	*Number of customers disconnected* (N_C)	*Load curtailed,* L_C *(kW)*	*Duration of interruption, d (hours)*	*Customer hours curtailed,* $N_C d$	*Energy not supplied,* $L_C d$ *(kWh)*
1	2	800	3600	3	2400	10800
	3	600	1800	3	1800	8400
2	6	300	1800	2	600	3600
3	3	600	2800	1	600	2800
4	5	500	2400	1.5	750	3600
	6	300	1800	1.5	450	2700
Total		3100	15200		6600	31900

Number of customers affected = 800 + 600 + 300 + 500 = 2200 = N_a.

$$\text{SAIDI} = \frac{\Sigma N_C d}{\Sigma N} = \frac{6600}{3100} = 1.65 \text{ hours/customer}$$

$$= 99.0 \text{ minutes/customer}$$

$$\text{CAIDI} = \frac{\Sigma N_C d}{\Sigma N_C} = \frac{6600}{3100} = 2.13 \text{ hours/customer interruption}$$

$$= 127.7 \text{ minutes/customer interruption}$$

$$\text{ASAI} = \frac{\Sigma N \times 8760 - \Sigma N_C d}{\Sigma N \times 8760} = \frac{4000 \times 8760 - 6600}{4000 \times 8760}$$

$$= 0.999812$$

$$\text{ASUI} = 1 - 0.999812 = 0.000188$$

$$\text{ENS} = \Sigma L_C d = 31900 \text{ kWh}$$

$$\text{AENS} = \frac{\text{ENS}}{\Sigma N} = \frac{31900}{4000} = 7.98 \text{ kWh/customer}$$

$$\text{ACCI} = \frac{\text{ENS}}{N_a} = \frac{31900}{2200} = 14.5 \text{ kWh/customer affected}$$

A recent EPRI (Electric Power Research Institute) research project has established [8] that the most frequently used indices for assessing system performance were customer-related. The histograms [8] shown in Fig. 7.3 illustrate the popularity of the various performance indices among the utilities that responded to the survey associated with the EPRI project.

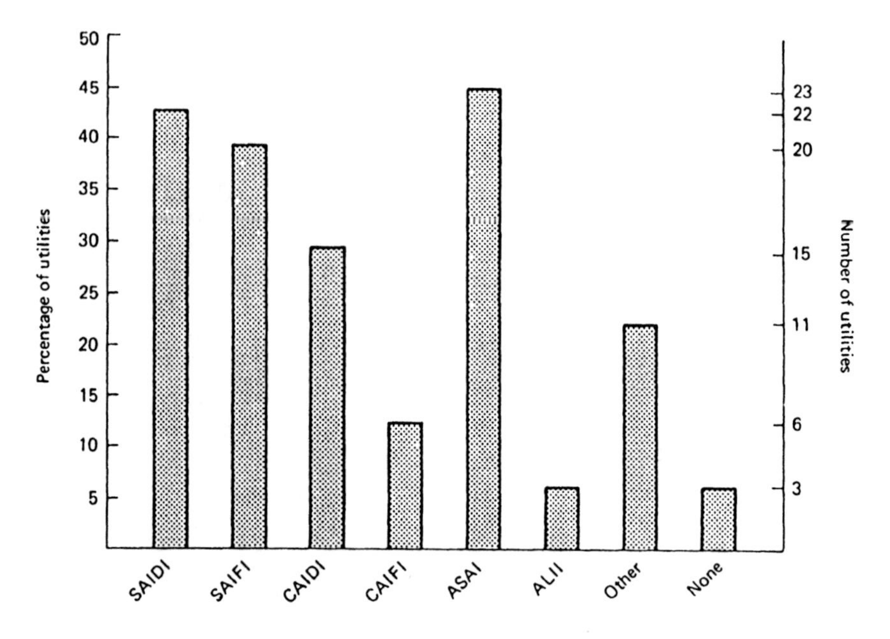

Fig. 7.3 Frequency of use of various performance indices
SAIDI Average interruption duration per customer served
SAIFI Average interruptions per customer served
CAIDI Average interruption duration per customer interrupted
CAIFI Average interruptions per customer interrupted
ASAI Customer hours available/customer hours demand
ALII Average connected kVA interrupted per kVA of connected load served

7.3.5 System prediction

In order to illustrate the evaluation of the customer- and load-orientated indices during the assessment of future performance, known as system prediction, reconsider the system shown in Fig. 7.1 and the primary reliability indices shown in Table 7.3. Let the number of customers and average load demand at each busbar be as shown in Table 7.6.

Table 7.6 Details of the system of Fig. 7.1

Load point	*Number of customers*	*Average load demand (kW)*
L1	200	1000
L2	150	700
L3	100	400
Total	450	2100

The customer- and load-orientated indices can now be evaluated as follows:

$$\text{SAIFI} = \frac{0.2 \times 200 + 0.3 \times 150 + 0.45 \times 100}{200 + 150 + 100}$$

$$= 0.289 \text{ interruptions/customer yr.}$$

$$\text{SAIDI} = \frac{1.2 \times 200 + 1.7 \times 150 + 2.9 \times 100}{450}$$

$$= 1.74 \text{ hours/customer yr.}$$

$$\text{CAIDI} = \frac{1.2 \times 200 + 1.7 \times 150 + 2.9 \times 100}{0.2 \times 200 + 0.3 \times 150 + 0.45 \times 100}$$

$$= 6.04 \text{ hours/customer interruption}$$

$$\text{ASAI} = \frac{450 \times 8760 - (1.2 \times 200 + 1.7 \times 150 + 2.9 \times 100)}{450 \times 8760}$$

$$= 0.999801$$

$$\text{ASUI} = 1 - 0.999801$$

$$= 0.000199$$

$$\text{ENS} = 1000 \times 1.2 + 700 \times 1.7 + 400 \times 2.9$$

$$= 3550 \text{ kWh/yr}$$

$$\text{AENS} = \frac{3550}{450}$$

$$= 7.89 \text{ kWh/customer yr.}$$

7.4 Application to radial systems

Many distribution systems are designed and constructed as single radial feeder systems. There are additionally many other systems which, although constructed as meshed systems, are operated as single radial feeder systems by using normally open points in the mesh. The purpose of these normally open points is to reduce the amount of equipment exposed to failure on any single feeder circuit and to ensure that, in the event of a system failure or during scheduled maintenance periods, the normally open point can be closed and another opened in order to minimize the total load that is disconnected.

The techniques described in this chapter can be used to evaluate the three primary indices and the additional customer- and load-orientated indices for all of these systems. Additional techniques are required, however, if a more rigorous

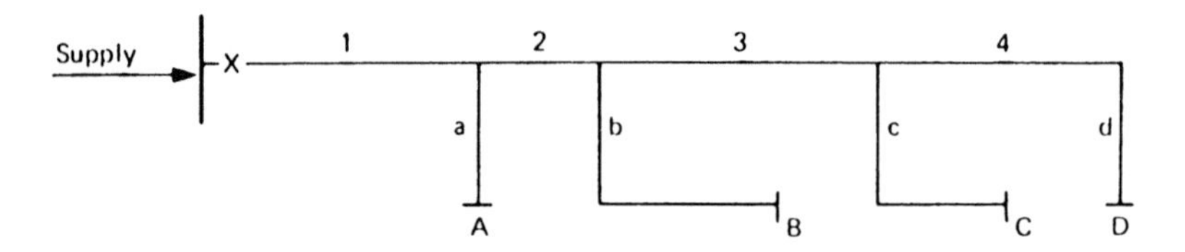

Fig. 7.4 Typical radial distribution network

analysis is desired of parallel systems and systems that are meshed. These additional techniques are presented in Chapter 8.

Consider now the system shown in Fig. 7.4. This is a single line representation of the system and the following discussion assumes that any fault, single-phase or otherwise, will trip all three phases.

It is normally found in practice that lines and cables have a failure rate which is approximately proportional to their length. For this example let the main feeder (Sections 1, 2, 3, 4) have a failure rate of 0.1 f/km yr and the lateral distributors (a, b, c, d) have a failure rate of 0.2 f/km yr. Using these basic data and the line lengths given in Table 7.7 gives the reliability parameters also shown in Table 7.7.

If all component failures are short circuits then each failure will cause the main breaker to operate. If there are no points at which the system can be isolated then each failure must be repaired before the breaker can be reclosed. On the basis of this operating procedure, the reliability indices of each load point (A, B, C, D) can be evaluated using the principle of series systems as shown in Table 7.8.

In this example, the reliability of each load point is identical. The operating policy assumed for this system is not very realistic and additional features such as isolation, additional protection and transferable loads can be included. These features are discussed in the following sections.

Table 7.7 Reliability parameters for system of Fig. 7.4

Component	*Length (km)*	λ *(f/yr)*	*r (hours)*
Section			
1	2	0.2	4
2	1	0.1	4
3	3	0.3	4
4	2	0.2	4
Distributor			
a	1	0.2	2
b	3	0.6	2
c	2	0.4	2
d	1	0.2	2

Table 7.8 Reliability indices for the system of Fig. 7.4

Component failure	Load pt A			Load pt B			Load pt C			Load pt D		
	λ (*f/yr*)	*r* (*hours*)	*U* (*hours/yr*)	λ (*f/yr*)	*r* (*hours*)	*U* (*hours/yr*)	λ (*f/yr*)	*r* (*hours*)	*U* (*hours/yr*)	λ (*f/yr*)	*r* (*hours*)	*U* (*hours/yr*)
Section												
1	0.2	4	0.8	0.2	4	0.8	0.2	4	0.8	0.2	4	0.8
2	0.1	4	0.4	0.1	4	0.4	0.1	4	0.4	0.1	4	0.4
3	0.3	4	1.2	0.3	4	1.2	0.3	4	1.2	0.3	4	1.2
4	0.2	4	0.8	0.2	4	0.8	0.2	4	0.8	0.2	4	0.8
Distributor												
a	0.2	2	0.4	0.2	2	0.4	0.2	2	0.4	0.2	2	0.4
b	0.6	2	1.2	0.6	2	1.2	0.6	2	1.2	0.6	2	1.2
c	0.4	2	0.8	0.4	2	0.8	0.4	2	0.8	0.4	2	0.8
d	0.2	2	0.4	0.2	2	0.4	0.2	2	0.4	0.2	2	0.4
Total	2.2	2.73	6.0	2.2	2.73	6.0	2.2	2.73	6.0	2.2	2.73	6.0

(where $\lambda_{total} = \Sigma\, \lambda$, $U_{total} = \Sigma\, U$ and $r_{total} = \Sigma\, u / \Sigma\, \lambda$).

Table 7.9 Customers and load connected to the system of Fig. 7.4

Load point	*Number of customers*	*Average load connected (kW)*
A	1000	5000
B	800	4000
C	700	3000
D	500	2000

If the average demand and number of customers at each load point is known, the primary indices shown in Table 7.8 can be extended to give the customer- and load-orientated indices. Let the average load and number of customers at A, B, C and D be as shown in Table 7.9.

The additional indices for this system can now be evaluated as

SAIFI = 2.2 interruptions/customer yr

SAIDI = 6.0 hours/customer yr

CAIDI = 2.73 hours/customer interruption

ASUI= 0.000685 ASAI = 0.999315

ENS = 84.0 MWh/yr AENS = 28.0 kWh/customer yr

7.5 Effect of lateral distributor protection

Additional protection is frequently used in practical distribution systems. One possibility in the case of the system shown in Fig. 7.4 is to install fusegear at the tee-point in each lateral distributor. In this case a short circuit on a lateral distributor causes its appropriate fuse to blow; this causes disconnection of its load point until the failure is repaired but does not affect or cause the disconnection of any other load point. The system reliability indices are therefore modified to those shown in Table 7.10.

In this case the reliability indices are improved for all load points although the amount of improvement is different for each one. The most unreliable load point is B because of the dominant effect of the failures on its lateral distributor. The additional indices for this system are:

SAIFI = 1.15 interruptions/customer yr

SAIDI = 3.91 hours/customer yr

CAIDI = 3.39 hours/customer interruption

ASUI = 0.000446 ASAI = 0.999554

ENS = 54.8 MWh/yr AENS = 18.3 kWh/customer yr

Table 7.10 Reliability indices with lateral protection

	Load pt A			*Load pt B*			*Load pt C*			*Load pt D*		
Component failure	λ (*f/yr*)	r (*hours*)	U (*hours/yr*)	λ (*f/yr*)	r (*hours*)	U (*hours/yr*)	λ (*f/yr*)	r (*hours*)	U (*hours/yr*)	λ (*f/yr*)	r (*hours*)	U (*hours/yr*)
Section												
1	0.2	4	0.8	0.2	4	0.8	0.2	4	0.8	0.2	4	0.8
2	0.1	4	0.4	0.1	4	0.4	0.1	4	0.4	0.1	4	0.4
3	0.3	4	1.2	0.3	4	1.2	0.3	4	1.2	0.3	4	1.2
4	0.2	4	0.8	0.2	4	0.8	0.2	4	0.8	0.2	4	0.8
Distributor												
a	0.2	2	0.4									
b				0.6	2	1.2						
c							0.4	2	0.8			
d										0.2	2	0.4
Total	1.0	3.6	3.6	1.4	3.14	4.4	1.2	3.33	4.0	1.0	3.6	3.6

7.6 Effect of disconnects

A second or alternative reinforcement or improvement scheme is the provision of disconnects or isolators at judicious points along the main feeder. These are generally not fault-breaking switches and therefore any short circuit on a feeder still causes the main breaker to operate. After the fault has been detected, however, the relevant disconnect can be opened and the breaker reclosed. This procedure allows restoration of all load points between the supply point and the point of isolation before the repair process has been completed. Let points of isolation be installed in the previous system as shown in Fig. 7.5 and let the total isolation and switching time be 0.5 hour.

The reliability indices for the four load points are now modified to those shown in Table 7.11.

In this case, the reliability indices of load points A, B, C are improved, the amount of improvement being greater for those near to the supply point and less for those further from it. The indices of load point D remain unchanged because isolation cannot remove the effect of any failure on this load point. The additional customer- and load-orientated indices for this configuration are

SAIFI = 1.15 interruptions/customer yr

SAIDI = 2.58 hours/customer yr

CAIDI = 2.23 hours/customer interruption

ASUI = 0.000294 ASAI = 0.999706

ENS = 35.2 MWh/yr AENS = 11.7 kWh/customer yr

7.7 Effect of protection failures

The reliability indices for each load point in Sections 7.5 and 7.6 were evaluated assuming that the fuses in the lateral distributor operated whenever a failure occurred on the distributor they were supposed to protect. Occasionally, however, the primary protection system fails to operate. In these cases, the back-up protection functions. In order to illustrate this aspect and its effect on the reliability indices,

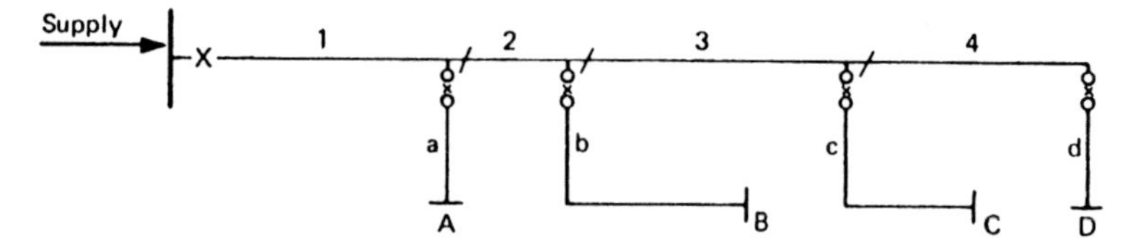

Fig. 7.5 Network of Fig. 7.4 reinforced with disconnects and fusegear

Table 7.11 Reliability indices with lateral protection and disconnects

	Load pt A			*Load pt B*			*Load pt C*			*Load pt D*		
Component failure	λ *(f/yr)*	*r* *(hours)*	*U* *(hours/yr)*	λ *(f/yr)*	*r* *(hours)*	*U* *(hours/yr)*	λ *(f/yr)*	*r* *(hours)*	*U* *(hours/yr)*	λ *(f/yr)*	*r* *(hours)*	*U* *(hours/yr)*
Section												
1	0.2	4	0.8	0.2	4	0.8	0.2	4	0.8	0.2	4	0.8
2	0.1	0.5	0.05	0.1	4	0.4	0.1	4	0.4	0.1	4	0.4
3	0.3	0.5	0.15	0.3	0.5	0.15	0.3	4	1.2	0.3	4	1.2
4	0.2	0.5	0.1	0.2	0.5	0.1	0.2	0.5	0.1	0.2	4	0.8
Distributor												
a	0.2	2	0.4									
b				0.6	2	1.2						
c							0.4	2	0.8			
d										0.2	2	0.4
Total	1.0	1.5	1.5	1.4	1.89	2.65	1.2	2.75	3.3	1.0	3.6	3.6

Table 7.12 Reliability indices if the fuses operate with a probability of 0.9

Component failure	Load pt A			Load pt B			Load pt C			Load pt D		
	λ (f/yr)	r (hours)	U (hours/yr)	λ (f/yr)	r (hours)	U (hours/yr)	λ (f/yr)	r (hours)	U (hours/yr)	λ (f/yr)	r (hours)	U (hours/yr)
Section												
1	0.2	4	0.8	0.2	4	0.8	0.2	4	0.8	0.2	4	0.8
2	0.1	0.5	0.05	0.1	4	0.4	0.1	4	0.4	0.1	4	0.4
3	0.3	0.5	0.15	0.3	0.5	0.15	0.3	4	1.2	0.3	4	1.2
4	0.2	0.5	0.1	0.2	0.5	0.1	0.2	0.5	0.1	0.2	4	0.8
Distributor												
a	0.2	2	0.4	0.02	0.5	0.01	0.02	0.5	0.01	0.01	0.5	0.01
b	0.06	0.5	0.03	0.6	2	1.2	0.06	0.5	0.03	0.06	0.5	0.03
c	0.04	0.5	0.02	0.04	0.5	0.02	0.4	2	0.8	0.04	0.5	0.02
d	0.02	0.5	0.01	0.02	0.5	0.01	0.02	0.5	0.01	0.2	2	0.4
Total	1.12	1.39	1.56	1.48	1.82	2.69	1.3	2.58	3.35	1.12	3.27	3.66

consider the system shown in Fig. 7.5 and assume that the fusegear operates with a probability of 0.9, i.e. the fuses operate successfully 9 times out of 10 when required. In this case the reliability indices shown in Table 7.11 are modified because, for example, failures on distributors b, c and d also contribute to the indices of load point A. Similarly for load points B, C and D. The contribution to the failure rate can be evaluated using the concept of expectation.

$$\text{Failure rate} = (\text{failure rate} \mid \text{fuse operates}) \times P(\text{fuse operates}) + (\text{failure rate} \mid \text{fuse fails}) \times P(\text{fuse fails})$$

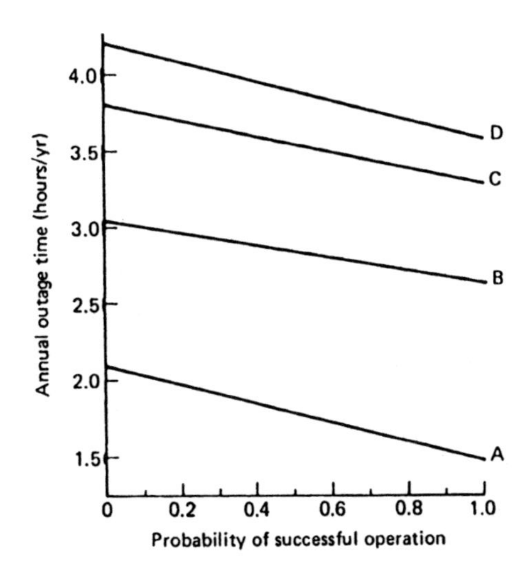

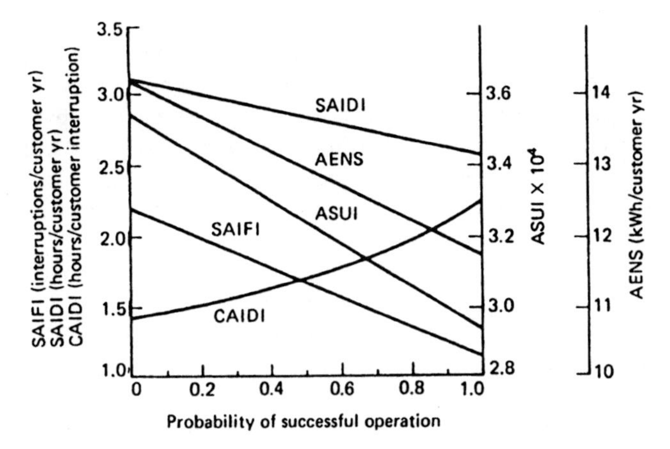

Fig. 7.6 Effect of protection failures on load point indices

Therefore the contribution to the failure rate of load point A by distributor b is:

failure rate = 0 × 0.9 + 0.6 × 0.1 = 0.06

The modified indices are shown in Table 7.12 assuming that all failures can be isolated within 0.5 hour.

The results shown in Table 7.12 indicate that the reliability of each load point degrades as expected, the amount of degradation being dependent on the probability that the fusegear operates successfully and the relative effect of the additional failure events compared with those that occur even if the fuses are 100% reliable in operation. This effect is illustrated in Fig. 7.6, which shows the change in load point annual outage time as a function of the probability that the fuses operate successfully. In this figure, the unavailability associated with success probabilities of 1.0 and 0.9 correspond to the results shown in Table 7.11 and Table 7.12 respectively, and a success probability of 0.0 corresponds to the results that would be obtained if the fusegear did not exist in the distributors.

The additional customer- and load-orientated indices are

SAIFI = 1.26 interruptions/customer yr

SAIDI = 2.63 hours/customer yr

CAIDI = 2.09 hours/customer interruption

ASUI = 0.000300 ASAI = 0.999700

ENS = 35.9 MWh/yr AENS = 12.0 kWh/customer yr

7.8 Effect of transferring loads

7.8.1 No restrictions on transfer

As described in Section 7.4, many distribution systems have open points in a meshed configuration so that the system is effectively operated as a radial network but, in the event of a system failure, the open points can be moved in order to recover load that has been disconnected. This operational procedure can have a marked effect on the reliability indices of a load point because loads that would otherwise have been left disconnected until repair had been completed can now be transferred onto another part of the system.

Consider the system shown in Fig. 7.5 and let feeder section 4 be connected to another distribution system through a normally open point as shown in Fig. 7.7. In this case, the reliability indices of each load point are shown in Table 7.13 assuming that there is no restriction on the amount of load that can be transferred through the backfeed.

The results shown in Table 7.13 indicate that the failure rate of each load point does not change, that the indices of load point A do not change because load transfer

Table 7.13 Reliability indices with unrestricted load transfers

	Load pt A			*Load pt B*			*Load pt C*			*Load pt D*		
Component failure	λ *(f/yr)*	*r* *(hours)*	*U* *(hours/yr)*	λ *(f/yr)*	*r* *(hours)*	*U* *(hours/yr)*	λ *(f/yr)*	*r* *(hours)*	*U* *(hours/yr)*	λ *(f/yr)*	*r* *(hours)*	*U* *(hours/yr)*
Section												
1	0.2	4	0.8	0.2	0.5	0.1	0.2	0.5	0.1	0.2	0.5	0.1
2	0.1	0.5	0.05	0.1	4	0.4	0.1	0.5	0.05	0.1	0.5	0.05
3	0.3	0.5	0.15	0.3	0.5	0.15	0.3	4	1.2	0.3	0.5	0.15
4	0.2	0.5	0.1	0.2	0.5	0.1	0.2	0.5	0.1	0.2	4	0.8
Distributor												
a	0.2	2	0.4									
b				0.6	2	1.2						
c							0.4	2	0.8			
d										0.2	2	0.4
Total	1.0	1.5	1.5	1.4	1.39	1.95	1.2	1.88	2.25	1.0	1.5	1.5

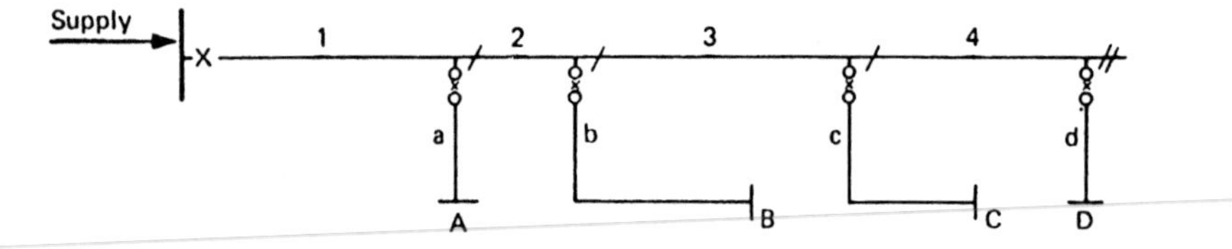

Fig. 7.7 Network of Fig. 7.5 connected to a normally open point

cannot recover any load lost, and that the greatest effect occurs for the load point furthest from the supply point and nearest to the normally open transfer point.

In this case the additional reliability indices are

SAIFI = 1.15 interruptions/customer yr

SAIDI = 1.80 hours/customer yr

CAIDI = 1.56 hours/customer interruption

ASUI = 0.000205 ASAI = 0.999795

ENS = 25. MWh/yr AENS = 8.4 kWh/customer yr

7.8.2 Transfer restrictions

It is not always feasible to transfer all load that is lost in a distribution system onto another feeder through a normally open point. This restriction may exist because the failure occurs during the high load period and either the feeder to which the load is being transferred or the supply point feeding the second system has limited capacity. In this case the outage time associated with a failure event is equal to the isolation time if the load can be transferred, or equal to the repair time if the load cannot be transferred. The average of these values can be evaluated using the concept of expectation, since

$$\text{outage time} = (\text{outage time} \mid \text{transfer}) \times P(\text{of transfer}) + (\text{outage time} \mid \text{no transfer}) \times P(\text{of no transfer})$$

As an example, consider the outage time of load point B of Fig. 7.7 due to failure of feeder section 1 if the probability of being able to transfer load is 0.6; then

$$\text{outage time} = 0.5 \times 0.6 + 4.0 \times 0.4 = 1.9 \text{ hours.}$$

The complete set of reliability indices is shown in Table 7.14.

The additional reliability indices are

SAIFI = 1.15 interruptions/customer yr

SAIDI = 2.11 hours/customer yr

Table 7.14 Reliability indices with restricted load transfers

	Load pt A			Load pt B			Load pt C			Load pt D		
Component failure	λ (*f/yr*)	*r* (*hours*)	*U* (*hours/yr*)	λ (*f/yr*)	*r* (*hours*)	*U* (*hours/yr*)	λ (*f/yr*)	*r* (*hours*)	*U* (*hours/yr*)	λ (*f/yr*)	*r* (*hours*)	*U* (*hours/yr*)
Section												
1	0.2	4	0.8	0.2	1.9	0.38	0.2	1.9	0.38	0.2	1.9	0.38
2	0.1	0.5	0.05	0.1	4	0.4	0.1	1.9	0.19	0.1	1.9	0.19
3	0.3	0.5	0.15	0.3	0.5	0.15	0.3	4	1.2	0.3	1.9	0.57
4	0.2	0.5	0.1	0.2	0.5	0.1	0.2	0.5	0.1	0.2	4	0.8
Distributor												
a	0.2	2	0.4									
b				0.6	2	1.2						
c							0.4	2	0.8			
d										0.2	2	0.4
Total	1.0	1.5	1.5	1.4	1.59	2.23	1.2	2.23	2.67	1.0	2.34	2.34

CAIDI = 1.83 hours/customer interruption

ASUI = 0.000241 ASAI = 0.999759

ENS = 29.1MWh/yr AENS = 9.7 kWh/customer yr.

The indices shown in Table 7.14 lie between those of Table 7.11 (no transfer possible) and those of Table 7.13 (no restrictions on transfers). The results shown in Fig. 7.8 illustrate the variation of load point annual outage time as the probability of transferring loads increases from 0.0 (Table 7.11) to 1.0 (Table 7.13).

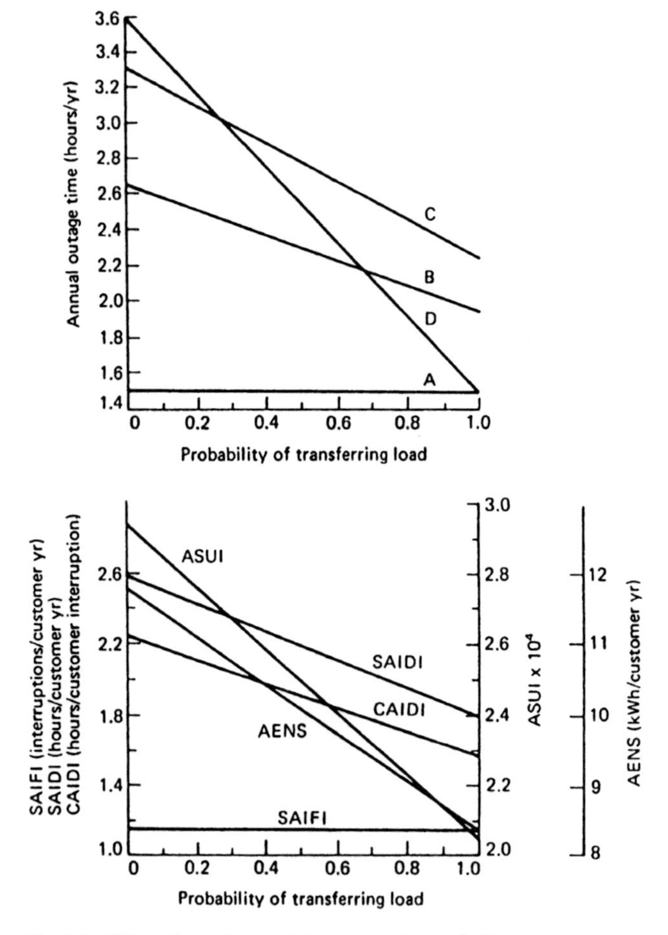

Fig. 7.8 Effect of transfer restrictions on load point indices

It may be thought unrealistic to consider load transfers related to a probability of making the transfer. Instead it may be preferable to consider the amount of load that can be recovered based on the load that has been disconnected and the available transfer capacity of the second system at that particular loading level on the system. This requires a more exhaustive analysis and is discussed in Chapter 9. It also requires knowledge of the load–duration curves for each load point, although in practice the shapes of these are usually assumed to be identical.

A summary of all the indices evaluated in Sections 7.4–8 is shown in Table 7.15.

Table 7.15 Summary of indices

	Case 1	*Case 2*	*Case 3*	*Case 4*	*Case 5*	*Case 6*
Load point A						
λ f/yr	2.2	1.0	1.0	1.12	1.0	1.0
r hours	2.73	3.6	1.5	1.39	1.5	1.5
U hours/yr	6.0	3.6	1.5	1.56	1.5	1.5
Load point B						
λ f/yr	2.2	1.4	1.4	1.48	1.4	1.4
r hours	2.73	3.14	1.89	1.82	1.39	1.59
U hours/yr	6.0	4.4	2.65	2.69	1.95	2.23
Load point C						
λ f/yr	2.2	1.2	1.2	1.3	1.2	1.2
r hours	2.73	3.33	2.75	2.58	1.88	2.23
U hours/yr	6.0	4.0	3.3	3.35	2.25	2.67
Load point D						
λ f/yr	2.2	1.0	1.0	1.12	1.0	1.0
r hours	2.73	3.6	3.6	3.27	1.5	2.34
U hours/yr	6.0	3.6	3.6	3.66	1.5	2.34
System indices						
SAIFI	2.2	1.15	1.15	1.26	1.15	1.15
SAIDI	6.0	3.91	2.58	2.63	1.80	2.11
CAIDI	2.73	3.39	2.23	2.09	1.56	1.83
ASAI	0.999315	0.999554	0.999706	0.999700	0.999795	0.999759
ASUI	0.000685	0.000446	0.000294	0.000300	0.000205	0.000241
ENS	84.0	54.8	35.2	35.9	25.1	29.1
AENS	28.0	18.3	11.7	12.0	8.4	9.7

Case 1. Base case shown in Fig. 7.4.
Case 2. As in Case 1, but with perfect fusing in the lateral distributors.
Case 3. As in Case 2, but with disconnects on the main feeder as shown in Fig. 7.5.
Case 4. As in Case 3, probability of successful lateral distributor fault clearing of 0.9.
Case 5. As in Case 3, but with an alternative supply as shown in Fig. 7.7.
Case 6. As in Case 5, probability of conditional load transfer of 0.6.

7.9 Probability distributions of reliability indices

7.9.1 Concepts

The reliability indices evaluated in the previous sections are average or expected values. Due to the random nature of the failure and restoration processes, the indices for any particular year deviate from these average values. This deviation is represented by probability distributions and a knowledge of these distributions can be beneficial in the reliability assessment of present systems and future reinforcement schemes. This problem has been examined in recent papers [9–11] in order to estimate the distributions which adequately represent the failure rate and restoration time of a load.

7.9.2 Failure rate

The failure times can normally be assumed to be exponentially distributed because the components operate in their useful or operating life period (see Section 6.5 of *Engineering Systems*). Also the system failure rate for radial networks depends only on the component failure rates and not the restoration times. Consequently, non-exponential restoration times do not affect the failure rate distribution. Under these circumstances, it has been shown [9,10] that the load point failure rate of a radial system obeys a Poisson distribution (Section 6.6 of *Engineering Systems*). From Equation (6.19) of *Engineering Systems*, the probability of n failures in time t is

$$P(n) = \frac{(\lambda t)^n e^{-\lambda t}}{n!} \tag{7.15}$$

Equation (7.15) can be used to evaluate the probability of any number of failures per year at each load point knowing only the average value of failure rate λ. As an example, consider the failure rates given in Table 7.11 in order to evaluate the probability of 0, 1, 2, 3, 4 failures occurring in a year at the load points. These results are shown in Table 7.16.

Table 7.16 Probability that n failures occur in a year

	Probability of n failures/yr at load point			
n	A	B	C	D
0	0.368	0.247	0.301	0.368
1	0.368	0.345	0.361	0.368
2	0.184	0.242	0.217	0.184
3	0.061	0.113	0.087	0.061
4	0.015	0.039	0.026	0.015

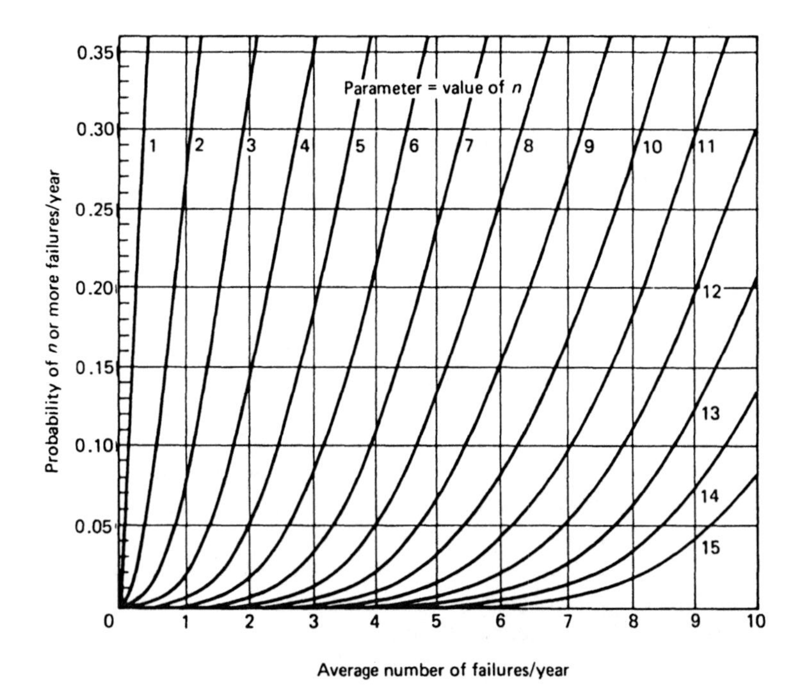

Fig. 7.9 Probability of n or more failures/year given average failure rate

The results shown in Table 7.16 indicate that the ratio between the probabilities for a given number of failures in a year is not constant between the load points and, consequently, is not equal to the ratio between the respective average failure rates.

An alternative to the repetitive use of Equation (7.15) is to construct parametric graphs from which the relevant probabilistic information can be deduced. One such set of graphs [11] is shown in Fig. 7.9, from which the probability of n or more failures per year for a given average failure rate can be ascertained.

7.9.3 Restoration times

It has been suggested [9] that the load point outage duration can be approximated by a gamma distribution (see Section 6.10 of *Engineering Systems*) if the restoration times are exponentially distributed. This suggestion has been confirmed [10, 11] by a series of Monte Carlo simulations (see Chapter 12 for more detail regarding this approach). In this case a relatively simple approach [9] can be used to evaluate the probabilities of outage durations.

The main problem is that practical restoration times are not usually exponentially distributed. It has been shown [11] from Monte Carlo simulations that, when non-exponential distributions are used to represent the restoration times, the load point duration cannot generally be represented by a gamma distribution and may not be described by any known distribution. In these cases, the only solution is to perform Monte Carlo simulations, which can be rather time consuming. It should be noted, however, that although the underlying distribution may not be known, the average values of outage duration evaluated using the techniques of the previous sections are still valid; it is only the distribution around these average values that is affected.

7.10 Conclusions

This chapter has described the basic techniques needed to evaluate the reliability of distribution systems. These techniques are perfectly adequate for the assessment of single radial systems and meshed systems that are operated as single radial systems. The techniques must be extended, however, in order to assess more complex systems such as parallel configurations and systems that are operated as a mesh. The extended techniques, described in Chapter 8, are extensions of those discussed in this chapter and therefore the underlying concepts of this chapter should be assimilated and understood before proceeding to the next chapter.

It is not possible to assess realistically the reliability of a system without a thorough understanding of the relevant operational characteristics and policy. The reliability indices that are evaluated are affected greatly by these characteristics and policy. This relationship has been illustrated in this chapter using a number of case studies. These do not exhaust the possibilities, however—many other alternatives are feasible. Some of these alternatives are included in the following set of problems and it is believed that, by studying the examples included in the text and solving the following problems, a reader should be in a sound position to use the techniques for his own type of systems and to study the effect of reinforcement and various operating policies.

7.11 Problems

1 Let the system shown in Fig. 7.10 have the reliability parameters shown in Table 7.7 and the load point details shown in Table 7.9. Assume that an isolator can be operated

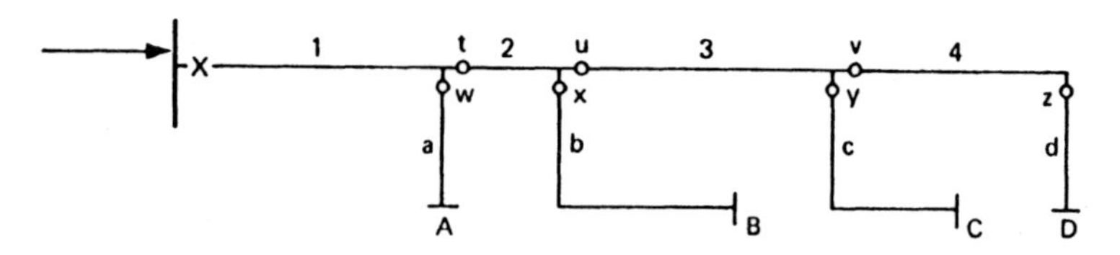

Fig. 7.10

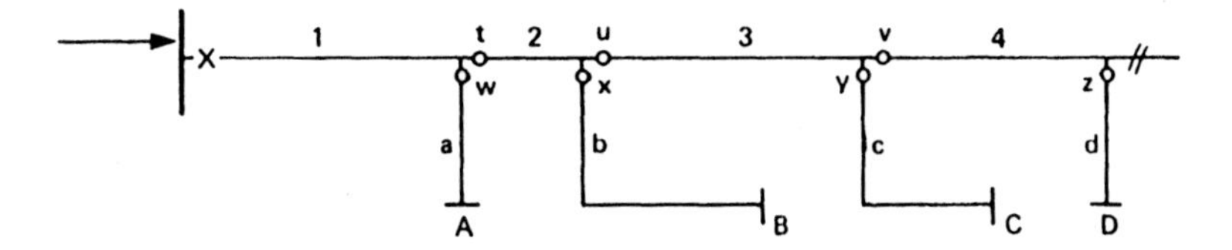

Fig. 7.11

Table 7.17

Case number	Solid links at	Isolators at	Fuses: At	Fuses: Prob. of success	Breakers: At	Breakers: Prob. of success
(i)	w–z	t–v	—	—	—	—
(ii)	—	t–z	—	—	—	—
(iii)	—	t–v	w–z	1.0	u	1.0
(iv)	—	t–v	w–z	0.9	u	0.95

in 0.5 hour. The points t–z are places at which the installation of additional components is being considered as shown in the case studies given in Table 7.17. For each case, evaluate the failure rate, outage time and unavailability of each load point and the values of SAIFI, SAIDI, CAIDI, ASAI and AENS.

2 The system shown in Fig. 7.10 is reinforced by connecting it through a normally open point to another feeder as shown in Fig. 7.11. Re-evaluate the indices specified in Problem 1 for all four case studies if:
(a) there are no transfer restrictions;
(b) the probability of being able to transfer load is 0.7.

7.12 References

1. Dixon, G. F. L., Hammersley, H., *Reliability and its Cost on Distribution Systems.* International Conference on Reliability of Power Supply Systems (1977), IEE Conference Publication No. 148.
2. Canadian Electrical Association, *Distribution System Reliability Engineering Guide*, CEA publication (March 1976).
3. Electricity Council, *Reliability Engineering and Cost–Benefit Techniques for Use in Power System Planning and Design*, Electricity Council Research Memorandum, ECR/M966 (October 1976).
4. IEEE Committee Report, 'Bibliography on the application of probability methods in power system reliability evaluation', *IEEE Trans. on Power Apparatus and Systems*, **PAS-91** (1972), pp. 649–60.

5. IEEE Committee Report, 'Bibliography on the application of probability methods in power system reliability evaluation, 1971–1977', *IEEE Trans. on Power Apparatus and Systems*, **PAS-97** (1978), pp. 2235–42.
6. Gaver, D. P., Montmeat, F. E., Patton, A. D., 'Power system reliability: I—Measures of reliability and methods of calculation', *IEEE Trans. on Power Apparatus and Systems*, **PAS-83** (1964), pp. 727–37.
7. Montmeat, F. E., Patton, A. D., Zemkoski, J., Cumming, D. J., 'Power system reliability: II—Applications and a computer program', *IEEE Trans. on Power Apparatus and Systems*, **PAS-84** (1965), pp. 636–43.
8. Northcote-Green, J. E. D., Vismor, T. D., Brooks, C. L., Billinton, R., *Integrated distribution system reliability evaluation: Part I—Current practices*, CEA Engineering and Operating Division Meeting (March 1980).
9. Patton, A. D., 'Probability distribution of transmission and distribution reliability performance indices', *Reliability Conference for the Electric Power Industry* (1979), pp. 120–3.
10. Billinton, R., Wojczyuski, E., Rodych, V., 'Probability distributions associated with distribution system reliability indices', *Reliability Conference for the Electric Power Industry* (1980).
11. Billinton, R., Wojczynski, E., Godfrey, M., 'Practical calculations of distribution system reliability indices and their probability distributions', *Trans. Canadian Electrical Association Paper* 81-D-41 (1981).

8 Distribution systems—parallel and meshed networks

8.1 Introduction

Chapter 7 described the basic techniques used to evaluate the reliability of distribution systems and applied these techniques to simple radial networks. These basic techniques have been used in practice for some considerable time but are restricted in their application because they cannot directly be used for systems containing parallel circuits or meshed networks.

Distribution reliability evaluation techniques have been enhanced and rapidly developed [1–6] in recent years and have reached a stage at which a comprehensive evaluation is now possible. These extended techniques permit a complete analysis of parallel and meshed systems with or without transfer facilities existing and can account for all failure and restoration procedures that are known to the system planner and operator. Clearly the techniques cannot account for unknown events but the structure of the techniques is sufficiently flexible and convenient that these additional events can be included as and when they manifest themselves.

Although the techniques in this and the preceding chapter are described in relation to distribution systems, they are equally applicable to any part of the power system network including both transmission and subtransmission. It should be noted, however, that the techniques neglect the reliability effects of the generation and if this is required, an analysis based on the concepts of composite systems (Chapter 6) must be used. It should also be noted that a complete representation of the power system network is not viable and it must be divided into groups or subsystems in order for it to be solved practically. If, for example, the group being analyzed is a subtransmission system, the evaluated load point reliability indices are not those of the customer but, instead, are those of the supply point that interfaces with the next lower level of the network. Consequently these evaluated indices can be used as the reliability indices of the input to the next level. This procedure can be continued sequentially until the customer load points have been reached.

8.2 Basic evaluation techniques

8.2.1 State space diagrams

One method that can be used to evaluate the reliability of a continuously operated system is based on the construction of state space diagrams as described in Section 9.3 of *Engineering Systems*. Although this method is accurate, it becomes infeasible for large distribution networks. It does, however, have an important role to play in power system reliability evaluation. Firstly, it can be used as the primary evaluation method in certain applications, some of which are described in Chapter 11. Secondly, it is frequently used as a means of deducing approximate evaluation techniques. Thirdly, it is extremely useful as a standard evaluation method against which the accuracy of approximate methods can be compared.

In order to illustrate the development and consequent complexity of state space diagrams, consider the dual transformer feeder system shown in Fig. 8.1.

It is assumed in this example that the two busbars and the circuit breakers are 100% reliable. This is not restrictive, however, and both types of components can be taken into account. This will be seen in Sections 8.3, 8.9 and Chapter 10. The reliability of the system is therefore governed by the two lines (components 1 and 2) and the two transformers (3 and 4).

If each component can reside in one of two states (up and down), there are $2^4 = 16$ system states to consider, these being shown as a state space diagram in Fig. 8.2.

This state space diagram is the input information required for a Markov technique and the system reliability can be evaluated using the techniques described in Chapters 8 and 9 of *Engineering Systems*. This method becomes impractical for large distribution systems since the construction of the state space diagram becomes very cumbersome as well as fraught with difficulties. The alternative is to use approximate techniques as discussed in Chapter 11 of *Engineering Systems*. The remaining sections of this chapter will therefore be concerned only with these alternative methods, although the concept of state space diagrams will be used either as a means of developing alternative methods, illustrating a technique, or as a secondary solution method.

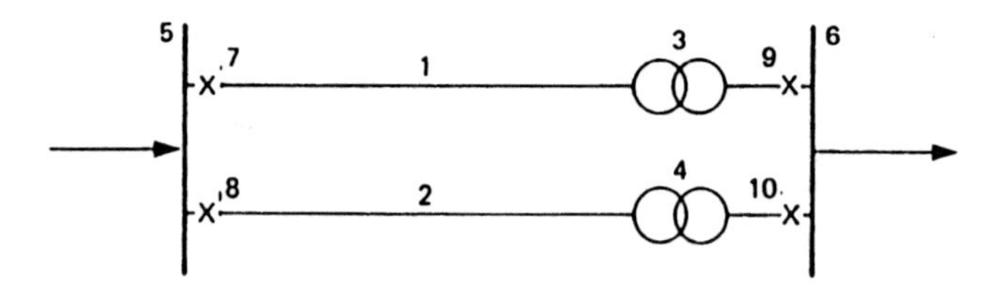

Fig. 8.1 Dual transformer feeder

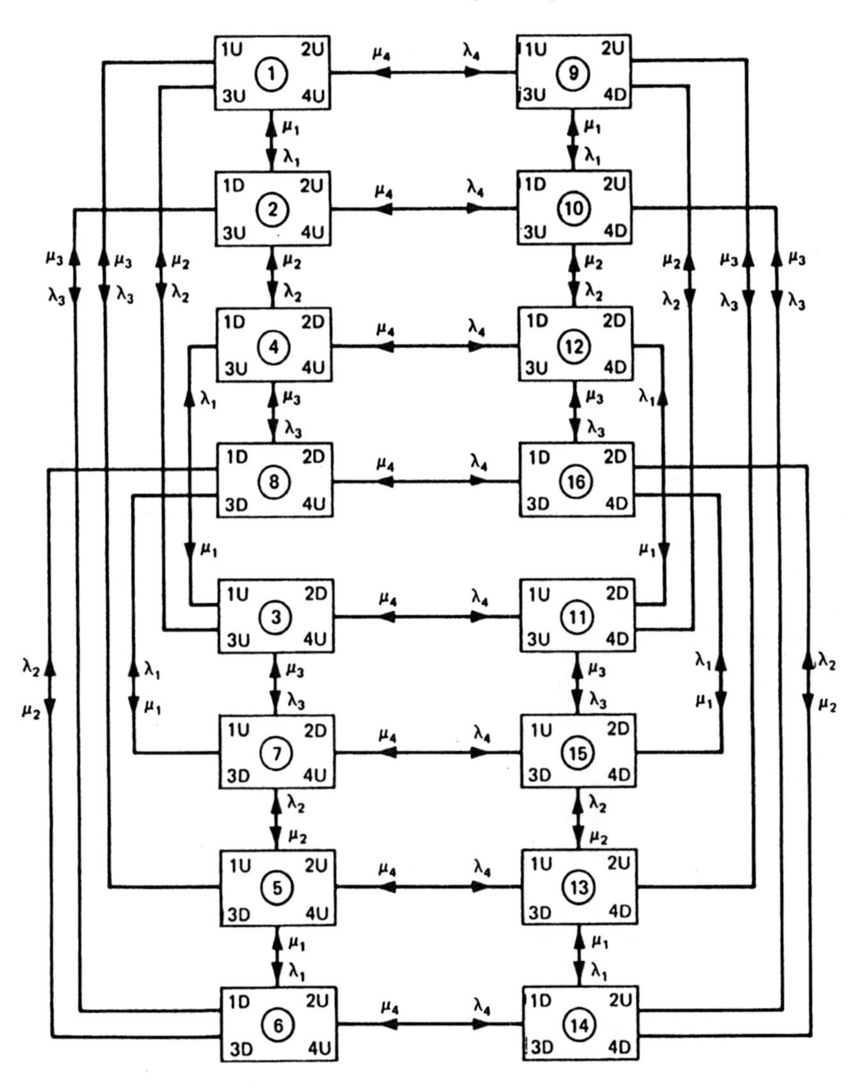

Fig. 8.2 State space diagram for system of Fig. 8.1

8.2.2 Approximate methods

An alternative method to state space diagrams is a method based on a set of appropriate equations for evaluating the failure rate, outage duration and annual outage time or unavailability. This alternative method is described fully in Chapter 11 of *Engineering Systems* and leads to the following set of equations

(a) for two components in parallel

$$\lambda_{pp} = \frac{\lambda_1\lambda_2(r_1 + r_2)}{1 + \lambda_1 r_1 + \lambda_2 r_2} \tag{8.1a}$$

$$\simeq \lambda_1\lambda_2(r_1 + r_2) \quad \text{when } \lambda_i r_i << 1 \tag{8.1b}$$

$$r_{pp} = \frac{r_1 r_2}{r_1 + r_2} \tag{8.2}$$

$$U_{pp} = f_{pp} r_{pp} \tag{8.3a}$$

$$\simeq \lambda_{pp} r_{pp} = \lambda_1\lambda_2 r_1 r_2 \tag{8.3b}$$

The approximations introduced in Equations (8.1) and (8.3) are generally valid for component failures in transmission and distribution networks. Consequently the approximate equations (Equations (8.1b) and (8.3b)) are used almost universally in the reliability evaluation of these systems. The following equations for three components in parallel and all subsequent equations are based on these approximations.

(b) for three components in parallel

$$\lambda_{pp} = \lambda_1\lambda_2\lambda_3(r_1 r_2 + r_2 r_3 + r_3 r_1) \tag{8.4}$$

$$r_{pp} = \frac{r_1 r_2 r_3}{r_1 r_2 + r_2 r_3 + r_3 r_1} \tag{8.5}$$

$$U_{pp} = \lambda_{pp} r_{pp} = \lambda_1\lambda_2\lambda_3 r_1 r_2 r_3 \tag{8.6}$$

As described in Chapter 11 of *Engineering Systems*, Equations (8.1)–(8.6), together with similar equations for higher-order events, can be used either as part of a network reduction process or in conjunction with a minimal cut set approach.

8.2.3 Network reduction method

The network reduction method creates a sequence of equivalent components obtained by gradually combining series and parallel components. In the case of the system shown in Fig. 8.1, this method would combine components 1 and 3 in series, components 2 and 4 in series and finally combine these two equivalent components in parallel. The numerical result can therefore be evaluated using Equations (7.1)–(7.3) and Equations (8.1)–(8.3) as appropriate. If the reliability data for each component of Fig. 8.1 is that shown in Table 8.1, the load point reliability indices can be evaluated as follows. Combining components 1 and 3 in series (Equations (7.1)–(7.3)) gives

$\lambda_{13} = 0.51$ f/yr $\quad r_{13} = 11.76$ hours $\quad U_{13} = 6$ hours/yr

Table 8.1 Reliability data for system of Fig. 8.1

Component	λ *(f/yr)*	*r (hours)*
1	0.5	10
2	0.5	10
3	0.01	100
4	0.01	100

Combining components 2 and 4 will give the same values; the load point reliability indices can therefore be evaluated using Equations (8.1)–(8.3) to give

$$\lambda_{pp} = 0.51 \times 0.51(11.76 + 11.76)/8760 = 6.984 \times 10^{-4} \text{ f/yr}$$

$$r_{pp} = \frac{11.76 \times 11.76}{11.76 + 11.76} = 5.88 \text{ hours}$$

$$U_{pp} = \lambda_{pp} r_{pp} = 4.107 \times 10^{-3} \text{ hours/yr}$$

There are three main disadvantages of this method. These are

(a) It cannot be used to analyze a system in which the components are not simply in series or parallel.

(b) Critical or unreliable areas and components become absorbed into equivalent components and their effect becomes increasingly impossible to identify as the amount of reduction increases. Essential attributes of a properly structured reliability analysis are to identify the events causing a system to fail and the contribution made by each event in addition to the overall values of the load point indices.

(c) The technique is not amenable to further development in order to include different modes of failure, maintenance, weather effects, etc. These aspects will be described in later sections.

Despite these disadvantages, the network reduction method can be useful in practice, particularly in the case of simple hand calculations when extra analytical refinements are not desired.

8.2.4 Failure modes and effects analysis

The alternative to network reduction is a failure modes and effects analysis. The failure modes are directly related to the minimal cut sets of the system and therefore the latter are used to identify the failure modes. The minimal cut set methods and their deduction were described in detail in Section 5.3 of *Engineering Systems* and therefore will not be discussed in depth at this point. It should be noted, however, that the formal application of a minimal cut set algorithm is not always necessary, since it is often possible to identify the failure modes of most systems from a visual inspection. The exceptions are when a digital computer is being used for the analysis, when the system is complex or when high-order events are desired.

The failure modes that are identified in this way represent component outages that must overlap to cause a system outage. The events are therefore defined as overlapping outages and the associated outage time is defined as the overlapping outage time. At this point, only component failures will be considered. These are defined as forced outages. A list of related definitions is included in Appendix 1.

Each overlapping outage is effectively a set of parallel elements and its effect can be evaluated using the equations for parallel components (Equations (8.1)–(8.6)). Also, since each of these overlapping outages will cause system failure, all the overlapping outages are effectively in series from a reliability point of view. The system indices can therefore be evaluated by applying the equations for series components (Equations (7.1)–(7.3)) in order to combine all the overlapping outages.

In order to illustrate this technique, reconsider the system shown in Fig. 8.1 and the reliability data shown in Table 8.1. A failure modes and effects analysis will give the results shown in Table 8.2.

Small differences can be seen between the results shown in Table 8.2 and those shown in Section 8.2.3. The reason for this is that the failure modes analysis is based on approximate equations and differences are therefore expected. The differences will be negligible, however, provided that $\lambda r \ll 1$ for each component; this is normally the case for power system networks. It should be noted that the value of system unreliability is evaluated using a summation rule. This rule gives the upper bound to system unreliability (see Section 5.3.3 of *Engineering Systems*). The lower bound is evaluated by subtracting the product of all pairs of event unavailabilities. In the present example this would give 4.106×10^{-3} hours/yr, a value which is identical to that in Section 8.2.3 to three decimal places and which is sufficiently close to the upper bound for all practical purposes. A similar comparison for most systems would show that the upper and lower bounds are virtually the same within

Table 8.2 Reliability indices for system of Fig. 8.1

Overlapping outages (failure events)	λ_{pp} *(f/yr)*	r_{pp} *(hours)*	U_{pp} *(hours/yr)*
1 and 2	5.708×10^{-4}	5	2.854×10^{-3}
1 and 4	6.279×10^{-5}	9.09	5.708×10^{-4}
2 and 3	6.279×10^{-5}	9.09	5.708×10^{-4}
3 and 4	2.283×10^{-6}	50	1.142×10^{-4}
Total	6.986×10^{-4}	5.88	4.110×10^{-3}

$\lambda_{total} = \Sigma\lambda_{pp}$, $U_{total} = \Sigma U_{pp}$, $r_{total} = \dfrac{\Sigma U_{pp}}{\Sigma\lambda_{pp}}$.

practical limits and fully justifies the omission of using more accurate evaluation methods.

An appraisal of the results shown in Table 8.2 shows that considerably more information is given by the failure modes approach. For instance, in the present example the results show that the system failure rate and unavailability are mainly due to the overlapping forced outage of the two lines but that the system outage duration is mainly due to the overlapping forced outage of the two transformers. This type of information can be vital in assessing critical areas and deducing those areas in which investment will give the greatest reliability improvement. This information is not readily deducible from the network reduction method, particularly when the system increases in size.

8.3 Inclusion of busbar failures

The results shown in Sections 8.2.3 and 8.2.4 were evaluated assuming the breakers and busbars were 100% reliable. Failures of these components can be taken into account using both network reduction and failure modes analyses. The effect of busbars is considered in this section and that of breakers in Section 8.9. Assume the busbars in Fig. 8.1 have the reliability data shown in Table 8.3.

(a) *Network reduction*

The two busbars are effectively single components in series with the two parallel branches. Their effect on the load point reliability indices can therefore be evaluated by using Equations (7.1)–(7.3) and the results previously obtained in Section 8.2.3, i.e.

$$\lambda_{pp} = 6.984 \times 10^{-4} + 0.01 + 0.02 = 3.070 \times 10^{-2} \text{ f/yr}$$

$$U_{pp} = 4.107 \times 10^{-3} + 0.01 \times 5 + 0.02 \times 2 = 9.411 \times 10^{-2} \text{ hours/yr}$$

$$r_{pp} = \frac{U_{pp}}{\lambda_{pp}} = 3.07 \text{ hours}$$

(b) *Failure modes analysis*

The additional failure modes (minimal cut sets) are component 5 and component 6, giving the results shown in Table 8.4.

Table 8.3 Reliability indices of busbars in Fig. 8.1

Component	*λ (f/yr)*	*r (hours)*
5	0.01	5
6	0.02	2

Table 8.4 Reliability indices including busbar failures

Failure event	λ_{pp} *(f/yr)*	r_{pp} *(hours)*	U_{pp} *(hours/yr)*
Subtotal from Table 8.2	6.986×10^{-4}	5.88	4.110×10^{-3}
5	1×10^{-2}	5	5×10^{-2}
6	2×10^{-2}	2	4×10^{-2}
Total	3.070×10^{-2}	3.07	9.411×10^{-2}

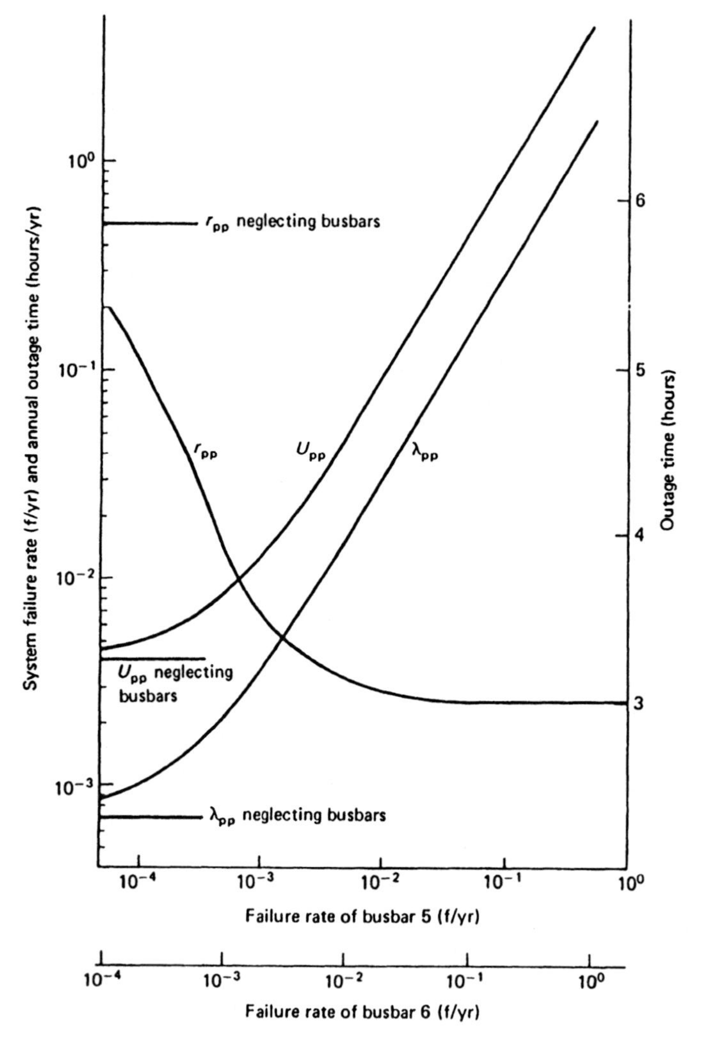

Fig. 8.3 Effect of busbar failures

It is seen from these results that the load point reliability is dominated by failures of the two busbars. This is expected because these two components are series components and therefore counteract much of the benefit provided by the redundancy aspect of the parallel branches. Recognition of series components can be an important aspect in system design because they frequently possess the dominant role. This must not be assumed to be a general conclusion, however, because their contribution depends on their own individual reliability indices. For instance, if the failure rate of each busbar was very small compared with that of the parallel branches, the contribution by the parallel branches could still be significant. This effect is illustrated by the results of Fig. 8.3, which shows the variation in the values of λ_{pp}, r_{pp} and U_{pp} of the load point as the failure rate of the busbars are increased, the repair times being those shown in Table 8.3.

8.4 Inclusion of scheduled maintenance

8.4.1 General concepts

A detailed discussion of scheduled maintenance together with the modelling and evaluation techniques was given in Section 11.6 of *Engineering Systems*. The major part of this discussion will therefore not be repeated at this point. A summary is included, however, together with the application of the techniques to the power system problem.

A scheduled maintenance outage, as defined in Appendix 1, is an outage that is planned in advance, is deferable if necessary and involves the removal of a component or components in order to perform preventive maintenance.

As discussed in Section 11.6 of *Engineering Systems*, a scheduled outage is not usually included in the reliability evaluation of a load point if, by this action alone, the load point is disconnected. Consequently scheduled maintenance of single radial systems or of single series components in a parallel or meshed system, e.g. busbars 5 and 6 of Fig. 8.1, are not considered. This does not mean that such outages do not occur in practice, but instead it acknowledges the fact that customers can be notified in advance or alternative arrangements can be made and therefore the load point is outaged deliberately, which cannot be considered as a randomly occurring event.

These concepts lead to the conclusion that scheduled maintenance is simulated only on overlapping events associated with parallel and meshed networks. In addition, however, as discussed in Section 11.6 of *Engineering Systems*, only the sequence 'maintenance followed by a forced outage', i.e. a forced outage overlapping an (existing) maintenance outage is considered, because the reverse sequence would cause disruption of the load point due to the maintenance outage alone.

8.4.2 Evaluation techniques

Using the concepts of Section 8.4.1 and the discussion of Section 11.6 of *Engineering Systems*, the equations for forced outages overlapping a maintenance outage previously given in Section 11.6 of *Engineering Systems* can be derived. Let λ_i'' and r_i'' be the maintenance outage rate and maintenance time of component i, λ_j and r_j be the forced outage failure rate and repair time of component j and λ_{pm}, r_{pm} and U_{pm} be the rate, duration and annual outage time of the load point due to forced outages overlapping a maintenance outage. From Equations (11.23)–(11.26) of *Engineering Systems*

(a) for two components in parallel or a second-order failure event (minimal cut set)

$$\lambda_{pm} = \lambda_1''(\lambda_2 r_1'') + \lambda_2''(\lambda_1 r_2'') \tag{8.7}$$

$$U_{pm} = \lambda_1''(\lambda_2 r_1'')\frac{r_1'' r_2}{r_1'' + r_2} + \lambda_2''(\lambda_1 r_2'')\frac{r_1 r_2''}{r_1 + r_2''} \tag{8.8}$$

$$r_{pm} = U_{pm}/\lambda_{pm} \tag{8.9}$$

(b) for three components in parallel or a third-order failure event

$$\lambda_{pm} = \lambda_a + \lambda_b + \lambda_c + \lambda_d + \lambda_e + \lambda_f \tag{8.10}$$

$$U_{pm} = \lambda_a r_a + \lambda_b r_b + \lambda_c r_c + \lambda_d r_d + \lambda_e r_e + \lambda_f r_f \tag{8.11}$$

and

$$r_{pm} = U_{pm}/\lambda_{pm} \tag{8.12}$$

where

$$\lambda_a = \lambda_1''(\lambda_2 r_1'')\left(\lambda_3 \frac{r_1'' r_2}{r_1'' + r_2}\right) \qquad \lambda_b = \lambda_1''(\lambda_3 r_1'')\left(\lambda_2 \frac{r_1'' r_3}{r_1'' + r_3}\right)$$

$$\lambda_c = \lambda_2''(\lambda_1 r_2'')\left(\lambda_3 \frac{r_2'' r_1}{r_2'' + r_1}\right) \qquad \lambda_d = \lambda_2''(\lambda_3 r_2'')\left(\lambda_1 \frac{r_2'' r_3}{r_2'' + r_3}\right)$$

$$\lambda_e = \lambda_3''(\lambda_1 r_3'')\left(\lambda_2 \frac{r_3'' r_1}{r_3'' + r_1}\right) \qquad \lambda_f = \lambda_3''(\lambda_2 r_3'')\left(\lambda_1 \frac{r_3'' r_2}{r_3'' + r_2}\right)$$

$$r_a = r_b = \frac{r_1'' r_2 r_3}{r_1'' r_2 + r_2 r_3 + r_3 r_1''} \qquad r_c = r_d = \frac{r_1 r_2'' r_3}{r_1 r_2'' + r_2'' r_3 + r_3 r_1}$$

$$r_e = r_f = \frac{r_1 r_2 r_3''}{r_1 r_2 + r_2 r_3'' + r_3'' r_1}$$

Similar equations can be deduced for any number of parallel components or any order of failure event using the same logic.

Since the system can fail either due to forced outages overlapping forced outages or forced outages overlapping maintenance outages, the two effects can be combined together to give the overall reliability indices of the load point, i.e.

$$\lambda = \lambda_{pp} + \lambda_{pm} \tag{8.13}$$

$$U = \lambda_{pp} r_{pp} + \lambda_{pm} + r_{pm} \tag{8.14}$$

$$r = U/\lambda \tag{8.15}$$

where λ_{pm} and r_{pm} are given by Equations (8.7)–(8.12) and λ_{pp} and r_{pp} (forced outages overlapping forced outages) are given by Equations (8.1)–(8.6).

8.4.3 Coordinated and uncoordinated maintenance

An implied assumption in Equations (8.7)–(8.12) is that each component is removed from service for scheduled maintenance quite separately and independently of all others. This is a maintenance policy that is prevalent in many utilities and can be described as an uncoordinated maintenance policy. An alternative maintenance policy is a coordinated one in which each component of a given branch is maintained simultaneously. The merit of this policy is that the total exposure time in a year during which a component in another branch may fail is reduced and therefore the probability of system failure is also reduced. The disadvantage of the policy is that it may require additional manpower in order to maintain several components at the same time which consequently increases the operational costs. The improvement in reliability, however, may justify this increased expenditure particularly if increased reliability is necessary and the alternative is increased capital expenditure to improve the system redundancy. There can be no general conclusion on this point since it depends on relative values of labor costs, equipment costs and outage costs. An economic appraisal is therefore required of all these factors and their effect on the incremental cost of reliability.

The evaluation of the contribution to the load point indices due to coordinated maintenance is essentially the same as for uncoordinated maintenance. The only difference is that, instead of considering component forced outages overlapping component maintenance outages, component forced outages overlapping branch maintenance outages are considered. Referring to Fig. 8.1, assuming busbars and breakers are 100% reliable and defining branch 1 as that containing components 1, 3, 7 and 9 and branch 2 as that containing components 2, 4, 8 and 10, the following failure events associated with component forced outages overlapping branch maintenance outages can be deduced:

—branch 1 on maintenance and component 2 forced out
—branch 1 on maintenance and component 4 forced out

—branch 2 on maintenance and component 1 forced out
—branch 2 on maintenance and component 3 forced out.

A similar but longer list would be deduced if failures of the breakers were also considered. After deducing these failure events, the reliability indices can be evaluated using modified forms of Equations (8.7)–(8.12). The modification requires only the first term of Equations (8.7) and (8.8) and only the first two terms of Equations (8.10) and (8.11). This modification is necessary because, for example, λ_1'' in Equation (8.7) represents the maintenance rate of branch 1 and λ_2 represents the failure rate of the component. This completely defines the relevant failure event. The second term of Equation (8.7) therefore has no practical meaning. A similar conclusion can be made for the last four terms of Equation (8.10).

8.4.4 Numerical example

(a) *Uncoordinated maintenance*

Consider first an uncoordinated maintenance policy in the system of Fig. 8.1 and assume that each component is individually removed for scheduled maintenance once a year for a period of 8 hours.

In order to illustrate the method of evaluation, consider the second-order event involving line 1 and transformer 4. Using the above maintenance data and the data in Table 8.1 with Equations (8.7)–(8.9) gives

$$\lambda_{pm}(1,4) = [1 \times (0.01 \times 8) + 1 \times (0.5 \times 8)]/8760 = 4.658 \times 10^{-4} \text{ f/yr}$$

$$U_{pm}(1,4) = \frac{1 \times 0.01 \times 8}{8760}\left(\frac{8 \times 100}{8 + 100}\right) + \frac{1 \times 0.5 \times 8}{8760}\left(\frac{8 \times 10}{8 + 10}\right)$$
$$= 2.097 \times 10^{-3} \text{ hours/yr}$$

$$r_{pm}(1,4) = 4.50 \text{ hours}$$

The complete set of results is shown in Table 8.5, assuming that busbars and breakers are 100% reliable.

The results shown in Table 8.5 are increased significantly due to the contribution of scheduled maintenance. The difference could be much less significant and sometimes negligible for systems in which one or more dominant series components exist. This would occur, for instance, if the breakers and busbars were not considered to be 100% reliable.

(b) *Coordinated maintenance*

Consider now a coordinated maintenance policy in the system of Fig. 8.1 and assume that each branch (as defined in Section 8.4.3) is maintained once a year for 8 hours. As an example of evaluation, consider the event, branch 1 on maintenance and component 2 forced out. The contribution by this event is

Table 8.5 Reliability indices neglecting busbars and breakers

Failure event	λ_{pm} *(f/yr)*	r_{pm} *(hours)*	U_{pm} *(hours/yr)*
1 + 2	9.132×10^{-4}	4.44	4.059×10^{-3}
1 + 4	4.658×10^{-4}	4.50	2.097×10^{-3}
3 + 2	4.658×10^{-4}	4.50	2.097×10^{-3}
3 + 4	1.826×10^{-5}	7.41	1.353×10^{-4}
Subtotal	1.863×10^{-3}	4.50	8.388×10^{-3}
	λ_{pp} *(f/yr)*	r_{pp} *(hours)*	U_{pp} *(hours/yr)*
Indices from Table 8.2	6.986×10^{-4}	5.88	4.110×10^{-3}
	λ *(f/yr)*	*r (hours)*	*U (hours/yr)*
Total load-point indices	2.562×10^{-3}	4.88	1.25×10^{-2}

$$\lambda_{pm}(\text{branch 1, comp.2}) = 1 \times (0.5 \times 8)/8760 = 4.566 \times 10^{-4} \text{ f/yr}$$

$$r_{pm}(\text{branch 1, comp. 2}) = \frac{8 \times 10}{8 + 10} = 4.44 \text{ hours}$$

$$U_{pm}(\text{branch 1, comp. 2}) = \lambda_{pm} r_{pm} = 2.029 \times 10^{-3} \text{ hours/yr}$$

A complete list of the events and indices is shown in Table 8.6 for the case when breakers and busbars are considered 100% reliable. These results can be compared with those of Table 8.5 which shows that the annual outage time due to

Table 8.6 Reliability indices assuming coordinated maintenance

Failure event	λ_{pm} *(f/yr)*	r_{pm} *(hours)*	U_{pm} *(hours/yr)*
Branch 1 +			
2 out	4.566×10^{-4}	4.44	2.029×10^{-3}
4 out	9.132×10^{-6}	7.41	6.765×10^{-5}
Branch 2 +			
1 out	4.566×10^{-4}	4.44	2.029×10^{-3}
3 out	9.132×10^{-6}	7.41	6.765×10^{-5}
Subtotal	9.315×10^{-4}	4.50	4.193×10^{-3}
	λ_{pp} *(f/yr)*	r_{pp} *(hours)*	U_{pp} *(hours/yr)*
Indices from Table 8.2	6.986×10^{-4}	5.88	4.110×10^{-3}
	λ *(f/yr)*	*r (hours)*	*U (hours/yr)*
Total load-point indices	1.630×10^{-3}	5.09	8.303×10^{-3}

maintenance effects, U_{pm}, is reduced by a factor of 2, since the total exposure time during which each branch is out of service due to maintenance has been halved by the coordinated maintenance policy.

8.5 Temporary and transient failures

8.5.1 Concepts

The techniques and equations in Sections 8.2–4 consider all failures to be grouped together with an overall failure rate and average repair time. This may be justified for expediency and simplification of the calculations but it overshadows the significance and effects of the different types of failures that can arise in a power system network.

Two particular types of failures that can occur are those that cause damage to the component which must then be repaired and those that do not damage the component. An example of the second type is a lightning strike which trips the protection breakers or blows the protection fuses but does not damage the shorted components. Service is then restored by closing the breakers automatically or manually or replacing the fuses. In either case the outage time is relatively small and may be negligible with automatic reclosure. The effect on the customer of these types of failures is therefore significantly different from that of failures that require components to be repaired, and it is beneficial to separate them in the reliability evaluation of the network.

The IEEE standard [7] defining forced outages (see Appendix 1) designates these two types of failures as permanent forced outages and transient forced outages respectively. In the UK they are generally known as damaged faults and non-damaged faults, respectively. The actual term used is not particularly fundamental provided it is clearly understood which failures are included in the two categories and their effect on the system behaviour. In this book, the non-damaged type of failure will be further subdivided into transient forced outages which are restored by automatic switching and temporary forced outages which are restored by manual switching or fuse replacement. The reason is that the outage time of transient outages is negligible and frequently ignored, whereas that of temporary outages may be quite long, particularly if the restoration action must be performed in rural areas. The basic method of analysis, however, is the same for both of these sub-categories.

8.5.2 Evaluation techniques

The method for evaluating the reliability indices of a load point to include the effect of transient and temporary failures is an extension of the concepts already described in Sections 8.2–4.

The failure modes of the system are identified either visually or using the minimal cut set method. This will give failure events of first order, second order and, if required, higher orders.

(a) *First order*

The first-order events are evaluated using directly the concepts of Equations (7.1)–(7.3), in which the values λ_i and r_i are the appropriate failure rate and outage duration respectively of the transient or temporary outage. In the case of transient failures (automatic restoration) the values of r_i may be negligible, in which case the contribution of such failures to the annual outage time can be neglected. This may not be valid in the case of temporary failures, since the outage time may be several hours if detection of the cause of failure is difficult and the site of failure is in a remote rural area.

(b) *Second order*

The second-order events can be evaluated using the concepts of Equations (8.1)–(8.3) and of overlapping failure events. The overlapping events that could be considered are

(i) a temporary or transient failure overlapping a temporary or transient failure;
(ii) a temporary or transient failure overlapping a permanent failure;
(iii) a temporary or transient failure overlapping a scheduled maintenance outage.

(i) *Temporary/transient failures overlapping temporary/transient failures*

These events are frequently neglected in practice because the probability is small and contributes very little to the overall result. They can be included, however, using a set of equations similar to those described below for type (ii).

(ii) *Temporary/transient failures overlapping a permanent failure*

The reliability indices associated with these events are evaluated using the concepts of Equations (8.1)–(8.3)

$$\lambda_{pt} = \lambda_{t1}(\lambda_2 r_{t1}) + \lambda_2(\lambda_{t1} r_2) + \lambda_{t2}(\lambda_1 r_{t2}) + \lambda_1(\lambda_{t2} r_1)$$

$$= \lambda_{t1}\lambda_2(r_{t1} + r_2) + \lambda_1\lambda_{t2}(r_1 + r_{t2}) \qquad (8.16)$$

$$= \lambda_a + \lambda_b$$

$$U_{pt} = \lambda_a \frac{r_{t1} r_2}{r_{t1} + r_2} + \lambda_b \frac{r_1 r_{t2}}{r_1 + r_{t2}} \qquad (8.17)$$

$$r_{pt} = U_{pt}/\lambda_{pt} \qquad (8.18)$$

where λ_{ti} = transient or temporary failure rate of component i

r_{ti} = restoration or reclosure time following temporary or transient failure of component i

λ_j, r_j = permanent failure rate and repair time of component j respectively

When the reclosure time is small or negligible, this being particularly the case for transient failures, Equations (8.16)–(8.18) reduce to

$$\lambda_{pt} = \lambda_{t1}\lambda_2 r_2 + \lambda_1\lambda_{t2} r_1 \tag{8.19}$$

$$U_{pt} = \lambda_{t1}\lambda_2 r_{t1} r_2 + \lambda_1\lambda_{t2} r_1 r_{t2} \tag{8.20}$$

$$r_{pt} = U_{pt}/\lambda_{pt} \tag{8.21}$$

If it is felt necessary to include the effect of temporary or transient failures overlapping transient or temporary failures, i.e. type (i) above, then Equations (8.16)–(8.21) are modified by considering the λ_a terms only and replacing λ_2 and r_2 by λ_{t2} and r_{t2} respectively.

(iii) *Temporary/transient failures overlapping a maintenance outage*

The reliability indices associated with these events are evaluated using the concepts of Equations (8.7)–(8.9)

$$\lambda_{tm} = \lambda_1''(\lambda_{t2} r_1'') + \lambda_2''(\lambda_{t1} r_2'') \tag{8.22}$$

$$U_{tm} = \lambda_1''(\lambda_{t2} r_1'')\frac{r_1'' r_{t2}}{r_1'' + r_{t2}} + \lambda_2''(\lambda_{t1} r_2'')\frac{r_{t1} r_2''}{r_{t1} + r_2''} \tag{8.23}$$

$$r_{tm} = U_{tm}/\lambda_{tm} \tag{8.24}$$

Similar equations can be derived for third-order events. These are shown in Appendix 3 for the case when a temporary or transient failure overlaps two permanent outages and a temporary or transient failure overlaps one permanent outage and one maintenance outage. Further equations can be deduced for other combinations of overlapping outages using the same basic concepts.

The overall indices are then given by

$$\lambda = \lambda_{pp} + \lambda_{pm} + \lambda_{pt} + \lambda_{tm} \tag{8.25a}$$

$$U = \lambda_{pp} r_{pp} + \lambda_{pm} r_{pm} + \lambda_{pt} r_{pt} + \lambda_{tm} r_{tm} \tag{8.25b}$$

$$r = U/\lambda \tag{8.25c}$$

Table 8.7 Data of temporary forced outages

Component	*Temporary failure rate,* λ_t *(f/yr)*	*Reclosure time,* r_t *(hours)*
1	2	0.25
2	2	0.25
3	1	0.5
4	1	0.5

8.5.3 Numerical example

Reconsider the system shown in Fig. 8.1 and assume that, in addition to the data given in the previous sections, the lines and transformers suffer temporary failures having the data shown in Table 8.7.

Table 8.8 Reliability indices including temporary failures

Failure event	λ_{pt} *(f/yr)*	r_{pt} *(hours)*	U_{pt} *(hours/yr)*
1 + 2	2.340×10^{-3}	0.24	5.708×10^{-4}
1 + 4	8.282×10^{-4}	0.41	3.425×10^{-4}
3 + 2	8.282×10^{-4}	0.41	3.425×10^{-4}
3 + 4	2.295×10^{-4}	0.50	1.142×10^{-4}
Subtotal 1	4.226×10^{-3}	0.32	1.370×10^{-3}
	λ_{tm} *(f/yr)*	r_{tm} *(hours)*	U_{tm} *(hours/yr)*
1 + 2	3.653×10^{-3}	0.24	8.856×10^{-4}
1 + 4	2.740×10^{-3}	0.32	8.725×10^{-4}
3 + 2	2.740×10^{-3}	0.32	8.725×10^{-4}
3 + 4	1.826×10^{-3}	0.47	8.595×10^{-4}
Subtotal 2	1.096×10^{-2}	0.32	3.490×10^{-3}
Subtotal 1 & Subtotal 2	1.519×10^{-2}	0.32	4.860×10^{-3}
	λ_{pp} *(f/yr)*	r_{pp} *(hours)*	U_{pp} *(hours/yr)*
Indices from Table 8.2	6.986×10^{-4}	5.88	4.110×10^{-3}
	λ_{pm} *(f/yr)*	r_{pm} *(hours)*	U_{pm} *(hours/yr)*
Indices from Table 8.5	1.863×10^{-3}	4.50	8.388×10^{-3}
Subtotal 3	2.562×10^{-3}	4.88	1.250×10^{-2}
Total load-point indices	1.775×10^{-2}	0.98	1.736×10^{-2}

If the effects of breakers and busbars are again neglected, the modified reliability indices using Equations (8.16)–(8.18) and (8.22)–(8.24) are as shown in Table 8.8.

The results shown in Table 8.8 indicate an effect frequently found in the analysis of distribution networks, that is, the annual outage time is increased by a relatively small margin (compare Tables 8.5 and 8.8) but the failure rate is increased sharply and the average duration is decreased. For this reason it is often a reasonable approximation to neglect temporary and particularly transient failures if only the annual outage time is considered important, but this type of failure should be included if all three indices are to be evaluated.

It should be noted that the summation of the indices associated with different types of events leads to a set of indices that have no real physical meaning. Instead they are simply the long-run average values that would be expected given the random sequence of failure events that are possible. They are useful, however, in assessing the average behavior of a system as a function of alternative reinforcement schemes and as a set of input data to an economic evaluation of such schemes.

8.6 Inclusion of weather effects

8.6.1 Concepts

All power system networks are exposed to varying weather conditions. This in itself would not pose any problems but it is found from experience that the failure rate of most components is a function of the weather to which they are exposed. In some weather conditions, the failure rate of a component can be many times greater than that found in the most favorable weather conditions. For these reasons, the effect of weather (and any other environmental condition) has been considered for several years and techniques have been developed that permit its effect to be included in the analysis.

The weather conditions that cause high failure rates of components are generally infrequent and of short duration. During these periods, however, the failure rates increase sharply and the probability of overlapping failures is very much greater than that in favorable weather. This creates what is known as the bunching effect due to the fact that component failures are not randomly distributed throughout the year but are more probable in constrained short periods in the year. If this fact is neglected, the reliability indices evaluated for a load point can be over-optimistic and consequently very misleading.

It should be noted that the techniques used to account for failure bunching do not imply that there is dependence between the failures of components. Although the components may reside within a common environment which affects the failure rates of the components, the actual failure process of overlapping outages still assumes the component failures to be independent. There is no suggestion therefore that the process is a common mode or dependent failure, only that the independent failure rates are enhanced because of the common environment.

It is also worth noting that, although the following techniques are described in relation to a common weather environment, they are equally applicable to failure processes in other types of varying environment such as temperature, stress, etc.

8.6.2 Weather state modelling

The failure rate of a component is a continuous function of weather, which suggests that it should be described either by a continuous function or by a large set of discrete states. This proves impossible in practice due to difficulties in system modelling, data collection and data validation. The problem must therefore be restricted to a limited number of states, a number which is sufficient to represent the problem of failure bunching but small enough to make the solution tractable.

The IEEE Standard [7] subdivides the weather environment into three classifications: normal, adverse and major storm disaster. These are defined in Appendix 1. Although techniques have been developed [8, 9] to evaluate the effect of these three weather states, the problems are still great and therefore only the first two—normal and adverse—are generally considered. The third state, major storms, is usually reserved for consideration of major system disturbances.

The large range of weather conditions must therefore be classified as either normal or adverse. This frequently causes concern and is one reason why two-state weather modelling has been seldom used in the past. The criterion for deciding into which category each type of weather must be placed is dependent on its impact on the failure rate of components. Those weather conditions having little or no effect on the failure rate should be classed as normal and those having a large effect should be classed as adverse. Examples of adverse weather include lightning storms, gales, typhoons, snow and ice.

One important feature in the collection of weather durations is that all periods of normal and adverse weather must be collated even if no failures occur during any given period. This point cannot be stressed too greatly since it is of little use allocating a particular failure event to normal weather or adverse weather after it has occurred if the starting and finishing times of the weather periods have not been

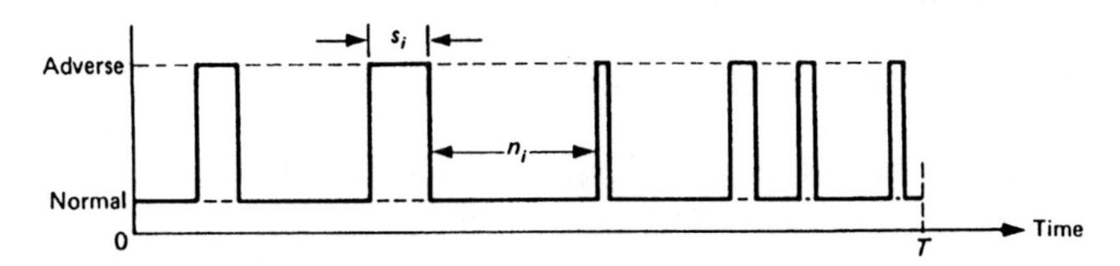

Fig. 8.4 Chronological variation of weather

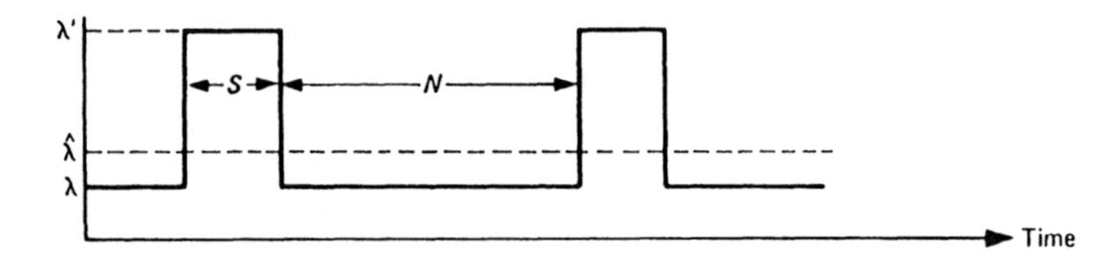

Fig. 8.5 Average weather duration profile

ascertained. This aspect requires cooperation between the utility and the appropriate weather bureau or weather center. Failure to collect such statistics comprehensively will mean significant errors, not only in the statistics themselves, but also in subsequent reliability analyses.

Data collected for durations of weather will produce a chronological variation as depicted in Fig. 8.4.

The pattern of durations of weather can be considered a random process which can then be described by expected values, i.e. expected duration of normal weather is given by $N = \Sigma_i n_i/T$ and expected duration of adverse weather is given by $S = \Sigma_i s_i/T$. These expected values produce the average weather profile shown in Fig. 8.5.

8.6.3 Failure rates in a two-state weather model

When a decision has been made concerning which weather conditions contribute to the constrained two-state model, all subsequent failures should be allocated to one of these states depending on the prevailing weather at the time of failure. This then permits the failure rate in each of the weather states to be ascertained. It should be noted that these failure rates must be expressed as the number of failures per year of that particular weather condition and not as the number of failures in a calendar year. This requirement follows from the concepts and definition of a transition rate as described in Section 9.2.1 of *Engineering Systems*. It is evident therefore that, because adverse weather is generally of short duration, several calendar years of operation may be necessary to achieve one year of adverse weather.

Define

λ — component failure rate in normal weather expressed in failures/year of normal weather

λ' — component failure rate in adverse weather expressed in failures/year of adverse weather

An average value of failure rate $\hat{\lambda}$ expressed in failures per calendar year can be derived from λ, λ', N and S using the concept of expectation, i.e.

$$\hat{\lambda} = \frac{N}{N+S}\lambda + \frac{S}{N+S}\lambda' \tag{8.26a}$$

Table 8.9 Relative magnitude of λ and λ'

F	λ (*f/yr of normal weather*)	λ' (*f/yr of adverse weather*)
0	0.600	0.000
0.5	0.300	30.0
1.0	0.000	60.0

Since generally $N >> S$, the value of $\hat{\lambda}$ is approximately equal to λ. These values of λ, λ' and $\hat{\lambda}$ are also shown in Fig. 8.5.

At the present time, most data collection schemes do not recognize λ and λ' but instead are only responsive to $\hat{\lambda}$. This is now gradually changing and more and more utilities are recognizing the need to identify this data. As this development continues, both the quality of the fault reporting scheme and the quality of the reliability analysis will benefit. The values of λ and λ' can, however, be evaluated from $\hat{\lambda}$ using Equation (8.26a) if the values of N, S and the proportion of failures (F) occurring in adverse weather are known, since

$$\lambda = \hat{\lambda}\frac{N+S}{N}(1-F) \tag{8.26b}$$

$$\lambda' = \hat{\lambda}\frac{N+S}{S}F \tag{8.26c}$$

Even if the value of F is unknown, a complete sensitivity analysis can be made using $0 \leq F \leq 1$ to establish the effect of adverse failures on the behaviour of the system.

The relative magnitude of λ and λ' can be illustrated by considering a realistic numerical example in which $\hat{\lambda} = 0.594$ f/yr, $N = 200$ hours, $S = 2$ hours. These values are shown in Table 8.9 for values of $F = 0$, 0.5 and 1.0, i.e. no failures, 50% of failures and all failures occur in adverse weather, respectively.

The results shown in Table 8.9 clearly indicate that the failure rate during the short adverse weather periods is very large, much greater than the overall average value and would significantly increase the probability of overlapping failures during these periods.

It is worth noting at this point the significance of λ, λ' and $\hat{\lambda}$. Although a data collection scheme may identify and store $\hat{\lambda}$, this is not a physical parameter of the components but is only a statistical quantity that relates λ, λ', N and S. It therefore does not truly represent the behaviour of the components. The real physical parameters determining failure of components are the values of λ and λ'. In order to illustrate this effect, consider two identical components subjected to different two-state weather patterns; the first has durations N_1 and S_1, the second has durations N_2 and S_2 and the adversity of each type of weather is the same for both components. It follows therefore that two entirely different values of $\hat{\lambda}$ would be obtained. The physical failure mechanism for both components would be identical, however, and so would the values of λ and λ'. Consequently the consistency of and confidence in the data would increase if λ and λ' are collected instead of $\hat{\lambda}$.

8.6.4 Evaluation methods

The first contribution [10,11] to the evaluation of a two-state weather model proposed a set of approximate equations for use with a network reduction method. These, although a major step forward, contained certain weaknesses which were identified from a Markov analysis [12] of the same problem. Subsequently a modified set of equations were proposed [8] which now form the basis of most evaluation methods.

These equations, which will be described in the next sections, can be used as part of a network reduction process or, more fruitfully, in association with a failure modes (minimal cut set) analysis. One fundamentally important feature is that the equations for second-order events must not be used to combine sequentially three or more parallel components. Considerable errors would accrue which does not occur in the case of a single-state weather model. The appropriate set of equations must therefore be used for each order of event being evaluated.

The equations deduced in the following sections are considered only for the general case of a second-order event involving a forced outage overlapping a forced outage and a forced outage overlapping a maintenance outage. These equations can be enhanced by subdividing the forced outage into permanent, temporary and transient using the concepts described in Section 8.5 and by extending the concepts of second-order events to third- and higher-order events. (These extended equations are shown in Appendix 3.)

8.6.5 Overlapping forced outages

The effect of weather on the reliability indices associated with overlapping outages is established by considering four separate cases. These are:

(a) initial failure occurs during normal weather, second failure occurs during normal weather;
(b) initial failure occurs during normal weather, second failure occurs during adverse weather;
(c) initial failure occurs during adverse weather, second failure occurs during normal weather;
(d) initial failure occurs during adverse weather, second failure occurs during adverse weather.

These four cases are mutually exclusive and exhaustive. The indices evaluated for each case can therefore be combined using conditional probability and the concepts of overlapping events described by Equation (11.19) of *Engineering Systems*.

Two constraints are imposed on the evaluation process: repair can be done during adverse weather; repair cannot be done during adverse weather.

(i) *Repair can be done during adverse weather*

(a) *Both failures occur during normal weather*

The contribution of this case to the overall failure rate is given by

λ_a = (probability of normal weather) × [(failure rate of component 1) × (probability of component 2 failing during the exposure time created by the failure of component 1)
+ (failure rate of component 2) × (probability of component 1 failing during the exposure time created by the failure of component 2)]

In this case the 'exposure time' is not simply the repair time of the failed component because repair can proceed into the adverse weather period. The second failure must occur during the proportion of the repair time that takes place in the normal weather period, i.e. the 'exposure time' is the time associated with the overlapping event of repair and normal weather and this is equal to $Nr/(N+r)$.

Therefore

$$\lambda_a = \frac{N}{N+S}\left[\lambda_1\left(\lambda_2 \frac{Nr_1}{N+r_1}\right) + \lambda_2\left(\lambda_1 \frac{Nr_2}{N+r_2}\right)\right] \tag{8.27a}$$

and if $r_i << N$, then

$$\lambda_a = \frac{N}{N+S}\left[\lambda_1(\lambda_2 r_1) + \lambda_2(\lambda_1 r_2)\right]$$

$$= \frac{N}{N+S}\left[\lambda_1\lambda_2(r_1 + r_2)\right] \tag{8.27b}$$

Equation (8.27b) therefore reduces to Equation (8.1) if only one weather state is considered.

(b) *Initial failure in normal weather, second failure in adverse weather*

The same principle is used in this case with the addition that the second failure can occur only if the weather changes before the second failure occurs. Consequently the failure rate of the second component is weighted by the probability that, during the repair time of the first component, the weather changes from normal to adverse. Also the 'exposure time' during which the second component must fail is the overlapping time associated with the repair of the first component and the duration of adverse weather. Therefore

$$\lambda_b = \frac{N}{N+S}\left[\lambda_1\left(\frac{r_1}{N}\right)\left(\lambda_2'\frac{Sr_1}{S+r_1}\right)+\lambda_2\left(\frac{r_2}{N}\right)\left(\lambda_1'\frac{Sr_2}{S+r_2}\right)\right] \tag{8.28}$$

where (r_1/N) represents the probability that the weather changes from normal to adverse during the repair of component 1. This can be deduced assuming exponential distributions since, in general,

$$\text{Prob(event)} = 1 - e^{-\lambda t}$$

and in this case, $t = r_1$ and $\lambda = 1/N$. Thus

$$\text{Prob(weather changing during repair of component 1)} = 1 - e^{-r_1/N}$$

which, if $r_1 << N$, reduces to

$$\text{Prob} \simeq 1 - \left(1 - \frac{r_1}{N}\right) + \cdots \simeq \frac{r_1}{N}$$

Similarly for the second term of Equation (8.28). This equation cannot be reduced further because the assumption $r_i << S$ is not generally valid.

(c) *Initial failure in adverse weather, second failure in normal weather*

This case is evaluated similarly to the second. In this case however

$$\text{Prob(weather changing during repair of component 1)} = 1 - e^{-r_1/S}.$$

Since r_1 and S are comparable, the simplification used in part (b) is not valid. Therefore either the probability of weather changing is used in its full form as shown above or this value of probability is assumed to be unity. Using the second assumption,

$$\lambda_c = \frac{S}{N+S}\left[\lambda_1'\left(\lambda_2\frac{Nr_1}{N+r_1}\right)+\lambda_2'\left(\lambda_1\frac{Nr_2}{N+r_2}\right)\right] \tag{8.29a}$$

which, if $r_i << N$, then

$$\lambda_c = \frac{S}{N+S}[\lambda_1'\lambda_2 r_1 + \lambda_2'\lambda_1 r_2] \tag{8.29b}$$

(d) *Both failures occur during adverse weather*

This case is similar to (a) and gives

$$\lambda_d = \frac{S}{N+S}\left[\lambda_1'\left(\lambda_2'\frac{Sr_1}{S+r_1}\right)+\lambda_2'\left(\lambda_1'\frac{Sr_2}{S+r_2}\right)\right] \tag{8.30}$$

(e) *Overall reliability indices*

The overall failure rate is given by

$$\lambda_{pp} = \lambda_a + \lambda_b + \lambda_c + \lambda_d \tag{8.31}$$

and since there is no restriction on repair,

$$r_{pp} = \frac{r_1 r_2}{r_1 + r_2} \tag{8.32}$$

$$U_{pp} = \lambda_{pp} r_{pp} \tag{8.33}$$

(ii) *Repair cannot be done during adverse weather*

In this case the concepts of deduction are identical to those described for case (i) with two modifications. Firstly, when the second failure occurs in normal weather, the 'exposure time' is the repair time of the first component since the whole of the repair is done during normal weather. Secondly, when the second failure occurs in adverse weather, the 'exposure time' is the duration of adverse weather since no repair is involved in this weather condition. Consequently

$$\lambda_a = \frac{N}{N+S}[\lambda_1 \lambda_2 (r_1 + r_2)] \tag{8.34}$$

$$\lambda_b = \frac{N}{N+S}\left[\lambda_1 \frac{r_1}{N} \lambda_2' S + \lambda_2 \frac{r_2}{N} \lambda_1' S\right] \tag{8.35}$$

$$\lambda_c = \frac{S}{N+S}[\lambda_1' \lambda_2 r_1 + \lambda_2' \lambda_1 r_2] \tag{8.36}$$

$$\lambda_d = \frac{S}{N+S}[\lambda_1'(\lambda_2' S) + \lambda_2'(\lambda_1' S)]$$

$$= \frac{S}{N+S}[2\lambda_1' \lambda_2' S] \tag{8.37}$$

In case (i) there were no restrictions on repair and therefore the average outage time was identical for all four cases. In the present case, repair cannot be done in adverse weather and therefore, when the second failure occurs in adverse weather, cases (b) and (d), the outage time will be increased by the duration of adverse weather, giving

$$\lambda_{pp} = \lambda_a + \lambda_b + \lambda_c + \lambda_d \tag{8.38}$$

$$U_{pp} = (\lambda_a + \lambda_c)\left(\frac{r_1 r_2}{r_1 + r_2}\right) + (\lambda_b + \lambda_d)\left(\frac{r_1 r_2}{r_1 + r_2} + S\right) \tag{8.39}$$

$$r_{pp} = U_{pp}/\lambda_{pp} \tag{8.40}$$

8.6.6 Numerical examples

(a) *System and data*

The application of the equations to consider the effect of the overlapping forced outages as derived in Section 8.6.5 is illustrated using the simple parallel network shown in Fig. 8.6. This system may represent either a real parallel circuit or a second-order failure event (minimal cut set) of a more complicated network. The process of analysis is identical in both cases.

It is assumed that both components are identical and each has the following numerical data:

$$\lambda = 0.20 \text{ f/year of normal weather}$$

$$\lambda' = 40.0 \text{ f/year of adverse weather}$$

$$r = 10 \text{ hours}$$

$$\lambda'' = 1 \text{ outage/calendar year}$$

$$r'' = 8 \text{ hours}$$

In addition it is assumed that the weather states have the following average durations:

$$N = 200 \text{ hours}$$

$$S = 2 \text{ hours}$$

(b) *Single weather state*

If the weather is not considered in the analysis, the average failure rate $\hat{\lambda}$ can be evaluated using Equation (8.26a)

$$\hat{\lambda} = \frac{200}{202} \times 0.20 + \frac{2}{202} \times 40 = 0.594 \text{ f/yr}$$

This value of $\hat{\lambda}$ is the failure rate which would be identified by a data collection scheme if the weather state were not associated with each system failure. It is evident that the value of $\hat{\lambda}$ is much closer to the failure rate during normal weather because the value of N is much greater than the value of S.

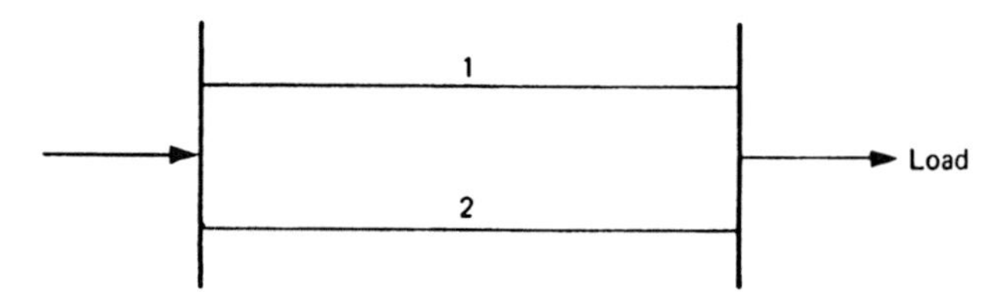

Fig. 8.6 Simple parallel transmission circuit

Using this value of $\hat{\lambda}$, the system reliability indices can be evaluated using Equations (8.1)–(8.3) which gives

$$\lambda_{pp} = 0.594 \times 0.594(10 + 10)/8760 = 8.06 \times 10^{-4} \text{ f/yr}$$

$$r_{pp} = \frac{10 \times 10}{10 + 10} = 5 \text{ hours}$$

$$U_{pp} = \lambda_{pp} r_{pp} = 4.03 \times 10^{-3} \text{ hours/yr}$$

(c) *Two weather states—repair possible in adverse weather*

This contribution can be evaluated from the data given in (a) above and Equations (8.27b), (8.28), (8.29b), (8.30)–(8.33).

$$\lambda_a = \frac{200}{202}[0.20 \times 0.20(10 + 10)]/8760 = 9.04 \times 10^{-5} \text{ f/yr}$$

$$\lambda_b = \frac{200}{202}\left[0.20\left(\frac{10}{200}\right)\left(40 \times \frac{2 \times 10}{2 + 10}\right) \times 2\right] \Big/ 8760 = 1.51 \times 10^{-4} \text{ f/yr}$$

$$\lambda_c = \frac{2}{202}[40 \times 0.20 \times 10 \times 2]/8760 = 1.81 \times 10^{-4} \text{ f/yr}$$

$$\lambda_d = \frac{2}{202}\left[40\left(40 \times \frac{2 \times 10}{2 + 10}\right) \times 2\right] \Big/ 8760 = 6.03 \times 10^{-3} \text{ f/yr}$$

$$\lambda_{pp} = \lambda_a + \lambda_b + \lambda_c + \lambda_d = 6.45 \times 10^{-3} \text{ f/yr}$$

$$r_{pp} = \frac{10 \times 10}{10 + 10} = 5 \text{ hours}$$

$$U_{pp} = \lambda_{pp} r_{pp} = 3.23 \times 10^{-2} \text{ hours/yr}$$

A similar set of results would be obtained if repair were not possible in adverse weather. In this case Equations (8.34)–(8.40) would be used.

(d) *Sensitivity analyses*

It is seen, by comparing the previous results for a single-state and a two-state weather model, that the failure rate and annual outage time is much greater for the two-state weather model. This shows the importance of recognizing the effect of the environment and identifying in which weather state the failures occur. It is very useful therefore to establish the system reliability indices as a function of the number of failures occurring in adverse weather. This type of sensitivity analysis is illustrated considering the system shown in Fig. 8.6 and assuming $N = 200$ hours, $S = 2$ hours, $r = 10$ hours and $\hat{\lambda} = 0.594$ f/yr, i.e. as evaluated in (b) above. The values of λ and λ' can be evaluated using Equation (8.26) for values of F between zero and unity and the system indices evaluated as illustrated in (c) above.

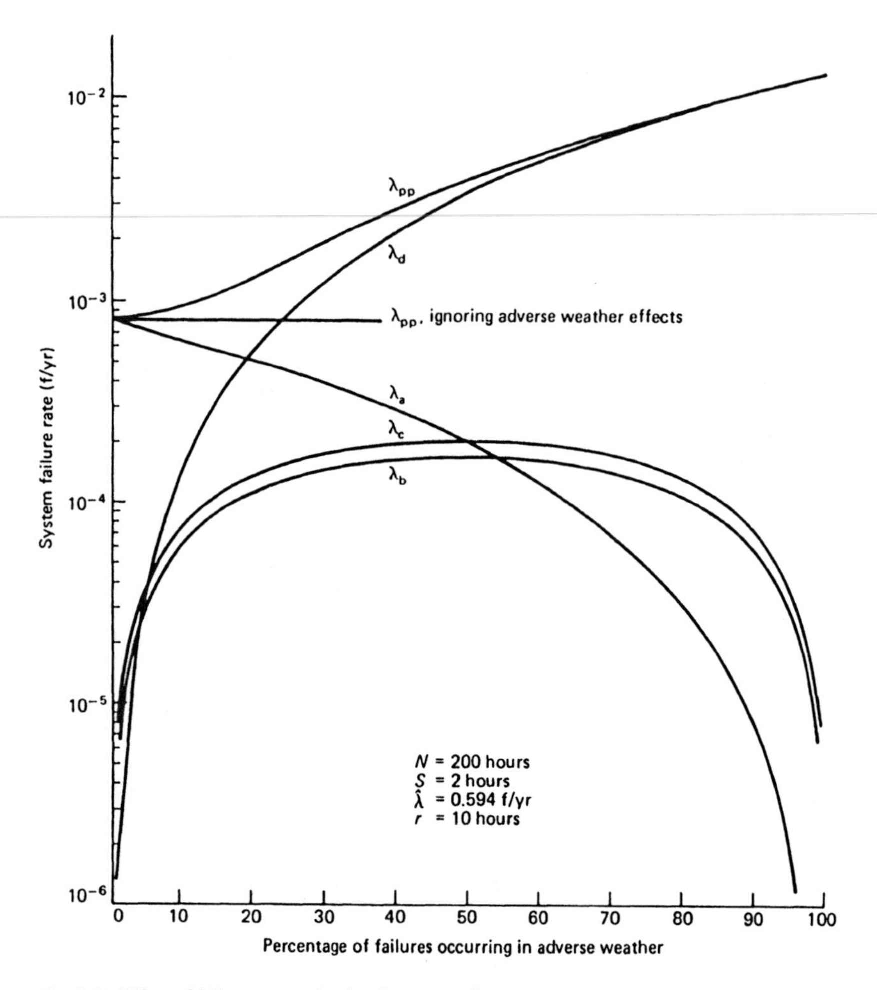

Fig. 8.7 Effect of failures occurring in adverse weather

These results are shown in Fig. 8.7, from which it is very evident that, as the number of failures occurring in adverse weather increases, the system failure rate also increases very sharply. The ratio between the failure rate if all failures occur in adverse weather and that when all failures occur in normal weather is about 17 to 1. This ratio can be defined as an error factor since it defines the error that will be introduced in the evaluation of failure rate if the effect of weather is neglected. The variation in the value of this error factor is shown in Fig. 8.8 as a function of the percentage of failures that occur in adverse weather. It is seen that the error increases rapidly as the percentage of adverse weather failures increases, and consequently a very optimistic evaluation would be obtained if the effect of weather were ignored.

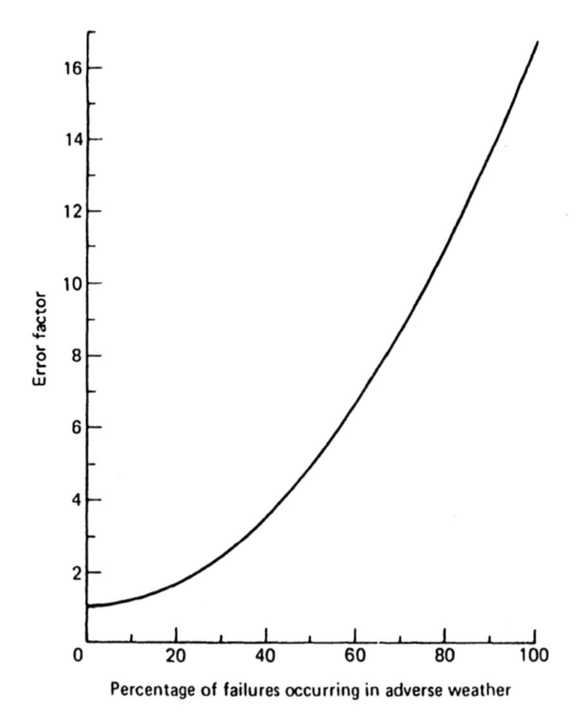

Fig. 8.8 Error factor in value of failure rate

The results shown in Fig. 8.7 also show the four contributions to the system failure rate. These indicate that when most failures occur in normal weather the system failure rate is dominated by λ_a, and when most failures occur in adverse weather the system failure rate is dominated by λ_d. The contribution by λ_b and λ_c is small in all cases.

Similar sensitivity results are shown in Figs. 8.9–11 for the same system and basic reliability data. These show the sensitivity of the system failure rate to average duration of adverse weather (Fig. 8.9), average duration of normal weather (Fig. 8.10) and average repair time (Fig. 8.11). In all cases the system failure rate is mainly affected by the value of λ_d for the component data chosen and shown on the figures. Also shown in Figs. 8.9 and 8.10 is the variation of $\hat{\lambda}$ with changes in the duration of weather.

8.6.7 Forced outage overlapping maintenance

A forced outage overlapping a maintenance outage can be considered in a similar way to that for a single weather state (Section 8.4) and overlapping forced outages (Section 8.6.5). There are, however, three cases to be considered. One further

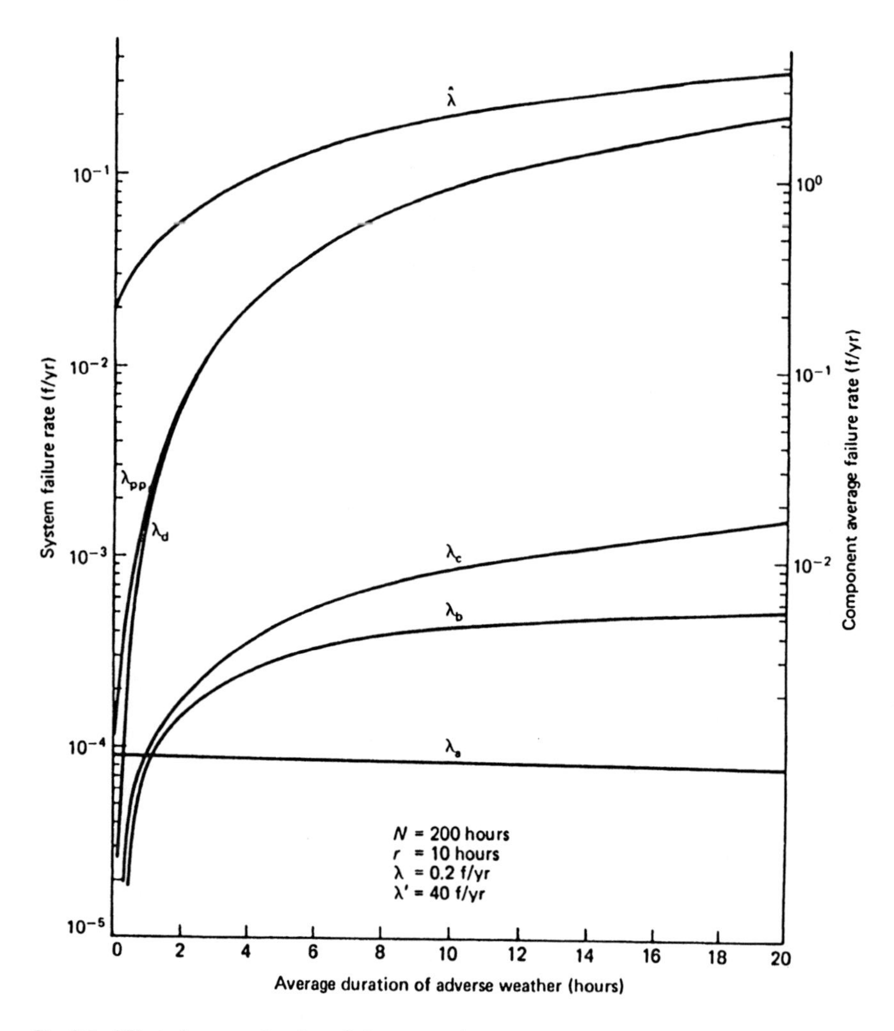

Fig. 8.9 Effect of average duration of adverse weather

constraint is generally imposed in addition to that considered previously in which a component is not removed for maintenance if this action alone would cause a system outage. The additional constraint is

maintenance is not commenced during adverse weather.

(i) *Maintenance not permitted if adverse weather is probable*

If commencement of maintenance is not permitted when adverse weather is probable, the equations are identical to those for a single weather state (Equations (8.7)–(8.9)) since adverse weather has no effect on the reliability indices.

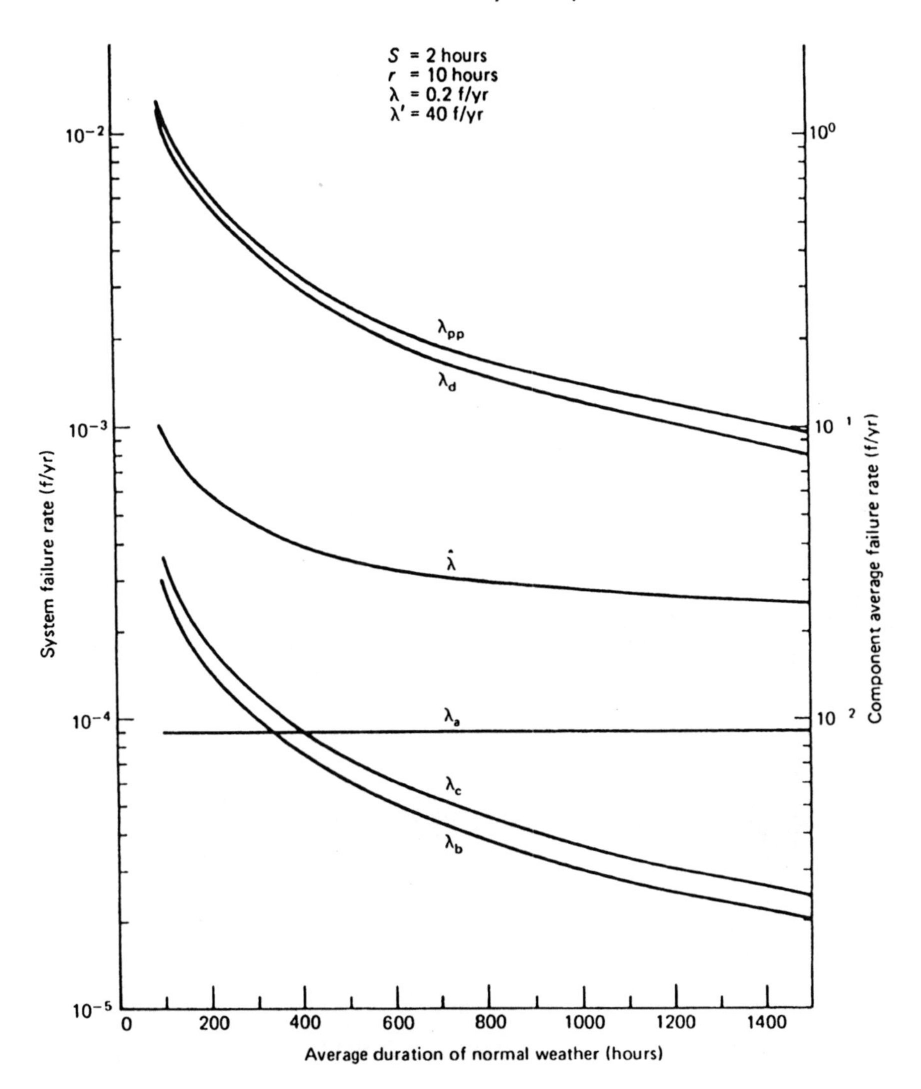

Fig. 8.10 Effect of average duration of normal weather

(ii) *Maintenance continued into adverse weather*

This case assumes that maintenance is commenced only during normal weather but that the weather may change during maintenance. It also assumes that both maintenance and repair can be continued during the adverse weather period. This is similar to case (i) of overlapping forced outages (Section 8.6.5).

The equations associated with this case can be derived from Equations (8.27) and (8.28) since the initial outage (maintenance) can only occur in normal weather.

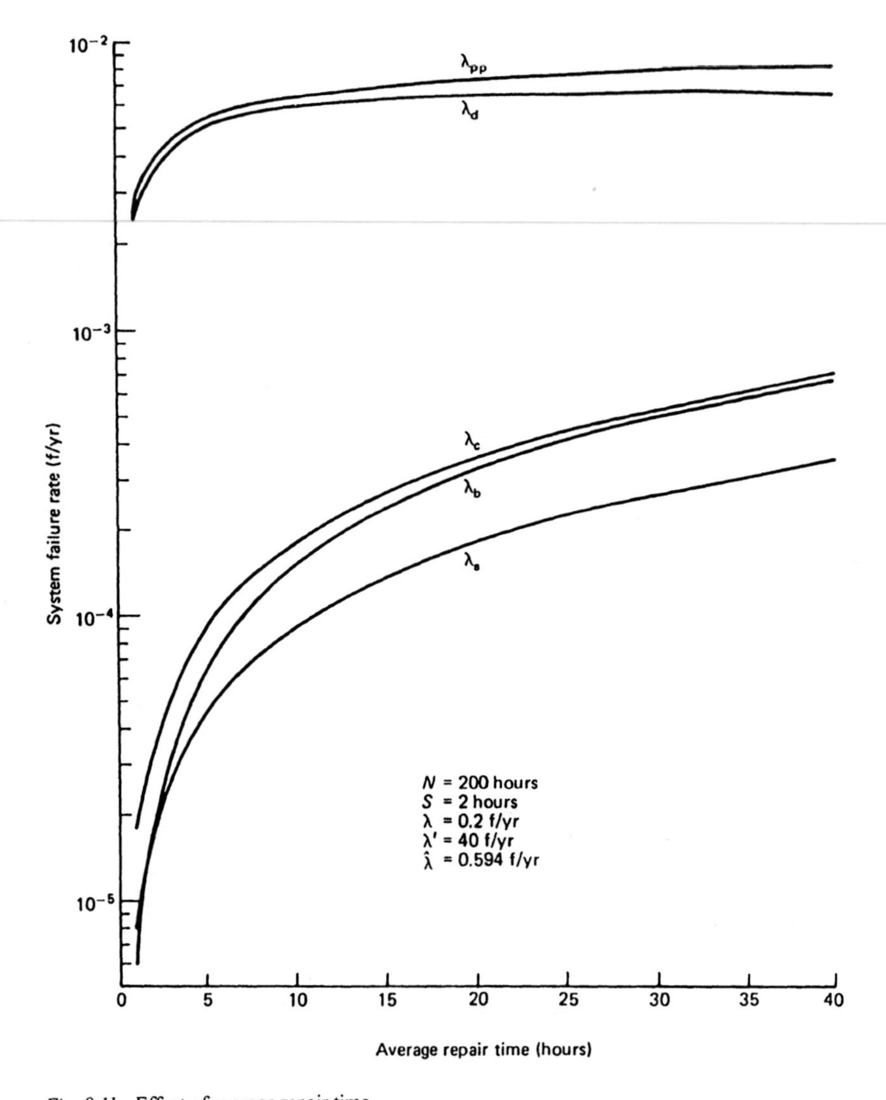

Fig. 8.11 Effect of average repair time

For this reason, the probability of normal weather prior to the maintenance outage is unity and therefore the term $N/(N+S)$ in these equations is inappropriate. On the basis of these principles, the contribution to the failure rate is

$$\lambda_{pm} = \lambda_1''\lambda_2 \frac{Nr_1''}{N+r_1''} + \lambda_2''\lambda_1 \frac{Nr_2''}{N+r_2''} + \lambda_1''\frac{r_1''}{N}\lambda_2' \frac{Sr_1''}{S+r_1''} + \lambda_2''\frac{r_2''}{N}\lambda_1' \frac{Sr_2''}{S+r_2''} \quad (8.41a)$$

which, if $r_i << N$, gives

$$\lambda_{pm} = \lambda_1''\lambda_2 r_1'' + \lambda_2''\lambda_1 r_2'' + \lambda_1''\frac{r_1''}{N}\lambda_2'\frac{Sr_1''}{S+r_1''} + \lambda_2''\frac{r_2''}{N}\lambda_1'\frac{Sr_2''}{S+r_2''} \tag{8.41b}$$

If the four terms in Equation (8.41) are defined as

λ_a'', λ_b'', λ_c'', λ_d'',

then

$$U_{pm} = (\lambda_a'' + \lambda_c'')\frac{r_1''r_2}{r_1''+r_2} + (\lambda_b'' + \lambda_d'')\frac{r_1 r_2''}{r_1+r_2''} \tag{8.42}$$

$$r_{pm} = U_{pm}/\lambda_{pm} \tag{8.43}$$

(iii) *Maintenance not continued into adverse weather*

This case also assumes that maintenance is commenced only during normal weather but that the weather may change. It further assumes, similarly to case (ii) of overlapping forced outages (Section 8.6.5), that neither maintenance nor repair is continued into adverse weather. Consequently, Equations (8.34), (8.35) and (8.39) can be adapted to give

$$\lambda_{pm} = \lambda_1''\lambda_2 r_1'' + \lambda_2''\lambda_1 r_2'' + \lambda_1''\frac{r_1''}{N}\lambda_2' S + \lambda_2''\frac{r_2''}{N}\lambda_1' S \tag{8.44}$$

If these four terms are again defined as λ_a'', λ_b'', λ_c'' and λ_d'', then

$$U_{pm} = \lambda_a''\left(\frac{r_1''r_2}{r_1''+r_2}\right) + \lambda_b''\left(\frac{r_1 r_2''}{r_1+r_2''}\right) + \lambda_c''\left(\frac{r_1''r_2}{r_1''+r_2} + S\right) + \lambda_d''\left(\frac{r_1 r_2''}{r_1+r_2''} + S\right) \tag{8.45}$$

$$r_{pm} = U_{pm}/\lambda_{pm} \tag{8.46}$$

8.6.8 Numerical examples

The application of the equations that take into account the effect of maintenance and derived in Section 8.6.7 can be illustrated using the previous parallel network shown in Fig. 8.6.

(a) *Single weather state*

If the effect of weather is neglected, the contribution due to forced outages overlapping maintenance can be evaluated using Equations (8.7)–(8.8) and the data given in Section 8.6.6.

$\lambda_{pm} = 1(0.594 \times 8)/8760 + 1(0.594 \times 8)/8760$

$$= 1.08 \times 10^{-3} \text{ f/yr}$$

$$r_{pm} = \frac{10 \times 8}{10 + 8} = 4.44 \text{ hours}$$

$$U_{pm} = 4.82 \times 10^{-3} \text{ hours/yr}$$

The total indices for the system can now be evaluated from the above indices and those derived in Section 8.6.6 using Equations (8.13)–(8.15) as

$$\lambda = 8.06 \times 10^{-4} + 1.08 \times 10^{-3} = 1.89 \times 10^{-3} \text{ f/yr}$$

$$U = 4.03 \times 10^{-3} + 4.82 \times 10^{-3} = 8.88 \times 10^{-3} \text{ hours/yr}$$

$$r = U/\lambda = 4.71 \text{ hours}$$

(b) *Two weather states*

(i) *Maintenance not permitted if adverse weather is probable*

In this case (see Section 8.6.7(i)), Equations (8.7)–(8.9) can be used with the data given in Section 8.6.6 to give

$$\lambda_{pm} = 1(0.2 \times 8) \times 2/8760 = 3.65 \times 10^{-4} \text{ f/yr}$$

$$r_{pm} = \frac{10 \times 8}{10 + 8} = 4.44 \text{ hours}$$

$$U_{pm} = \lambda_{pm} r_{pm} = 1.62 \times 10^{-3} \text{ hours/yr}$$

These can be combined with the values for overlapping forced outages (Section 8.6.6) to give the total indices

$$\lambda = 6.45 \times 10^{-3} + 3.65 \times 10^{-4} = 6.82 \times 10^{-3} \text{ f/yr}$$

$$U = 3.23 \times 10^{-2} + 1.62 \times 10^{-3} = 3.39 \times 10^{-2} \text{ hours/yr}$$

$$r = U/\lambda = 4.98 \text{ hours}$$

(ii) *Maintenance not continued into adverse weather*

In this case Equations (8.44)–(8.46) can be used to give

$$\lambda_{pm} = [1 \times 0.2 \times 8 + 1 \times 0.2 \times 8$$

$$+ 1 \times \frac{8}{200} \times 40 \times 2 + 1 \times \frac{8}{200} \times 40 \times 2]/8760$$

$$= 1.83 \times 10^{-4} + 1.83 \times 10^{-4} + 3.65 \times 10^{-4} + 3.65 \times 10^{-4}$$

$$= 1.10 \times 10^{-3} \text{ f/yr}$$

$$U_{pm} = \left(1.83 \times 10^{-4} \times \frac{8 \times 10}{8 + 10}\right) \times 2 + 3.65 \times 10^{-4} \times \left(\frac{8 \times 10}{8 + 10} + 2\right) \times 2$$

$$= 6.33 \times 10^{-3} \text{ hours/yr}$$

$$r_{pm} = 5.75 \text{ hours}$$

which can again be combined with the appropriate values for overlapping forced outages.

A similar set of results can be obtained if maintenance is continued into adverse weather using Equations (8.41)–(8.43).

8.6.9 Application to complex systems

The techniques to consider normal and adverse weather have been applied to a simple parallel system in the previous sections. Most systems, however, are clearly more complex than this particular example. The techniques can be applied to more complex systems with little difficulty. Two basic methods can be used.

The first method requires the failure modes or minimal cut sets to be deduced, in which case the previous techniques and equations can be applied to each of these failure modes. The load point indices can then be evaluated by combining the indices given by each contributing failure event.

The second method, which may be useful as a partial solution, particularly in the case of hand calculations, is to use wholly or partly a network reduction solution. This requires equivalent component indices to be evaluated.

These methods can be illustrated by means of the ring distribution system shown in Fig. 8.12.

In this example, assume that each distributor, 1, 2, 3 and 4, has the same component reliability data specified in Section 8.6.6(a) and again let $N = 200$ hours and $S = 2$ hours.

(a) *Failure modes method*

The failure modes for each load point of Fig. 8.12 are shown in Table 8.10. Each failure mode is a second-order event and the previous equations and techniques can be applied directly to each. Consider only load point L2. The indices for

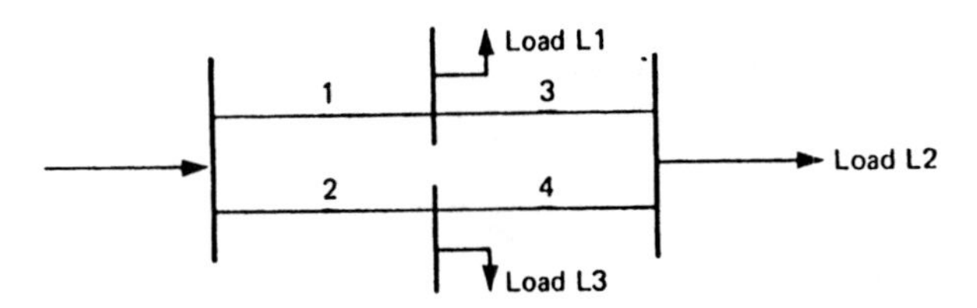

Fig. 8.12 Ring distribution system

Table 8.10 Failure modes for system of Fig. 8.12

Failure modes of load point		
L1	*L2*	*L3*
1 + 2	1 + 2	1 + 2
1 + 3	1 + 4	2 + 3
1 + 4	2 + 3	2 + 4
	3 + 4	

each failure event are identical to those evaluated in Section 8.6.6(c). Therefore the total load point indices are

$$\lambda = 4 \times 6.45 \times 10^{-3} = 2.58 \times 10^{-2} \text{ f/yr}$$

$$U = 4 \times 3.23 \times 10^{-2} = 1.29 \times 10^{-1} \text{ hours/yr}$$

$$r = U/\lambda = 5 \text{ hours}$$

(b) *Including network reduction*

The number of times that the equations must be applied is reduced if the series components of Fig. 8.12 are first reduced to an equivalent component. The overlapping failure events of the reduced network can then be deduced and the appropriate equations applied. The reduced failure events for Fig. 8.12 are shown in Table 8.11.

There is now only one failure event for each load point and therefore the equations need only be applied once for each of the load points. Again consider load point L2. The indices of the equivalent components can be evaluated from the principle of series systems, i.e.

$$\begin{aligned} \lambda \text{ (equivalent component 1, 3)} &= \Sigma\, \lambda \\ &= 0.20 + 0.20 \\ &= 0.40 \text{ f/yr of normal weather} \\ \lambda' \text{ (equivalent component 1, 3)} &= \Sigma\, \lambda' \\ &= 40 + 40 \\ &= 80.0 \text{ f/yr of adverse weather} \end{aligned}$$

Table 8.11 Reduced failure events for system of Fig. 8.12

Load point	*Failure events*
L1	1 + (2, 3, 4)
L2	(1, 3) + (2, 4)
L3	(1, 3, 4) + 2

U (equivalent component 1, 3) $= \Sigma\, \lambda r$

$= 0.20 \times 10 + 0.20 \times 10$

$= 4$ hours/yr of normal weather

U' (equivalent component 1, 3) $= \Sigma\, \lambda' r$

$= 40 \times 10 + 40 \times 10$

$= 800$ hours/yr of adverse weather

$$r \text{ (equivalent component 1, 3)} = \frac{\Sigma\, \lambda r}{\Sigma\, \lambda} = \frac{\Sigma\, \lambda' r}{\Sigma\, \lambda'}$$

$= 10$ hours

The indices are the same for equivalent component (2, 4) since all components are considered identical. It the above values are substituted into Equations (8.27b), (8.28), (8.29b), (8.30)–(8.33), the reliability indices of load point L2 would be evaluated as:

$$\lambda = 2.58 \times 10^{-2} \text{ f/yr}, \qquad r = 5 \text{ hours}, \qquad U = 1.29 \times 10^{-1} \text{ hours/yr}$$

i.e. they would be identical to those evaluated above.

The same principle can be applied in the case of load points L1 and L3.

8.7 Common mode failures

8.7.1 Evaluation techniques

The concepts and evaluation techniques for common mode failures have been described at length in Section 11.7 of *Engineering Systems* and will therefore not be repeated in detail in this section. It is useful, however, to recall some of the aspects before attempting to combine common mode failures and weather effects which is the subject of Section 8.8.

Two of the most useful models [13] for representing a second-order overlapping failure event including common mode failures are shown in Figs 8.13 and 8.14.

The significance of the values of repair rates in these two models should be appreciated and understood. In the case of the model with separate down states, Fig. 8.14, all repair rates are non-zero and equal to the reciprocal of the appropriate component repair times. In addition all common mode failures are restored by an equivalent common mode repair process.

In the case of the model with a single down state, Fig. 8.13, the interpretation is rather different. In this case the value of μ_{12} is zero if all repairs are conducted independently and each component is returned to service separately. Consequently, μ_{12} is not strictly the reciprocal of a repair time but instead represents the proportion of repairs that involve both components being returned to service simultaneously.

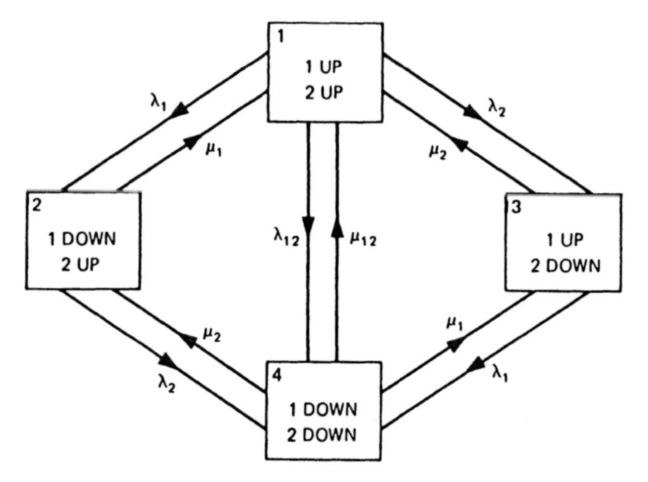

Fig. 8.13 Model for two components with single down state

As an example, consider a double circuit transmission line for which 10% of all double circuit outages are restored by a common mode restoration process. Therefore μ_{12} is 10% of μ_1 or of μ_2. Having evaluated μ_{12} in this way, it can be reciprocated to give a value of r_{12}, which represents an equivalent repair time.

It was shown in Section 11.7.2 of *Engineering Systems* that the set of equations [5] representing these models is

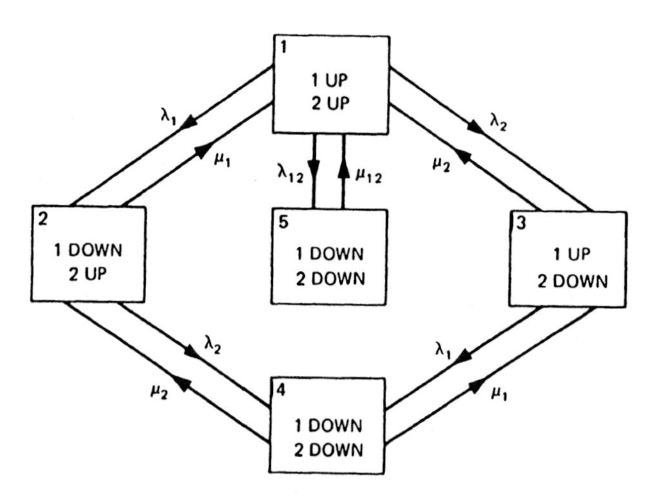

Fig. 8.14 Model for two components with separate down states

(a) *Failure rate*

For both models,

$$\lambda_{pp} = \lambda_1\lambda_2(r_1 + r_2) + \lambda_{12} \tag{8.47}$$

(b) *Outage time*

If

$$r_1 = \frac{1}{\mu_1}, \quad r_1 = \frac{1}{\mu_2}, \quad r_{12} = \frac{1}{\mu_{12}}$$

then, for the model of Fig. 8.13,

$$r_{pp} = \frac{r_1 r_2 r_{12}}{r_1 r_2 + r_2 r_{12} + r_{12} r_1} \tag{8.48a}$$

and if $\mu_{12} = 0$

$$r_{pp} = \frac{r_1 r_2}{r_1 + r_2} \tag{8.48b}$$

and for the model of Fig. 8.14

$$r_{pp} = \frac{\lambda_1\lambda_2 r_1 r_2 + \lambda_{12} r_{12}}{\lambda_{pp}} \tag{8.49}$$

8.7.2 Application and numerical examples

(a) *Effect of model on outage duration*

The discussion in Section 11.7 of *Engineering Systems* and Equations (8.48) and (8.49) show that the two models give the same value for failure rate but different values for the outage duration. It is important therefore to assess the most suitable model for any particular practical situation. In order to illustrate this effect on outage duration, consider a two-component parallel system and the following data

$$\lambda_1 = \lambda_2 = 0.5 \text{ f/yr}$$

$$\lambda_{12} = 0.05 \text{ f/yr}$$

$$r_1 = r_2 = 2.5 \text{ hours}$$

$$r_{12} = 10 \text{ hours and } 20 \text{ hours}$$

The values of outage duration can be evaluated from Equations (8.48) and (8.49) and are shown in Table 8.12. The results shown in Table 8.12 indicate that the outage time for the model shown in Fig. 8.13 is approximately equal to the overlapping outage time associated with independent failures only, whereas that of

Table 8.12 Outage durations including common mode failures

r_{12} (hours)	*Outage duration (hours) for model of Fig.* 8.14	8.13	
	Eqn (8.49)	*Eqn (8.48a)*	*Eqn (8.48b)*
10	9.98	1.11	1.25
20	19.95	1.18	1.25

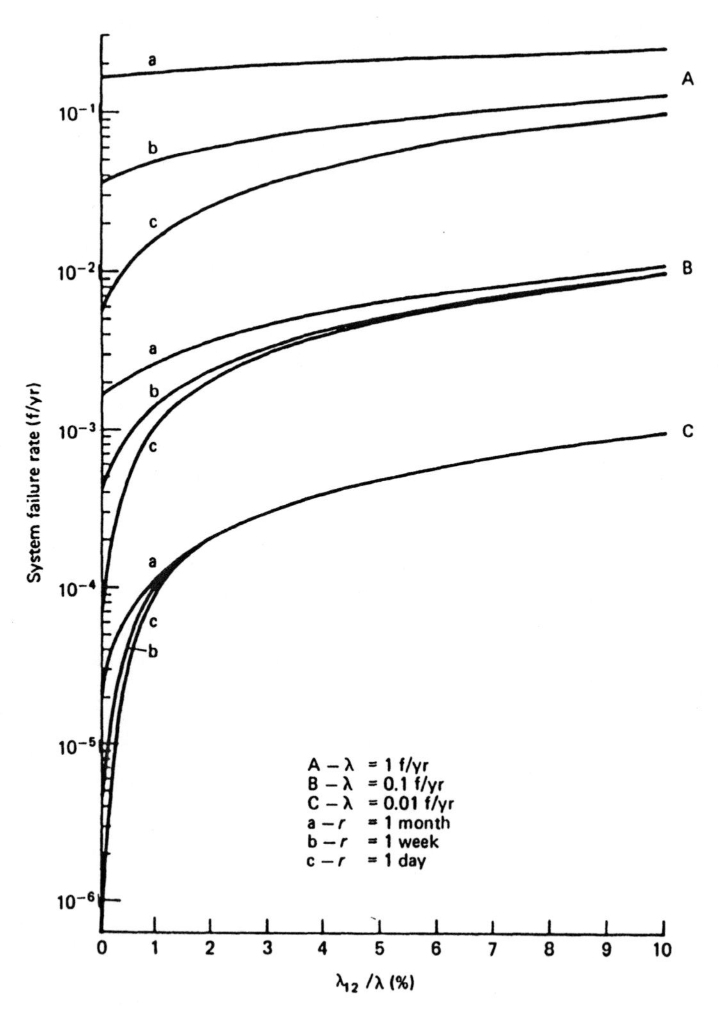

Fig. 8.15 Effect of common mode failures

Fig. 8.14 is approximately equal to the outage time associated with the common repair process.

(b) *Sensitivity analysis of common mode failures*

Consider again a two-component parallel system represented by the model of Fig. 8.13 and assume the following data

$$\lambda_1 = \lambda_2 = 0.01, 0.1, 1.0 \text{ f/yr}$$

$$r_1 = r_2 = 1 \text{ day } (= 24 \text{ hours}),\ 1 \text{ week } (= 168 \text{ hours}),\ 1 \text{ month } (= 720 \text{ hours})$$

$$\mu_{12} = 0$$

$$\lambda_{12} = 0 \text{ to } 10\% \text{ of the independent failure rate } \lambda_1$$

The results for the failure rate of the system are shown in Fig. 8.15, which again stresses the points made in Section 11.7 of *Engineering Systems* that only a small percentage of common mode failures causes a significant increase in the evaluated failure rate of a load point and the effect of common mode failures should be included in the analysis if they can be identified as a potential cause of failure.

8.8 Common mode failures and weather effects

8.8.1 Evaluation techniques

The examples and discussion in Sections 8.6 and 8.7 demonstrate that common mode failures and the effect of weather can independently increase the failure rate of a load point very significantly compared with the value that would be evaluated if these two effects were neglected. It follows therefore that if these two effects were considered simultaneously, it is possible that the evaluated indices would be even greater. This possibility has been considered [6] fairly recently, and the combined evaluation methods and effects have been described and discussed.

The relevant equations for these combined effects can be deduced very readily from the concepts used to derive the weather related equations (Section 8.6) and the common mode equations (Section 8.7).

(a) *Repair can be done during adverse weather*

Equation (8.47) shows that the load point failure rate is given by the summation of that due to independent failures and that due to common mode failures. The same principle applies when a two-state weather model is used. Consequently, if λ_{12} and λ'_{12} are defined as the common mode failure rate in normal and adverse weather respectively,

$$\lambda_{pp} = \lambda_a + \lambda_b + \lambda_c + \lambda_d + \frac{N}{N+S}\lambda_{12} + \frac{S}{N+S}\lambda'_{12} \tag{8.50}$$

where λ_a, λ_b, λ_c and λ_d are defined by Equations (8.27b), (8.28), (8.29b) and (8.31) respectively.

If Equation (8.50) is compared with Equation (8.26a), it can be seen that the summation of the last two terms of Equation (8.50) give the average failure rate per calendar year due to common mode failures, i.e.

$$\hat{\lambda}_{12} = \frac{N}{N+S}\lambda_{12} + \frac{S}{N+S}\lambda'_{12} \tag{8.51}$$

Therefore

$$\lambda_{pp} = \lambda_a + \lambda_b + \lambda_c + \lambda_d + \hat{\lambda}_{12} \tag{8.52}$$

Equations (8.50)–(8.52) apply equally to the single down state model (Fig. 8.13) and the separate down state model (Fig. 8.14).

The outage time associated with the two models can be evaluated as follows:

(i) *Single down state model*

In this case the outage time is given by Equation (8.48) with no modification.

(ii) *Separate down state model*

In this case, the principle of Equations (8.32), (8.33) and (8.49) can be used to give

$$\begin{aligned} U_{pp} &= (\lambda_a + \lambda_b + \lambda_c + \lambda_d)\frac{r_1 r_2}{r_1 + r_2} + \frac{N}{N+S}\lambda_{12} r_{12} + \frac{S}{N+S}\lambda'_{12} r_{12} \\ &= (\lambda_a + \lambda_b + \lambda_c + \lambda_d)\frac{r_1 r_2}{r_1 + r_2} + \left(\frac{N}{N+S}\lambda_{12} + \frac{S}{N+S}\lambda'_{12}\right) r_{12} \\ &= (\lambda_a + \lambda_b + \lambda_c + \lambda_d)\frac{r_1 r_2}{r_1 + r_2} + \hat{\lambda} r_{12} \end{aligned} \tag{8.53}$$

$$r_{pp} = U_{pp}/\lambda_{pp} \tag{8.54}$$

where λ_a, λ_b, λ_c and λ_d are given by Equations (8.27b), (8.28), (8.29b) and (8.31), respectively.

(b) *Repair cannot be done during adverse weather*

Equations (8.50)–(8.52) also apply to the case when repair cannot be done during adverse weather with the exception that λ_a, λ_b, λ_c and λ_d are given by Equations (8.34)–(8.37) respectively.

The outage times associated with the two models are as follows:

(i) *Single down state model*

In this case, as in Section 8.6.5(ii), the outage time of failures occurring in adverse weather must be increased by the duration of adverse weather. Therefore the outage duration is evaluated using the concepts of Equations (8.39), (8.40) and (8.48) to give

$$U_{pp} = \left(\lambda_a + \lambda_c + \frac{N}{N+S}\lambda_{12}\right) r_e + \left(\lambda_b + \lambda_d + \frac{S}{N+S}\lambda'_{12}\right)(r_e + S) \tag{8.55}$$

where

$$r_e = \frac{r_1 r_2 r_{12}}{r_1 r_2 + r_2 r_1 + r_{12} r_1}$$

and λ_a, λ_b, λ_c and λ_d are given by Equations (8.34)–(8.37) respectively. Thus

$$r_{pp} = U_{pp}/\lambda_{pp} \tag{8.56}$$

(ii) *Separate down state model*

In this case, the outage time for some of the failure modes must also be increased by the duration of adverse weather. Therefore Equations (8.39) and (8.53) are modified to:

$$U_{pp} = (\lambda_a + \lambda_c)\frac{r_1 r_2}{r_1 + r_2} + (\lambda_b + \lambda_c)\left(\frac{r_1 r_2}{r_1 + r_2} + S\right)$$

$$+ \frac{N}{N+S}\lambda_{12} r_{12} + \frac{S}{N+S}\lambda'_{12}(r_{12} + S) \tag{8.57}$$

$$r_{pp} = U_{pp}/\lambda_{pp} \tag{8. 58}$$

where λ_a, λ_b, λ_c, λ_d are given by Equations (8.34)–(8.37).

8.8.2 Sensitivity analysis

In order to illustrate the effect of combining common mode failures and weather modelling on the system reliability indices, reconsider the system shown in Fig. 8.6, the data shown in Section 8.6.6(a), the single down state common mode failure model and assume that repair can be done in adverse weather. A failure rate sensitivity analysis can now be made using Equations (8.50) or (8.52). These results are shown in Fig. 8.16 as a function of common mode failures and in Fig. 8.17 as a function of number of failures occurring in adverse weather. The curve labelled zero in Fig. 8.17 is identical to that of λ_{pp} in Fig. 8.7.

These results show that both adverse weather effects and common mode failures can significantly affect the system failure rate. They also show that, if the percentage of common mode failures is relatively large, above about 4–5%, this contribution is much more significant than that of adverse weather effects.

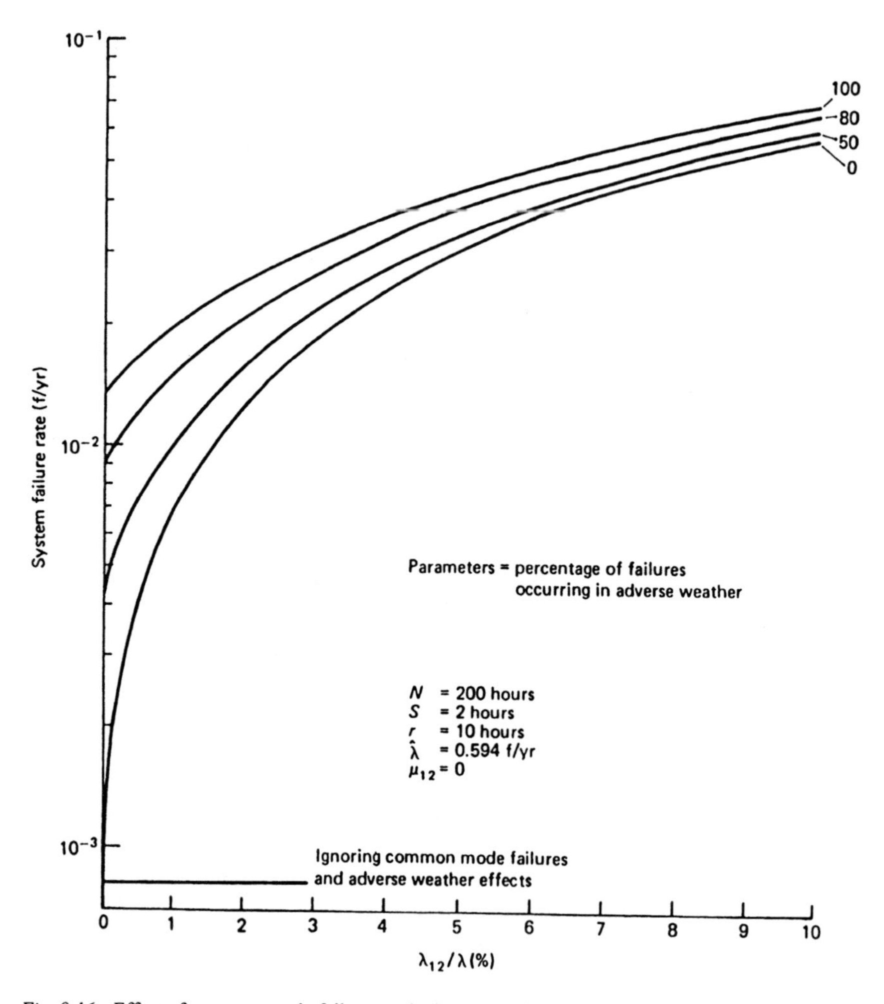

Fig. 8.16 Effect of common mode failures and adverse weather

8.9 Inclusion of breaker failures

8.9.1 Simplest breaker model

The simplest way to include the effect of breakers is to treat them identically to the components considered in Sections 8.2 and 8.3. This introduces no complexities and the previous techniques can be applied directly.

In order to illustrate this, assume that all breakers of Fig. 8.1 have a failure rate of 0.05 f/yr and a repair time of 20 hours. The new load point reliability indices using a failure modes analysis are shown in Table 8.13.

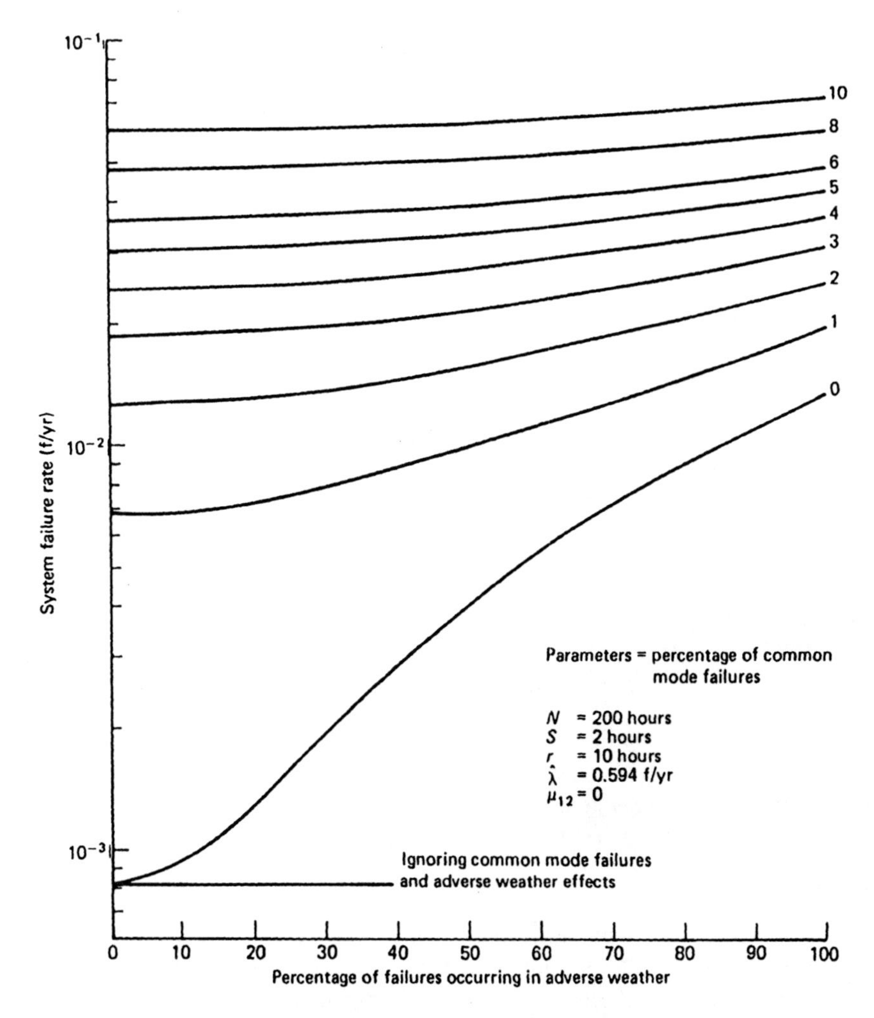

Fig. 8.17 Effect of common mode failures and adverse weather

The results shown in Table 8.13 differ only marginally from those in Table 8.4 owing to the dominant effect of the two busbars. If these busbars were 100% reliable, however, a significant increase would be observed when breaker failures are included.

8.9.2 Failure modes of a breaker

Most power system components can be represented by a two-state model that identifies the operating (up) state and the failure (down) state. This is not true for a breaker, however, because such a model ignores its switching function during

Table 8.13 Reliability indices using simple breaker model

Failure event	λ_{pp} *(f/yr)*	r_{pp} *(hours)*	U_{pp} *(hours/yr)*
Subtotal from Table 8.4	3.070×10^{-2}	3.07	9.411×10^{-2}
7 + 8	1.142×10^{-5}	10	1.142×10^{-4}
7 + 2	8.562×10^{-5}	6.67	5.708×10^{-4}
7 + 4	6.849×10^{-6}	16.7	1.142×10^{-4}
7 + 10	1.142×10^{-5}	10	1.142×10^{-4}
1 + 8	8.562×10^{-5}	6.67	5.708×10^{-4}
1 + 10	8.562×10^{-5}	6.67	5.708×10^{-4}
3 + 8	6.849×10^{-6}	16.7	1.142×10^{-4}
3 + 10	6.849×10^{-6}	16.7	1.142×10^{-4}
9 + 8	1.142×10^{-5}	10	1.142×10^{-4}
9 + 2	8.562×10^{-5}	6.67	5.708×10^{-4}
9 + 4	6.849×10^{-6}	16.7	1.142×10^{-4}
9 + 10	1.142×10^{-5}	10	1.142×10^{-4}
Total	3.112×10^{-2}	3.13	9.731×10^{-2}

fault conditions. The breaker model should therefore recognize an increased number of states. For a normally closed breaker, the complete set of states consists of the following:

(a) operates successfully in its closed state;
(b) opens successfully when required to do so;
(c) fails to open when required to do so;
(d) opens inadvertently when not requested to do so;
(e) suffers an open circuit;
(f) suffers a short circuit on the busbar side;
(g) suffers a short circuit on the line side.

The previous model does not recognize all of the states (a)–(g). A detailed modelling procedure for breakers is described in Chapter 10 in which substations and switching stations are discussed. The techniques and models of Chapter 10 are equally applicable to distribution systems and can be used if so desired. The more enhanced models are not always necessary, however, and simplifications can be made when evaluating the load point reliability indices of distribution systems. These are discussed in Section 8.9.3.

8.9.3 Modelling assumptions

The main assumptions for a simplified breaker model are:

(i) probability of opening successfully (state b) is unity;
(ii) probability of not opening successfully (state c) is zero;
(iii) probability of an open circuit (state e) is negligible.

Assumptions (i) and (ii) imply that states (b) and (c) can be neglected although the effect was discussed in Section 7.7 in relation to fusegear failures in simple radial systems. The justification for the assumption is that, in distribution and transmission systems, the probability of not opening successfully is usually very small and the contribution due to this malfunction is negligible compared with other significant contributions.

Breakers are normally located at the sending end of a single radial feeder and at both ends of a branch in a parallel or meshed system. It follows from assumptions (i) and (ii) and this method of design that short circuit faults on any branch component, other than the breakers themselves, will be isolated by their protection breakers which will therefore limit the effect of the fault to the branch in which it occurs. This permits the previous network reduction or failure modes analysis to be used without modification.

Assumption (iii) is justified because the probability of an open circuit on any power system component is usually very small and is negligible in comparison with short circuits. State (e) can therefore be neglected. This leaves only states (d), (f) and (g) to be considered as failure states.

State (d) usually manifests itself due to false signals being developed by or in the protection system. Its effect is similar to that of an open circuit and will only affect the branch in which the breaker exists.

A short circuit on a breaker, states (f) and (g), can cause different switching effects depending on the operational design of the protection system, i.e. whether a short circuit on a breaker can be isolated by its own switching action. For example, a fault on the line side of breaker 7, state (g), in Fig. 8.1 could be protected by itself and breaker 9. Similarly a fault on the busbar side of breaker 7, state (f), could be protected by itself and the breaker (not shown) protecting busbar 5. This can be defined as 'short circuits cleared by itself'. On the other hand, if the fault cannot be cleared by its own operation, then in both of the above examples, the breaker protecting busbar 5 and breaker 9 must operate. This can be defined as 'short circuits not cleared by itself'.

The modelling techniques described in the next section are derived from the above assumptions and concepts.

8.9.4 Simplified breaker models

(a) Inadvertent opening

Inadvertent opening failures can be modelled in either of two ways:

(i) The breaker is identified as a system component and the inadvertent opening indices allocated to it.

(ii) The breaker is neglected as a component but its inadvertent opening indices are combined with the reliability indices of the next component in the branch in which the breaker exists. The indices are combined as for series components.

For example, the inadvertent opening indices of breaker 7 in Fig. 8.1 can be combined with the reliability indices of line 1.

(b) Short circuits not cleared by itself

In this case, the breaker is neglected as a component and its short-circuit indices are combined with the reliability indices of the busbar to which it is connected. These indices are combined as for series components. For example, the indices of breakers 7 and 9 are combined with those of busbars 5 and 6 respectively.

(c) Short circuits cleared by itself

In this case, the short-circuit indices associated with the busbar side of the breaker are combined with the reliability indices of the busbar and the short-circuit indices associated with the line side of the breaker are either:

(i) allocated to the breaker being considered, as in (a(i)), or
(ii) combined with the reliability indices of the line, as in (a(ii)).

It should be noted that the model in (b) above is also applicable to any additional terminal equipment that exists in a branch between the breaker protecting that branch and the relevant busbar to which it is connected.

8.9.5 Numerical example

The concepts described in the previous sections can be illustrated by again considering the system shown in Fig. 8.1. Assume that the breaker failure rate of 0.05 f/yr specified in Section 8.9.1 is due to

inadvertent opening—20% ≡ 0.01 f/yr

short circuits on busbar side—40% ≡ 0.02 f/yr

short circuits on line side—40% ≡ 0.02 f/yr

and that the repair time is 20 hours for each failure mode.

Furthermore, assume that each breaker can clear its own short circuits The failures of the breakers can now be modelled as follows

inadvertent opening: as in (a(i))

short circuits on busbar side: as in (c)

short circuits on line side: as in (c(i))

The modified reliability indices of the busbars and breakers are

busbar 5: $\lambda = 0.01 + 0.02 = 0.03$ f/yr
$U_5 = 0.01 \times 5 + 0.02 \times 20 = 0.45$ hours/yr
$r_5 = U_5/\lambda_5 = 15$ hours

busbar 6: $\lambda_6 = 0.02 + 0.02 = 0.04$ f/yr
$U_6 = 0.02 \times 2 + 0.02 \times 20 = 0.44$ hours/yr
$r_6 = U_6/\lambda_6 = 11$ hours

breakers: $\lambda = 0.01 + 0.02 = 0.03$ f/yr
$r = 20$ hours.

Using the above component data and that given in Table 8.1 gives the load point reliability indices shown in Table 8.14.

The results shown in Table 8.14 are significantly greater than those of Table 8.13, which indicates the importance of modelling the system components, particularly breakers, in the most realistic way. In both cases (Tables 8.13 and 8.14) the basic component data was the same but the method of using this data was different. It is evident that no single method for processing or using the data can be given since both aspects are a function of the mode of failure and the type of protection scheme being used. The discussion, however, indicates how these aspects may be considered once the mode of failure and protection scheme have been identified. As noted previously, a more detailed and comprehensive description of modelling different modes of failure and the subsequent switching procedures are given in Chapter 10 in relation to substations and switching stations. These additional concepts can also be used, if desired, in the analysis of distribution systems.

8.10 Conclusions

This chapter has extended the basic techniques described in Chapter 7 for simple radial systems to more complicated parallel and meshed systems. It has also described evaluation methods that can consider a wider range of failure modes including permanent, temporary, transient and common mode outages as well as the effect of maintenance and weather.

The evaluation problems and concepts of breakers and the complexities associated with their switching effects have been introduced. The models used for including these effects have been simplified in this chapter but these will be extended quite considerably in Chapter 10. The extended techniques are not frequently necessary in the case of distribution systems but, if desired, the techniques of Chapter 10 can be used for these systems.

The only reliability indices considered in this chapter were the failure rate, repair time and annual outage time. Additional indices such as average load disconnected and energy not supplied can also be evaluated. These indices will be discussed, however, in Chapter 9, when other reliability criteria are also imposed on the system behavior. In this chapter, the only criterion has been one of continuity between the load and supply point. This therefore assumes that parallel paths are fully redundant, a criterion which is not strictly relevant to many practical systems.

The main conclusion that can be drawn from the discussion and results of the present chapter is that all features known to affect the reliability of a power system network and all realistic failure and restoration modes should be taken into account. This frequently means that fault reporting schemes must be developed and evolved to produce reliability data of the appropriate form and detail that is commensurate

Table 8.14 Reliability indices using modified breaker models

Failure event	λ_{pp} *(f/yr)*	r_{pp} *(hours)*	U_{pp} *(hours/yr)*
Subtotal from Table 8.2	6.986×10^{-4}	5.88	4.110×10^{-3}
5	3×10^{-2}	15.0	4.5×10^{-1}
6	4×10^{-2}	11.0	4.4×10^{-1}
7 + 8	4.110×10^{-6}	10.0	4.110×10^{-5}
7 + 2	5.137×10^{-5}	6.67	3.425×10^{-4}
7 + 4	4.110×10^{-6}	16.7	6.849×10^{-5}
7 + 10	4.110×10^{-6}	10.0	4.110×10^{-5}
1 + 8	5.137×10^{-5}	6.67	3.425×10^{-4}
1 + 10	5.137×10^{-5}	6.67	3.425×10^{-4}
3 + 8	4.110×10^{-6}	16.7	6.849×10^{-5}
3 + 10	4.110×10^{-6}	16.7	6.849×10^{-5}
9 + 8	4.110×10^{-6}	10.0	4.110×10^{-5}
9 + 2	5.137×10^{-5}	6.67	3.425×10^{-4}
9 + 4	4.110×10^{-6}	16.7	6.849×10^{-5}
9 + 10	4.110×10^{-6}	10.0	4.110×10^{-5}
Total	0.0709	12.6	0.8960

with the reliability models. It follows therefore that these two aspects, reliability data and evaluation models, must evolve together; as one or other is developed, the other must follow.

8.11 Problems

1 Evaluate the failure rate, average outage duration and annual outage time of load points L1 and L2 in Fig. 8.18:

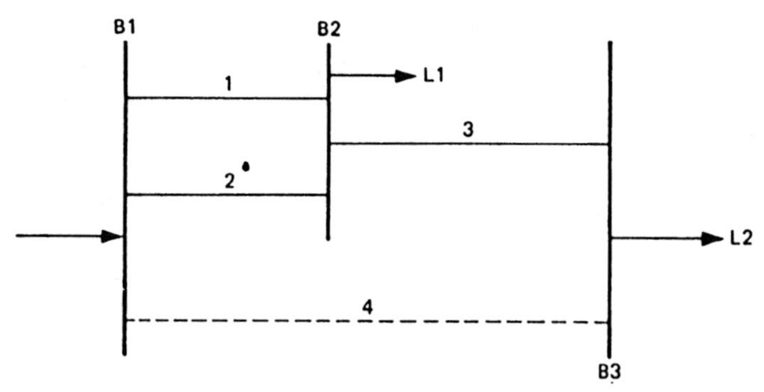

Fig. 8.18

Table 8.15

Component	λ *(f/yr)*	*r (hours)*
line 1, 2	0.5	8
line 3	0.1	2
line 4	0.25	10
busbars	0.001	1

(a) the busbars are assumed to be 100% reliable and
   (i) without line 4;
   (ii) with line 4;
(b) the busbars are not 100% reliable and
   (i) without line 4;
   (ii) with line 4.

The component reliability data is shown in Table 8.15.

2 Re-evaluate the load point indices of Fig. 8.18 for case (a) only in Problem 1 when the lines are maintained once per year for a duration of 8 hours. Assume that maintenance is not done if maintenance alone will cause load point failure and that maintenance is performed on one line only at a time.

3 Evaluate the failure rare, average outage duration and annual outage time of load point L1 in Fig. 8.19 if the breakers always operate successfully when requested to do so and can isolate their own short circuits. Assume that 10% of breaker failures are due to inadvertent opening and that 45% of the failures occur on each side of the breaker. The component reliability data is shown in Table 8.16.

4 Re-evaluate the indices of L1 in Problem 3 if coordinated maintenance is performed on each branch once per year for a period of 8 hours.

5 Evaluate the failure rate, average outage time and annual outage time of the load point L in Fig. 8.20 if each busbar is 100% reliable and each line has a permanent failure rate

Table 8.16

Component	λ *(f/yr)*	*r (hours)*
busbars	0.01	3
lines	0.50	5
breakers	0.05	10

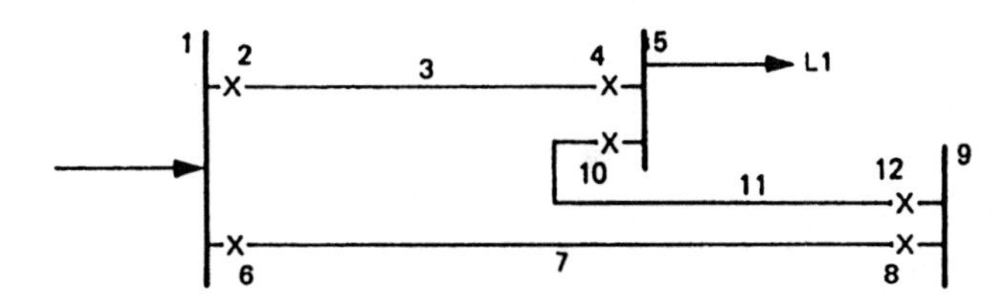

Fig. 8.19

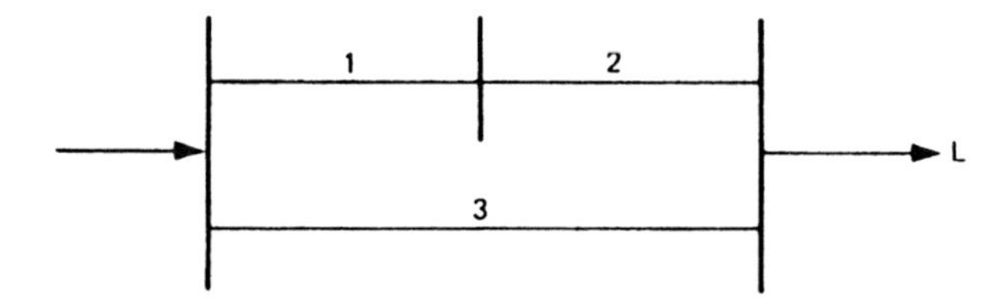

Fig. 8.20

of 0.5 f/yr, a temporary failure rate of 2 f/yr, a repair time of 8 hours and a reclosure time of 10 minutes.

6 The three lines of a system similar to that in Fig. 8.20 have the average reliability indices shown in Table 8.17.
Given that the average durations of normal and adverse weather are 200 hours and 1.5 hours respectively, evaluate the failure rate, average outage duration and annual outage time of load point L for the conditions:
(a) no failures occur in adverse weather;
(b) 50% of failures occur in adverse weather;
(c) all failures occur in adverse weather.
Assume:
(i) repairs can be done during adverse weather;
(ii) no repairs can be done during adverse weather.

7 Re-evaluate the reliability indices of Problem 6 if repairs and maintenance cannot be done in adverse weather and maintenance is performed on each line separately once per year for a period of 8 hours.

8 A data collection scheme shows that each line of a system similar to that of Fig. 8.20 has a failure rate/year of normal weather of 0.25 and a failure rate/year of adverse weather of 50. Given that $N = 250$ hours, $S = 2$ hours, evaluate:
(a) the average failure rate of each line;
(b) the percentage of failures occurring during adverse weather;
(c) the reliability indices of the load point L if repair can be done during adverse weather and each line has a repair time of 8 hours.

9 Re-evaluate the reliability indices of Problem 8 if lines 1 and 3 and lines 2 and 3 can suffer common mode failures.
Assume:
(a) the common mode failure rate to be 5% of the independent failure rate of each line;
(b) the system can be modelled with a single down state as illustrated in Fig. 8.13;
(c) all lines are restored to service independently, i.e. $\mu_{12} = 0$.

Table 8.17

Line	*Average failure rate (f/yr)*	*Repair time (hours)*
1	0.5	10
2	0.5	10
3	1.0	10

8.12 References

1. IEEE Committee Report, 'Bibliography on the application of probability methods in power system reliability evaluation', *IEEE Trans. on Power Apparatus and Systems*, **PAS-91** (1972), pp. 649–60.
2. IEEE Committee Report, 'Bibliography on the application of probability methods in power system reliability evaluation, 1971–1977', *IEEE Trans. on Power Apparatus and Systems*, **PAS-97** (1978), pp. 2235–42.
3. Billinton, R., Medicherla, T. K. P., Sachdev, M. S., 'Common cause outages in multiple circuit transmission lines', *IEEE Trans. on Reliability*, **R-27** (1978), pp. 128–131.
4. Allan, R. N., Dialynas, E. N., Homer, I. R., 'Modelling and evaluating the reliability of distribution systems', *IEEE Trans. on Power Apparatus and Systems*, **PAS-98** (1979), pp. 2181–9.
5. Allan, R. N., Dialynas, E. N., Homer, I. R., *Modelling Common Mode Failures in the Reliability Evaluation of Power System Networks*, IEEE PES Winter Power Meeting (1979), paper A79 040-7.
6. Billinton, R., Kumar, Y., *Transmission Line Reliability Models Including Common Mode and Adverse Weather Effects*, IEEE PES Winter Power Meeting (1980), paper A80 080-2.
7. IEEE Standard 346: 1973, *Terms for Reporting and Analyzing Outages of Electrical Transmission and Distribution Facilities and Interruptions to Customer Service.*
8. Billinton, R., Grover, M. S., 'Reliability assessment of transmission and distribution schemes', *IEEE Trans. on Power Apparatus and Systems*, **PAS-94** (1975), pp. 724–33.
9. Billinton, R., Grover, M. S., 'Quantitative evaluation of permanent outages in distribution systems', *IEEE Trans. on Power Apparatus and Systems*, **PAS-94** (1975), pp. 733–42.
10. Gaver, D. P., Montmeat, F. E., Patton, A. D., 'Power system reliability: I—Measures of reliability and methods of calculation', *IEEE Trans. on Power Apparatus and Systems*, **PAS-83** (1964), pp. 727–37.
11. Montmeat, F. E., Patton, A. D., Zemkoski, J., Cumming, D. J., 'Power system reliability: II—Applications and a computer program', *IEEE Trans. on Power Apparatus and Systems*, **PAS-84** (1965), pp. 636–43.
12. Billinton, R., Bollinger, K. E., 'Transmission system reliability evaluation using Markov processes', *IEEE Trans. on Power Apparatus and Systems*, **PAS-87** (1968), pp. 538–47.
13. Billinton, R., 'Transmission system reliability models', *Proceedings EPRI Conference on Power System Reliability Research Needs and Priorities* (March 1978), EPRI WS-77-60.

9 Distribution systems — extended techniques

9.1 Introduction

The models and techniques described in Chapter 8 allow the three basic reliability indices, expected failure rate (λ), average outage duration (r), and average annual outage time (U), to be evaluated for each load point of any meshed or parallel system. These three basic indices permit a measure of reliability at each load point to be quantified and allow subsidiary indices such as the customer interruption indices (see Section 7.3.2) to be found. They have three major deficiencies, however:

(a) they cannot differentiate between the interruption of large and small loads;
(b) they do not recognize the effects of load growth by existing customers or additional new loads;
(c) they cannot be used to compare the cost–benefit ratios of alternative reinforcement schemes nor to indicate the most suitable timing of such reinforcements.

These deficiencies can be overcome by the evaluation of two additional indices, these being:

(i) the average load disconnected due to a system failure, measured in kW or MW and symbolized by L;
(ii) the average energy not supplied due to a system failure, measured in kWh or MWh and symbolized by E.

These are not new indices since they were developed for generating capacity reliability evaluation using the loss of energy method (see Chapter 2). The recent application [1, 2] to distribution systems, however, allows a more complete analysis of these systems. This was demonstrated in Section 7.3.3, in which the load and energy indices for simple radial systems were evaluated. This was a simple exercise for these basic systems because the load and energy indices are easily deduced from the average load and the annual outage time of each load point. This chapter describes how these additional indices can be evaluated [2] for more complex systems.

The criterion used in Chapter 8 for determining a load point failure event was 'loss of continuity', i.e. a load point fails only when all paths between the load point and all sources are disconnected. This assumes that the system is fully redundant and any branch is capable of carrying all the load demanded of it. This clearly is unrealistic. For this reason, the previous 'loss of continuity' criterion is best described as 'total loss of continuity' (TLOC). In addition, a system outage or failure event may not lead to TLOC but may cause violation of a network constraint, e.g. overload or voltage violation, which necessitates that the load of some or all of the load points be reduced.

This type of event was initially defined [3, 4] as loss of quality. These initial considerations have now been considerably developed [2] and the event defined as partial loss of continuity (PLOC). The evaluation of PLOC events becomes of great significance if the load and energy indices are to be evaluated. The relevant techniques needed to evaluate these events and the load and energy indices are described in this chapter.

Many systems have interconnections which allow the transfer of some or all the load of a failed load point to other neighboring load points through normally open points. This concept was previously described in a simplistic way for radial systems in Section 7.8. A more realistic discussion of available techniques [2] is given in this chapter.

Finally, management decisions of the most appropriate reinforcement or expansion scheme cannot be based only on the knowledge of the reliability indices of each scheme. It is also necessary to know the economic implications of each of these schemes. This aspect is briefly discussed in the final section of this chapter.

9.2 Total loss of continuity (TLOC)

The criterion of TLOC is that used consistently in Chapter 8, and therefore the values of λ, r and U can be evaluated exactly as described in Chapter 8. Furthermore the values of L and E are readily evaluated knowing only the average load connected to each load point since, as described in Section 7.3.3,

$$\begin{aligned} L &= L_a \\ &= L_p f \end{aligned} \tag{9.1}$$

where

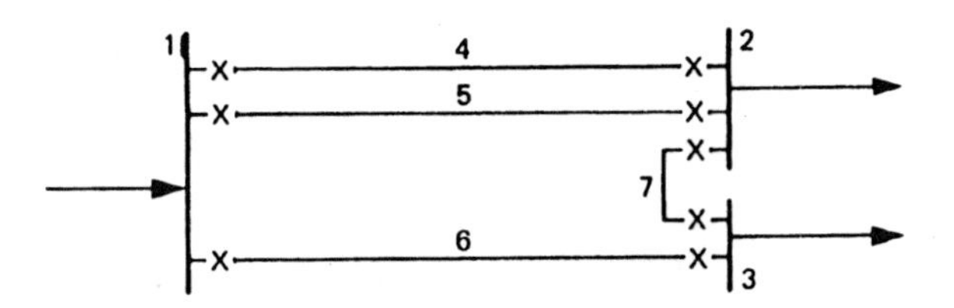

Fig. 9.1 Typical ring system with two load points

Table 9.1 Reliability data for the system of Fig. 9.1

Component	*Failure rate (f/yr)*	*Repair time (hours)*
lines 4–7	0.02	10
busbars 1–3	0.01	5

Table 9.2 Loading data for the system of Fig. 9.1

Load Point	*Peak load (MW)*	*Number of customers*
2	20	2000
3	10	1000

$$\left.\begin{aligned} L_a &= \text{average load at load point} \\ L_p &= \text{peak load at load point (maximum demand)} \\ f &= \text{load factor} \\ E &= LU \text{ where } U = \text{annual outage time of load point.} \end{aligned}\right\} \quad (9.2)$$

Finally the additional customer-orientated indices described in Section 7.3.2 can be obtained if so desired.

Table 9.3 Total loss of continuity indices for system of Fig. 9.1

Event	λ *(f/yr)*	*r (hours)*	*U (hours/yr)*	*L (MW)*	*E (MWh/yr)*
Load point 2					
1	0.01	5	0.05	15	0.75
2	0.01	5	0.05	15	0.75
4 + 5 + 6	3.13×10^{-7}	3.33	1.04×10^{-10}	15	1.56×10^{-9}
4 + 5 + 3	1.14×10^{-11}	2.5	2.60×10^{-11}	15	3.90×10^{-10}
4 + 5 + 7	3.13×10^{-11}	3.33	1.04×10^{-10}	15	1.56×10^{-9}
Total	2.00×10^{-2}	5	1.00×10^{-1}	15	1.5
Load point 3					
1	0.01	5	0.05	7.5	3.75×10^{-1}
3	0.01	5	0.05	7.5	3.75×10^{-1}
6 + 2	3.42×10^{-7}	3.33	1.14×10^{-6}	7.5	8.55×10^{-6}
6 + 7	9.13×10^{-7}	5	4.57×10^{-6}	7.5	3.43×10^{-5}
4 + 5 + 6	3.13×10^{-11}	3.33	1.04×10^{-10}	7.5	7.8×10^{-10}
Total	2.00×10^{-2}	5	1.00×10^{-1}	7.5	0.75

SAIFI = 0.02 interruptions/customer yr
SAIDI = 0.10 hours/customer yr
CAIDI = 5.0 hours/customer interruption
ASAI = 0.999989
ASUI = 1.142×10^{-5}
ENS = 2.25 MWh/yr
AENS = 0.75 kWh/customer yr

As an example, consider the system shown in Fig. 9.1, the reliability data shown in Table 9.1 and the load data shown in Table 9.2. In Table 9.1 it is assumed that the data for the lines and busbars have been modified to account for breaker failures as described in Section 8.9. Although only permanent failures and a single-state weather model are used in this example, these can be extended to include all the aspects discussed in Chapter 8. It is also assumed, for simplicity only, that the load–duration curve for each load point follows a straight line with a load factor of 0.75.

The reliability indices (λ, r, U, L, E) can be evaluated using the techniques of Chapter 8, and the additional customer-orientated indices can be deduced using the techniques of Sections 7.3.2 and 7.3.3. These results are shown in Table 9.3.

9.3 Partial loss of continuity (PLOC)

9.3.1 Selecting outage combinations

A partial loss of continuity event could potentially occur for any combination of branch and busbar outages except those that cause a total loss of continuity. In order to be rigorous, it would therefore be necessary to simulate all possible outage combinations except those that are known to lead to a TLOC event. This may be feasible for very small systems but it becomes impractical for large ones. Consequently the outage combinations to be studied must be restricted.

It is usually feasible to study all first-order outages and usually reasonable to neglect third- and higher-order outages. The second-order outages can be selected by one of the following methods:

(a) select all second-order outages if the number is small;
(b) manually determine, from experience, those second-order outages that could cause concern;
(c) since the minimal cut sets identify weak links of the system, the third-order cuts can be used [2] to identify potential second-order PLOC events. These are obtained by taking all second-order combinations from each third-order minimal cut set obtained for the load point of interest. For the system of Fig. 9.1 and using the information given in Table 9.3, this would mean simulating the following second-order events:
load point 2—(4 + 5), (4 + 6), (5 + 6), (4 + 3), (5 + 3), (4 + 7), (5 + 7)
load point 3—(4+5), (4+6), (5+6).

9.3.2 PLOC criteria

After determining the outage combinations to be considered, it is necessary to deduce whether any or all of these form a PLOC event. This can only be achieved using a load flow and establishing whether a network constraint has been violated.

The most realistic load flow is an a.c. one [5], although others can be used if preferred, e.g. approximations such as a d.c. load flow can be used if deemed sufficiently satisfactory.

The purpose of the load flow routine, which should be performed with the peak loads at each load point, is to identify whether any network constraints are violated under certain loading conditions and therefore to ascertain whether the outage combination being considered leads to a PLOC event at one or more load points. Two possible network constraints are line overloads and busbar voltage violations.

9.3.3 Alleviation of network violations

The only network violation to be considered in this book is line overload. If required, other violations can be considered using similar techniques.

If any of the outage combinations causes a line overload, it may be necessary to disconnect sufficient load at one or more load points in order to remove this overload. If the load is shed at a load point of interest, then the outage condition being simulated causes a PLOC event at that load point. If the overload is considered acceptable, the outage condition does not lead to a PLOC event and can be ignored.

A given overload can often be alleviated by reducing the load at a number of load points, either individually or in combination. Each of the possible ways produces different PLOC indices for the load points. It is therefore not possible to define a single method of achieving the objective of overload alleviation that would give absolutely consistent results. The decision as to which method should be used must rest with the particular utility performing the analysis and this decision should be based on their accepted load shedding policy. Amongst others, the following are possible methods for load shedding:

(a) load is shed at those load points which alleviate the overload with a minimum shedding of load;
(b) load is reduced proportionately at all load points that can affect the overload;
(c) load is shed [2, 6] at the receiving end of the overloaded line.

9.3.4 Evaluation of PLOC indices

The reliability indices of each load point of interest due to a PLOC event can be evaluated using the model shown in Fig. 9.2 in which it is assumed that the system can satisfy all load demands without the outage condition (base case) and

λ_e = rate of occurrence of outage condition; this is evaluated using the techniques of Chapter 8 and can include all modes of failure, two-state weather model and maintenance;
μ_e = reciprocal of the average duration r_e of the outage condition;
L_s = maximum load that can be supplied to the load point of interest during the outage condition;
P = probability of load being greater than L_s;

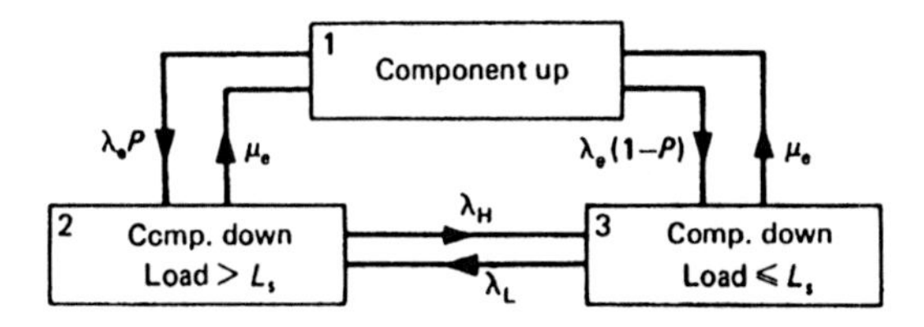

Fig. 9.2 State space diagram for PLOC

λ_H = transition rate from load levels > L_s to load levels ≤ L_s
= reciprocal of average duration (r_H) of load level > L_s;
λ_L = transition rate from load levels ≤ L_s to load levels > L_s
= reciprocal of average duration (r_L) of load level ≤ L_s.

The down state or failure state of the load point is state 2 in which the outage condition and a load level greater than L_s has occurred.

It has been shown [2] from an analysis of the state space diagram in Fig. 9.2 that the rate of occurrence of a PLOC event is

$$\lambda = \lambda_e P + \lambda_e(1 - P)\,\lambda_L \frac{r_e r_L}{r_e + r_L} \qquad (9.3)$$

A rigorous Markov analysis is not necessary, however, because Equation (9.3) can be explained in words as follows:

A PLOC event occurs if EITHER [a failure occurs during a high load level (first term of Equation (9.3))] OR [a failure occurs during a low load level AND a transition to a high load level occurs during the overlapping time associated with component repair and the duration of a low load level (second term of Equation (9.3))].

The average duration of the PLOC event is

$$r = \frac{r_e r_H}{r_e + r_H} \qquad (9.4)$$

or

$$r = r_e \qquad (9.5)$$

Equation (9.4) applies to the operating situation in which excess load is connected and disconnected each time a load transition between states 2 and 3 occur. This is a likely operating policy when load switching can be performed easily, e.g. when remote control is used.

Equation (9.5) applies when the excess load, once disconnected, remains disconnected until the repair has been completed. This is a likely operating policy in remote or rural areas to prevent many manual load transfer operations.

In the case of PLOC, the average load disconnected is the average load in excess of the maximum level L_s that can be sustained. This is deduced by evaluating the area under the load-duration curve for load levels greater than L_s to give the energy that cannot be supplied *given* the outage condition and dividing by the time for which load L_s is exceeded. Essentially this is an application of conditional probability and the details are shown in Fig. 9.3.

Using the above concepts, the complete set of expressions for evaluating λ, r, U, L, E associated with a PLOC event are

$$\lambda = \lambda_e P + \lambda_e (1 - P) \lambda_L \frac{r_e r_L}{r_e + r_L} \tag{9.6}$$

$r = r_e$ if excess load remains disconnected until repair is complete (9.7)

$r = \dfrac{r_e r_H}{r_e + r_H}$ if load is disconnected and reconnected each time a load transition occurs (9.8)

$$U = \lambda r \tag{9.9}$$

$$L = \left[\int_0^{t_1} L(t)\mathrm{d}t - L_s t_1 \right] \Big/ t_1 \tag{9.10}$$

$$E = LU \tag{9.11}$$

where $L(t)$ represents the load–duration curve and t_1 is the time for which the load level > L_s.

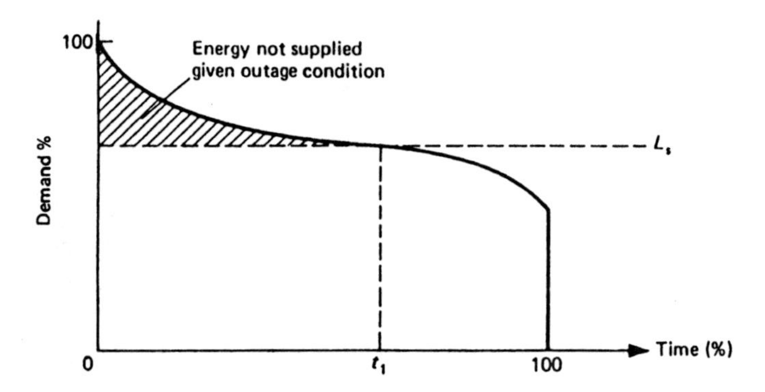

Fig. 9.3 Load-duration curve for evaluating L and E

If a number of PLOC events exist, which is likely for most systems, the overall PLOC indices for the load point can be evaluated using the concept of series systems to give

$$\lambda_p = \sum \lambda \tag{9.12a}$$

$$U_p = \sum U \tag{9.12b}$$

$$E_p = \sum E \tag{9.12c}$$

$$r_p = U_p/\lambda_p \tag{9.12d}$$

$$L_p = E_p/U_p \tag{9.12e}$$

9.3.5 Extended load–duration curve

In order to evaluate the load disconnected and energy not supplied in the case of a PLOC event, it is necessary to know the load–duration curve of each load point. These are often not known with great accuracy at distribution load points but it is usually possible to estimate them with sufficient accuracy for the purposes of reliability evaluation. Load–duration curves themselves (see Fig. 9.3) were discussed in Chapter 2. These previous concepts remain the same and can be used to evaluate the value of P in Equation (9.6).

It is also necessary to know the values of λ_L, r_L and r_H in order to apply Equations (9.6)–(9.11). These values are interrelated since

$$\lambda_H P = \lambda_L(1 - P)$$

i.e.

$$\lambda_L = \frac{P}{1-P}\lambda_H = \frac{P}{1-P}\frac{1}{r_H} \tag{9.13}$$

and since

$$r_L = \frac{1}{\lambda_L} = \frac{1-P}{P} r_H \tag{9.14}$$

It follows from Equations (9.13) and (9.14) that, if r_H is deduced from a data collection scheme, the other values can be evaluated. These values of r_H can be deduced from the same empirical data that is used to deduce the load–duration curves. This empirical data is usually obtained by integrating the load demand over short intervals of time, e.g. $\frac{1}{2}$ hour, 1 hour, etc. Consider the data shown in Fig. 9.4 and the particular load demand L.

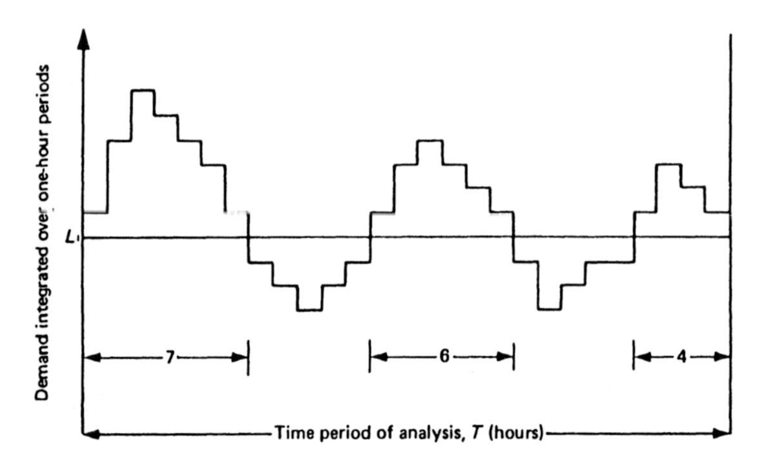

Fig. 9.4 Variation of demand in hourly intervals

If n is defined as the number of hourly intervals, f is the number of occasions that the load demand exceeds L, and T is the time period (in hours) of analysis, then

$$r_{\mathrm{H}} = \frac{n}{f} = \frac{7+6+4}{3} = 5.67 \text{ hours} \tag{9.15}$$

$$P = \frac{n}{T} = \frac{7+6+4}{7+5+6+5+4} = 0.62963 \tag{9.16}$$

If this process is repeated for all load levels, then the resultant variations of P and r_{H} are obtained. These together can be defined as an extended load–duration curve.

9.3.6 Numerical example

In order to illustrate the effect of including a PLOC criterion, reconsider the system shown in Fig. 9.1 and the data shown in Tables 9.1 and 9.2. The outages that could lead to PLOC events were identified in Section 9.3.1 as

load point 2—3, 4, 5, 6, 7, (4 + 5), (4 + 6), (5 + 6), (4 + 3), (5 + 3), (4 + 7), (5 + 7)

load point 3—2, 4, 5, 6, 7, (4 + 5), (4 + 6), (5 + 6)

In order to simplify the process and permit relatively easy hand calculations, assume that the load flow is inversely proportional to the line reactance. This clearly is an over-simplification and would not be done in a real evaluation exercise. Because of this assumption, only the relative line reactances are needed and these, together with the assumed line capacities, are shown in Table 9.4.

Table 9.4 Line reactances and capacities of Fig. 9.1

Line	*Relative reactance*	*Capacity (MW)*
4, 5	1.0	18
6	1.5	8
7	0.5	8

The load–duration curves for the two load points were defined as straight lines in Section 9.2 with maximum and minimum loads of 20 and 10 MW for load point 2 and 10 and 5 MW for load point 3. The reliability results for the PLOC events can now be evaluated using this loading information, the data shown in Tables 9.1, 9.2 and 9.4, the assumed values of r_H shown in Table 9.5, the load shedding criterion described in Section 9.3.3 and assuming the load remains disconnected until repair is completed, i.e. Equation (9.7). The full details of this analysis are shown in Table 9.5.

It can be seen from Table 9.5 that the PLOC indices associated with load point 2 are very small, whereas those associated with load point 3 are very significant. When these PLOC indices are combined with the TLOC given previously in Table 9.3, the overall indices shown in Table 9.6 are obtained.

It is evident from the results shown in Table 9.6 that, although PLOC may be insignificant for some load points of a system, they may dominate those of other load points. It follows therefore that PLOC should be included in the reliability analysis of distribution systems in order to ensure accuracy of the evaluation and the most reliable set of information necessary for the decision-making process of expansion and reinforcement.

9.4 Effect of transferable loads

9.4.1 General concepts

The loads on a distribution system are not usually connected directly to the load-point busbar but are distributed along feeders which themselves are fed by the load-point busbar. If the system does not have transferable facilities, the individual loads are lumped together and considered as a single point load connected to the load point of interest. This is illustrated in Fig. 9.5(a) and is the conventional technique used in previous sections and chapters.

If the effect of transferring load through normally open points in the load feeder (Fig. 9.5b) is to be considered, however, this single point load representation is less useful and extended techniques should be implemented. The importance of modelling transfer facilities exists because, in the event of a TLOC or PLOC failure event at a load point, it may be possible to recover some or all the disconnected

Table 9.5 PLOC indices for system of Fig. 9.1

Event	L_s (MW)	P	r_H (hours)	λ_e (f/yr)	r_e (hours)	λ (f/yr)	U (hours/yr)	L (MW)	E (MWh/yr)
Load point 2									
3	20	0	—	—	—	—	—	—	—
4	20	0	—	—	—	—	—	—	—
5	20	0	—	—	—	—	—	—	—
6	20	0	—	—	—	—	—	—	—
7	20	0	—	—	—	—	—	—	—
4 + 5	8	1	8760	9.13×10^{-7}	5	9.13×10^{-7}	4.57×10^{-6}	7	3.20×10^{-5}
4 + 6	10	1	8760	9.13×10^{-7}	5	9.13×10^{-7}	4.57×10^{-6}	5	2.28×10^{-5}
5 + 6	10	1	8760	9.13×10^{-7}	5	9.13×10^{-7}	4.57×10^{-6}	5	2.28×10^{-5}
4 + 3	18	0.2	5	3.42×10^{-7}	3.33	1.07×10^{-7}	3.58×10^{-7}	1	3.58×10^{-7}
5 + 3	18	0.2	5	3.42×10^{-7}	3.33	1.07×10^{-7}	3.58×10^{-7}	1	3.58×10^{-7}
4 + 7	18	0.2	5	9.13×10^{-7}	5	3.29×10^{-7}	1.64×10^{-6}	1	1.64×10^{-6}
5 + 7	18	0.2	5	9.13×10^{-7}	5	3.29×10^{-7}	1.64×10^{-6}	1	1.64×10^{-6}
Total					4.90	3.61×10^{-6}	1.77×10^{-5}	4.61	8.16×10^{-5}

2	8	0.4	10	0.01	5	5.50×10^{-3}	2.75×10^{-2}	1	2.75×10^{-2}
4	2.6	1	8760	0.02	10	2.00×10^{-2}	2.00×10^{-1}	4.9	9.80×10^{-1}
5	2.6	1	8760	0.02	10	2.00×10^{-2}	2.00×10^{-1}	4.9	9.80×10^{-1}
6	8	0.4	10	0.02	10	1.28×10^{-2}	1.28×10^{-1}	1	1.28×10^{-1}
7	8	0.4	10	0.02	10	1.28×10^{-2}	1.28×10^{-1}	1	1.28×10^{-1}
4 + 5	0	1	8760	9.13×10^{-7}	5	9.13×10^{-7}	4.57×10^{-6}	7.5	3.42×10^{-5}
4 + 6	8	0.4	10	9.13×10^{-7}	5	5.02×10^{-7}	2.51×10^{-6}	1	2.51×10^{-6}
5 + 6	8	0.4	10	9.13×10^{-7}	5	5.02×10^{-7}	2.51×10^{-6}	1	2.51×10^{-6}
Total					9.62	7.11×10^{-2}	6.84×10^{-1}	3.27	2.24

SAIFI = 0.0237 interruptions/customer yr
SAIDI = 0.2280 hours/customer yr
CAIDI = 9.62 hours/customer interruption
ASAI = 0.999974

ASUI = 2.603×10^{-5}
ENS = 2.24 MWh/yr
AENS = 0.747 kWh/customer yr

Table 9.6 Overall indices for system of Fig. 9.1

Load point	Criterion	λ (f/yr)	r (hours)	U (hours/yr)	L (MW)	E (MWh/yr)
2	TLOC	2.00×10^{-2}	5	1.00×10^{-1}	15	1.5
	PLOC	3.61×10^{-6}	4.90	1.77×10^{-5}	4.61	8.16×10^{-5}
	Total	0.020	5	0.100	15	1.50
3	TLOC	2.00×10^{-2}	5	1.00×10^{-1}	7.5	0.75
	PLOC	7.11×10^{-2}	9.62	6.84×10^{-1}	3.27	2.24
	Total	0.091	8.62	0.784	3.81	2.99

SAIFI = 0.0437 interruptions/customer yr
SAIDI = 0.3280 hours/customer yr
CAIDI = 7.51 hours/customer interruption
ASAI = 0.999963
ASUI = 3.744×10^{-5}
ENS = 4.49 MWh/yr
AENS = 1.497 kWh/yr

load by transferring it to neighboring load points through the normally open points. This problem was discussed simply in Section 7.8 and more deeply in recent publications [2, 6].

9.4.2 Transferable load modelling

The state space diagram for a transferable load system is shown in Fig. 9.6. State 1 represents the system operating normally. State 2 represents the system after the initiating failure has occurred and all affected feeders have been disconnected

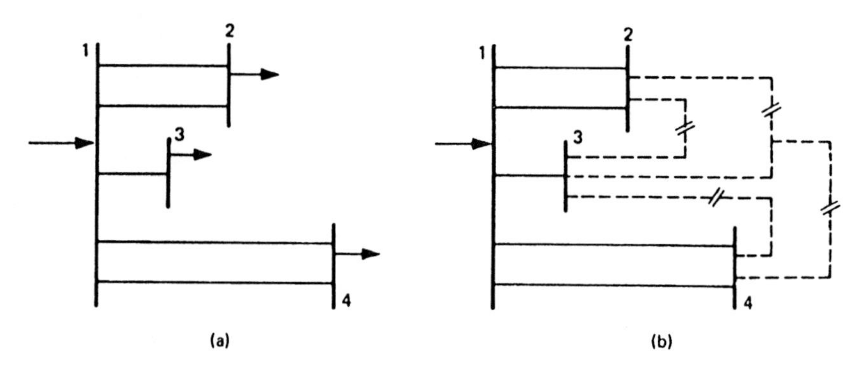

Fig. 9.5 Distribution system (a) without and (b) with transferable loads

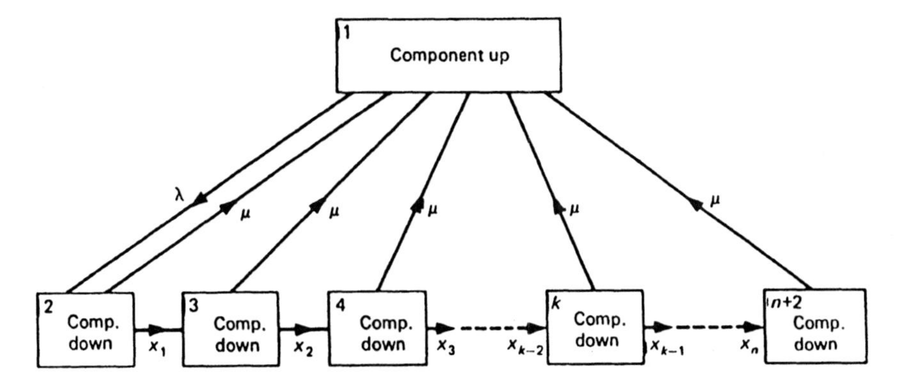

Fig. 9.6 State space diagram of transferable load model

(TLOC) or sufficient feeders have been disconnected to relieve network violations (PLOC). Feeders can now be sequentially transferred, wherever possible, to neighboring load points, and these sequential transfer operations are represented by States 3 to $(n + 2)$, where n is the number of individual transfers that can be performed.

In the case of TLOC events, λ is the rate of occurrence and μ is the reciprocal of the average outage time of the initiating event. These are evaluated using the techniques of Chapter 8. In the case of PLOC, λ is given by Equation (9.6) and μ is given by the reciprocal of r in Equation (9.7) if it is assumed that the excess load remains disconnected from the normal source of supply until repair of the initiating event is completed.

In both the TLOC and PLOC cases, x_k represents the switching rate from State $(k + 1)$ to State $(k + 2)$, i.e. it is the reciprocal of the average switching time. These times and rates can be ascertained knowing the importance or priority of each feeder and the order in which they will be transferred.

All the states of Fig. 9.6 except State 1 are associated with the failure state of the initiating event and represent the down state of the load point. Consequently, the transition rate into the down state (failure rate of the load point) and the average down time remain the same as that if transferable facilities were not available. The problem therefore is to evaluate the probability of residing in each of the down substates, the load that can be transferred and therefore the energy that cannot be supplied given residence in the substates and then, using a conditional probability approach, to evaluate the average load disconnected and energy not supplied during the repair time of the initiating event.

It has been shown [2] that the probability of residing in each substate of Fig. 9.6 is given by the following recursive formulae

$$P_1 = \frac{\mu}{\lambda + \mu} \tag{9.17a}$$

$$P_2 = \frac{\lambda\mu}{(\lambda + \mu)(\mu + x_1)} \tag{9.17b}$$

$$P_k = \frac{P_{k-1}x_{k-2}}{(\mu + x_{k-1})} \quad \text{for } k = 3, \ldots, (n+1) \tag{9.17c}$$

$$P_{n+2} = 1 - \sum_{i=1}^{n+1} P_i \tag{9.17d}$$

9.4.3 Evaluation techniques

The amount of load that can be recovered during each of the transfer states in Fig. 9.6 may be less than the maximum demand on the feeders being transferred due to limitations caused by the feeder capacity itself, by the capacity of feeders to which it is being transferred or by the capacity of the new load point to which it is to be connected.

The first two restrictions can be assessed by comparing [2] the loading profiles, i.e. how the loads are distributed along the feeders, with the capacity profiles of the feeders. An example of this comparison is given in Section 9.4.4. The third restriction can be assessed by performing a load flow on the system after the load has been transferred. In all cases, the maximum load of each feeder that can be transferred (L_{pt}) can be evaluated and the maximum load that cannot be transferred, i.e. remains disconnected, is then given by

$$L_{pd} = L_p - L_{pt} \tag{9.18}$$

where L_p = peak load or maximum demand on the feeder.

The energy not supplied (E_{ij}) to feeder j given state i of Fig. 9.6 is evaluated by the area under the load–duration curve having a peak value L_p before the feeder is transferred and a peak value L_{pd}, after the feeder is transferred. The total energy not supplied is then found by summating E_{ij} weighted by P_i for all feeders ($j = 1, \ldots, f$) and all down substates ($i = 2, \ldots, n + 2$), i.e.

$$E = \sum_{i=2}^{n+2} \sum_{j=1}^{f} E_{ij} P_i \tag{9.19}$$

and

$$L = E/U \tag{9.20}$$

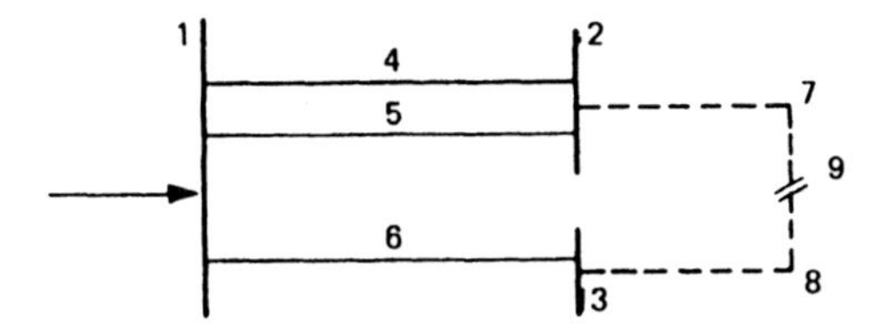

Fig. 9.7 Radial system with transferable loads

9.4.4 Numerical example

Consider the system shown in Fig. 9.7 and the reliability, loading and capacity data shown in Table 9.7. The loading and capacity profiles for this system are shown in Fig. 9.8. In this example the capacity profiles are constant. This may not necessarily be so in practice, in which case the combined load profile must be reduced, if necessary to prevent overload at any restricted part of the feeder capacity. The profiles in Fig. 9.8 indicate that only 1 MW of the load on feeder 8 can be transferred to load point 2 and only 3 MW of the load on feeder 7 can be transferred to load point 3.

The reliability results are shown in Tables 9.8–9.10.

The results shown in Table 9.10 indicate the significant improvement that can be gained using transferable load systems and demonstrate the need to include the feature in a reliability analysis if such transferable facilities exist.

Table 9.7 Data for system shown in Fig. 9.7

(a) *Component data*

Component	λ *(f/hr)*	*r (hours)*	*Capacity (MW)*	*Switching time (hours)*
1–3	0.01	5	—	—
4–5	0.02	10	3	—
6	0.02	10	5	—
9	—	—	—	0.5

(b) *Loading data*

Load feeder	*Peak load L_p (MW)*	*Min. load (MW)*	*Load factor*	*Capacity (MW)*
7	4	2	0.75	5
8	2	1	0.75	5

Assume (i) straight-line load-duration curves;
(ii) load is uniformly distributed along each feeder;
(iii) cable capacity is uniform.

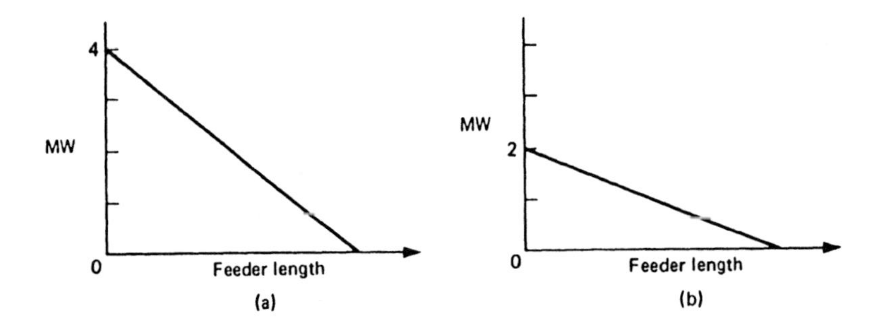

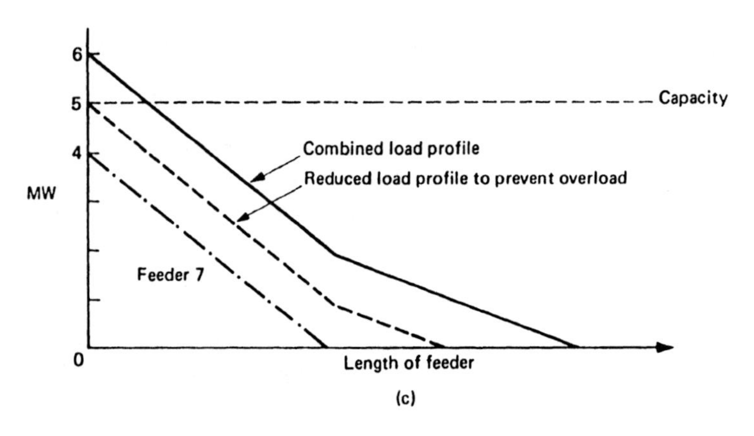

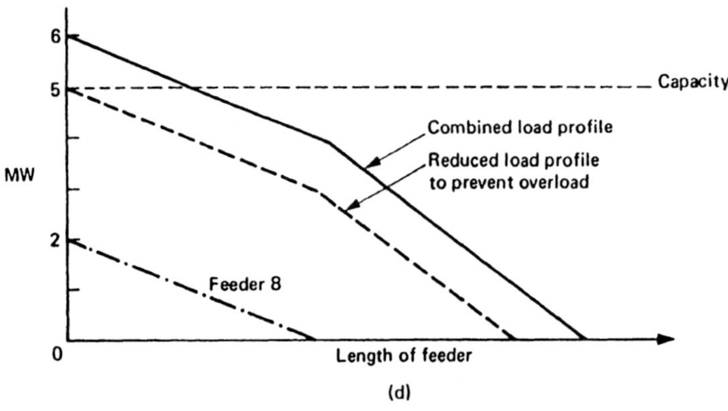

Fig. 9.8 Load and capacity profiles
(a) Load profile of feeder 7 seen from load point 2
(b) Load profile of feeder 8 seen from load point 3
(c) Combined load profile seen from load point 2
(d) Combined load profile seen from load point 3

Table 9.8 Reliability indices without transferring loads

(A) *TLOC*

Event	λ (*f/yr*)	*r* (*hours*)	*U* (*hours/yr*)	*L* (*MW*)	*E* (*MWh/yr*)
Load point 3					
1	0.01	5	0.05	1.5	0.075
3	0.01	5	0.05	1.5	0.075
6	0.02	10	0.20	1.5	0.300
Total	0.04	7.5	0.30	1.5	0.450
Load point 2					
1	0.01	5	0.05	3	0.15
2	0.01	5	0.05	3	0.15
4 + 5	9.13×10^{-7}	5	4.57×10^{-6}	3	1.37×10^{-5}
Total	0.02	5	0.10	3	0.30

(B) *PLOC*

Event	L_s (*MW*)	*P*	r_H (*hours*)	λ_e (*f/yr*)	r_e (*hours*)	λ (*f/yr*)	*U* (*hours/yr*)	*L* (*MW*)	*E* (*MWh/yr*)
Load point 3									
No PLOC events									
Load point 2									
4	3	0.5	10	0.02	10	0.015	0.15	0.5	0.075
5	3	0.5	10	0.02	10	0.015	0.15	0.5	0.075
Total					10	0.03	0.30	0.5	0.15

9.5 Economic considerations

9.5.1 General concepts

Power systems exist in order to provide, as economically and as reliably as possible, electrical energy to the customer. It is implicit in this philosophy that it is not justifiable to increase reliability for its own sake; supply reliability should only be increased if some benefit will economically accrue to society. The benefit, measured in terms of an increase in conventional reliability indices, of any proposed reinforcement can be evaluated using the reliability techniques of this chapter and Chapters 7 and 8. When associated with the cost of such a scheme, the result means nothing other than a cost per unit increase in reliability and gives no indication of the likely return on investment or real benefit to the customer or utility. Several authors have considered this problem but, to date, there is no universal acceptance of the most suitable evaluation method or of the cost data.

Table 9.9 Reliability indices with load transfers

Event	Sub state	λ (f/yr)	r (hours)	U (hours/yr)	P_t	L_{pd} (MW)	L (MW)	E (MWh/yr)
(A) *TLOC*								
Load point 2								
1	—	0.01	5	0.05	—	4	3	0.15
2	2				5.19×10^{-7}	4	3	1.36×10^{-2}
	3				5.19×10^{-6}	1	0.75	3.41×10^{-2}
	Total	0.01	5	0.05			0.95	4.77×10^{-2}
4 + 5	2				4.74×10^{-4}	4	3	1.25×10^{-6}
	3				4.74×10^{-10}	1	0.75	3.11×10^{-6}
	Total	9.13×10^{-7}	5	4.57×10^{-6}			0.95	4.36×10^{-6}
Total		0.02	5	0.10			1.98	0.198
Load point 3								
1	—	0.01	5	0.05	—	2	1.5	0.075
3	2				5.19×10^{-7}	2	1.5	6.82×10^{-3}
	3				5.19×10^{-6}	1	0.75	3.41×10^{-2}
	Total	0.01	5	0.05			0.82	4.09×10^{-2}

6	2				1.09×10^{-6}	2	1.5	1.43×10^{-2}
	3				2.17×10^{-5}	1	0.75	1.43×10^{-1}
	Total	0.02	10	0.2			0.79	0.157
Total		0.04	7.5	0.3			0.91	0.273
(B) *PLOC*								
Load point 2								
4	2				8.15×10^{-7}	4	3	2.14×10^{-2}
	3				1.63×10^{-5}	0	0	0
	Total	0.015	10	0.15			0.14	2.14×10^{-2}
5	2				8.15×10^{-7}	4	3	2.14×10^{-2}
	3				1.63×10^{-5}	0	0	0
	Total	0.015	10	0.15			0.14	2.14×10^{-2}
Total		0.03	10	0.3			0.14	4.28×10^{-2}

N.B. (i) With outage event 1, no load can be transferred.
(ii) In the case of PLOC, it is assumed that the feeder is disconnected in state 2 and load is either reconnected to the busbar or transferred in state 3.

Table 9.10 Comparison of reliability indices

Load point	λ (f/yr)	r (hours)	U (hours/yr)	L (MW)	E (MWh/yr)
2—no transfers	0.05	8	0.4	1.13	0.450
— with transfers	0.05	8	0.4	0.60	0.241
3—no transfers	0.04	7.5	0.3	1.50	0.450
—with transfers	0.04	7.5	0.3	0.91	0.273

It is universally accepted that the reliability of a system can be increased by increased investment. At the same time, the outage costs of the system will decrease. This leads to the concept of an optimum reliability as depicted in Fig. 9.9.

The essential problem in applying the concept of optimum reliability is lack of knowledge of true outage costs and the features that should be included.

9.5.2 Outage costs

The outage costs have two parts: that seen by the utility and that seen by society or the customer.

The utility outage costs include:

(a) loss of revenue from customers not served;
(b) loss of customer goodwill;
(c) loss of future potential sales due to adverse reaction;
(d) increased expenditure due to maintenance and repair.

These costs, however, usually form only a very small part of the total outage costs. A greater part of the costs comprises those seen by the customer and most of these are extremely difficult, if not impossible, to quantify. They include:

(a) costs imposed on industry due to lost manufacture, spoiled products, damaged equipment, extra maintenance, etc.
(b) costs imposed on residential customers due to spoiled deep frozen foods, alternative heating and lighting costs, etc.

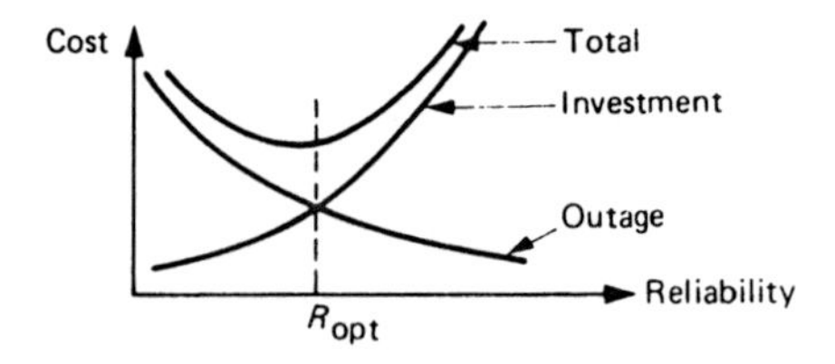

Fig. 9.9 Relationship between costs and reliability

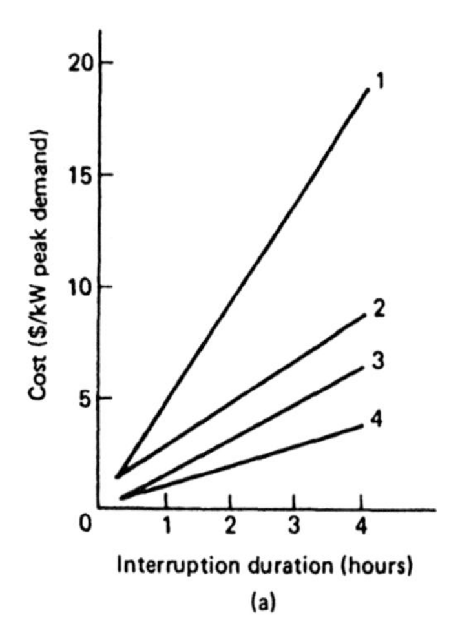

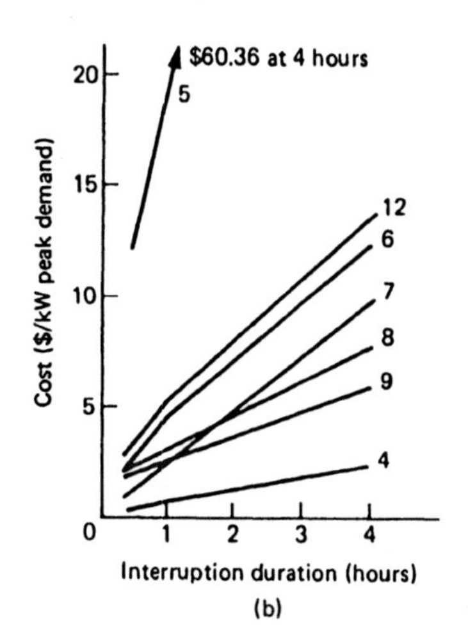

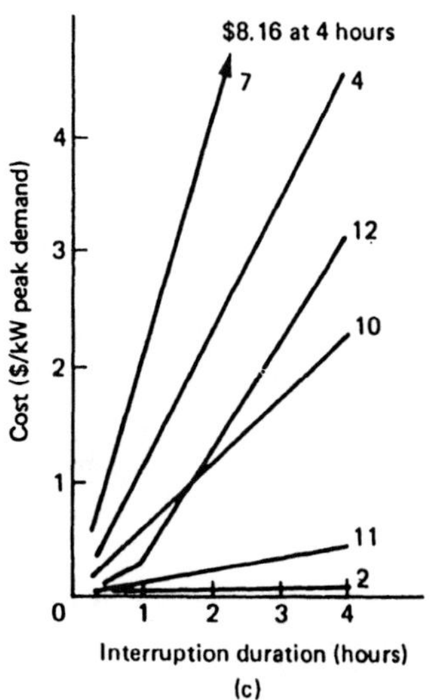

Fig. 9.10 Interruption costs estimates: (a) total; (b) industrial; (c) residential

1 New York blackout
2 Ontario Hydro
3 France
4 Sweden
5 IEEE small industrial
6 Ontario Hydro small industrial
7 Great Britain
8 IEEE large industrial
9 Ontario Hydro large industrial
10 Myers
11 Markel
12 University of Saskatchewan

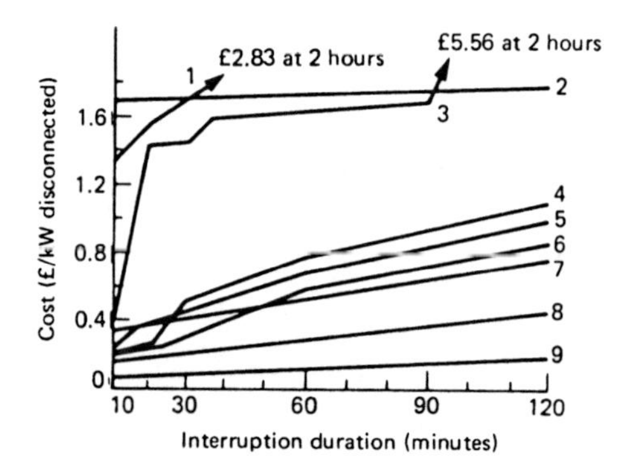

Fig. 9.11 Interruption costs estimates for manufacturing industries [9]

1 Potato crisps
2 Plastic sheeting
3 Paper machine rolls
4 Confectionery
5 Rubber types
6 Glass bottles
7 Paper
8 Trucks and tractors
9 Spun fiber

(c) costs which become difficult to quantify and which range from loss of convenience, inability to enjoy one's hobbies and pastimes, to severe situations occurring during blackouts such as looting, rioting, failure of hospital services, etc.

The costs per kilowatt interrupted found in a number of previous studies are shown in Figs. 9.10 and 9.11 as a function of the interruption duration. Several interesting features can be discerned in these results.

(a) The suggested costs vary over a wide range (many are also now very out of date).
(b) The costs depend very much on the country of origin.
(c) The costs depend on the type of customer and are very different for small and large industries, type of industry, residential customers, and would also be different for commercial customers and agriculture customers.
(d) The costs are not necessarily linear as a function of interruption duration for a given type of customer. This is a particular problem because a simple cost/kW or per other index would be an approximation and could give rise to misleading economic appraisals. The non-linear cost relation should be ascertained and then convolved with the appropriate outage time probability distribution (see Section 7.9 and Chapter 12).

Most of the problems described above are being actively studied. These and related aspects are discussed in Chapter 13.

9.6 Conclusions

This chapter has considered a number of techniques for extending the reliability evaluation of distribution systems. The importance of including a partial loss of continuity (PLOC) criterion has been demonstrated. In addition it has been shown that, if it is possible to transfer loads from one load point to another, these facilities should be included in the assessment when the load and energy indices are to be evaluated with reasonable accuracy.

Finally, the economic implications associated with reinforcement and expansion of distribution systems has been discussed briefly. This area still remains relatively open and more work is required, particularly that concerning outage costs, before a full, accurate and consistent economic assessment can be made. The current status regarding reliability cost and worth is discussed in Chapter 13.

9.7 Problems

1 Two load points are presently fed individually by single transformer feeders as shown in Fig. 9.12. Each of the present transformers has a capacity of 4 MW. The maximum demand at each load point is expected to grow to 5 MW and four alternative reinforcement schemes are to be considered. These are:
(i) the present 4 MW transformers to be replaced by 8 MW transformers;
(ii) an additional 4 MW transformer to be connected in parallel with each of the existing ones;
(iii) as reinforcement (i) plus a 3 MW link connected between the two load points;
(iv) as reinforcement (i) and the two low-voltage feeders, each of constant 8 MW capacity, connected together through a normally open point.

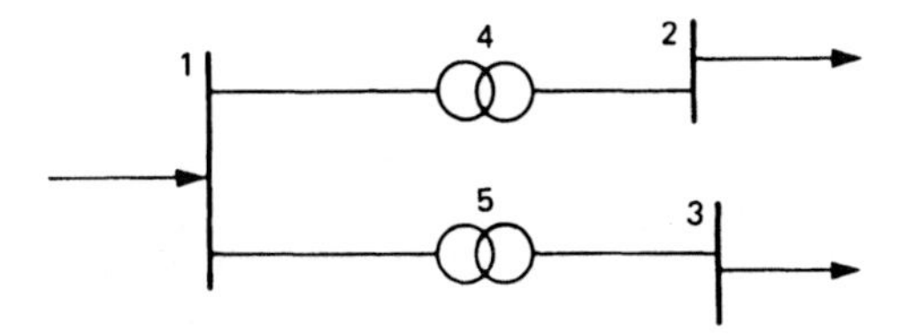

Fig. 9.12

Table 9.11

Component	*λ (f/yr)*	*r (hours)*	*Switching time (hours)*
Busbars	0.01	5	—
Transformer feeders	0.02	10	—
Interconnector	0.01	5	—
Normally open point	—	—	0.5

Table 9.12

Load point	*Peak load (MW)*	*Minimum load (MW)*	*Load factor*	*Number of customers*
2	5	2.5	0.75	500
3	5	2.5	0.75	500

Evaluate the values of λ, r, U, L, E including, when appropriate, both TLOC and PLOC indices at each load point
Assume
—load flow inversely proportional to reactance and all reactances equal
—load-duration curves are straight lines
—average duration of load level > maximum demand = 0
—average duration of load level > 60% of maximum demand = 20 hours
—the load level-duration curve is a straight line between the above two limits

9.8 References

1. Allan, R. N., Dialynas, E. N., Homer, I. R., 'Reliability indices and reliability worth in distribution systems', EPRI Workshop on Power System Reliability Research Needs and Priorities, Asilomar, California, March 1978, *EPRI Pub. WS-77-60*, pp. 6.21–8.
2. Allan, R. N., Dialynas, E. N., Homer, I. R., 'Modelling and evaluating the reliability of distribution systems', *IEEE Trans. on Power Apparatus and Systems*, **PAS-98** (1979), pp. 2181–9.
3. Billinton, R., Grover, M. S., 'Reliability evaluation in transmission and distribution systems', *Proc. IEE*, **122** (1975), pp. 517–23.
4. Allan, R. N., de Oliveira, M. F., 'Reliability modelling and evaluation of transmission and distribution systems', *Proc. IEE*, **124** (1977), pp. 535–41.
5. Stott, B., Alsac, O., 'Fast decoupled load flow', *IEEE Trans. on Power Apparatus and Systems*, **PAS-93** (1974), pp. 859–67.
6. Allan, R. N., Dialynas, E. N., Homer, I. R., *Partial Loss of Continuity and Transfer Capacity in the Reliability Evaluation of Power System Networks*. Power System Computational Conference, **PSCC 6** (1978), Darmstadt, West Germany.

10 Substations and switching stations

10.1 Introduction

The main difference between the power system networks discussed in Chapters 7–9 and substations and switching stations is that the latter systems comprise switching arrangements that are generally more complex. For this reason, several papers [1–3] have specifically considered techniques that are suitable for evaluating the reliability of such systems. It is recognized, however, that the reliability techniques described in Chapters 7–9 for power system networks are equally applicable to substations and switching stations (subsequently referred to only as substations). The reason for discussing substations as a separate topic is that the effect of switching is much more significant and the need for accurate models is greater.

The basic concepts of breakers in reliability evaluation were discussed in Section 8.4 and various models were described which permit their effect to be included in the analysis of distribution systems. These concepts are extended in this chapter, additional concepts are introduced and the relationship between substation design and its reliability is discussed.

It should be noted that, as stated in Chapter 8, all the concepts and modelling techniques described in the present chapter can be used in the evaluation of distribution systems and, conversely, the previous techniques can be used in the evaluation of substations. This is important because it means that, if required, the boundary of the system can be extended to encompass substations and a distribution system as one entity, and common evaluation methods can be used.

10.2 Effect of short circuits and breaker operation

10.2.1 Concepts

In order to illustrate the importance of recognizing the switching operation of breakers following short circuit faults, consider the two simple substations shown in Fig. 10.1.

Consider only short circuit failures of the transformers and the subsequent effect on the indices of the two load points. When T1 fails in the system of Fig.

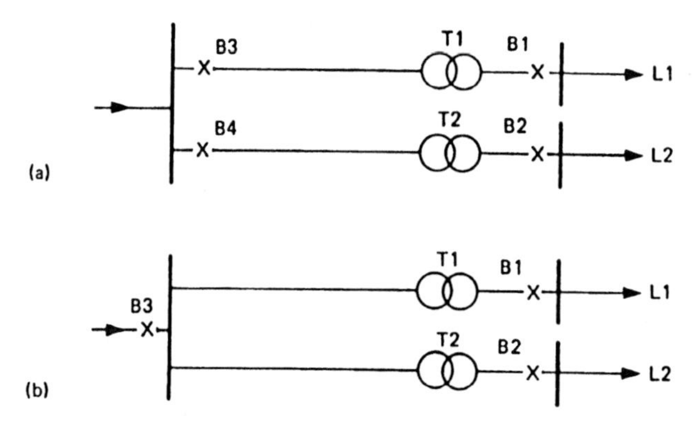

Fig. 10.1 Two simple substations

10.1(a), the input breaker B3 should operate, causing an interruption of load point L1 only. Similarly failure of T2 will interrupt load point L2 only. In both cases, the outage time of the load points will be the repair or replacement time for the appropriate transformer, giving the following load point reliability indices:

Load L1: $\lambda(\mathrm{T1}), r(\mathrm{T1}), U(\mathrm{T1})$

Load L2: $\lambda(\mathrm{T2}), r(\mathrm{T2}), U(\mathrm{T2})$

When T1 fails in the system of Fig. 10.1(b), however, the input breaker B3, which protects both transformers, should operate, causing interruption of both load points. Similarly failure of T2 will interrupt both load points. In this case the indices will be dependent on the subsequent operational procedures.

(a) *Isolation of failed component not possible*

If it is not possible or practical to isolate the failed component, breaker B3 will remain open until the relevant component has been repaired or replaced. In this case both load points will remain disconnected until this has been achieved and the indices will be:

loads L1 and L2 $\lambda = \lambda(\mathrm{T1}) + \lambda(\mathrm{T2})$

$$U = \lambda(\mathrm{T1})r(\mathrm{T1}) + \lambda(\mathrm{T2})r(\mathrm{T2})$$

$$r = U/\lambda$$

(b) *Isolation of failed component is possible*

In practice it is usually possible to isolate a failed component either using physically existing disconnects (isolators) or by disconnecting appropriate connections. In either case the protection breaker that has operated can be reclosed after the component is isolated. For example, after T1 has failed, it is isolated, B3 is reclosed and load L2 is reconnected. This procedure means that L1 is interrupted for the repair or replacement time of T1 but L2 is interrupted only for the relevant isolation or switching time. A similar situation occurs if T2 fails. The indices now become:

$$\text{load L1:}\quad \lambda(\text{L1}) = \lambda(\text{T1}) + \lambda(\text{T2})$$

$$U(\text{L1}) = \lambda(\text{T1})r(\text{T1}) + \lambda(\text{T2})s(\text{T2})$$

$$r(\text{L1}) = U(\text{L1})/\lambda(\text{L1})$$

$$\text{load L2:}\quad \lambda(\text{L2}) = \lambda(\text{T1}) + \lambda(\text{T2})$$

$$U(\text{L2}) = \lambda(\text{T1})s(\text{T1}) + \lambda(\text{T2})r(\text{T2})$$

$$r(\text{L2}) = U(\text{L2})/\lambda(\text{L2})$$

where $s(\ )$ is the switching or isolation time of the failed component.

10.2.2 Logistics

Misconceptions occasionally arise in regard to the numerical values associated with the outage times, particularly that concerning the switching or isolation time.

When the relevant information is being collected or assessed, the appropriate outage time must be measured from the instant the failure occurs to the instant at which the load is reconnected. Consequently both repair and switching times contain several logistic aspects including:

(a) time for a failure to be noted (in rural distribution systems without telemetry this includes the time it takes for a customer to notify the utility of supply failure);
(b) time to locate the failed component;
(c) time to travel to the location of the failed component and the relevant disconnects (isolators) and breakers;
(d) time required to make the appropriate operating decisions;
(e) time to perform the required action itself.

The summation of these items can mean that switching times in particular are very much greater than the actual time needed to complete the switching sequence itself.

10.2.3 Numerical examples

Example 1

Consider first the systems shown in Fig. 10.1 and let each transformer have a failure rate of 0.1 f/yr, a repair time of 50 hours and a switching time of 2 hours.

(a) *System of Fig.* 10.1*(a)*

$$\lambda(\text{L1}) = \lambda(\text{L2}) = 0.1 \text{ f/yr}$$

$r(\text{L1}) = r(\text{L2}) = 50$ hours

$U(\text{L1}) = U(\text{L2}) = 5$ hours/yr

(b) *System of Fig.* 10.1*(b)*

(i) *Isolation not possible*

$\lambda(\text{L1}) = \lambda(\text{L2}) = 0.1 + 0.1 = 0.2$ f/yr

$U(\text{L1}) = U(\text{L2}) = 0.1 \times 50 + 0.1 \times 50 = 10$ hours/yr

$r(\text{L1}) = r(\text{L2}) = 50$ hours

(ii) *Isolation is possible*

$\lambda(\text{L1}) = \lambda(\text{L2}) = 0.1 + 0.1 = 0.2$ f/yr

$U(\text{L1}) = U(\text{L2}) = 0.1 \times 50 + 0.1 \times 2 = 5.2$ hours/yr

$r(\text{L1}) = r(\text{L2}) = 26$ hours

These results indicate that it is important to recognize the switching effects of protection breakers, the failure modes of the load point and the mode by which service to the load point is restored.

Example 2

Consider now the system shown in Fig. 10.2 and the reliability data shown in Table 10.1.

The reliability indices of load point A (identical for load points B and C) are shown in Table 10.2. It is seen from these results that the annual outage time is dominated by that of transformer 3. This effect can be reduced by using a spare transformer rather than repairing the failed one.

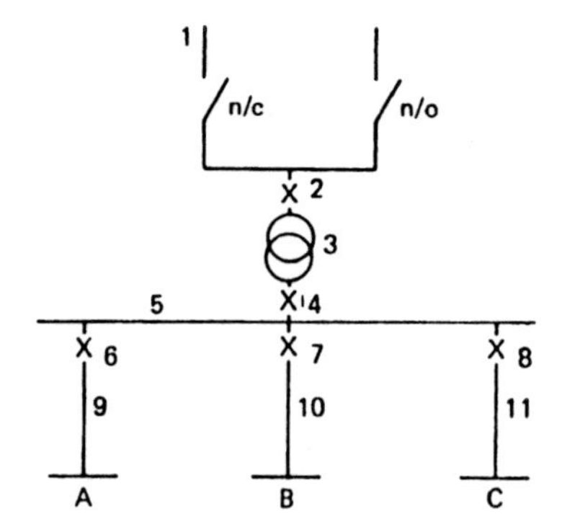

Fig. 10.2 Substation feeding three radial loads

Table 10.1 Reliability data for system of Fig. 10.2

Component	*Short circuit failure rate (f/yr)*	*Repair time (hours)*	*Switching time (hours)*
h.v. line 1	0.05	20	0.5
h.v. breaker 2	0.002	50	1.0
transformer 3	0.01	200	1.0
l.v. breakers, 4, 6–8	0.005	20	1.0
l.v. busbar 5	0.005	5	1.0
l.v. feeders 9–11	0.05	5	1.0

If, in this example, a spare transformer is available and can be installed in 10 hours, the values of r and U for this failure event reduce to 10 hours and 0.1 hour/yr. The system indices then become:

$\lambda = 0.137$ f/yr, $r = 5.2$ hours, $U = 0.710$ hours/yr

Generally one spare transformer, or indeed spares for other system components, are used to service several operating transformers, and an economic appraisal is necessary to judge the merits of investing in one or more spares and the benefits accruing from being able to reduce the system outage times of those parts of the system for which the spares are being carried. If each part of the system for which spares are available is analyzed independently of the others, then it is assumed that the number of failures is less than or equal to the number of spares available and that spares can be restocked or components repaired before further failures occur. This assumption is usually valid since the failure rate of system components is very small, although it does mean that each spare must not be expected to service a large

Table 10.2 Reliability indices for load point A

Failed component	λ *(f/yr)*	r *(hours)*	U *(hours/yr)*
h.v. line 1	0.05	0.5	0.025
h.v. breaker 2	0.002	50	0.100
transformer 3	0.01	200	2.000
l.v. breaker 4	0.005	20	0.100
l.v. busbar 5	0.005	5	0.025
l.v. breaker 6	0.005	20	0.100
l.v. breaker 7	0.005	1	0.005
l.v. breaker 8	0.005	1	0.005
l.v. feeder 9	0.05	5	0.250
Total	0.137	19.1	2.610

number of system components. A more exact method of analysis can be made using the techniques of sparing discussed in Section 11.3.

10.3 Operating and failure states of system components

The previous discussion and examples illustrate the need to consider short circuits and their effect on subsequent switching. There are other failure modes which must be considered in a practical system. It is not necessary, however, to identify separately each individual mode of failure. Instead, modes of failure which have an identical effect on the system behavior can be grouped together and represented by a single set of component indices in the reliability evaluation. For most system components, the pooled system states are:

(a) operates successfully;
(b) suffers an open circuit;
(c) suffers a short circuit.

In the case of breakers, however, several other states must be considered because of their switching actions. Seven operating and failure states were identified in Section 8.9.2 for normally closed breakers. These were:

(a) operates successfully in its closed state;
(b) opens successfully when required to do so;
(c) fails to open when required to do so;
(d) opens inadvertently when not requested to do so;
(e) suffers an open circuit;
(f) suffers a short circuit on the busbar side;
(g) suffers a short circuit on the line side.

In the case of a normally open breaker, these states reduce to:

(a) closes successfully when required to do so;
(b) fails to close when required to do so;
(c) suffers a short circuit on the busbar side;
(d) suffers a short circuit on the line side.

At this stage, as in Section 8.9.3, it will be assumed that breakers open and close successfully when requested to do so. These failure modes are discussed in Sections 10.6–8. Therefore, the only failure modes to be considered at this stage are open circuits, short circuits and the inadvertent opening of normally closed breakers.

10.4 Open and short circuit failures

10.4.1 Open circuits and inadvertent opening of breakers

It is generally found in practical systems that, when an open circuit fault occurs on a component, the protection system does not operate. The fault manifests itself therefore only on the component that has failed and the other system components remain energized. If this is not the case and protection breakers do operate, then this type of failure should be treated in the same way as a short circuit failure of the component and the appropriate indices combined together.

Open circuits of power system components occur very infrequently and can usually be ignored. This does not apply to normally closed breakers which can suffer inadvertent opening due to malfunction of the protection system or the breaker itself. These malfunctions manifest themselves as an open circuit and are treated accordingly.

If it is known that open circuits and inadvertent opening of breakers do not cause protection breakers to operate, an appropriate set of indices must be estimated which represents the open circuit (including inadvertent opening) failure rate.

10.4.2 Short circuits

Short circuit faults are the dominant source of component failures and always cause the operation of the protection system. Since this is a different system effect from that of open circuits, it must be categorized separately and appropriate indices estimated.

The basic method for assessing the impact of short circuits on the reliability of a load point was used in Section 10.2. This intuitive approach is the basis of more formalized methods; these are discussed in Section 10.5. This basic method identifies the protection breakers that trip following a short circuit failure of a component and deduces whether these cause interruption of the load point being analyzed.

This method of evaluation is relatively simple for most system components, including breakers which do not or cannot clear their own short circuit faults (see Section 8.9.3). In the case of breakers which can clear some of their own short circuits, the method described in Section 8.9.4 should be used and the short circuit failure rate allocated to the breaker or the components on either side of the breaker as appropriate. The methods and analyses described in the following sections assume that either:

(a) the breakers cannot clear their own faults and therefore the indices represent the total short circuit failures of the respective breakers; or

Table 10.3 Reliability data of breakers of Fig. 10.2

	Failure rate				
Breaker	*Open circuit (f/yr)*	*Short circuit (f/yr)*	*Total (f/yr)*	*Repair time (hours)*	*Switching time (hours)*
2	0.003	0.002	0.005	50	1.0
4, 6–8	0.005	0.005	0.010	20	1.0

Table 10.4 Modified reliability indices for load point A

Failed component	λ *(f/yr)*	*r* *(hours)*	*U* *(hours/yr)*
h.v. line 1	0.05	0.5	0.025
h.v. breaker 2	0.005	50	0.250
transformer 3	0.01	10	0.100
l.v. breaker 4	0.01	20	0.200
l.v. busbar 5	0.005	5	0.025
l.v. breaker 6	0.01	20	0.200
l.v. breaker 7	0.005	1	0.005
l.v. breaker 8	0.005	1	0.005
l.v. feeder 9	0.05	5	0.250
Total	0.150	7.1	1.060

(b) the breakers can clear some of their own faults and therefore the indices have already been allocated to the appropriate components.

10.4.3 Numerical example

Consider again the system shown in Fig. 10.2 and, in addition to the short circuit information shown in Table 10.1, let the breakers have the open circuit failure rates shown in Table 10.3. The modified set of reliability indices for load point A (load points B and C are identical) is shown in Table 10.4. These results assume that the transformer can be replaced by a spare in 10 hours.

10.5 Active and passive failures

10.5.1 General concepts

It is evident from the previous sections that the switching actions must be modelled and simulated in the reliability evaluation process. This section describes the more formal approach to this modelling and simulation.

When switching actions occur, a three-state model is required [1, 2, 4–7], the three states being:

(i) state before the fault;
(ii) state after the fault but before isolation;
(iii) state after isolation but before repair is completed.

This set of states is shown in Fig. 10.3.

In some cases, however, breakers are not required to operate, e.g. open circuits and inadvertent operation of breakers. In these cases, a two-state model only is necessary. These states are:

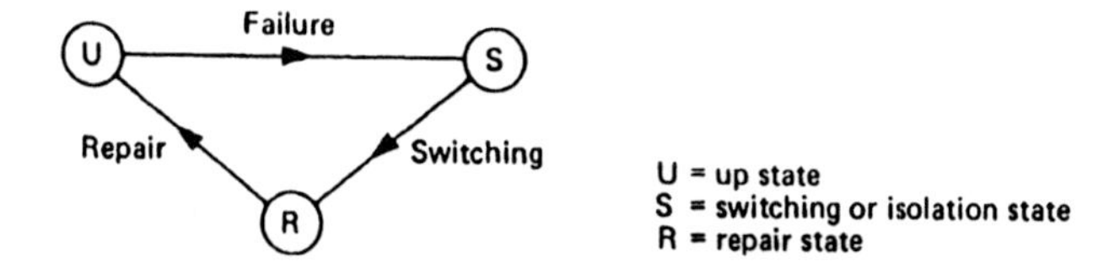

Fig. 10.3 Three-state component model

(i) state before the fault;
(ii) state after the fault but before repair is completed.
This set of states is shown in Fig. 10.4.

If the repair process is assumed to be the same in both cases, the two models can be superimposed to give the combined state space diagram shown in Fig. 10.5.

The two modes of failure, one leading to state R and the other to state S, have been designated [1, 7] as passive and active failures respectively. These are defined [1, 7] as:

(a) *Passive event*: A component failure mode that does not cause operation of protection breakers and therefore does not have an impact on the remaining healthy components. Service is restored by repairing or replacing the failed component. Examples are open circuits and inadvertent opening of breakers.

(b) *Active event*: A component failure mode that causes the operation of the primary protection zone around the failed component and can therefore cause the removal of other healthy components and branches from service. The actively failed component is isolated and the protection breakers are reclosed. This leads to service

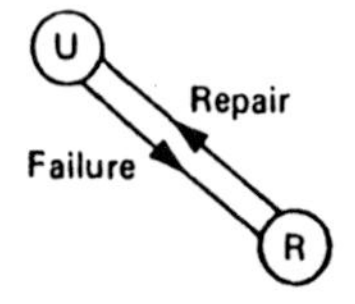

Fig. 10.4 Two-state component model

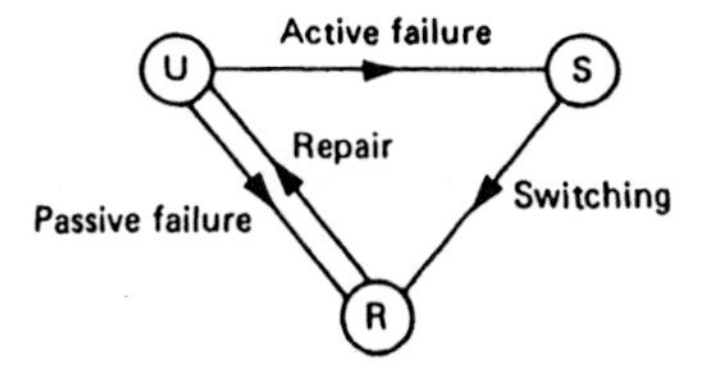

Fig. 10.5 State space diagram for active and passive failures

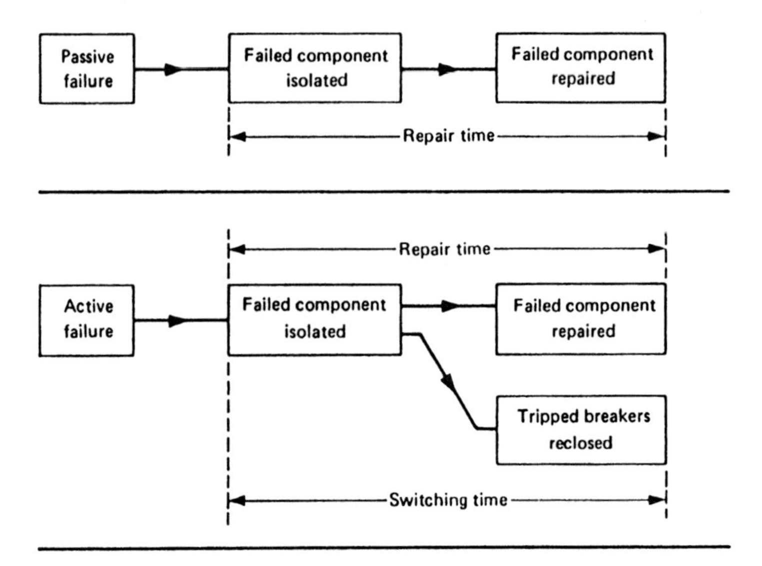

Fig. 10.6 Sequences following passive and active failures

being restored to some or all of the load points. It should be noted, however, that the failed component itself can be restored to service only after repair or replacement.

These concepts and definitions lead to the sequence flow chart shown in Fig. 10.6.

10.5.2 Effect of failure mode

The essential requirement of a reliability assessment is to identify whether the failure of a component or a combination of components causes the failure of the load point of interest. If it does, the event must be counted as a load point failure event; if it does not, the event can be disregarded at least as far as the load point of interest is concerned. Consequently it is necessary to determine whether state R and/or state S of Fig. 10.5 constitutes a load point failure event.

It can be seen from Fig. 10.5 that, if a passive event leads to the failure of a given load point, i.e. state R is a load point failure event, then an active event on the same component also leads to the failure of the load point. The reverse situation may not be true, however, because state R need not be a load point failure event when state S is. This is likely to happen in a practical system because state S represents a wider outage situation due to the protection breakers being open during the existence of this state.

These aspects are illustrated by the failure modes analysis of the system shown in Fig. 10.7.

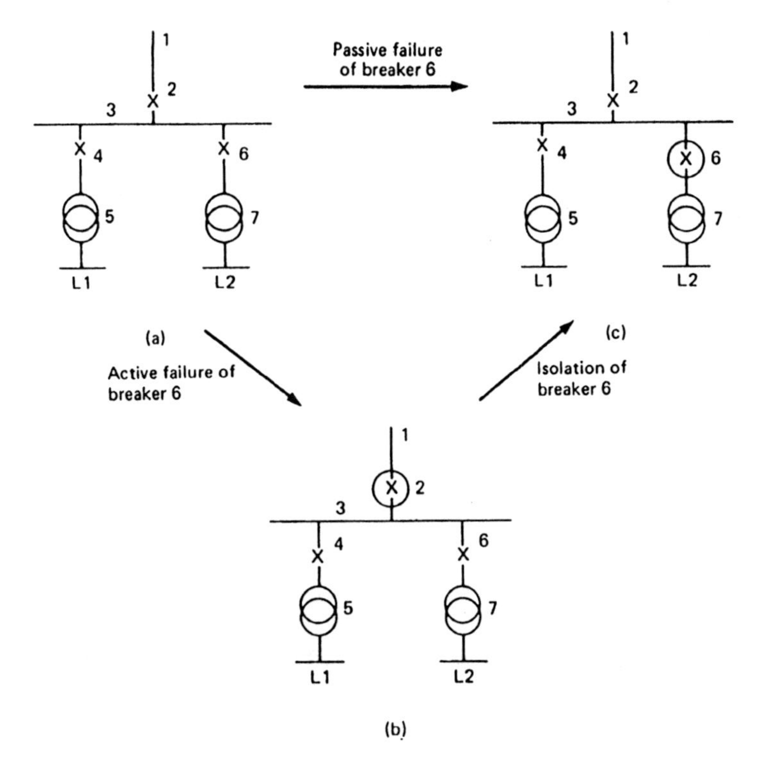

Fig. 10.7 The effect of failure modes
(a) System operating normally (state U of Fig. 10.5), L1 and L2 being supplied
(b) System state following active failure of breaker 6 (state S of Fig. 10.5), L1 and L2 disconnected
(c) System state following passive failure of breaker 6 or following switching after active failure of breaker 6 (state R of Fig. 10.5), L1 being supplied, L2 disconnected

The failure events of load point L1 are:

(a) passive events: 1P, 2P, 3P, 4P, 5P
(b) active events: 1A, 2A, 3A, 4A, 5A, 6A

It can be seen that, for each passive event, there is a corresponding active event. The failure events can therefore be grouped together to give:

(c) grouped events: 1(P + A), 2(P + A), 3(P + A), 4(P + A), 5(P + A), 6A
or: 1T, 2T, 3T, 4T, 5T, 6A

In the above events, P = passive, A = active and T = total (= P + A).

It follows therefore that the separate identity of the passive event is lost because it can be combined with its corresponding active event to produce a total failure event. Therefore the component reliability data required is:

λ—total failure rate
λ^a—active failure rate ($= \lambda$ if all failures are active events)
r—repair time
s—switching or isolation time

10.5.3 Simulation of failure modes

The deduction of the failure events of a load point can be made in two sequential steps [6]:

(a) a failure modes or minimal cut-set analysis [6, 8] is performed on the system. This simulates the total failure events including overlapping outages up to any desired order, effect of weather, maintenance, common mode failures, etc.

(b) active failure events are simulated using the following algorithm [6]:

   (i) Choose the first component on which an active failure event is to be simulated. Neglect those components which constitute a first-order total failure event because these include the appropriate active failure.

   (ii) Identify whether the actively failed component and/or one or more of its protection breakers appears in a path between the load point of interest and a source.

   (iii) If these components do not appear in any of the paths, the actively failed component does not constitute a load point failure event.

   (iv) If these components cause all paths to be broken, the active failure constitutes a first-order load point failure event.

   (v) If these components break some of the paths, the minimal cut sets of the remaining paths are evaluated. These cut sets together with the actively failed component form a second- or higher-order failure event, the order of the event being that of the minimal cut set plus one.

   (vi) Repeat the above steps until all components on which an active failure is to be simulated have been considered.

   (vii) Each time a load point failure event is identified by this process, it must be checked for minimality and, if it contains an existing lower order event, must be discarded.

In order to illustrate the identification and simulation of these failure modes, consider the system shown in Fig. 10.8.

The failure events of load-point L1 are

(a) total failure events:

4	1 + 5	2 + 5	3 + 5
	1 + 6	2 + 6	3 + 6
	1 + 7	2 + 7	3 + 7
	1 + 8	2 + 8	3 + 8
	1 + 9	2 + 9	3 + 9

The above second-order events are associated with overlapping forced outages and a forced outage overlapping a maintenance outage.

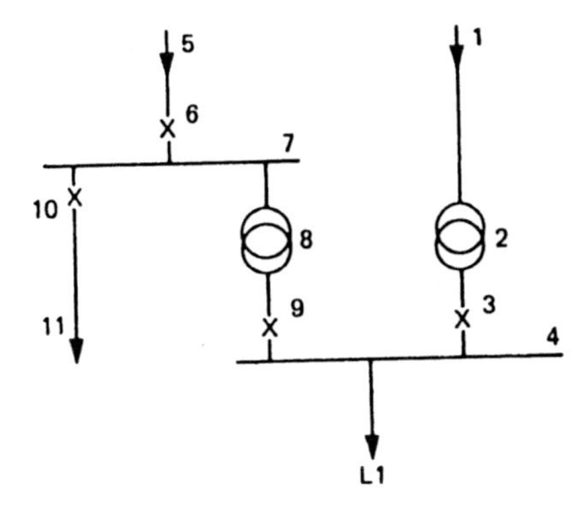

Fig. 10.8 System used to identify failure events

(b) first-order active failures:

3A

9A

(c) second-order active failures:

10A + 1

10A + 2

10A + 3

The total outage in the above failure events may be due to a forced outage or a maintenance outage.

10.5.4 Evaluation of reliability indices

(a) *Total failure events*

The indices for the total failure events are evaluated using the equations for overlapping outages described in Chapters 7–9. The evaluation can include the effect of normal and adverse weather considerations, common mode failures and maintenance.

(b) *Active failure events*

The indices for the active failure events are evaluated using the concept of overlapping outages but with the appropriate component data.

First-order events **(component 1 actively failed):**

$$\lambda_{ap} = \lambda_1^a \tag{10.1}$$

$$r_{ap} = s_1 \tag{10.2}$$

Second-order events

(i) Active failure of component 1 overlapping a total failure of component 2:

$$\lambda_{ap} = \lambda_1^a(\lambda_2 s_1) + \lambda_2(\lambda_1^a r_2) \tag{10.3(a)}$$

$$= \lambda_1^a \lambda_2 (s_1 + r_2) \tag{10.3(b)}$$

If $r_2 >> s_1$

$$\lambda_{ap} \simeq \lambda_1^a \lambda_2 r_2 \tag{10.3(c)}$$

$$r_{ap} = \frac{s_1 r_2}{s_1 + r_2} \tag{10.4(a)}$$

If $r_2 >> s_1$, then $r_{ap} \simeq s_1$. (10.4(b))

Equation (10.3(c)) is the second term of (10.3(a)) and implies that the overlapping sequence is 'active failure of component 1 during the repair time of component 2', i.e. the probability that component 2 fails during the switching time of component 1 is negligible.

(ii) Active failure of component 1 overlapping a maintenance outage of component 2.

$$\lambda_{am} = \lambda_2''(\lambda_1^a r_2'') \tag{10.5}$$

$$r_{am} = \frac{s_1 r_2''}{s_1 + r_2''} \tag{10.6(a)}$$

If $r_2'' >> s_1$, then $r_{am} \simeq s_1$. (10.6(b))

It can be seen that the forms of the equations are the same as those for overlapping outages described in Chapter 8. In addition, similar equations can be deduced for active failures overlapping temporary and transient outages and for active failures in a two-state weather environment. In practice, active failures overlapping transient and temporary outages are usually ignored because the duration of both active failures before switching and transient and temporary outages are small and can be considered negligible.

If it is considered necessary to include them, the equations are of the same form as Equations (8.16)–(8.18), in which the permanent failure rate and repair time are replaced by the active failure rate and switching time respectively.

The equations for a two-state weather model are again of the same form as Equations (8.27)–(8.40), in which the permanent failure rate and repair time of one component is replaced by its active failure rate and switching time respectively.

10.6 Malfunction of normally closed breakers

10.6.1 General concepts

It was assumed in Sections 10.4 and 10.5 that normally closed breakers open successfully when requested to do so. In practice they occasionally fail to respond due to malfunction of the protection system, the relaying system or the breaker itself. This is generally referred to as a stuck-breaker condition [1, 2, 6]. The probability of this situation can be evaluated from either a data collection scheme that records breaker operations and malfunctions or from a reliability analysis of the complete protection system.

The second method is described in Section 11.4. In the first method, the probability that a breaker fails to open when required (stuck-breaker probability) is given by:

$$P_c = \frac{\text{number of times breaker fails to operate when required}}{\text{number of times breaker is requested to operate}}$$

If a breaker fails to open, other protection devices respond, causing a breaker or breakers further from the faulted component to operate. This may cause a greater proportion of the system and additional load points to be disconnected.

10.6.2 Numerical example

Reconsider the system shown in Fig. 10.2. If stuck-breaker conditions are included in the analysis, the following additional events are found:

component 10 actively failed and breaker 7 stuck (10A + 7S);

component 11 actively failed and breaker 8 stuck (11A + 8S).

These events manifest themselves only if a breaker is stuck. Consequently the relevant failure rate must be weighted by the value of stuck probability. If the stuck-breaker probability of each of these two breakers is 0.05, then the reliability indices of load point A in Fig. 10.2 are modified to those shown in Table 10.5, again assuming that a spare transformer is available.

Table 10.5 Reliability indices of load point A including stuck breakers

Failure event	λ *(f/yr)*	r *(hours)*	U *(hours/yr)*
Subtotal from Table 10.4	0.150	7.1	1.060
10A + 7S	0.0025	1	0.0025
11A + 8S	0.0025	1	0.0025
Total	0.155	6.9	1.065

The results shown in Table 10.5 indicate that the effect of stuck breakers is small in this example and could have been neglected with little error. This is frequently the case, and therefore consideration of more than one breaker stuck simultaneously is usually ignored.

10.6.3 Deduction and evaluation

The deduction of failure events associated with stuck breakers can be achieved using a modified form [6] of the algorithm given in Section 10.5.3. In this case the simulation process also includes the breaker or breakers that operate if a primary protection breaker has not opened successfully. No other modifications are necessary.

The system shown in Fig. 10.8 can again be used to identify the events involving stuck breakers. These are

(a) First-order events:

1A + 3S	6A + 9S
2A + 3S	7A + 9S
	8A + 9S
	10A + 9S

(b) Second-order events:

11A + 10S + 1
11A + 10S + 2
11A + 10S + 3

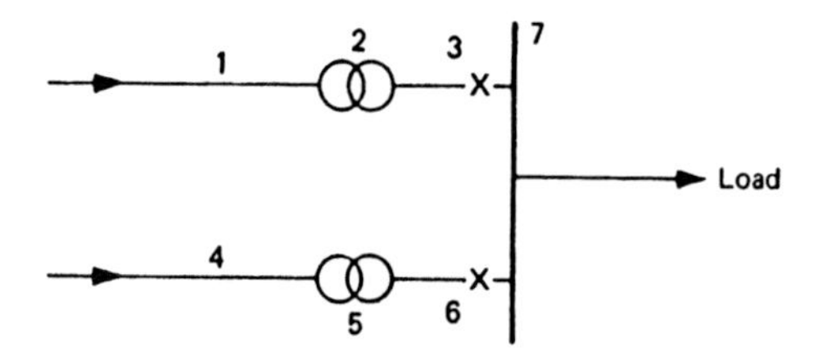

Fig. 10.9 Typical substation

Table 10.6 Reliability data for system of Fig. 10.9

Components	λ *(f/yr)*	r *(hours)*	λ^a *(f/yr)*	s *(hours)*	P_c
1, 4	0.09	7.33	0.09	1.0	—
2, 5	0.10	50	0.10	1.0	—
3, 6	0.02	3.0	0.01	1.0	0.06
7	0.024	2.0	0.024		—

The total outage in these events may again be due to a forced outage or a maintenance outage.

The equations used for evaluating the relevant reliability indices are also a modified set of those given in Section 10.5.4. The only difference is that the value of λ^a given in Equations (10.1), (10.3) and (10.5) is weighted by the value of stuck probability, P_c.

10.7 Numerical analysis of typical substation

(a) *Base case study*

Consider the system shown in Fig. 10.9 and the reliability data shown in Table 10.6. In addition assume that uncoordinated maintenance is performed on each component once a year for 8 hours. The reliability indices shown in Table 10.7 can then be evaluated.

The results obtained for this example show that the three indices of the load point are affected fairly equally by all three groups of events. Although these results are related to the assumed data, they illustrate the need to consider all types of events including the simulation of active failures and stuck-breaker conditions.

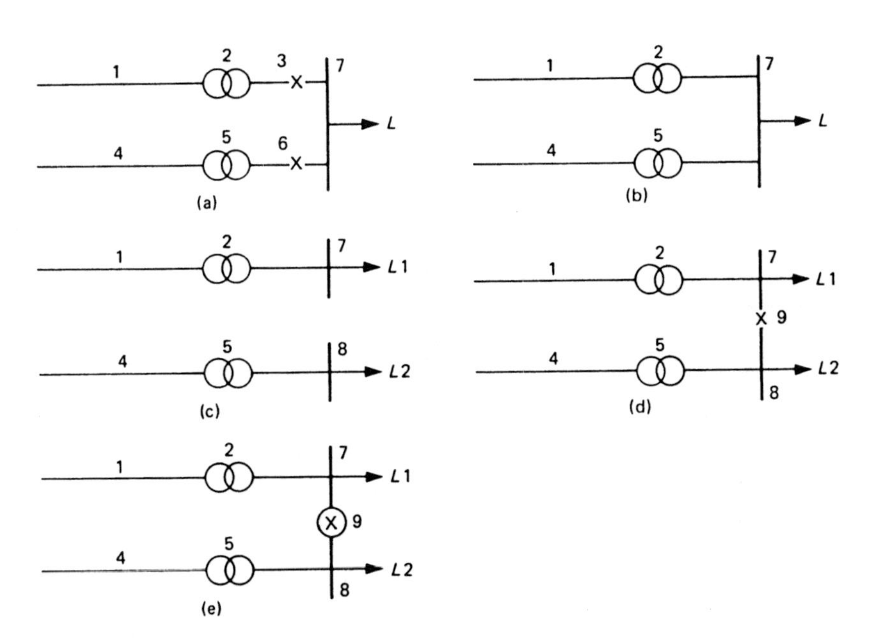

Fig. 10.10 Alternative substation arrangements

Table 10.7 Reliability indices for system of Fig. 10.9

(a) Overlapping total outages and a maintenance outage

Failure event	λ_{pp} *(f/yr)*	r_{pp} *(hours)*	U_{pp} *(hours/yr)*	λ_{pm} *(o/yr)*	r_{pm} *(hours)*	U_{pm} *(hours/yr)*
7	2.40×10^{-2}	2.0	4.80×10^{-2}	—	—	—
1 and 4	1.36×10^{-5}	3.67	4.97×10^{-5}	1.64×10^{-4}	3.83	6.29×10^{-4}
1 and 5	5.89×10^{-5}	6.39	3.77×10^{-4}	1.74×10^{-4}	5.44	9.44×10^{-4}
1 and 6	2.12×10^{-6}	2.13	4.52×10^{-6}	1.01×10^{-4}	3.53	3.54×10^{-4}
2 and 4	5.89×10^{-5}	6.39	3.77×10^{-4}	1.74×10^{-4}	5.44	9.44×10^{-4}
2 and 5	1.14×10^{-4}	25	2.85×10^{-3}	1.83×10^{-4}	6.90	1.26×10^{-3}
2 and 6	1.21×10^{-5}	2.83	3.42×10^{-5}	1.10×10^{-4}	6.11	6.70×10^{-4}
3 and 4	2.12×10^{-6}	2.13	4.52×10^{-6}	1.01×10^{-4}	3.53	3.54×10^{-4}
3 and 5	1.21×10^{-5}	2.83	3.42×10^{-5}	1.10×10^{-4}	6.11	6.70×10^{-4}
3 and 6	2.74×10^{-7}	1.50	4.11×10^{-7}	3.63×10^{-5}	2.18	7.97×10^{-5}
Subtotal	2.43×10^{-2}	2.13	5.17×10^{-2}	1.15×10^{-3}	5.14	5.91×10^{-3}

(b) Active failures

	λ_{ap}	r_{ap}	U_{ap}	λ_{am}	r_{am}	U_{am}
3A	1.00×10^{-2}	1.00	1.00×10^{-2}	—	—	—
6A	1.00×10^{-2}	1.00	1.00×10^{-2}	—	—	—
Subtotal	2.00×10^{-2}	1.00	2.00×10^{-2}	—	—	—

(c) Active failures and stuck breaker

1A + 3S	5.40×10^{-3}	1.00	5.40×10^{-3}	—	—	—
2A + 3S	6.00×10^{-3}	1.00	6.00×10^{-3}	—	—	—
4A + 6S	5.40×10^{-3}	1.00	5.40×10^{-3}	—	—	—
5A + 6S	6.00×10^{-3}	1.00	6.00×10^{-3}	—	—	—
Subtotal	2.28×10^{-2}	1.00	2.28×10^{-2}	—	—	—
Subtotals	6.71×10^{-2}	1.41	9.45×10^{-2}	1.15×10^{-3}	5.14	5.91×10^{-3}
Total	6.83×10^{-2} (f/yr)	1.47 (hours)	1.00×10^{-1} (hours/yr)			

(b) *Sensitivity studies*

It is important to perform sensitivity studies and to consider alternative schemes during the practical evaluation of systems. This is illustrated in this section by considering various alternatives to the scheme shown in Fig. 10.9 and by considering the effect of the stuck-breaker probability. These case studies are:

Case A: base case as described in (a) above (Fig. 10.10(a));
Case B: as case A but without the l.v. breakers (Fig. 10.10(b));
Case C: a split l.v. busbar with the load divided equally between the two busbars (Fig. 10.10(c));

Table 10.8 Reliability analysis of cases A–E

Event	λ *(f/yr)*	*r (hours)*	*U (hours/yr)*	*L (MW)*	*E (MWh/yr)*
Case A (as Table 10.7)					
	6.71×10^{-2}	1.41	9.45×10^{-2}	15	1.42
Case B					
7	2.40×10^{-2}	2	4.80×10^{-2}	15	
1 + 4	1.36×10^{-5}	3.67	4.97×10^{-5}	15	
1 + 5	5.89×10^{-5}	6.39	3.77×10^{-4}	15	
2 + 4	5.89×10^{-5}	6.39	3.77×10^{-4}	15	
2 + 5	1.14×10^{-4}	25	2.85×10^{-3}	15	
1A	9.00×10^{-2}	1	9.00×10^{-2}	15	
2A	1.00×10^{-1}	1	1.00×10^{-1}	15	
4A	9.00×10^{-2}	1	9.00×10^{-2}	15	
5A	1.00×10^{-1}	1	1.00×10^{-1}	15	
Total	4.04×10^{-1}	1.07	4.32×10^{-1}	15	6.48
Case C—load *L*1 (load *L*2 identical)					
7	2.40×10^{-2}	2	4.80×10^{-2}	7.5	
1	9.00×10^{-2}	7.33	6.60×10^{-1}	7.5	
2	1.00×10^{-1}	50	5.00	7.5	
Total	2.14×10^{-1}	26.7	5.71	7.5	42.8
Case D—load *L*1 (load *L*2 identical)					
7	2.40×10^{-2}	2	4.80×10^{-2}	7.5	
1 + 4	1.36×10^{-5}	3.67	4.97×10^{-5}	7.5	
1 + 5	5.89×10^{-5}	6.39	3.77×10^{-4}	7.5	
1 + 8	2.30×10^{-6}	1.57	3.61×10^{-6}	7.5	
1 + 9	2.12×10^{-6}	2.13	4.52×10^{-6}	7.5	
2 + 4	5.89×10^{-5}	6.39	3.77×10^{-4}	7.5	
2 + 5	1.14×10^{-4}	25	2.85×10^{-3}	7.5	
2 + 8	1.42×10^{-5}	1.92	2.74×10^{-5}	7.5	
2 + 9	1.21×10^{-5}	2.83	3.42×10^{-5}	7.5	
1A	9.00×10^{-2}	1	9.00×10^{-2}	7.5	
2A	1.00×10^{-1}	1	1.00×10^{-1}	7.5	
9A	1.00×10^{-2}	1	1.00×10^{-2}	7.5	
4A + 9S	5.40×10^{-3}	1	5.40×10^{-3}	7.5	
5A + 9S	6.00×10^{-3}	1	6.00×10^{-3}	7.5	
8A + 9S	1.44×10^{-3}	1	1.44×10^{-3}	7.5	
Total	2.37×10^{-1}	1.12	2.65×10^{-1}	7.5	1.99
Case E—load *L*1 (load *L*2 identical)					
7	2.40×10^{-2}	2	4.80×10^{-2}	7.5	
1	9.00×10^{-2}	1	9.00×10^{-2}	7.5	
2	1.00×10^{-1}	1	1.00×10^{-1}	7.5	
9A	1.00×10^{-2}	1	1.00×10^{-2}	7.5	
Total	2.24×10^{-1}	1.11	2.48×10^{-1}	7.5	1.86

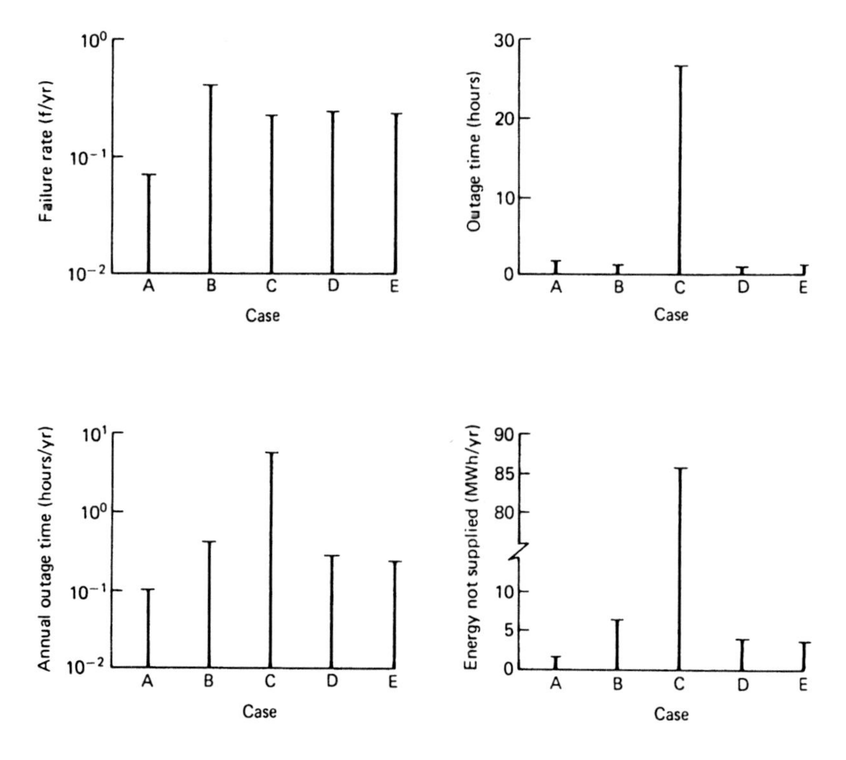

Fig. 10.11 Comparison between cases A–E

Case D: as case C but with a normally closed bus section breaker between the two busbars (Fig. 10.10(d));

Case E: as case D but with the bus section breaker normally open (Fig. 10.10(e));

Case F: as case A but with the stuck-breaker probability varying between 0.0 and 1.0;

Case G: as case D but with the stuck-breaker probability varying between 0.0 and 1.0.

Maintenance is neglected in the evaluation of these systems and the following data is assumed:

Reliability data—as given in Table 10.6 assuming the normally open breaker has the same indices as the normally closed breakers.

Loading data—the average load L connected to busbar 7 in cases A, B and F is 15 MW and the number of customers is 1500. The average load $L1$ and $L2$ connected to busbars 7 and 8 in cases C, D, E and G are each equal to 7.5 MW and the number of customers is each equal to 750.

The detailed analysis for cases A–E are shown in Table 10.8, a comparative summary of these results are shown in Fig. 10.11 and the results of cases F and G are shown in Fig. 10.12. These analyses assume that failure of the alternative supply in case E is neglected and that the normally open breaker always closes when required. These assumptions are discussed in Section 10.8.

The results shown in Fig. 10.11 clearly demonstrate the merits of case A and the demerits of case C, these being the most expensive and least expensive respectively in terms of capital investment. This conclusion relates directly to the data used and a different conclusion may be reached with different data and designs. The important point, however, is the need for such an assessment in order to arrive at an objective engineering judgement.

The results of Fig. 10.12, which illustrate the effect of stuck-breaker probability in cases F and G, show that case F is superior to case G at small values of stuck-breaker probability but that the two sets of results converge at large values of probability. It is most unlikely in practice for the value of stuck probability to be very large, but the results again demonstrate the need to perform sensitivity studies associated with the component data before a final choice of system design or reinforcement is made.

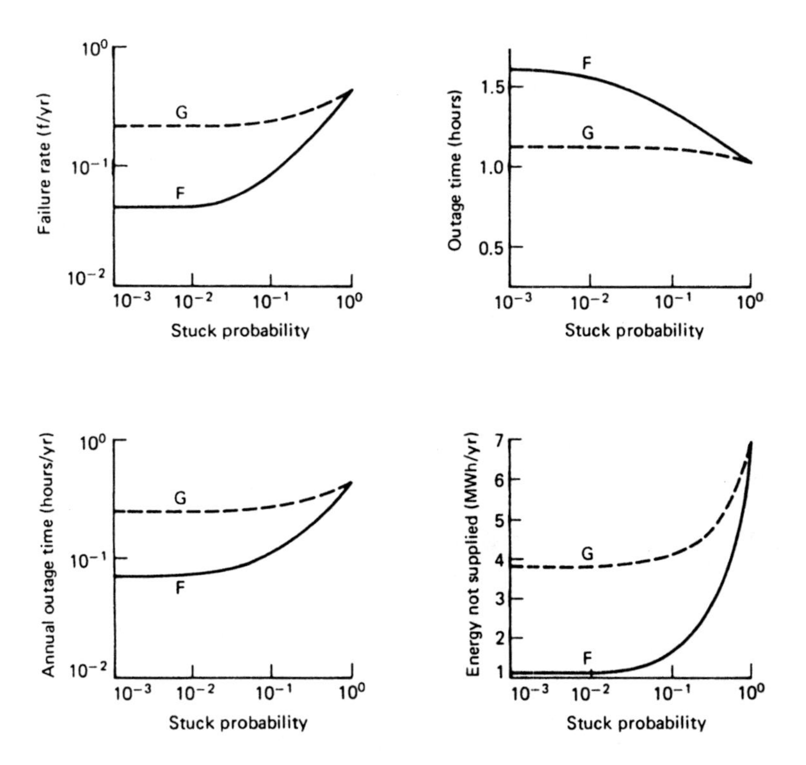

Fig. 10.12 Reliability analysis of cases F and G

10.8 Malfunction of alternative supplies

10.8.1 Malfunction of normally open breakers

Some of the breakers in a power system network, particularly bus-section breakers, are left open in order to reduce busbar fault levels. This can sometimes degrade the reliability of a load point because it reduces the number of closed paths leading to the load point and therefore reduces the number of components that must exist in an overlapping outage in order to fail the load point. This problem must be accepted, however, because of competing practical aspects.

When a load point failure event occurs, the normally open breaker permits service to be restored by closing the breaker. This reduces the outage time to a switching time provided the breaker responds. This concept was used in Section 10.7 in order to evaluate the reliability of case E. If the breaker fails to close, however, then the load point remains disconnected until the originally failed component is repaired or the normally open breaker can be made to respond. The implication in this concept is one of conditional probability associated with the outage time. The appropriate indices of a given failure event are

λ = failure rate of the event;

$$\begin{aligned} r &= \text{(switching time of normally open breaker)} \\ &\quad \times \text{(probability of successful closing)} \\ &\quad + \text{(repair time of component)} \\ &\quad \times \text{(probability of not closing or stuck probability)} \\ &= s(1-P_0) + rP_0 \end{aligned} \tag{10.7}$$

Equation (10.7) assumes two possibilities only, which is usually sufficient for practical purposes. If repeated attempts are made to close the breaker, Equation (10.7) can be extended to include a sequence of switching times, i.e.:

$$r = \sum_i s_i P_i + rP_0 \tag{10.8}$$

where

Table 10.9 Modified reliability indices of case E

Event	λ *(f/yr)*	r *(hours)*	U *(hours/yr)*	L *(MW)*	E *(MWh/yr)*
7	0.024	2.0	0.048	7.5	
1	0.09	1.38	0.124	7.5	
2	0.10	3.94	0.394	7.5	
9A	0.01	1.0	0.01	7.5	
Total	0.224	2.57	0.576	7.5	4.32

$$P_0 + \sum_i P_i = 1.0$$

In order to illustrate the application of this technique, reconsider case E in Section 10.7 and let the stuck probability (P_0) of the normally open breaker be 0.06. The detailed reliability analysis of load point $L1$ ($L2$ being identical) is shown in Table 10.9.

10.8.2 Failures in alternative supplies

When a load point failure occurs and an alternative supply is created by closing a normally open breaker, it is possible for the load point to subsequently fail due to a failure event in the alternative supply. The probability of such a subsequent event is likely to be small in practice and it is often reasonable to neglect the possibility, particularly if the number of alternative supplies is greater than one. If, however, it is considered desirable to include the effect in the evaluation exercise, a conditional probability approach similar to that described in Section 10.8.1 can be used [7, 9].

In the present case:

λ = failure rate of the original event plus failure rate of the alternative supply when in use

$$= \lambda_1 + \lambda_e \text{ (probability of alternative supply being used)}$$
$$\simeq \lambda_1 + \lambda_e(\lambda_1 r_1)$$
$$= \lambda_1(1 + \lambda_e r_1) \tag{10.9}$$

where λ_1 is the failure rate of the original event and λ_e is the failure rate of the alternative supply. This latter value can be found from the minimal cut sets (or failure events) of this alternative supply, ensuring that any events already included in λ_1 are ignored.

U = (unavailability of the load point given the alternative supply is available) × (probability that alternative supply is available) + (unavailability of the load point given that the alternative supply is unavailable) × (probability that alternative supply is unavailable)

$$= U_1[1 - (P_0 + U_e)] + U_2(P_0 + U_e) \tag{10.10}$$

where U_1 and U_2 are the unavailabilities of the load point given that the alternative supply are and are not available respectively.

U_e is the unavailability of the alternative supply and P_0 is the stuck probability of the normally open breaker as defined in Section 10.8.1.

$$r = U/\lambda \tag{10.11}$$

which, if failures of the alternative supply are not considered ($\lambda_e = U_e = 0.0$), becomes identical to Equation (10.7).

In order to illustrate the application of this technique, reconsider the same example used in Section 10.8.1, i.e. case E of Section 10.7.

The only events that can be recovered by closing the normally open breaker in the case of load point $L1$ are failures of line 1 and transformer 2. While the alternative supply is being used, failures of components 4, 5, 8 and 9 could occur. Therefore,

$$\lambda_e = \lambda(4) + \lambda(5) + \lambda(8) + \lambda(9)$$

$$= 0.09 + 0.10 + 0.024 + 0.02 = 0.234 \text{ f/yr}$$

$$U_e = \lambda(4)r(4) + \lambda(5)r(5) + \lambda(8)r(8) + \lambda(9)r(9)$$

$$= (0.09 \times 7.33) + (0.10 \times 50) + (0.024 \times 2) + (0.02 \times 3)$$

$$= 5.768 \text{ hours/yr}$$

and, from Equations (10.9)–(10.11) and Tables 10.8 and 10.9:

$$\lambda(1) = 0.09\left(1 + 0.234 \times \frac{7.33}{8760}\right) \simeq 0.09 \text{ f/yr}$$

$$\lambda(2) = 0.10\left(1 + 0.234 \times \frac{50}{8760}\right) \simeq 0.10 \text{ f/yr}$$

$$U(1) = 0.09\left[1 - \left(0.06 + \frac{5.768}{8760}\right)\right] + 0.660\left(0.06 + \frac{5.768}{8760}\right)$$

$$= 0.125 \text{ hour/yr}$$

$$U(2) = 0.10\left[1 - \left(0.06 + \frac{5.768}{8760}\right)\right] + 5.0\left(0.06 + \frac{5.768}{8760}\right)$$

$$= 0.397 \text{ hour/yr}$$

Consequently the modified reliability indices for load point $L1$ ($L2$ being identical) are shown in Table 10.10.

The results shown in Table 10.10 differ only marginally from those shown in Table 10.9, indicating that, in this case, the effect of the stuck breaker dominates over the effect of subsequent failures in the alternative supply.

One interesting comparison that can be made is the variation in reliability indices due to changes in stuck-breaker probability for case D (normally closed bus section breaker) and case E (normally open bus section breaker). This comparison is shown in Fig. 10.13. These results indicate that, although the indices are nominally the same when the stuck-breaker probabilities are zero, the indices of the substation having a normally open breaker degrade very much more significantly as the stuck-breaker probability increases. This indicates that it is preferable to have the bus section breaker closed if other system constraints permit.

Table 10.10 Modified reliability indices of case E

Event	λ (f/yr)	r (hours)	U (hours/yr)	L (MW)	E (MWh/yr)
7	0.024	2	0.048	7.5	
1	0.09	1.39	0.125	7.5	
2	0.10	3.97	0.397	7.5	
9A	0.01	1.0	0.01	7.5	
Total	0.224	2.59	0.580	7.5	4.35

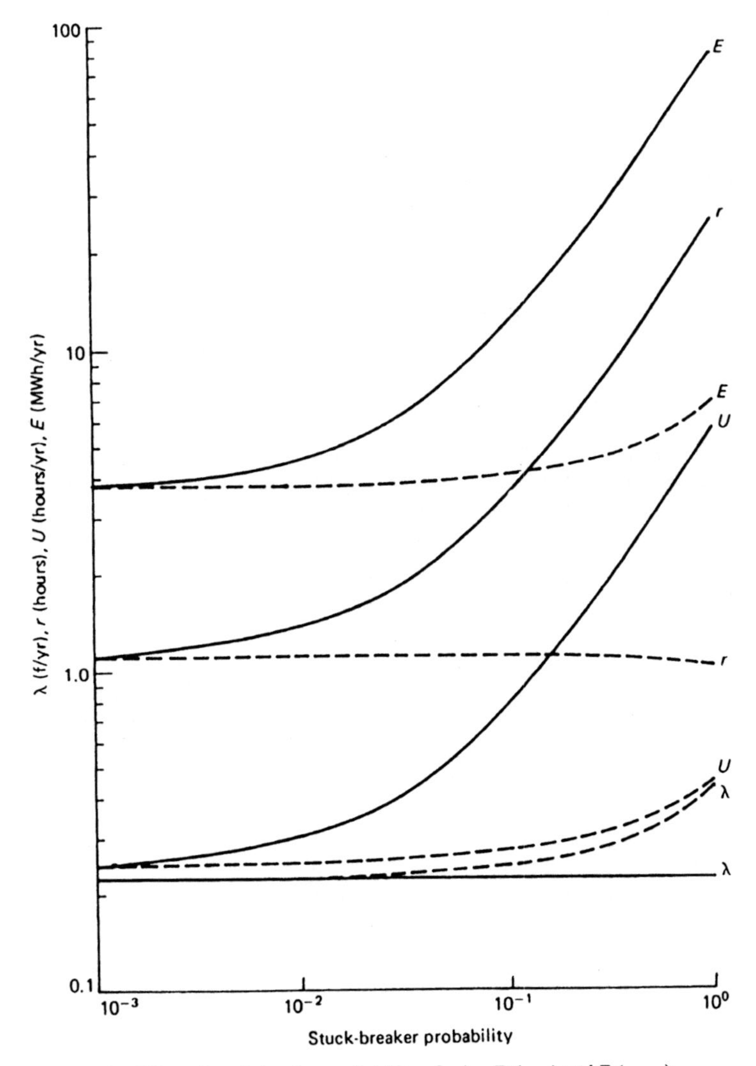

Fig. 10.13 Effect of stuck-breaker probability of cases D (- - -) and E (——)

10.9 Conclusions

This chapter has considered the concepts of open and short circuits, their impact on the operation of circuit breakers and the consequential effect on the reliability of load points. This has been done primarily in relation to substations and switching stations, but all these considerations are equally applicable to other parts of the power system, including transmission and distribution networks. The only reason for concentrating on substations and switching stations is that the switching effects of breakers are usually more dominant in such situations.

The numerical and sensitivity analyses that have been included demonstrate the marked effect that switching actions can and do have in real systems. The results, however, are directly a consequence of the data that was used and care must be taken not to assume too many general conclusions from the results of a specific solution.

The concepts and techniques discussed in this chapter assumed a single-state weather model and independent overlapping outages only. These are not natural limitations, however, and the concepts of two-state weather models, common mode failures and transient and temporary outages can be included in the present techniques with no additional complexities.

10.10 Problems

1 Seven possible designs for a particular substation are shown in Fig. 10.14. The average load on the busbar is 15 MW and the number of customers is 1500. Evaluate the values of failure rate, average outage duration, annual outage time and energy not supplied using the data given in Table 10.11.
Note: Consider overlapping forced outages up to second-order, first-order active failures and first-order active failures overlapping a stuck breaker, and ignore subsequent failures of a normally open path when used as the alternative supply.

2 Evaluate the failure rate, average outage time and annual outage time of load point A of the two substations shown in Fig. 10.15. Assume the same reliability data as in Problem 1. Consider overlapping forced outages up to second-order and first-order active failures only and neglect stuck-breaker considerations.

Table 10.11 Reliability data

Component	λ *(f/yr)*	λ^a *(f/yr)*	*r (hours)*	*s (hours)*	P_c or P_0
H.V. lines	0.1	0.1	10	0.5	—
Breakers	0.05	0.02	20	0.5	0.1
Transformers	0.01	0.01	50	0.5	—

(busbars are assumed 100% reliable)

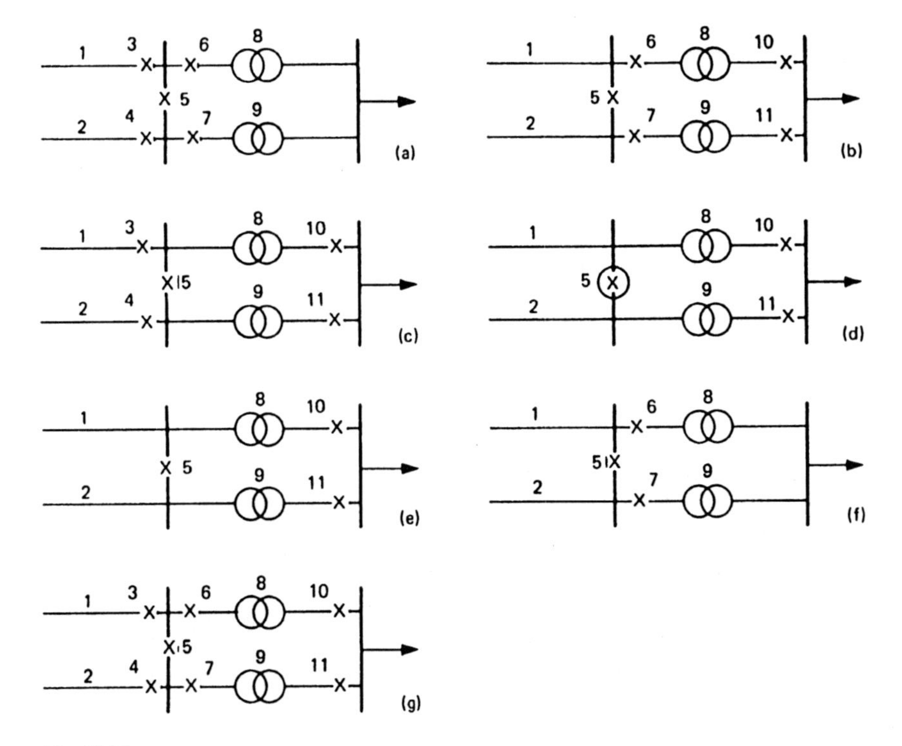

Fig. 10.14

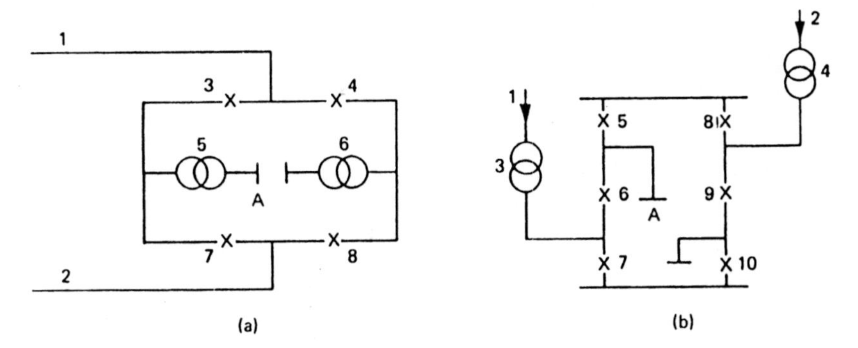

Fig. 10.15

10.11 References

1. Grover, M. S., Billinton, R., 'A computerized approach to substation and switching station reliability evaluation', *IEEE Trans. on Power Apparatus and Systems*, **PAS-93** (1974), pp. 1488–97.
2. Grover, M. S., Billinton, R., 'Substation and switching station reliability evaluation', *CEA Trans.*, **13** (pt 3), (1974), paper 74-SP-153.
3. Albrecht, P. F., 'Reliability evaluation of substation bus arrangements', *CEA Trans.*, **13** (pt 3), (1974), paper 74-SP-152.
4. Endrenyi, J., 'Three state models in power system reliability evaluations', *IEEE Trans. on Power Apparatus and Systems*, **PAS-90** (1971), pp. 1909–16.
5. Endrenyi, J., Maenhaut, P. C., Payne, L. E., 'Reliability evaluation of transmission systems with switching after faults—Approximations and a computer program', *IEEE Trans. on Power Apparatus and Systems*, **PAS-92** (1973), pp. 1863–75.
6. Allan, R. N., Billinton, R., DeOliveira, M. F., 'Reliability evaluation of electrical systems with switching actions', *Proc. IEE*, **123** (1976), pp. 325–30.
7. Allan, R. N., DeOliveira, M. F., 'Reliability modelling and evaluation transmission and distribution systems', *Proc. IEE*, **124** (1977), pp. 534–41.
8. Allan, R. N., Billinton, R., DeOliveira, M. F., 'An efficient algorithm for deducing the minimal cuts and reliability indices of a general network configuration', *IEEE Trans. on Reliability*, **R-25** (1976), pp. 226–33.
9. Allan, R. N., DeOliveira, M. F., Billinton, R., 'Reliability evaluation of the auxiliary electrical systems of power stations', *IEEE Trans. on Power Apparatus and Systems*, **PAS-96** (1977), pp. 1441–9.

11 Plant and station availability

11.1 Generating plant availability

11.1.1 Concepts

The models used in Chapters 2–6 represented generating units by a single component, the reliability indices of which were convolved together to form the generation model for evaluating system risk. This single-component representation is necessary in the assessment of large systems in order to reduce both computer time and computer storage. Each of these single components, however, represents a system of its own, the composition of which has a marked effect on the unit availability. A separate reliability evaluation of the generating plant is therefore desirable for two reasons:

(a) The reliability of a given generating plant configuration can be evaluated using historical component data. This index can then be used in the evaluation techniques described in Chapters 2–6.

(b) Comparative studies can be made of alternative generating plant configurations in order to assess the economic benefits of these alternatives.

The reliability of generating plant can be assessed using relatively simple evaluation methods based on state enumeration or, if applicable, the binomial distribution. Particular examples are shown in the following sections.

11.1.2 Generating units

A generating unit consists of a turbo-alternator set and one or more boilers. The latter are associated either with one specific set (this arrangement is known as a unit system) or with more than one set (a range system). These are illustrated in Fig. 11.1. Although the first type is more common, both need to be considered.

(a) *Unit systems*

Unit systems are the easiest to analyze because they are essentially series/parallel systems.

Example 1

Consider the system shown in Fig. 11.1(a) for which each unit can output 60 MW, each boiler has an availability of 0.91 and each set has an availability of 0.88. Each

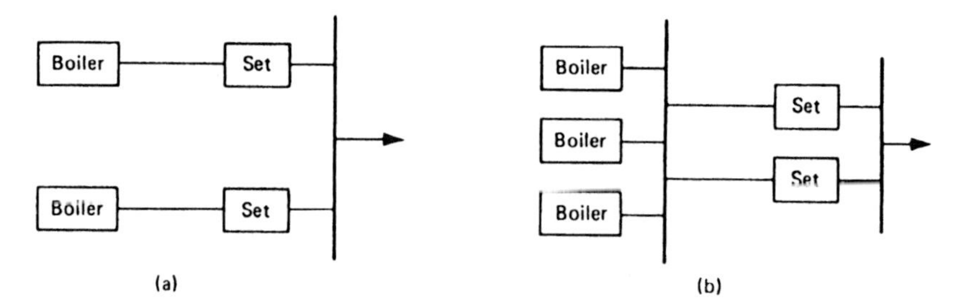

Fig. 11.1 Typical station configurations: (a) unit system; (b) range system

unit therefore has an availability of $0.91 \times 0.88 = 0.8008$. The reliability model for the complete system is shown in Table 11.1. The expected output of this station is

$$E(\text{MW}) = 120 \times 0.641281 + 60 \times 0.319038 = 96.10\,\text{MW}$$

Example 2

Consider the previous example but with two 40 MW boilers feeding each set, all other data remaining unchanged. The reliability models for each unit and the system are shown in Table 11.2. The expected output of this station is

$$E(\text{MW}) = 98.98\ \text{MW}$$

(b) *Range systems*

The reliability of range systems can be evaluated using a state enumeration method.

Example 3

Consider the system shown in Fig. 11.1(b). Let each set have a capacity of 60 MW and an availability of 0.88 and each boiler have a capacity of 40 MW and an availability of 0.91. The individual probabilities for each state in which the boilers and sets can reside are

Table 11.1 Reliability model for Example 1

Number of units on outage	*Available output (MW)*	*Probability*	
0	120	0.8008^2	$= 0.641281$
1	60	$2 \times 0.8008 \times 0.1992$	$= 0.319038$
2	0	0.1992^2	$= 0.039681$
			$\Sigma = 1.000000$

Table 11.2 Reliability models for Example 2

Available output (MW)	*State of boiler/sets*	*Probability*	
(a) *Each unit*			
60	both up (and) up	$0.91^2 \times 0.88$	= 0.728728
40	one up (and) up	$2 \times (0.91 \times 0.09) \times 0.88$	= 0.144144
0	both down (or) down	$0.09^2 + 0.12 - 0.09^2 \times 0.12$	= 0.127128
			1.000000
(b) *System*			
120	—	0.728728^2	= 0.531044
100	—	$2 \times 0.728728 \times 0.144144$	= 0.210084
80	—	0.144144^2	= 0.020778
60	—	$2 \times 0.728728 \times 0.127128$	= 0.185283
40	—	$2 \times 0.144144 \times 0.127128$	= 0.036649
0	—	0.127128^2	= 0.016162
			1.000000

P(3 boilers up) = 0.753571 P(2 sets up) = 0.7744

P(2 boilers up) = 0.223587 P(1 set up) = 0.2112

P(1 boiler up) = 0.022113 P(0 sets up) = 0.0144

P(0 boilers up) = 0.000729

The complete reliability model for this system can be deduced as shown in Table 11.3. The expected output of this station is

E(MW) = 97.13 MW

Table 11.3 Reliability model for Example 3

	Number of available			
Available output (MW)	*boilers*	*sets*	*Probability*	
120	3	2	0.583566	= 0.583566
80	2	2	0.173146	= 0.173146
60	3	1	0.159154	= 0.206376
	2	1	0.047222	
40	1	2	0.017124	= 0.021794
	1	1	0.004670	
0	3	0	0.010851	= 0.015118
	2	0	0.003220	
	1	0	0.000318	
	0	2	0.000565	
	0	1	0.000154	
	0	0	0.000010	

Table 11.4 Reliability model for Example 4

Available output (MW)	*Probability*
120	0.583566 + 0.173146 = 0.756712
60	0.206376 + 0.021794 = 0.228170
0	0.015118 = 0.015118

Example 4

If each of the boilers in the previous example were rated at 60 MW, the reliability model shown in Table 11.3 would be modified to that shown in Table 11.4. The expected output of this station is

$$E(\text{MW}) = 104.50 \text{ MW}$$

The previous examples show how the main components comprising a generating unit can be combined to give a reliability model for the station. The relative merits of each possible configuration can be evaluated by relating the reliability of each to the economic benefits that would accrue.

The analysis considered in this section need not and, in many cases, should not be limited to the main components of the generating system. Consideration should also be given to the configuration of the station transformers and the effect of the auxiliary systems feeding the boilers and the turbo-alternator sets. The first aspect is discussed in the following section and the second aspect is discussed in Section 11.2.

11.1.3 Including effect of station transformers

Generating plants are connected to the transmission network through one or more station transformers. There are two basic configurations that can be used in practice; either each generating unit is connected individually to one or more station transformers or a group of generating units shares one or more station transformers. These two configurations are illustrated in Fig. 11.2 for the unit generating system considered in Section 11.1.2.

The inclusion of the station transformers is a straightforward extension of the techniques described in Section 11.1.2.

Example 5

Consider the generating system used in Example 1 and let each unit be connected to an individual 60 MW transformer as shown in Fig. 11.2(a). Assume the unavailability of each transformer to be 0.01.

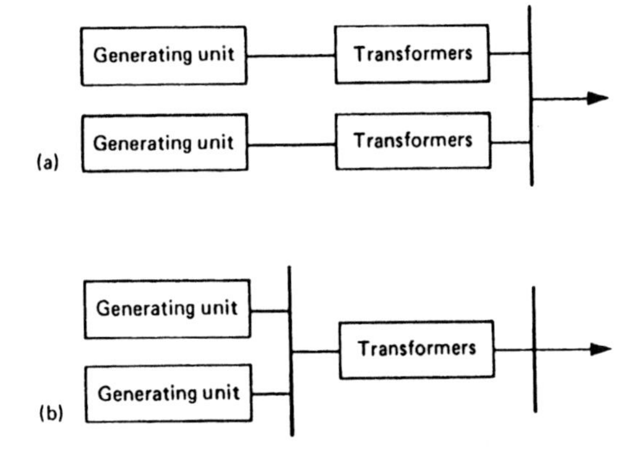

Fig. 11.2 Typical station transformer configurations: (a) individual transformers; (b) shared transformers

Each unit, including its associated transformer, now has an availability of $0.91 \times 0.88 \times 0.99 = 0.792792$ and the modified reliability model for the complete system is shown in Table 11.5.

Example 6

Consider the previous example but this time let the two generating units share a common 120 MW transformer as shown in Fig. 11.2(b). Assume the unavailability of the transformer to be 0.01.

In this example, the results shown in Table 11.1 must be modified as shown in Table 11.6.

Example 7

Reconsider Example 6, but replace the single 120 MW transformer by two 60 MW transformers having the same value of unavailability.

Table 11.5 Reliability model for Example 5

Available output (MW)	*Probability*	
120	0.792792^2	$= 0.628519$
60	$2 \times 0.792792(1 - 0.792792)$	$= 0.328546$
0	$(1 - 0.792792)^2$	$= \underline{0.042935}$
		1.000000
	$E(\text{MW}) = 95.14$	

Table 11.6 Reliability model for Example 6

Available output (MW)	*Probability*	
120	0.641281 × 0.99	= 0.634868
60	0.319038 × 0.99	= 0.315848
0	0.039681 × 0.99 + 0.01	= 0.049284
		1.000000
	E(MW) = 95.14	

In this case, the results shown in Table 11.1 are modified to those shown in Table 11.7.

From Examples 5–7, it is seen that the degradation in the reliability of these systems due to the inclusion of the station transformers is very small. This simply reflects the data being used and the outcome would be totally different for those cases in which the transformers have a greater unavailability and the generating units have a greater availability. The results illustrate the need for this type of analysis, however, in order to establish quantitatively the most significant contribution and hence to ensure that the system is reinforced, if deemed necessary, in a manner that gives the greatest economic benefit. In the case of Example 7, it would not be economically worthwhile to use two station transformers unless other technical reasons were important. Two particular instances that may necessitate two transformers are the need to improve the ability to maintain the transformers without a complete shut-down, and the need to use utility standard sizes of transformers which may include 60 MW but not 120 MW.

Although the unavailability of a transformer is usually quite small, this is due to a low value of failure rate. When a failure occurs, however, the outage time is usually very long unless a spare transformer is available. The application of sparing

Table 11.7 Reliability model for Example 7

Available output (MW)	*Probability*	
120	$0.641281 \times (0.99)^2$	= 0.628520
60	$0.641281 \times (2 \times 0.99 \times 0.01)$	
	$+ 0.319038 \times (0.99)^2$	
	$+ 0.319038 \times (2 \times 0.99 \times 0.01)$	= 0.331703
0	$0.641281 \times (0.01^2)$	
	$+ 0.319038 \times (0.01^2)$	
	$+ 0.039681 \times (1.00)$	= 0.039777
		1.000000
	E(MW) = 95.32	

allocation is therefore of considerable importance and this concept is discussed in Section 11.3.

11.2 Derated states and auxiliary systems

11.2.1 Concepts

Large modern generating plants are complex systems and contain large quantities of auxiliary equipment. Failure of any one, or possibly more than one item, of this equipment does not necessarily mean that the complete unit must be shut down. Instead, it can frequently be operated at a reduced output level. This concept is known as a partial output state or a derated state of a unit.

The inclusion of derated states in the risk assessment of a power system was described in Chapters 2 and 3. At that time, the implication of including (or neglecting) derated states was demonstrated but the underlying cause was deliberately omitted.

A typical thermal power station contains many auxiliaries; some of the most important are:

(a) forced draught fans; (b) induction draught fans;
(c) primary air fan for the pulverizer; (d) circulating water pumps;
(e) boiler feed water pumps; (f) condensate pumps;
(g) pulverizer drives; (h) soot blower air compressors;
(i) ash removal pumps; (j) lubricating oil pumps.

Most of these pumps and drives are electrically operated and a power station contains an electrical auxiliary system which resembles an elaborate distribution system. Failure of this electrical system leads to failure of the various pumps and drives, computer controls, instrumentation and safety devices. Consequently such failures can have a significant effect on the availability and safety of power stations and a number of technical papers [1–4] have been concerned with this problem area. Some of the auxiliaries in a typical installation may consist of two or more identical items, at least one of which may be fully redundant. In these cases, the failure of one set will have no effect on the output of the generating unit. In the operational phase (see Chapter 5), these items of equipment can be ignored as means of failure if it is assumed that, within the relevant lead time, the probability of more than one component failure can be neglected. In the planning phase, this assumption is generally invalid and multiple failure events should be considered.

Some of the other auxiliaries may not be fully redundant and the failure of the first item of a group of equipment leads to a derated state. The effect of these auxiliaries should be included therefore in both the operational and planning phases.

The system designer or operator can identify the effect of any auxiliary failure from his knowledge of the system and its operating requirements. Some failures may mean the unit output has to be reduced by a small amount, say 5%, whilst others may mean a reduction of up to 50%. Generally, failures that would cause a theoretical derating effect of greater than 50% lead to complete shut-down of the unit and can be classed as total unit failure.

11.2.2 Modelling derated states

Consider a 300 MW thermal unit for which the following components cause significant outages:

1 boiler;
1 turbine;
5 pulverizers (assume loss of each causes a 25% derating);
2 forced-draught fans (assume loss of each causes a 50% derating);
2 induction-draught fans (assume loss of each causes a 50% derating);
3 cooling water pumps (assume loss of each causes a 33% derating);
3 feed water pumps (assume loss of each causes a 33% derating).

If it is assumed that, during the lead time of the operational phase, more than one failure and the repair process can be neglected, the state space diagram [13] for this system is as shown in Fig. 11.3. This is now a very simple system to analyze since the probability of residing in each state at the end of a lead time T is obtained from Equation (5.4) as

$$P_i = \lambda_i T \tag{11.1}$$

where P_i is the probability of residing in state i ($i = 2, \ldots, 8$) and λ_i is the transition rate into state i.

This model is therefore an extension of the one shown in Fig. 5.1(c) and the effect can be incorporated into the techniques described in Chapter 5 without any modification.

Although the state space diagram shown in Fig. 11.3 is relatively simple, the model will grow rapidly in size as an increasing number of system components is included. The diagram can be considerably reduced since, from an operational risk point of view, it is not necessary to identify the cause of each state, only its effect on the system. Consequently, states of identical or near-identical capacity can be combined. In the case of the diagram in Fig. 11.3, this reduction will create the model shown in Fig. 11.4.

It is not realistic to neglect higher-order failure events and the repair process in the planning phase. On the other hand, a state space diagram of the form shown in Fig. 11.3 can become totally impractical if all states and all transitions between these states are included. Since the most important factor in system planning studies is to identify the effect on the system and not the cause, the relevant state space diagram can be constructed in terms of capacity states rather than component states. This type of technique was used to construct Fig. 11.4. It should be noted that for

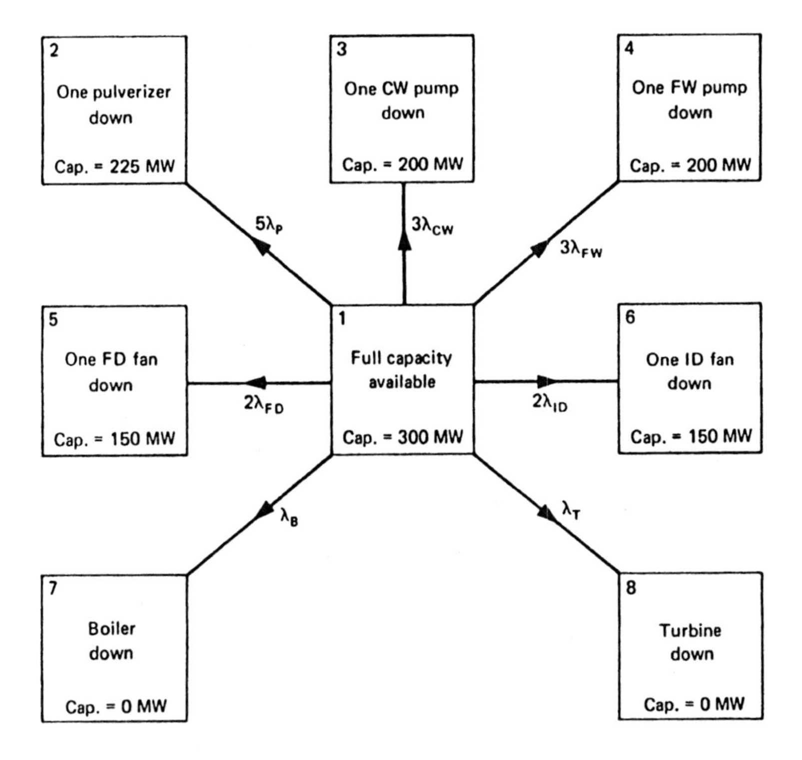

Fig. 11.3 State space diagram for operational studies

a detailed station design study, the cause of failure as well as its effect on the system is essential.

A generalized state space diagram that includes three derated states is shown in Fig. 11.5. In this model, it is only necessary to recognize each derating level (in practice several near-levels can be grouped together), the total period of time spent in each of the levels and the number of times each level is entered. The following indices can then be evaluated from this information:

$$P_i = \frac{T_i}{T} \tag{11.2}$$

$$\lambda_{ij} = \frac{N_{ij}}{T_i} \tag{11.3}$$

where P_i is the probability of residing in state i;
T_i is the total time spent in state i;
T is the total period of interest = $\Sigma_i\, T_i$;
λ_{ij} is the transition rate from state i to state j;
N_{ij} is the number of transitions that occur from state i to state j.

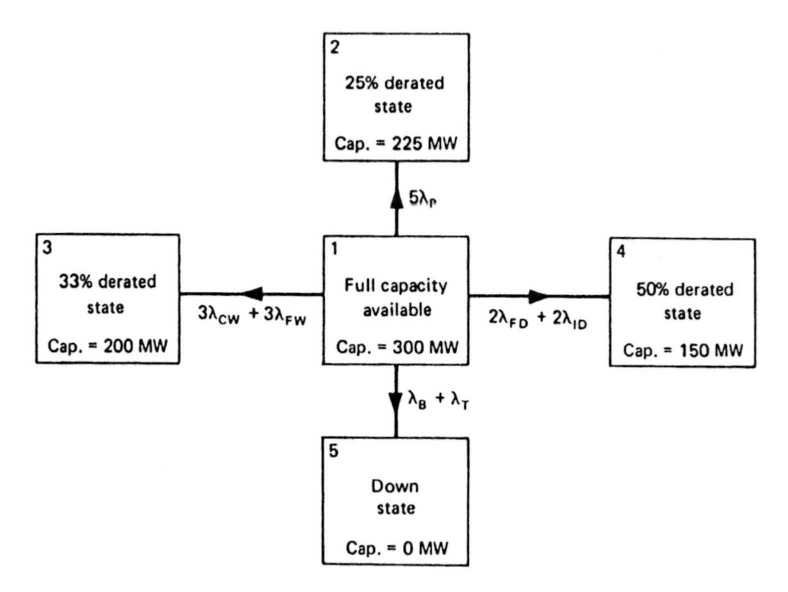

Fig. 11.4 Reduced state space diagram

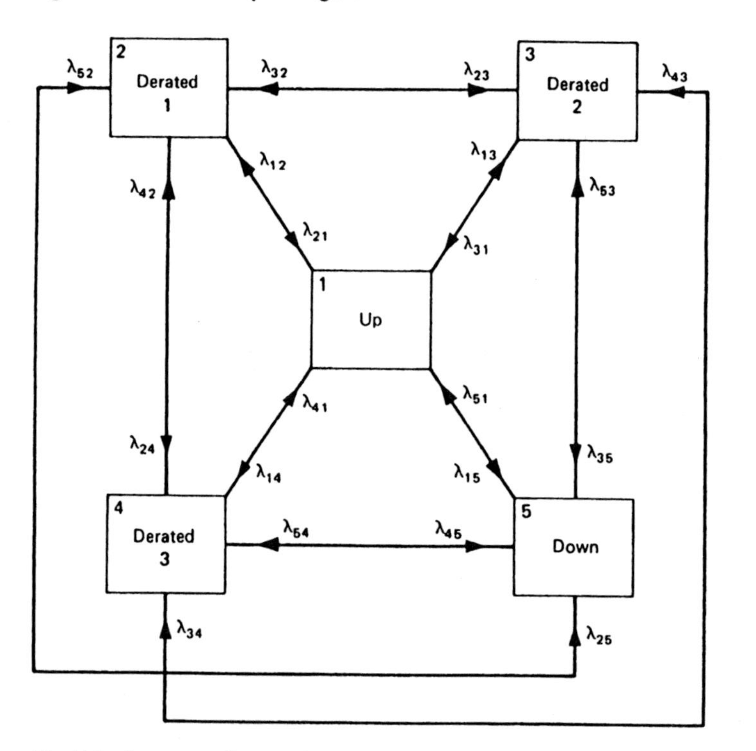

Fig. 11.5 State space diagram with three derated states

The only parameter required for the LOLP, LOLE and LOE methods (Chapter 2) is P_i. The values of λ_{ij} are required, however, for the frequency and duration approach (Chapter 3). It should be noted that several of the transitions shown in Fig. 11.5 may not exist in practice and can be ignored. Also, as shown in Section 2.4, the number of derated states that need to be modelled in order to obtain a realistic assessment of system risk is very small and rarely needs to exceed one such state.

11.3 Allocation and effect of spares

11.3.1 Concepts

The two most important concepts involved in a continuously operated and repairable system are the failure process and the restoration process. Most of the techniques in preceding chapters have assumed that the restoration process associated with a permanently failed component is achieved either by repair of the failed component or by replacing the failed component with a spare. In the latter case, it has generally been assumed that a spare is available as and when required. This assumption is reasonably justified in many cases, particularly those in which the failure rate is very small or the number of spares is relatively large. On the other hand, this assumption may be invalid and the availability of spares is an important criterion in the reliability assessment. In addition, it is frequently necessary to evaluate the number of spares that are required for a given application. The concept of modelling and evaluating the effect of spares is important in both of these situations.

The basic concepts associated with modelling and evaluating the effects of spares was discussed in Section 10.5 of *Engineering Systems*. These concepts will not be reiterated in this section and only the application of the concepts will be described.

There are many examples which can be used to describe sparing concepts and evaluation. The one chosen in this section relates to the same problem area described in the previous sections of this chapter, namely, the generating station and, in particular, the station transformer configuration. The discussion that follows, however, can easily be adapted and applied to other power system areas including the generating station auxiliaries, the transmission and distribution network and substation configurations. A further application is made in Section 11.5 in relation to HVDC systems.

11.3.2 Review of modelling techniques

It was shown in Section 10.5 of *Engineering Systems* how Markov techniques could be used to model and evaluate the probability of residing in, frequency of encoun-

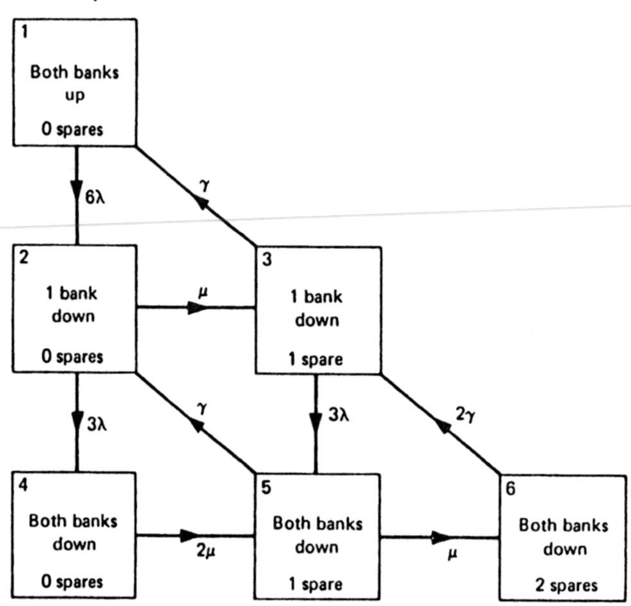

Fig. 11.6 Two transformer banks with no spares

tering and duration in each individual and cumulated state. This was achieved by first constructing the relevant state space diagram, inserting the appropriate failure rates λ, repair rates μ and installation rates γ and solving the resulting transition equations using Markov techniques.

As an example, consider a generating transformer substation that consists of two identical three-phase transformer banks, each bank consisting of three identical single-phase transformers. Each of these banks is considered to have failed totally and must be removed from service if any one of the three single-phase transformers in the bank fails. The bank can be returned to service when the failed transformer has been repaired or replaced. It is also assumed that no further failures can occur in the bank once it has been de-energized and removed from service. If any of these assumptions is not considered valid, the techniques can be readily adapted to suit the appropriate operating behavior.

Using the techniques described in Section 10.5 of *Engineering Systems*, the state space diagrams for this system when no spares are stored and when one spare is stored are shown in Figs. 11.6 and 11.7. It is assumed in these state space diagrams that there are no restrictions on the number of repair and installation processes that can be conducted simultaneously.

The concept shown in Figs. 11.6 and 11.7 and described in Section 10.5 of *Engineering Systems* can be extended to accommodate any number of service components and any number of spares.

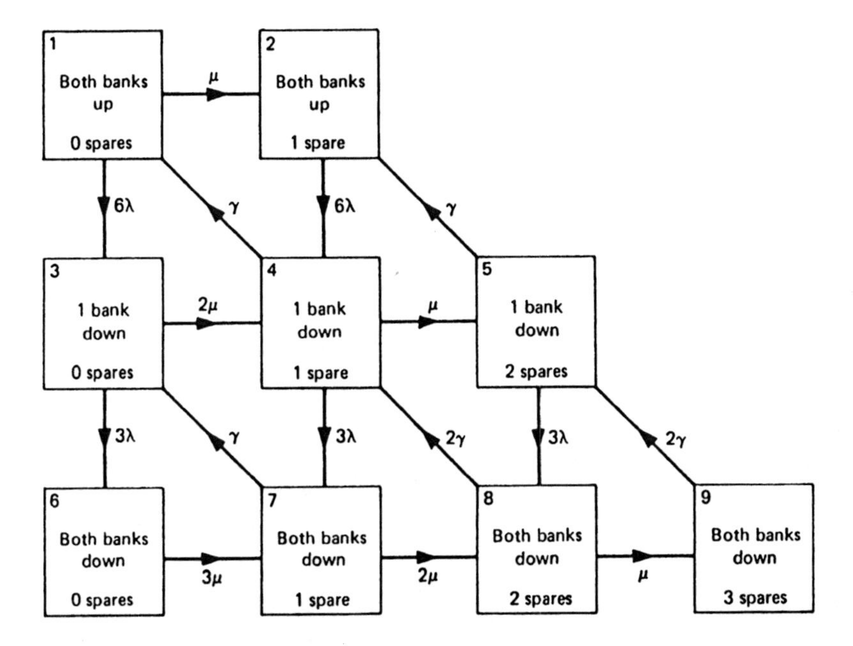

Fig. 11.7 Two transformer banks with one spare

11.3.3 Numerical examples

Consider a transformer substation which is to connect a 100 MW generating system to a transmission network. The following configurations are considered:

(a) a single transformer bank with no spares, one spare, two spares, three spares, four spares;

(b) two identical transformer banks with no spares, one spare, two spares.

The transformers of case (a) must be rated at least 100 MW. A range of sizes between 50 MW (no redundancy) and 100 MW (total redundancy) is considered in case (b).

The following data is used for each single-phase transformer:

failure rate $\lambda =$ 0.01, 0.1, 1.0 f/yr

repair rate $\mu =$ 4, 12, 52 repairs/yr (these being equivalent to repair times of about 3 months, 1 month and 1 week respectively)

installation rate $\gamma =$ installations/yr (this being equivalent to an installation time of about 2 days)

The results for case (a) are shown in Table 11.8, which includes the probability of the bank being in the up state, the probability of being in the down state and the

Table 11.8 Results for single transformer bank

λ (*f/yr*)	μ (*repairs/yr*)	*Probability*		*Expected MW capacity level*
		up state	*down state*	
No spares				
1	4	0.566125	0.433875	56.61
1	12	0.789644	0.210356	78.96
1	52	0.931024	0.068976	93.10
0.1	4	0.928816	0.071184	92.88
0.1	12	0.974052	0.025948	97.41
0.1	52	0.992646	0.007354	99.26
0.01	4	0.992394	0.007606	99.24
0.01	12	0.997343	0.002657	99.73
0.01	52	0.999260	0.000740	99.93
One spare				
1	4	0.852028	0.147972	85.20
1	12	0.961673	0.038327	96.17
1	52	0.982686	0.017314	98.27
0.1	4	0.995818	0.004182	99.58
0.1	12	0.998078	0.001922	99.81
0.1	52	0.998350	0.001650	99.84
0.01	4	0.999809	0.000191	99.98
0.01	12	0.999833	0.000167	99.98
0.01	52	0.999836	0.000164	99.98
Two spares				
1	4	0.953347	0.046653	95.33
1	12	0.982240	0.017760	98.22
1	52	0.983856	0.016144	98.39
0.1	4	0.998302	0.001698	99.83
0.1	12	0.998361	0.001639	99.84
0.1	52	0.998363	0.001637	99.84
0.01	4	0.999836	0.000164	99.98
0.01	12	0.999836	0.000164	99.98
0.01	52	0.999836	0.000164	99.98
Three spares				
1	4	0.978535	0.021465	97.85
1	12	0.983786	0.016214	98.38
Four spares				
1	4	0.983136	0.016864	98.31
1	12	0.983868	0.016132	98.39
Limiting values (*∞ spares*)				
1	—	0.983871	0.016129	98.39
0.1	—	0.998363	0.001637	99.84
0.01	—	0.999836	0.000164	99.98

expected MW capacity level of the transformer station. A similar set of results is shown in Tables 11.9 and 11.10 for case (b). Table 11.9 shows the state probabilities as a function of reliability indices and number of spares. Table 11.10 shows the expected MW capacity level of the transformer station as the rating of each bank is increased. The results for the single transformer bank as a function of number of

Table 11.9 State probabilities for parallel transformer bank

λ (f/yr)	μ (repairs/yr)	Probability of Both up	One up	Both down
No spares				
1	4	0.320498	0.491255	0.188247
1	12	0.623538	0.332213	0.044250
1	52	0.866806	0.128436	0.004758
0.1	4	0.862699	0.132233	0.005067
0.1	12	0.948777	0.050550	0.000673
0.1	52	0.985346	0.014600	0.000054
0.01	4	0.984847	0.015096	0.000058
0.01	12	0.994693	0.005300	0.000007
0.01	52	0.998520	0.001480	0.000001
One spare				
1	4	0.626282	0.299241	0.074476
1	12	0.891479	0.100994	0.007527
1	52	0.962990	0.036462	0.000547
0.1	4	0.986957	0.012784	0.000259
0.1	12	0.995575	0.004409	0.000016
0.1	52	0.996672	0.003323	0.000004
0.01	4	0.999562	0.000437	0.000000
0.01	12	0.999660	0.000340	0.000000
0.01	52	0.999672	0.000328	0.000000
Two spares				
1	4	0.823424	0.150992	0.025583
1	12	0.956365	0.042447	0.001188
1	52	0.967854	0.031874	0.000272
0.1	4	0.996254	0.003733	0.000013
0.1	12	0.996711	0.003286	0.000003
0.1	52	0.996729	0.003268	0.000003
0.01	4	0.999672	0.000328	0.000000
0.01	12	0.999672	0.000328	0.000000
0.01	52	0.999672	0.000328	0.000000
Limiting values (∞ spares)				
1	—	0.968254	0.031746	—
0.1	—	0.996732	0.003268	—
0.01	—	0.999672	0.000328	—

N.B. The values 0.000000 are precise to 6 decimal places and are not absolute zero.

Table 11.10 Expected capacity level of parallel transformer bank

λ (f/yr)	μ (repairs/yr)	*Expected MW capacity level when rating of each bank (MW) is* 50	60	70	80	90	100
No spares							
1	4	56.61	61.53	66.44	71.35	76.26	81.18
1	12	78.96	82.29	85.61	88.93	92.25	95.58
1	52	93.10	94.39	95.67	96.96	98.24	99.52
0.1	4	92.88	94.20	95.53	96.85	98.17	99.49
0.1	12	97.41	97.91	98.42	98.92	99.43	99.93
0.1	52	99.26	99.41	99.56	99.70	99.85	99.99
0.01	4	99.24	99.39	99.54	99.69	99.84	99.99
0.01	12	99.73	99.79	99.84	99.89	99.95	100.00
0.01	52	99.93	99.94	99.96	99.97	99.99	100.00
One spare							
1	4	77.59	80.58	83.58	86.57	89.56	92.55
1	12	94.20	95.21	96.22	97.23	98.24	99.25
1	52	98.12	98.49	98.85	99.22	99.58	99.95
0.1	4	99.33	99.46	99.59	99.72	99.85	99.97
0.1	12	99.78	99.82	99.87	99.91	99.95	100.00
0.1	52	99.83	99.87	99.90	99.93	99.97	100.00
0.01	4	99.98	99.98	99.99	99.99	100.00	100.00
0.01	12	99.98	99.99	99.99	99.99	100.00	100.00
0.01	52	99.98	99.99	99.99	99.99	100.00	100.00
Two spares							
1	4	89.89	91.40	92.91	94.42	95.93	97.44
1	12	97.76	98.18	98.61	99.03	99.46	99.88
1	52	98.38	98.70	99.02	99.34	99.65	99.97
0.1	4	99.81	99.85	99.89	99.92	99.96	100.00
0.1	12	99.84	99.87	99.90	99.93	99.97	100.00
0.1	52	99.84	99.87	99.90	99.93	99.97	100.00
0.01	4	99.98	99.99	99.99	99.99	100.00	100.00
0.01	12	99.98	99.99	99.99	99.99	100.00	100.00
0.01	52	99.98	99.99	99.99	99.99	100.00	100.00
Limiting values (∞ spares)							
1	—	98.41	98.73	99.05	99.37	99.68	100.00
0.1	—	98.84	99.87	99.90	99.93	99.97	100.00
0.01	—	99.98	99.99	99.99	99.99	100.00	100.00

N.B. The values 100.00 are precise to two decimal places and are not absolutely 100 MW.

spares are also shown in Figs. 11.8 and 11.9; Fig. 11.8 illustrates the unavailability of the bank and Fig. 11.9 illustrates the expected MW capacity level. The results in Tables 11.8–11.10 and Figs. 11.8 and 11.9 also include the limiting values which would occur if an infinite number of spares were available. Several important features and concepts can be deduced from these results:

(i) From Fig. 11.8, it is seen that the number of spares required in order for the unavailability to approximate to the limiting value is small but increases as the failure rate increases and the repair rate decreases (i.e. repair time increases).

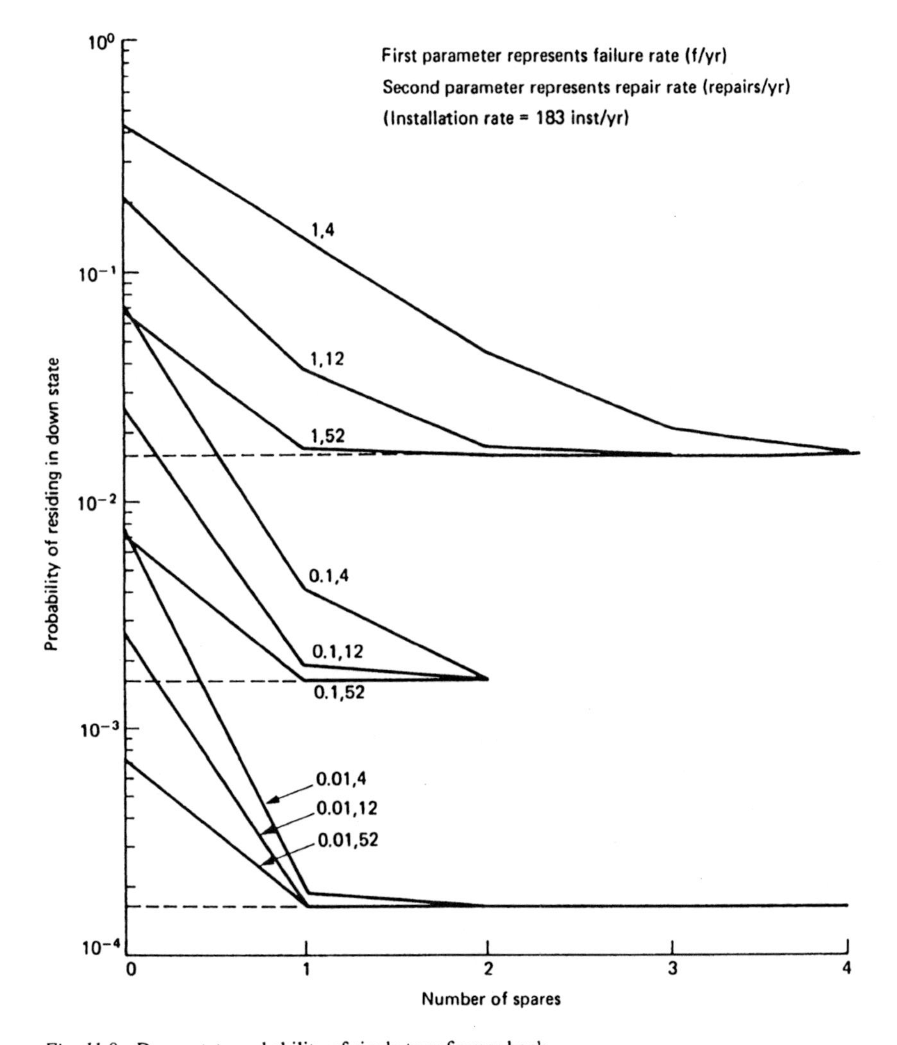

Fig. 11.8 Down state probability of single transformer bank

(ii) A similar effect to (i) can be seen in Fig. 11.9 which shows that a small number of spares increases the expected MW capacity level to the limiting value.

(iii) From Fig. 11.8, the unavailability when no spares are available can be smaller for a bank with high failure rate than for one with a small failure

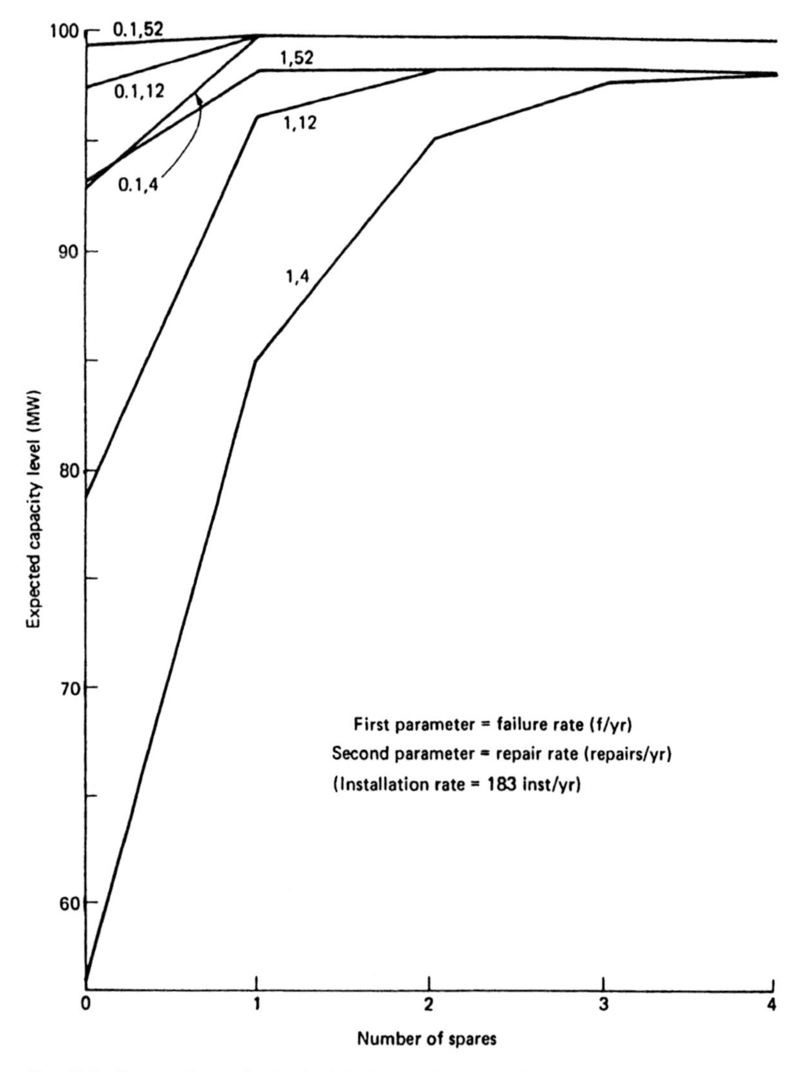

Fig. 11.9 Expected capacity level of single transformer bank

rate provided the repair rate is considerably greater (compare, for instance, the result for 1 f/yr and 52 repairs/yr with that for 0.1 f/yr and 4 repairs/yr). On the other hand, when spares are available, this observation can be reversed by a considerable margin (compare the same set of results with 1 and 2 spares available).

(iv) From Fig. 11.9, it is evident that the expected MW capacity level has a limiting value and, for a given set of reliability data for each transformer, cannot be exceeded no matter how many spares are carried. If transformers of a given quality only are available and the limiting MW capacity level is insufficient, the only alternative is to increase the number of transformer banks operating in parallel.

(v) Comparison of the results shown in Tables 11.8 and 11.10 indicates that the use of two transformer banks does not necessarily improve the performance of the station compared with using one. For instance:

(a) if no spares are carried, the rating of the parallel banks must be at least 60 MW each to derive some benefit;

(b) if one spare is carried, the rating of the parallel banks must be at least 80 MW for transformers having 1 f/yr and 4 repairs/yr, but only 60 MW for transformers having 1 f/yr and 52 repairs/yr in order to derive some benefit;

(c) the rating of the parallel banks must be at least 90 MW for 1 f/yr and 4 repairs/yr in order to derive some benefit when two spares are carried but only 50 MW when an infinite number of spares is available (limiting value).

These results clearly indicate the need to consider the values of failure rate, repair rate and component rating in any quantitative evaluation of sparing requirements. Many alternatives are possible in addition to increasing the number of spares, including investing in improved quality of components and therefore reducing the failure rate, investing in repair and installation resources and therefore increasing the relevant rates, using components of greater capacity, increasing the number of components operating in parallel. The most appropriate solution for a given requirement can only be established from a quantitative reliability assessment which should be used in conjunction with an economic appraisal of the various alternatives.

It should be noted that the results shown and discussed above were evaluated assuming that there were no restrictions on the number of repairs and installations that could be conducted simultaneously. If such restrictions existed due to lack of manpower or facilities, different results would be obtained and different conclusions might be reached. The analysis is performed in an identical manner, however, only the values of transition rates between states being changed. This point is discussed in Section 10.5 of *Engineering Systems*.

11.4 Protection systems

11.4.1 Concepts

The concept of a stuck breaker was introduced in Section 10.6 and its implication in network reliability evaluation was discussed. At that time, it was suggested that the value of stuck-breaker probability could be established from a data collection scheme by recording the number of requests for a breaker to open and the number of times the breaker failed to respond. In many practical applications, this method and the techniques described in Section 10.6 are sufficient.

The probability of a breaker responding to a failed component depends on the protection system, its construction and the quality of the components being used. This is a completely integrated system of its own and, as such, can be analyzed independently of the power system network which it is intended to protect. This independent analysis enables sensitivity and comparative studies to be made of alternative protection systems and also enables the index of stuck-breaker probability to be fundamentally derived.

A protection system can malfunction in two basic ways:

(a) It fails to operate when requested. A power system network is in a continually operating state and hence any failure manifests itself immediately. Such failures have been defined [5] as revealed faults. A protection system, however, remains in a dormant state until it is called on to operate. Any failures which occur in this system during the dormant state do not manifest themselves until the operating request is made when, of course, it will fail to respond. These failures have been defined [5] as unrevealed faults. In order to reduce the probability of an operating failure, the protection system should be checked and proof-tested at regular intervals.

(b) Spurious or inadvertent operation. This type of failure, which is due to a spurious signal being developed in the system, thus causing breakers to operate inadvertently, manifests itself immediately it occurs. Hence it is defined as a revealed fault. This type of failure was classed as a passive failure in Chapter 10 because it has an effect identical to an open-circuit fault.

11.4.2 Evaluation techniques and system modelling

Protection systems involve the sequential operation of a set of components and devices. For this reason, the network evaluation techniques described in previous chapters are not particularly suited to these systems. There are several alternative techniques available including fault trees [5], event trees [6, 7] and Markov modelling [8]. The event tree technique is particularly useful because it recognizes the sequential operational logic of a system and can be easily extended to include analysis of the system at increasing depth. For this reason, only this method will be discussed in this chapter.

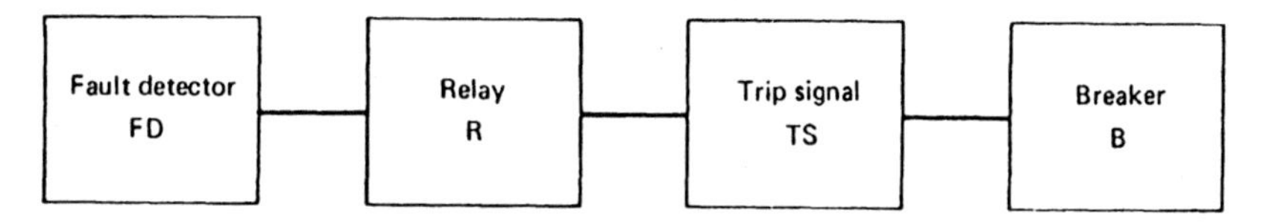

Fig. 11.10 Block diagram of a general protection scheme

The general principle of event trees, their deduction, application and associated probability evaluation was described in Section 5.7 of *Engineering Systems*. Therefore only the application of this method [14] to protection systems will be described here.

There are many types of protection systems and it is not possible to consider all of these within the scope of this chapter. Consequently the discussion relates only to a generalized form of protection system consisting of the blocks shown in Fig. 11.10.

These blocks can be related to most protection systems in which the fault detector includes appropriate CTs, VTs and comparators, the relay includes operating and restraint coils, the trip signal contains the trip signal device and associated power supply and finally the breaker is the actual device which isolates the faulted component.

11.4.3 Evaluation of failure to operate

(a) *System and basic event tree*

Consider a particular network component that is protected by two breakers B1 and B2. Assume that both breakers are operated by the same fault detector FD, relay R and trip signal device TS. The event tree, given the network component has failed, is therefore as shown in Fig. 11.11. This shows the sequence of events together with the outcomes of each event path, only one of which leads to complete success when both breakers open as requested.

(b) *Evaluating event probabilities*

The event probabilities needed are the probability that each device will and will not operate when required. These are time-dependent probabilities and will be affected by the time period between when they were last checked and the time when they are required to operate. The probability of operating when required increases as the period of time between checks is decreased provided the checking and testing is performed with skill and precision, i.e. the devices are left in an 'as-good-as-new' condition and not degraded by the testing procedure.

The time at which a device is required to operate is a random variable. The only single index that can be calculated to represent the probability of failing to

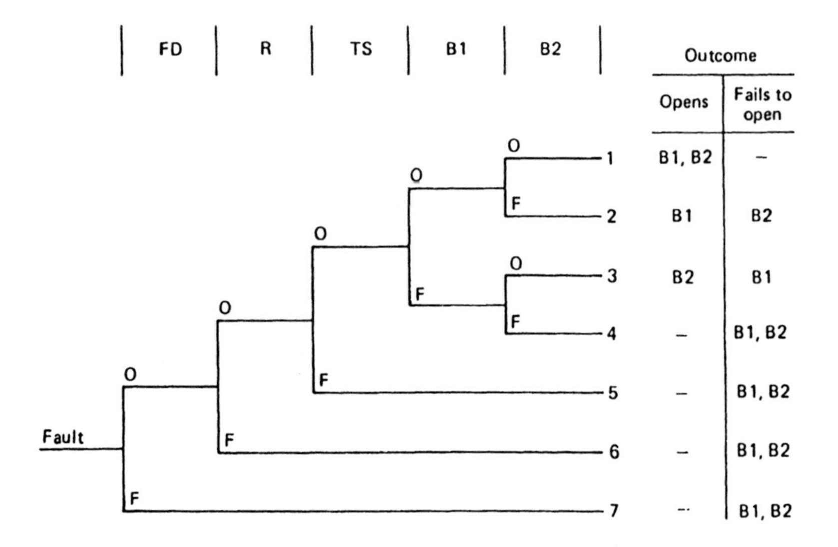

Fig. 11.11 Event tree of protection system: O—operates; F—fails to operate

respond is the average unavailability of the device between consecutive tests. This average unavailability has also been defined as the mean fractional dead time [5].

Assume that the times to failure of the device are exponentially distributed (a similar evaluation can be made for other distributions) and that the time period between consecutive tests is T_c. Then the average unavailability of the device is:

$$U = \frac{1}{T_c}\int_0^{T_c}(1 - e^{-\lambda t})dt = 1 - \frac{1}{\lambda T_c}(1 - e^{-\lambda T_c}) \tag{11.4}$$

If $\lambda T_c << 1$,

$$U \simeq 1 - \frac{1}{\lambda T_c}\left[1 - \left(1 - \lambda T_c + \frac{\lambda^2 T_c^2}{2} - \cdots\right)\right] = \frac{\lambda T_c}{2} \tag{11.5}$$

For the present system, assume for convenience that all of the devices (FD, R, TS and B) have a failure rate of 0.02 f/yr. The average unavailability, evaluated using Equations (11.4) and (11.5) for inspection intervals, T_c, of 3 months, 6 months and 1 year, is shown in Table 11.11. These results show that the error introduced by Equation (11.5) for this data is negligible.

(c) *Evaluating outcome probabilities*

The evaluation of the outcome probabilities is a simple exercise after the event tree has been deduced. First the paths leading to the required outcome are identified. The probability of occurrence of each relevant path is the product of the event

Table 11.11 Event probabilities of devices in Fig. 11.11

	Inspection interval of		
	3 months	*6 months*	*1 year*
Equation (11.4)	0.002496	0.004983	0.009934
Equation (11.5)	0.002500	0.005000	0.010000

probabilities in the path. The probability of occurrence of the outcome is then the sum of the probabilities of each path leading to that outcome.

In the present example:

Prob. (B1 not opening) $=\Sigma$ Prob. of paths 3 to 7

Prob. (B2 not opening) $=\Sigma$ Prob. of paths 2, 4 to 7

Prob. (B1 and B2 not opening) $=\Sigma$ Prob. of paths 4 to 7

Using the approximate data evaluated previously for event probabilities, the probability of each path for the event tree of Fig. 11.11 is shown in Table 11.12. Combining these probabilities appropriately gives the probability of B1 not opening, probability of B2 not opening and probability of B1 and B2 not opening on demand. These values are shown in Table 11.13.

The results shown in Table 11.13 indicate, as expected, that the probability of a breaker being stuck increases as the inspection interval increases. The results also indicate the more significant effect that the probability of both breakers not opening is almost the same value as the individual stuck-breaker probability. This can clearly have a significant impact on system operation. If, on the other hand, an assumption of independent overlapping failures was considered, the probability of B1 and B2 not opening would have been $(0.00996)^2 = 9.92 \times 10^{-5}$, which is 75 times smaller than the true value.

Table 11.12 Path probabilities for event tree of Fig. 11.11

	Probability for inspection interval of		
Path	*3 months*	*6 months*	*1 year*
1	0.987562	0.975248	0.950990
2	0.002475	0.004901	0.009606
3	0.002475	0.004901	0.009606
4	0.000006	0.000025	0.000097
5	0.002488	0.004950	0.009801
6	0.002494	0.004975	0.009900
7	0.002500	0.005000	0.010000

Table 11.13 Probability of breakers not opening on demand

Breaker not opening	*Probability for inspection interval of*		
	3 months	*6 months*	*1 year*
B1	0.00996	0.01985	0.03940
B2	0.00996	0.01985	0.03940
B1 and B2	0.00748	0.01495	0.02980

These results were evaluated assuming both breakers were actuated by exactly the same set of protection components. Redundancy can be included in this system, which can have a marked effect on the values of stuck probability. This is considered in the next subsections.

Although it was assumed in this section that the faulted system component is protected by two breakers, the concepts can be extended to any number of breakers. All that is required is the appropriate event tree for the system being considered.

(d) *Effect of sharing protection components*

It was assumed in the previous example that breakers B1 and B2 were both actuated by the same fault detector, relay and trip signal device. Consequently both breakers shared the same protection system and any failure in this system, other than the breaker itself, caused both breakers to malfunction. This possibility can be reduced by providing alternative channels to each of the breakers. In order to illustrate this, assume that two trip signal devices are used: one (TS1) actuates breaker B1 and the other (TS2) actuates breaker B2. In this case, the original event tree will be modified to that shown in Fig. 11.12.

The values of stuck-breaker probabilities can be evaluated using the previous data and evaluation techniques. These values are shown in Table 11.14 for an inspection interval of 6 months only. Comparison of these results with those in Table 11.13 shows that the stuck probability of B1 only (also B2 only) is unchanged. This is to be expected since the protection channel and the number of devices in the channel to each breaker is unchanged. The results also show, however, that the stuck probability of B1 and B2 together is considerably reduced, the ratio between the value shown in Table 11.14 and the independent overlapping failure probability being reduced to 25 to 1.

(e) *Effect of redundant protection components*

In the previous example, each trip signal device was assumed to actuate one of the breakers. This was shown to improve the probability that breakers B1 and B2 would not open. The system can be further improved by including redundancy in the protection channels. As an example, reconsider the previous system of two trip signal devices but this time assume that the operation of either of them causes the

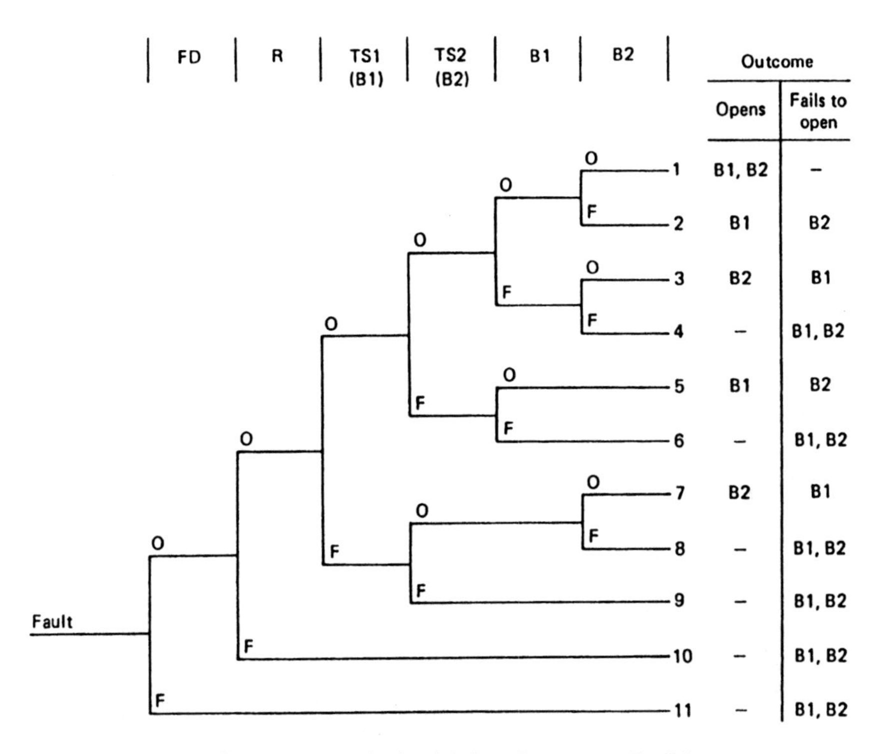

Fig. 11.12 Event tree for two separate trip signal devices: O—operates; F—fails to operate

operation of both breakers. These devices therefore are fully redundant and the associated event tree is shown in Fig. 11.13.

The values of stuck-breaker probabilities evaluated using the previous data are shown in Table 11.15 for an inspection interval of 6 months.

It is seen from these results that the probability that breakers (B1 and B2) do not open is reduced slightly compared with the results shown in Table 11.14, but the probability of breaker B1 (similarly for B2) is reduced considerably. These

Table 11.14 Stuck probability of breaker with separate trip signal devices

Breaker not opening	*Probability of not opening*
B1	0.01985
B2	0.01985
B1 and B2	0.01008

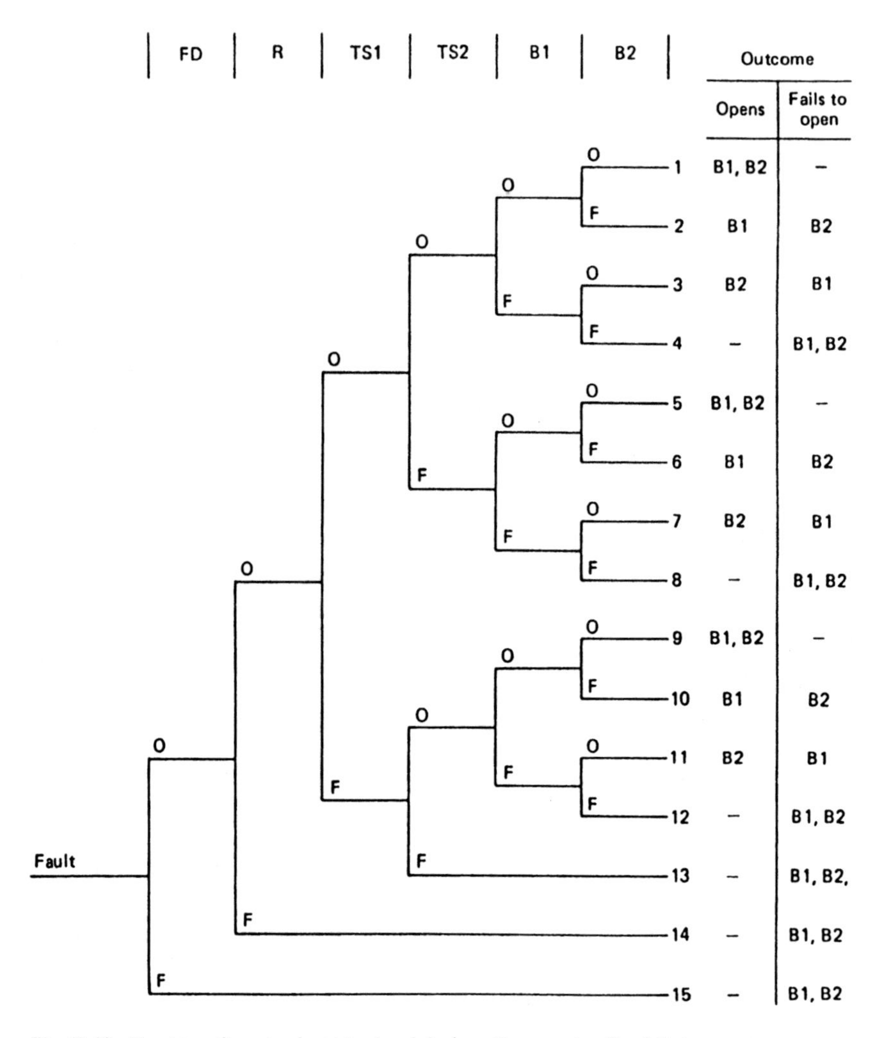

Fig. 11.13 Event tree for redundant trip signal devices: O—operates; F—fails to operate

Table 11.15 Stuck-breaker probability with redundancy

Breaker not opening	*Probability of not opening*
B1	0.01495
B2	0.01495
B1 and B2	0.01003

values would be affected by an even greater amount if further redundancy were incorporated in the system: for example, if redundancy were included in the fault detector (FD) and relay (R). This is not particularly necessary or even desirable from an economic point of view in terms of conventional power system operation. It is important, however, in safety applications, particularly those involving nuclear generator stations when considerable redundancy is used. The techniques, however, remain identical.

11.4.4 Evaluation of inadvertent operation

The modelling and evaluation of inadvertent operation is identical in concept to that for failure to operate. An event tree is constructed in a similar manner to Figs. 11.11–13, commencing from the point at which the false signal can occur. This initiating point is known as the initiating event and in Section 11.4.3 was the fault on the system component.

Consider, as an example, the protection system which, under a system fault condition, gives rise to the event tree shown in Fig. 11.11. If a false signal can be developed in the fault detection (FD) device, the event tree associated with this occurrence is the first six paths of Fig. 11.11. If the false signal develops in the trip signal (TS) device, the associated event tree is the first four paths of Fig. 11.11. The probability of one or more breakers inadvertently opening can therefore be evaluated using the previous technique and data with only two differences:

(i) the event tree will be smaller than those in Figs. 11.11–13 depending on the location in which the false signal is developed;

(ii) the value of probability associated with the device in which the false signal is developed is the occurrence probability of the false signal; the probabilities associated with all other subsequently operating components are identical to those used previously.

In order to illustrate this evaluation technique, consider that a false signal can originate in the trip signal device of Fig. 11.11 with a probability of 0.001. In this case, the first four paths of Fig. 11.11 are considered. Using the data of Section 11.4.3(b) for an inspection interval of 6 months, the following path probabilities can be evaluated *given* that the false signal has developed:

$$P\,(\text{path } 1) = 0.995 \times 0.995 = 0.990025$$

$$P\,(\text{path } 2) = 0.995 \times 0.005 = 0.004975$$

$$P\,(\text{path } 3) = 0.005 \times 0.995 = 0.004975$$

$$P\,(\text{path } 4) = 0.005 \times 0.005 = 0.000025$$

These values of path probabilities must be weighted by the probability that the false signal develops, in order to evaluate the probability of an inadvertent opening of a breaker. This gives the inadvertent opening probabilities shown in Table 11.16.

Table 11.16 Inadvertent opening probabilities

Breaker inadvertently opening	*Probability of opening*
B1 (path 2)	0.000050
B2 (path 3)	0.000050
B1 and B2 (path 1)	0.009900
None (path 4)	0.000000

The results shown in Table 11.16 relate to a false signal developing in the trip signal device. False signals can develop in other devices and a similar analysis should be done for each possible occurrence. These are mutually exclusive and therefore the probabilities of each contribution can be summated to give an overall probability of inadvertent opening.

The previous analysis enables probability of opening to be evaluated. A similar analysis can be made to determine the failure rate associated with inadvertent opening. In this case, the relevant path probabilities are weighted by the rate of occurrence of the false signal instead of its probability.

11.5 HVDC systems

11.5.1 Concepts

High voltage direct current (HVDC) power transmission has been the centre of many research studies, and considerable activity throughout the world is devoted to evaluating its technical benefits as part of the composite power system. To date, the number of HVDC schemes that exist or are being developed is minute in comparison with HVAC systems. This imbalance will always exist in the future since HVDC schemes are beneficial only in specific applications and are not useful for widespread power transmission. The specific applications include long-distance bulk power transmission, particularly between remote generation points and load centers, relatively long cable interconnections such as sea crossings, interconnection between two large isolated HVAC systems, and asynchronous tie-lines between or internal to HVAC systems.

The reliability evaluation of HVDC systems has received very little attention and only a few papers [9–12] have been published. This lack of interest simply reflects the relative size and application of HVAC and HVDC systems. This does not mean, however, that techniques for analyzing such systems do not exist: methods described in *Engineering Systems* as well as methods presented in previous sections of this book can be used very adequately. The purpose of this section is therefore to describe how these techniques can be used in the evaluation of HVDC schemes.

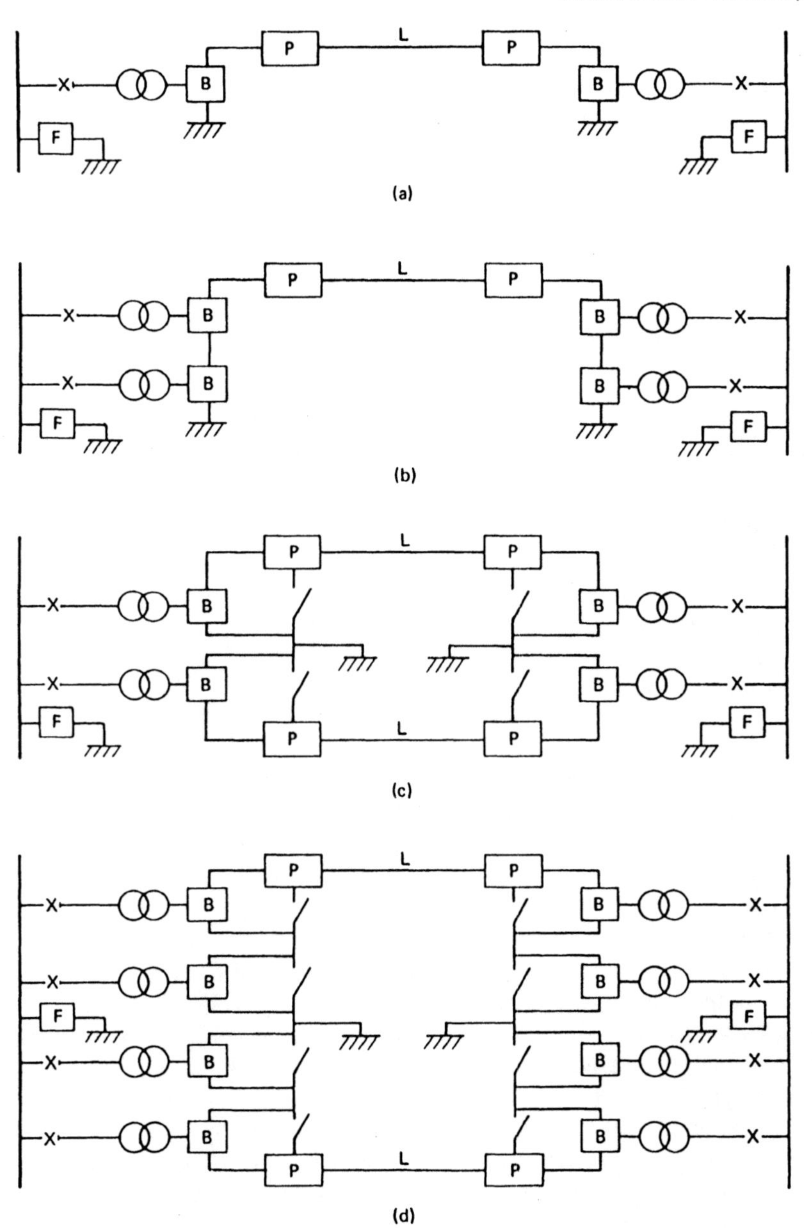

Fig. 11.14 Typical bridge configurations:
(a) dingle bridge, monopole (system 1);
(b) multibridge, monopole (system 2);
(c) single bridge, bipole (system 3);
(d) multibridge, bipole (system 4)
B—bridge, F—a.c. filter, P—pole equipment, L—transmission line

HVDC systems are only used as links between remote generation and the HVAC system or between two HVAC systems. This single link concept enables these systems to be solved using relatively simple techniques. The most important techniques consist of series systems and Markov modelling.

11.5.2 Typical HVDC schemes

An HVDC link consists of a rectifier station at the sending end, an inverter station at the receiving end and one or more transmission lines between them. The degree of complexity of the system increases when multibridges are used in the two converter stations and two poles are used for the transmission link. Typical, but simplified, configurations are shown in Fig. 11.14 for a single bridge, monopole link, a single bridge, bipole link, a multibridge, monopole link, and a multibridge, bipole link.

These systems can be divided into three main subsystems: the rectifier/inverter subsystem, the transmission line and pole equipment subsystem and the a.c. filters subsystem. These subsystems can be analyzed independently and finally combined to give the reliability of the complete HVDC link. These concepts are described in the following sections.

11.5.3 Rectifier/inverter bridges

The heart of the converter station is the bridge which includes the valves, damping, protection and control equipment. Some real systems still use mercury arc rectifiers, now steadily being replaced by the thyristor valve as the converting device. The evaluation concept, however, is essentially the same for both conversion methods.

All HVDC systems retain spare valves in the case of both mercury arc rectifiers and thyristors, and therefore the concept of sparing and allocation of spares is important in the reliability assessment of bridges. The techniques described in Section 10.5 of *Engineering Systems* and Section 11.3 of this book can be applied directly. For instance, Figs. 11.6 and 11.7 are applicable provided 'both banks' is replaced by 'both bridges' (i.e. a two-bridge device is being considered), 'spare' is replaced by 'spare valve' and the bridge failure rate λ_b is multiplied by the number of bridges that can fail in any of the system states. On this basis, all the previous concepts, evaluation techniques and conclusions remain equally valid.

In the case of multibridges, the number of bridges required for system success must be known. In order to increase voltage rating, bridges are connected in series. Failure of any one bridge in this case will fail the system. In order to increase power transmission, bridges can be connected in parallel. Failure of one or more bridges in this case does not necessarily fail the system but may send it into a derated state. This problem can be easily resolved by cumulating the relevant states of a state space diagram such as that shown in Fig. 11.7. In the case of a series system, 'both up' states represent system success and all others represent failure. In the case of a

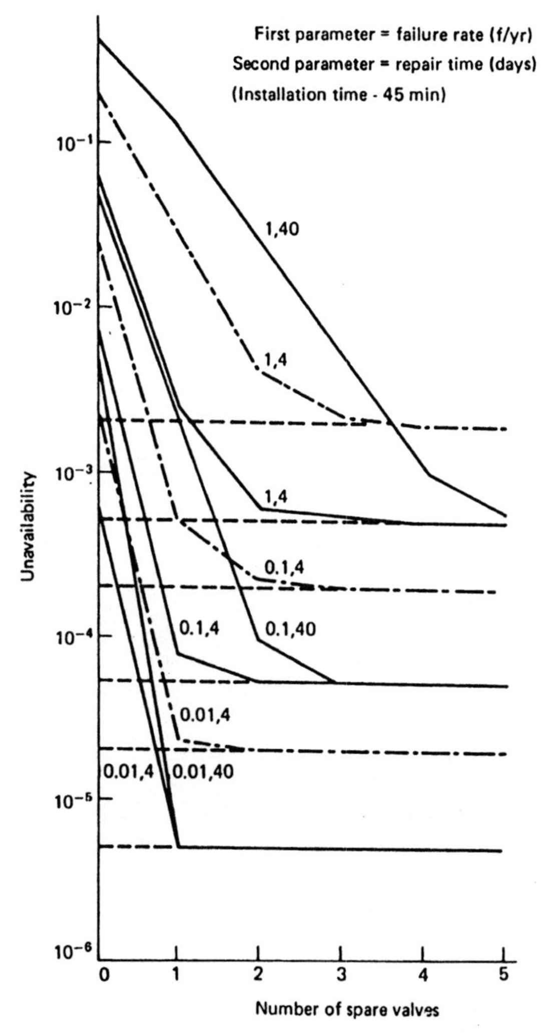

Fig. 11.15 Effect of spares on unavailability — Single bridge; — - — Two bridges in series; - - - Limiting probability

fully redundant system, 'both down' states represent system failure and all others represent success. In the case of a partially redundant system, 'both up' states represent a 100% capacity level, 'one down' states represent an intermediate capacity level, e.g. 50%, and 'both down' states represent complete failure.

As an illustrative example, the results shown in Fig. 11.15 were obtained [11] for a single mercury arc bridge consisting of six valves and for a system of two identical bridges connected in series. These results are similar to those shown in

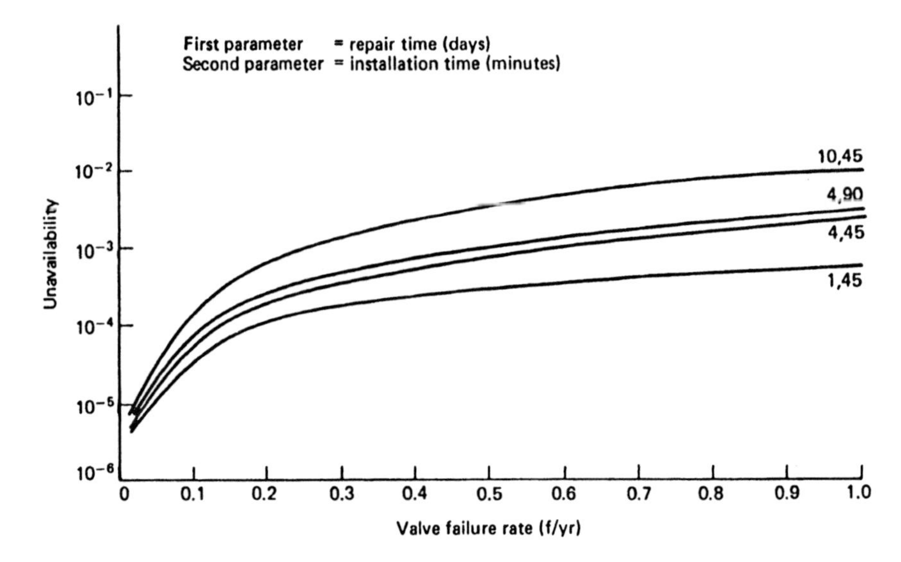

Fig. 11.16 Effect of valve failure rate on unavailability

Fig. 11.8. A similar set of results plotted as a function of valve failure rate is shown in Fig. 11.16 for the above single bridge having one spare valve.

11.5.4 Bridge equivalents

The full state space diagrams and the subsequent evaluation techniques described in Section 11.5.3 are an important component of converter station assessment and essential when considering spares and sparing allocation. The bridge, however, is only one component in an HVDC link and must be combined with the other components and subsystems in order to evaluate the reliability of the complete system.

This combination becomes rather difficult and certainly tedious if a complete representation is used for the bridge configuration. The bridge, however, can be considerably simplified during the analysis of the complete system using equivalent state space diagrams and models. The reason for this is that the bridge identities are no longer required at this stage, only the effect of bridge states on the operation of the system. This equivalencing was previously done in Section 11.2.2 in order to reduce the number of derated states in a generating station. The following two conditions [12] must apply in deducing the equivalent model:

(a) the mean time spent in the UP state of the equivalent model must be equal to the average duration of the UP states in the complete model;
(b) the availabilities of the various capacity levels in the equivalent and complete model must be equal.

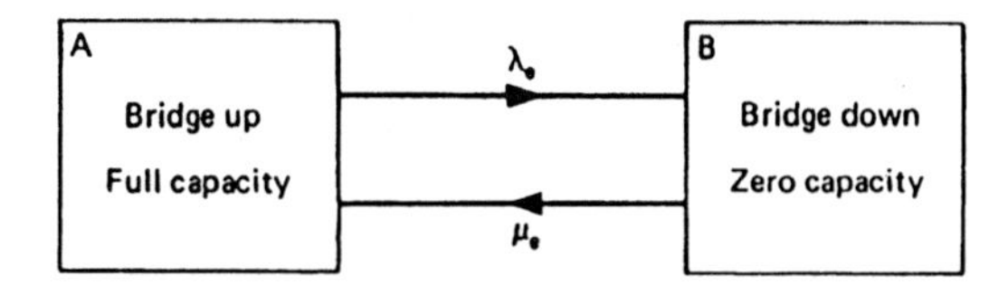

Fig. 11.17 Equivalent model for single bridge

(a) *Single bridge*

A single bridge can be represented in terms of capacity states by the state space diagram shown in Fig. 11.17.

The two transition rates in Fig. 11.17 are the equivalent failure rate and equivalent repair rate. For a single bridge [12] with no spare valves:

$$\text{Prob. (state A, up)} = \frac{\mu\gamma}{\lambda_b(\mu+\gamma)+\mu\gamma}$$

$$\text{Prob. (state B, down)} = \frac{\lambda_b(\mu+\gamma)}{\lambda_b(\mu+\gamma)+\mu\gamma}$$

$$\lambda_e = \lambda_b, \quad \mu_e = \frac{\mu\gamma}{\mu+\gamma}$$

where λ_b = bridge failure rate ($\equiv n\lambda$)
λ = valve failure rate
n = number of valves in bridge
μ = valve repair rate
γ = valve installation rate.

For a single bridge [12] with any number of spare valves,

$$\lambda_e = \lambda_b = n\lambda$$

$$\mu_e = \frac{\Sigma\,(\text{UP state probabilities})}{\Sigma\,(\text{DOWN state probabilities})} \times n\lambda$$

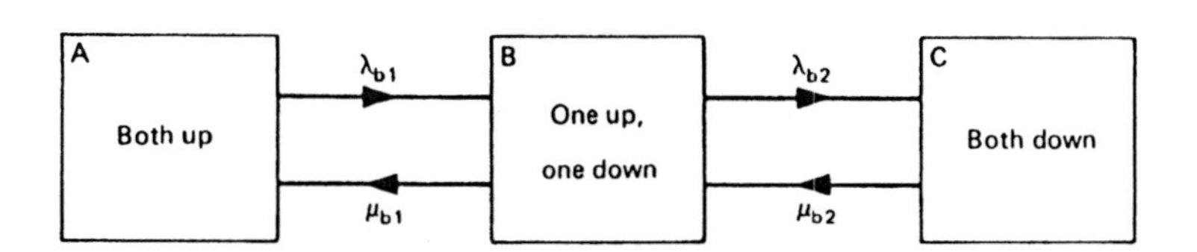

Fig. 11.18 Equivalent model for two identical bridges

The UP state and DOWN state cumulative probabilities can be evaluated by analyzing the appropriate state space diagrams using the techniques described in Chapters 9 and 10 of *Engineering Systems* and Section 11.3 of this book.

(b) *Multibridge*

The equivalent model for a multibridge depends on the number of capacity levels in which the bridge can exist. In the case of two identical bridges, the equivalent model will have a maximum of three states irrespective of the number of spares as shown in Fig. 11.18.

If both bridges are required for system success (series system), states B and C can be combined to give a two-state model, the indices of which can be evaluated as for the single bridge described in (a) above. Furthermore, if no further failures can occur when one bridge has failed and the system is de-energized, state C is not relevant and should be discarded. The indices of this equivalent model can again be evaluated as for a single bridge.

If the bridge contains redundancy, however, all three states of Fig. 11.18 are required. The cumulative probability of residing in each of these states can be evaluated using the concepts of Section 11.3 and the complete models such as shown in Figs. 11.6 and 11.7. In this case, state A of Fig. 11.18 represents full or 100% capacity, state B represents half or 50% capacity and state C represents zero capacity. Referring to the notation used in (a) above and to Fig. 11.6, the indices for the equivalent model of two bridges without spares can be deduced [12] as follows (other arrangements can be evaluated similarly):

$$\text{Prob. (full capacity)} = P_A = P_1$$

$$\text{Prob. (half capacity)} = P_B = P_2 + P_3$$

$$\text{Prob. (zero capacity)} = P_C = P_4 + P_5 + P_6$$

Frequency of transfer from state A to state B is

$$f_{AB} = f_{12}$$

$$f_{BC} = f_{24} + f_{35}$$

$$f_{BA} = f_{31}$$

$$f_{CB} = f_{52} + f_{63}$$

giving:

$$\lambda_{b1} = f_{AB}/P_A = 2\lambda_b = 2n\lambda \quad (n = \text{number of valves in each bridge})$$

$$\lambda_{b2} = f_{BC}/P_B = \lambda_b = n\lambda$$

$$\mu_{b1} = f_{BA}/P_B = \frac{\gamma P_3}{P_2 + P_3}$$

$$\mu_{b2} = f_{CB}/P_C = \frac{\gamma(P_5 + 2P_6)}{P_4 + P_5 + P_6}$$

11.5.5 Converter stations

Each converter station consists of not only one or more bridges, but also converter transformers and circuit breakers. These can be combined with the equivalent model for the bridges using the following method.

Generally each bridge is associated with its own converter transformer, circuit breaker and other relevant terminal equipment. These components operate as a series system and therefore a combined failure rate λ_a and repair rate μ_a can be deduced for these auxiliaries. These auxiliaries can be combined with the bridge to produce an equivalent model which can be used in subsequent evaluations. The complete state space diagram together with its equivalent model is shown in Fig. 11.19 for a single bridge system and in Fig. 11.20 for a two-bridge system.

It is assumed in the models shown in Figs. 11.19 and 11.20 that when a bridge fails its auxiliary is de-energized and cannot fail but remains in a standby mode until its associated bridge is repaired; similarly if an auxiliary fails. Consequently, state 6 of Fig. 11.20 is a failure state because, although one bridge and one auxiliary are operable, they are not associated with each other and cannot therefore be operated together but remain on standby. If they can be linked together, however, state 6 becomes a half-capacity state and further failures from this state become possible.

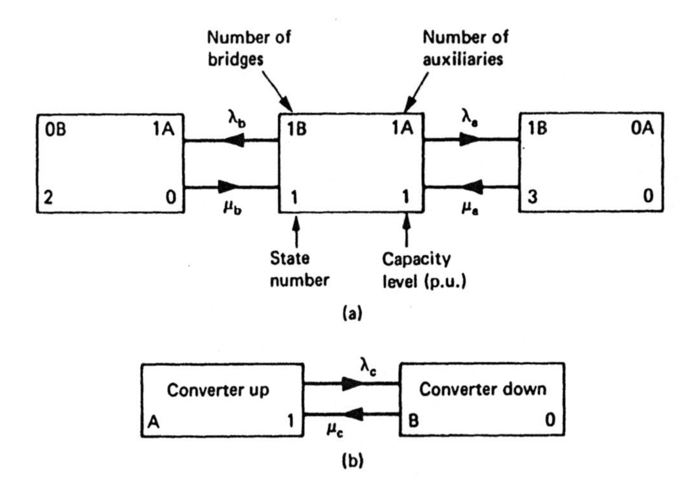

Fig. 11.19 Models for a converter station with a single bridge: (a) complete model; (b) equivalent model

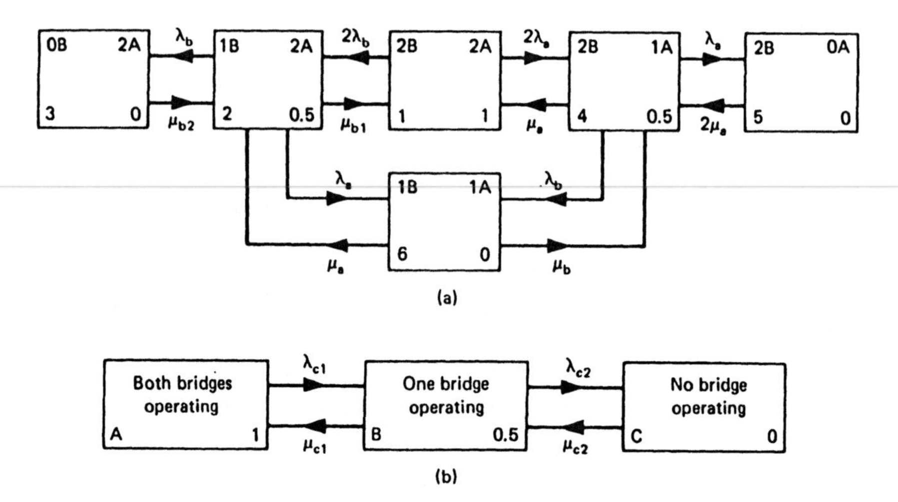

Fig. 11.20 Models for a converter station with two bridges: (a) complete model; (b) equivalent model

The equivalent indices λ_c and μ_c in the above models can be evaluated using the equivalencing concept described in Section 11.5.4. This analysis would show that:

$$\lambda_c = \lambda_a + \lambda_b$$

$$\lambda_{c1} = 2\lambda_{c2} = 2(\lambda_a + \lambda_b) = 2\lambda_c$$

The concept illustrated in Figs. 11.19 and 11.20 can be extended to any number of bridges and associated auxiliaries. In all cases, an equivalent model can be deduced in which each state represents a particular capacity level.

In order to simplify the analysis of the complete HVDC link, the two converter stations can be combined to create the next stage of equivalent models. This is again achieved using the previous principles. As an example, the state space diagrams shown in Fig. 11.21 represent the complete and equivalent models for identical sending end (SE) and receiving end (RE) converter stations, each containing two bridges. The notation used in Fig. 11.21 is the same as for Fig. 11.19 and the principle is similar to that of Fig. 11.20. Consequently, state 6 of Fig. 11.21 is a zero-capacity state since, although a bridge is available at each end of the link, they are connected to two different poles. If the system permits bridges to be connected to either pole, however, state 6 becomes a half-capacity state and further failures from this state can occur.

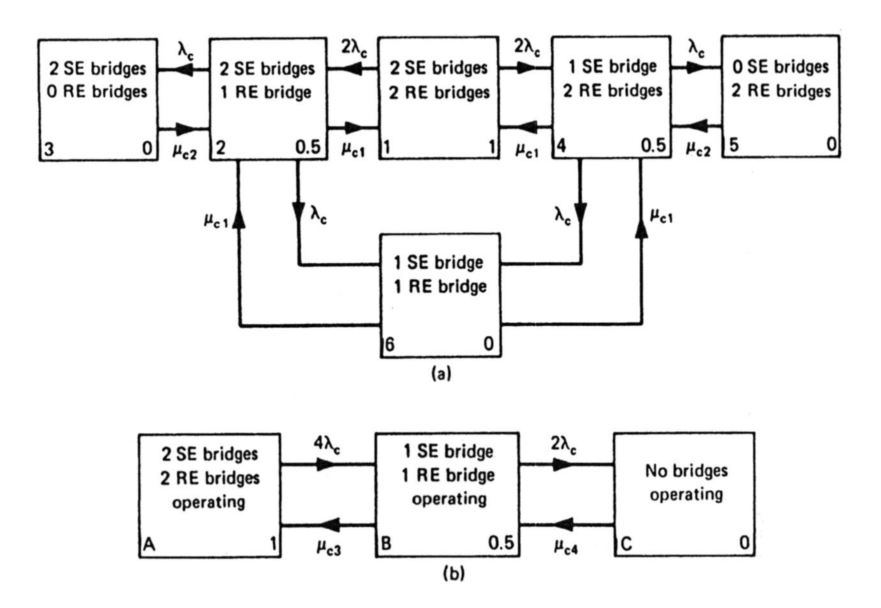

Fig. 11.21 Models for combined converter stations: (a) complete model; (b) equivalent model

11.5.6 Transmission links and filters

The two remaining subsystems of the complete HVDC link are the transmission lines and the filters.

If it is assumed that all the filters at each converter station are required for system success, the equivalent failure and repair rate, λ_f and μ_f, can be evaluated using the principle of series systems. Similarly, if both banks of filters at the two converter stations are required in order to operate the system, the equivalent filter model is as shown in Fig. 11.22. With the above assumptions the state in which both filter banks are out cannot exist.

The model for the transmission lines is similar to that for the filters except that the system can still be operated when one or more lines of a multi-line system are out of service. The model for a bipole link is therefore as shown in Fig. 11.23.

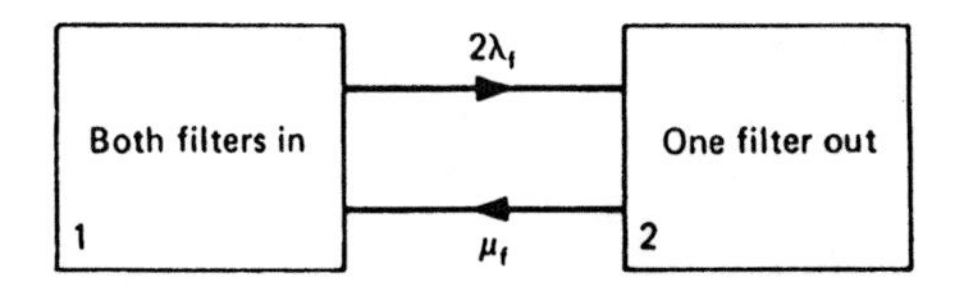

Fig. 11.22 Model for combined filter banks

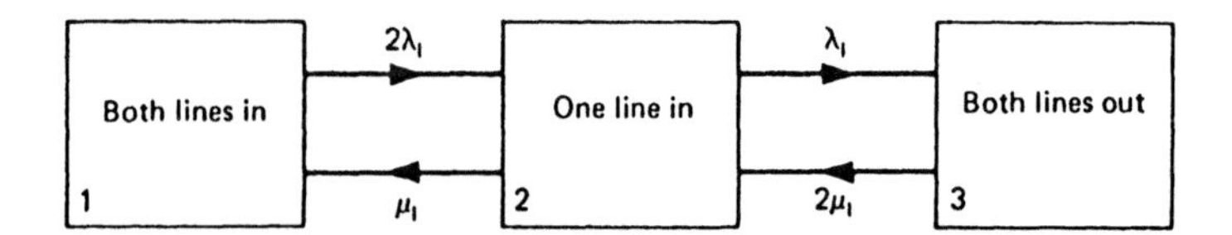

Fig. 11.23 Model for bipole transmission link

The indices used in Fig. 11.23 should represent the composite reliability indices of the actual transmission line and its associated pole equipment, all of which are effectively connected in series.

11.5.7 Composite HVDC link

The composite HVDC link can now be assessed by combining the individual equivalent models to form a complete state space diagram that represents the HVDC system. These composite models can become quite complex and only two examples are given in this section. Others can be created using similar logic. In these examples, it is assumed that, with the exception of the transmission line and its pole equipment, no further failures can occur when the system is in a de-energized state. In the case of the bipole example, it is also assumed that the system can operate in a monopole mode without encountering ground current problems. If this assumption is not valid, this state will create a zero capacity rather than half-capacity transmission level. Therefore the state space diagram is not altered, only the states which are cumulated together.

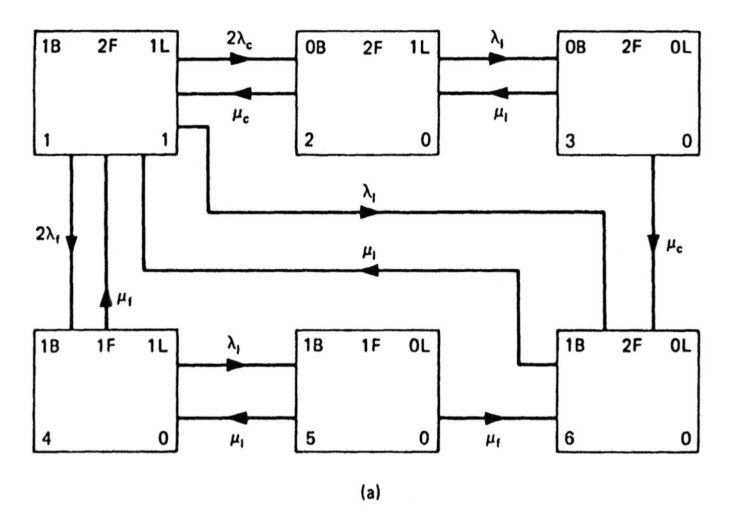

Fig. 11.24 Models for complete HVDC links:
(a) Single bridge, monopole

On the basis of the above discussion, the models for a single bridge, monopole system (system 1 of Fig. 11.14) and for a single bridge, bipole system (system 3 of Fig. 11.14) are shown in Figs. 11.24(a) and (b) respectively. The notation used is the same as in Fig. 11.19.

The models shown in Fig. 11.24 will reduce to a two-state model (100%, zero % capacity) for the single-bridge monopole system and a three-state model (100%, 50% and zero % capacity) for the single-bridge bipole system. Similar models can be constructed for other configurations.

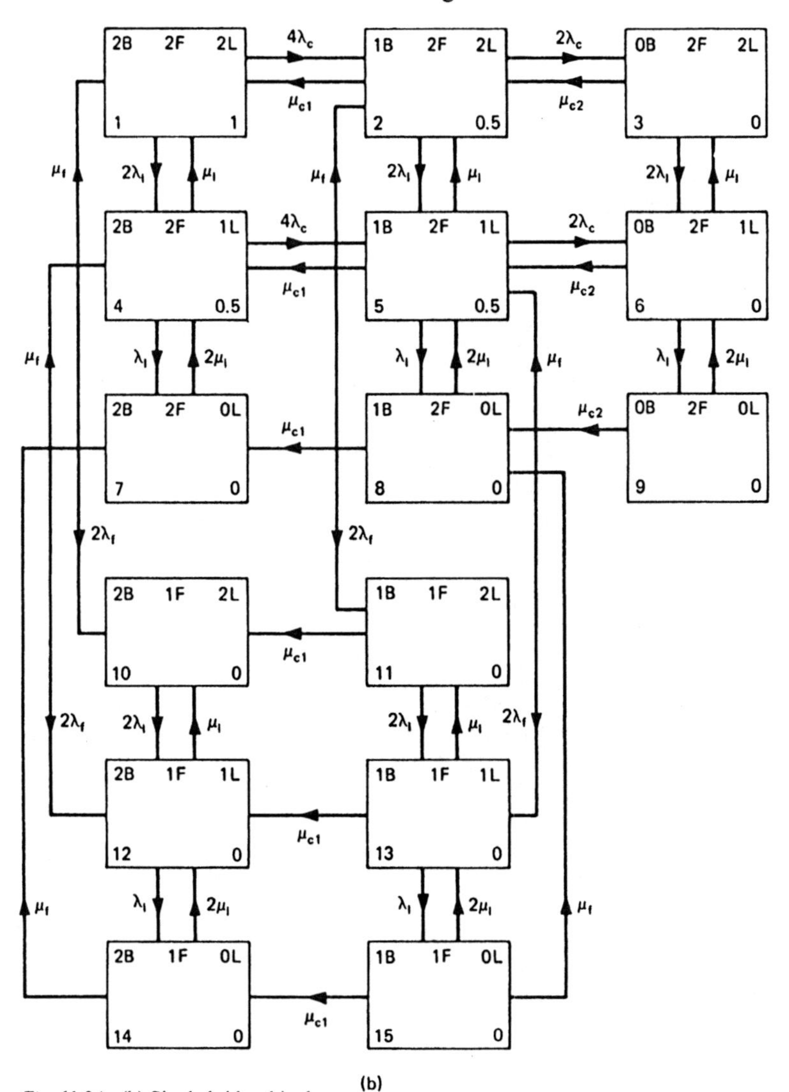

Fig. 11.24 (b) Single bridge, bipole

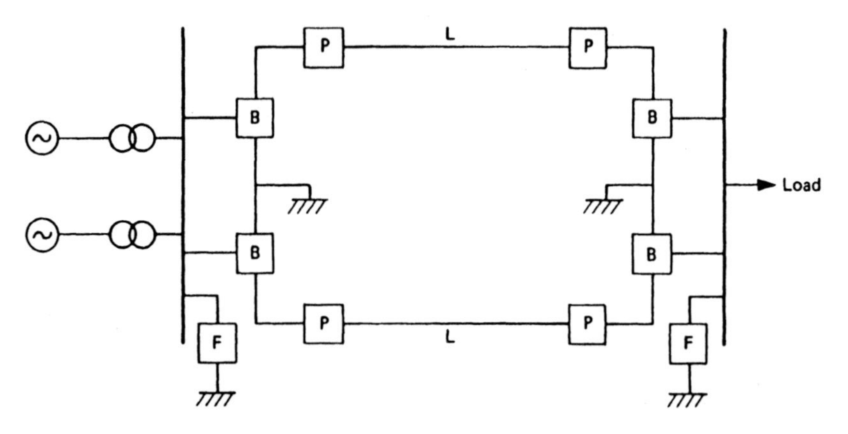

Fig. 11.25 Typical generation/HVDC system

Table 11.17 Reliability data used for system in Fig. 11.25

Component	*Failure rate (f/yr)*	Repair time (hours)	*Installation time (minutes)*
valves	0.25	96.0	45.0
generating units	0.50	87.6	—
transformers	0.012	168.0	—
transmission lines	1.50	4.0	—
pole equipment	0.04	8.0	—
filters	0.012	168.0	—

Table 11.18 Transmission capability probability tables

State	*Capacity (p.u.)*	*Probability*	*Frequency (per yr)*	*Duration (days)*	*Expected capacity (p.u.)*
No spare valves					
1	1.0	0.926473	3.7413	90.387	
2	0.5	0.071728	6.6109	3.963	0.9623
3	0.0	0.001799	0.2958	2.204	
One spare valve					
1	1.0	0.987850	3.9892	90.387	
2	0.5	0.011886	6.9466	0.625	0.9938
3	0.0	0.000264	0.0221	4.370	
Two spare valves					
1	1.0	0.988874	3.9933	90.387	
2	0.5	0.010868	6.9521	0.571	0.9943
3	0.0	0.000258	0.0210	4.490	
Three spare valves					
1	1.0	0.988885	3.9933	90.387	
2	0.5	0.010857	6.9522	0.570	0.9943
3	0.0	0.000258	0.0218	4.315	

11.5.8 Numerical examples

Consider the system [12] shown in Fig. 11.25 and the reliability data shown in Table 11.17. The models described in the previous sections can be combined with the generation capacity table to form a complete transmission capability table [10] as seen by the receiving end of the HVDC link. These transmission capability tables

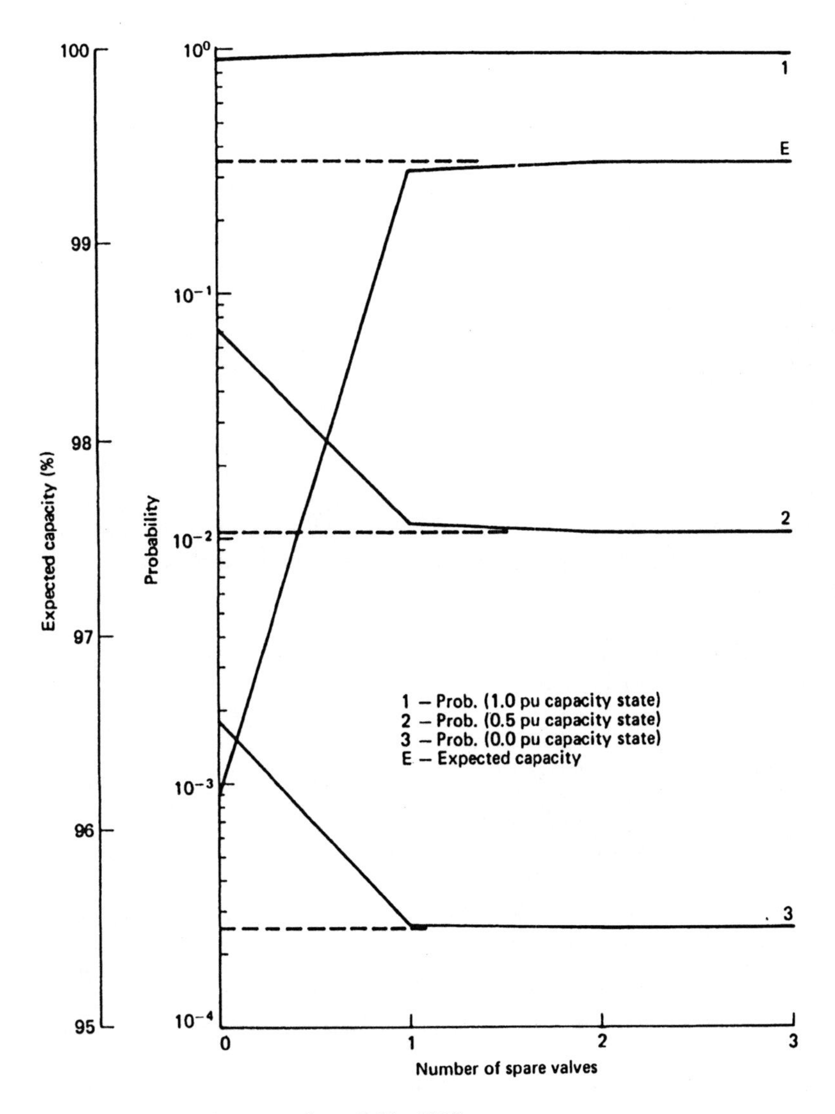

Fig. 11.26 Results for system shown in Fig. 11.25

[12] are shown in Table 11.18 when zero to three spare valves are retained for the bridges. The variation of state probabilities and expected capacity are also shown in Fig. 11.26.

These results clearly demonstrate the benefit of including spares, although it can be seen that little additional benefit is gained by having more than one spare. The average duration of remaining in the fully up state (State 1) is not affected by the number of spares, since these cannot affect the rate of failure, only the restoration process and therefore the duration of being in a DOWN state. The UP state duration would be affected, however, by the failure rate data used for the components, and sensitivity studies should be performed in practice that reflect possible ranges of failure data. In addition, other design changes, such as number of bridges, and the amount of system redundancy should be considered in practical applications and their effect on the transmission capability tables established. This can be achieved using the various models described in previous sections.

11.6 Conclusions

In this chapter we have considered a number of separate topic areas, all of which are related to the availability of plant and stations. In the reliability analysis of a complete power system, most of the plant considered in this chapter are simply represented as a single system component. This is quite adequate for most analyses, but it does neglect the fact that these single-component representations are frequently complex systems in their own right. It can be just as important to be able to analyze these systems in order to establish the effects of redesign, reinforcement and redundancy on their behavior and consequently their effect on the overall system performance.

The models and techniques described in this chapter enable the plant and station subsystems to be analyzed and also permit the very important concept of sparing to be considered in their design.

11.7 Problems

1. A base load generating station consists of two 60 MW turbo-alternators, each of which is fed by two 45 MW boilers. The average failure rate and average repair time of each boiler is 3 f/yr and 48 hours respectively and the average failure rate and average repair time of each turbo-alternator is 2 f/yr and 72 hours respectively. Evaluate the expected MW output of the complete station.
2. Repeat Problem 1 assuming that the boilers are (a) 30 MW, (b) 60 MW, the reliability indices of the boilers being the same as the 45 MW boilers.
3. Repeat Problem 1 assuming that both turbo-alternators can be fed by any of the four boilers.
4. Repeat Problem 1 assuming that each turbo-alternator is fed by a single 60 MW boiler having the same reliability indices as the 45 MW boilers.

Table 11.19

Component	Number	*MTTF (months)*	*Derating due to outage of one component (%)*
turbo-alternator	1	6	100
boiler	1	4	100
pulverizers	3	6	40
forced draught fan	2	9	50
induction draught fan	2	9	50
circulating water pump	4	12	30
feed water pump	4	12	30

5 Re-evaluate the MW output of the station in Problem 1 assuming that station transformers are included in the assessment. Each station transformer has a MTTF of 10 yr and a MTTR of 200 hours. The configurations to be analyzed are:
   (a) the output of each turbo-alternator is fed through one 60 MW transformer;
   (b) the output of both turbo-alternators are fed through a single 120 MW transformer;
   (c) the output of both turbo-alternators are fed through two parallel 90 MW transformers.

6 The main components of a particular generating station are as shown in Table 11.19. Evaluate the probability of residing in each capacity state during an operational phase if the lead time is 2 hours and failure of more than one component and the repair process are neglected.

7 During a particular 5-year period, a base load generating station was found to reside in the following derated states for the number of days shown in Table 11.20. At all other times it operated at its full capacity. Calculate the probability of residing in each of its capacity output levels and the value of equivalent forced outage rate (EFOR).

8 A substation transformer is protected by two breakers. The protection system consists of a fault detector (FD) having a failure rate of 0.01 f/yr, a combined relay/trip device (RT) having a failure rate of 0.04 f/yr and the breakers, which have a failure rate of 0.02 f/yr. The inspection interval is 6 months. Given that a fault occurs on the transformer, evaluate the probability of successful operation of the protection, the

Table 11.20

Derated state (%)	*Number of days in derated state*									
10	7	10	4	3	20	15	8	7	5	10
20	10	8	2	15	16	14	7	5	6	—
30	2	4	8	2	25	18	8	—	—	—
40	15	11	14	8	1	12	—	—	—	—
100	1	4	2	10	12	19	15	14	—	—

probability of exactly one breaker not operating, the probability of both breakers not operating, given that:
(a) the FD and RT is shared by both breakers;
(b) the FD is shared but separate RT serve the two breakers;
(c) separate FD and RT serve each breaker.

9 The system of Problem 8(a) is used to protect three breakers. Evaluate the probability of successful operation, the probability of exactly one breaker not operating, the probability of exactly two breakers not operating, the probability of all breakers not operating.

11.8 References

1. Allan, R. N., Billinton, R., de Oliveira, M. F., 'Reliability evaluation of the electrical auxiliary systems of power stations', *IEEE Trans. on Power Apparatus and Systems*, **PAS-96** (1977), pp. 1441–9.
2. Allan, R. N., de Oliveira, M. F., Billinton, R., 'Evaluating the reliability of auxiliary electrical systems of generating stations', *CEA Trans.* (1976), **15**, pt 3, paper 76-T-226.
3. Allan, R. N., de Oliveira, M. F., Chambers, J. A., Billinton, R., 'Reliability effects of the electrical auxiliary systems in power stations', *IEE Conf. Publ.* **148** (1977), pp. 28–31.
4. Allan, R. N., de Oliveira, M. F., Kozlowski, A., Williams, G. T., 'Evaluating the reliability of electrical auxiliary systems in multi-unit generating stations', *Proc. IEE*, **127**, pt C (1980), pp. 65–71.
5. Green, A. E., Bourne, A. J., *Reliability Technology*, John Wiley (1972).
6. Allan, R. N., Rondiris, I. L., Fryer, D. M., Tye, C., *Computational Development of Event Trees in Nuclear Reactor Systems*, Second National Reliability Conference, Birmingham, UK, March 1979, paper 3D/1.
7. Allan, R. N., Rondiris, I. L., Fryer, D. M., 'An efficient computational technique for evaluating the cut/tie sets and common-cause failures of complex systems', *IEEE Trans. on Reliability*, **R-30** (1981), pp. 101–9.
8. Singh, C., Patton, A. D., 'Protection system reliability modelling; unreadiness probability and mean duration of undetected faults', *IEEE Trans. on Reliability*, **R-29** (1980), pp. 339–40.
9. Heising, C. R., Ringlee, R. J., 'Prediction of reliability and availability of HVDC valve and HVDC terminal', *IEEE Trans. on Power Apparatus and Systems*, **PAS-89** (1970), pp. 619–24.
10. Billinton, R., Sachdev, M. S., 'Direct current transmission system reliability evaluation', *CEA Trans.*, **7**, pt 3 (1968), paper 68-SP-170.
11. Billinton, R., Prasad, V., 'Quantitative reliability analysis of HVDC transmission systems: I—Spare valve assessment in mercury arc bridge configurations', *IEEE Trans. on Power Apparatus and Systems*, **PAS-90** (1971), pp. 1034–46.

12. Billinton, R., Prasad, V., 'Quantitative reliability analysis of HVDC transmission systems: II—Composite system analysis', *IEEE Trans. on Power Apparatus and Systems*, **PAS-90** (1971), pp. 1047–54.
13. Billinton, R., Jain, A. V., 'Unit derating levels in spinning reserve studies', *IEEE Trans. on Power Apparatus and Systems*, **PAS-90** (1971), pp. 1677–87.
14. Allan, R. N., Adraktas, A. N., *Terminal Effects and Protection System Failures in Composite System Reliability Evaluation*, Paper No. 82 SM 428-1, presented at the IEEE Summer Power Meeting, 1982.

12 Applications of Monte Carlo simulation

12.1 Introduction

Previous chapters have centered almost exclusively on the use of analytical techniques. The only exceptions are the opening discussion in Chapter 1 and the brief use of simulation techniques in Section 9 of Chapter 7. However, as discussed in Chapter 1, there are two general approaches for assessing system reliability: the direct analytical method and simulation methods. It is therefore appropriate to describe the application of simulation techniques to power system reliability problems, to discuss the basic procedures used, and to illustrate the approach with some relatively simple examples in a similar way to that used for discussion of the analytical methods.

At this point it is worth noting that, from a teaching and tutorial point of view, simulation techniques may not be considered to be as intellectually stimulating as the analytical approach because there are no easily structured mathematical models and equations which can be determined either by logic or mathematical derivation: Instead the approach is entirely based on the use of random numbers generated by computer software. It is, however, important to recognize the principles involved, the processing logic required, and the interpretation of the output produced by the simulation algorithms.

It is also worth noting that, although the probabilistic analyses themselves may be based on a simulation approach, this does not override the fact that analytical models and equations, such as load flow, are still an essential ingredient in performing the required power system analyses and in determining the adequacy of system states.

A detailed description of simulation techniques is provided in Chapter 13 of *Engineering Systems*, and the reader is referred to that for a discussion of fundamental principles and general applications. Therefore only a review of the basic concepts is given in this chapter, sufficient for the reader to appreciate the principles needed to apply the concepts to power system problems.

12.2 Types of simulation

As discussed previously (Chapter 1 of this book and Chapter 13 of *Engineering Systems*), there are several types of simulation processes. They are all frequently and loosely referred to as Monte Carlo simulations (MCS). Strictly this is incorrect since MCS really relates to a process that is completely random in all respects. However, many processes are related to or with time and therefore do not possess all the random characteristics needed to use a true MCS technique. The process, however, is a stochastic process and can be analyzed using stochastic simulation.

Stochastic simulation itself can be used in one of two ways: random or sequential. The random approach simulates the basic intervals of the system lifetime by choosing intervals randomly. The sequential approach simulates the basic intervals in chronological order. The most appropriate of these two approaches depends on system effects and the objectives of the analyses.

There are some system problems for which one basic time interval has a significant effect on the next interval, and this can have a consequential significant impact on the reliability indices being evaluated. One example is the effect of hydrogeneration: the ability to use water in one interval of time can be greatly affected by how the water was used in previous intervals and the amount of rainfall and water infeed in these previous intervals. Another example is when the probability distributions of state durations and frequencies are required: these can only be evaluated explicitly if the chronology of the process is simulated. It follows from this discussion that the sequential approach will always work and the random approach is more restrictive. However, it is generally, but not universally, found that the random approach is less time consuming.

Although the above points regarding terminology are important to recognize, the term Monte Carlo simulation is used widely for all types of simulation processes and is generally understood in terms of its significance. Therefore, the term Monte Carlo simulation is used consistently in this book to refer to all simulation processes.

12.3 Concepts of simulation

The behavior pattern of n identical real systems operating in real time will all be different to varying degrees, including the number of failures, the time between failures, the restoration times, etc. This is due to the random nature of the processes involved. Therefore the specific behavior of a particular system could follow any of these behavior patterns. The simulation process is intended to examine and predict these real behavior patterns in simulated time, to estimate the expected or average value of the various reliability parameters, and to obtain, if required, the frequency/probability distribution of each of the parameters.

Some of the concepts and principles needed to achieve this can be established by considering the toss of a coin. The probability of getting a head or tail in a single throw is known to be $\frac{1}{2}$. However, this can also be estimated using the relative frequency interpretation of probability given by (see Equation (2.5) of *Engineering Systems*)

$$P\,(\text{of a head occurring}) = \lim_{N\to\infty}\left(\frac{H}{N}\right) \tag{12.1}$$

where H = number of heads
N = number of tosses

The outcomes obtained when a single coin was tossed 20 times are shown graphically in Fig. 12.1. These results indicate the following:

(a) A small number of tosses produce a very poor estimate of the probability, and therefore a large number of tosses are required.
(b) The value of probability oscillates but has a tendency toward the true value as the number of tosses is increased.
(c) The mean of the oscillations is not a good estimate of the true value since in the present case all values are equal to or greater than the true probability of getting a head.
(d) The true value sometimes occurs during the tossing process (three times in the present case), but this would not generally be known.
(e) Although it is not evident from these results, the sequence of outcomes would be different if the process was repeated, giving a completely different pattern of probabilities and a different estimate if the last value of probability in Fig. 12.1 is taken as the estimate.

All the above points are deduced from the behavior of a "real" system, i.e., the tossing of a coin. They are also pertinent in the simulation of the behavior of a real system and therefore need to be acknowledged and understood before commencing any MCS studies.

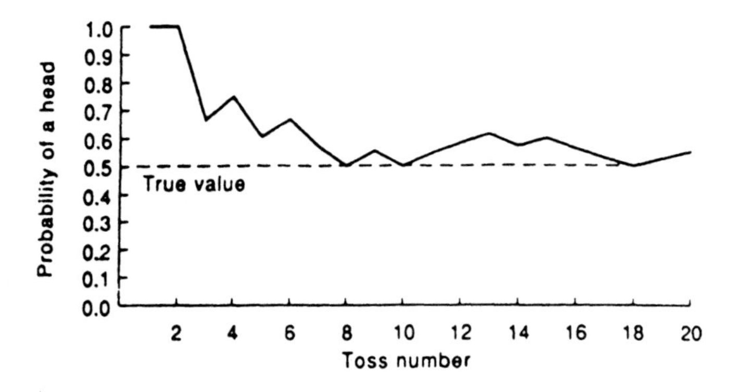

Fig. 12.1 Probability of tossing a head

12.4 Random numbers

When a real system is examined, the occurrence of events follows the inherent behavior of the components and variables contained in the system. When the system is simulated, however, the occurrence of the events depends upon models and probability distributions used to represent the components and variables. This is achieved using random numbers and converting these into density functions known to represent the behavior of the components and variables being considered. An understanding of random numbers, their generation, and conversion is therefore an essential part of MCS.

A detailed discussion and description of random numbers, their generation, and conversion is given in Chapter 13 of *Engineering Systems*. The reader is referred to that text for more details of this topic. It is sufficient to state at this point that uniform random numbers U (in the range 0 to 1) are generated computationally by a random number generator. These are then converted into values representing a nonuniform probability distribution using one of several alternative approaches.

In the case of an exponential distribution having a characteristic transition rate λ, the exponential variate T can be shown to be given by

$$T = -\frac{1}{\lambda} \ln U \tag{12.2}$$

12.5 Simulation output

Each simulation produces an estimate of each of the parameters being assessed. These estimates reflect the values of the random numbers selected for the process variables during that particular simulation. This procedure creates N estimates for each of the parameters, where N is the number of simulations performed.

There are many ways in which these estimates can be processed, including plots of the distributions such as frequency histograms or density functions, and point estimates such as means, modes, minima, maxima, and percentiles. A description of some of these is given in *Engineering Systems* together with numerical examples.

Plots of the estimates are extremely valuable and are one of the significant merits of MCS. These plots give a pictorial representation of the way the parameter can vary, including the very important tail areas, which, although perhaps occurring very infrequently, can have serious effects on the system behavior and consequences. A schematic representation of a very skewed distribution is shown in Fig. 12.2, which shows that, because of its extreme skewedness, the average value is very small and almost insignificant but that extremely high values can occur. This type of effect, which can easily occur in real systems, can be masked or ignored if only average values (or even standard deviations) are evaluated.

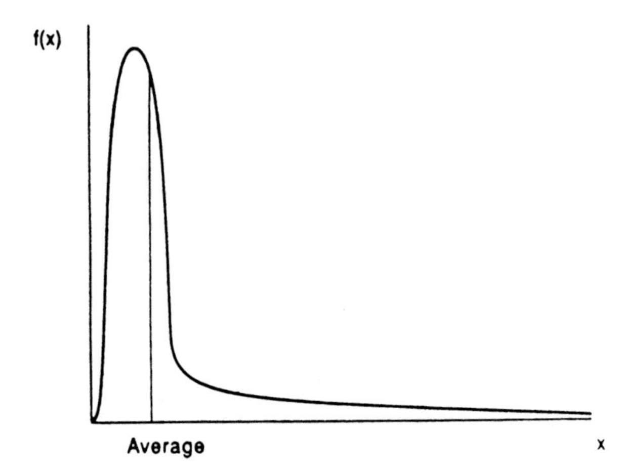

Fig. 12.2 Highly skewed probability density function

Two particular examples for which such distributions are useful are worth describing at this point. The first relates to the distribution associated with the state probabilities of a thermal generating unit. Although a discrete distribution, it is greatly skewed with the expected output generally close to its full capacity but clearly with a tail that reaches zero capacity. The second example relates to the trend in several countries which set standards for customer service. These require several things [1], one of which is that customers should not be disconnected for longer than a specified duration. An average value alone gives no indication whether this is likely or to what extent. Only a knowledge of the appropriate tail region can provide the objective information [2].

Histograms and density functions are easy to construct from the estimates obtained during the simulation process. This is described and illustrated in Chapter 13 of *Engineering Systems*. Also the expected value $E(x)$ and variance $V(x)$ of the observations can be found from:

$$E(x) = \frac{1}{N} \sum_{i=1}^{N} x_i \tag{12.3}$$

$$V(x) = \frac{1}{N-1} \sum_{i=1}^{N} (x_i - E(x))^2 \tag{12.4}$$

where N = number of simulations
x_i = value of parameter

These values only provide an estimate of the true values since, as indicated by Equation (12.1), the value of N must tend to infinity before the true value is obtained.

12.6 Application to generation capacity reliability evaluation

12.6.1 Introduction

The application of analytical techniques to generation capacity reliability evaluation is described in Chapter 2, together with a detailed discussion of concepts and a wide range of sensitivity analyses. It is not intended to repeat the discussion relating to these concepts nor to describe details of generation and load models. The reader is referred to Chapter 2 for these. Instead the intention is to illustrate three main points. The first is the general principles involved in applying MCS to the generation area, the second is to show that MCS can be applied to the same type of problem and produce the same results as the analytical approach, and the third is to demonstrate that the MCS approach can produce extended sets of results compared with the analytical method.

In order to achieve these three objectives, four types of studies are described in this chapter. These are:

(a) A reliability (LOLE) assessment of a generation system using a nonchronological load model. This is similar to the approach used in Chapter 2 and should produce similar results;
(b) A reliability (LOLE) assessment using a chronological load model. This shows the added benefits that can be achieved from MCS studies;
(c) A reliability assessment including LOEE/EENS evaluation as well as LOLE using a nonchronological load model;
(d) A similar reliability assessment as (c) but using a chronological load model.

12.6.2 Modelling Concepts

(a) *Generating unit states*

A generating unit can reside in one of a number of discrete mutually exclusive states. In the case of a two-state representation, the probabilities of residing in the up and down states are identified (see Chapter 2) as availability and unavailability (FOR) respectively. Additional state probabilities are needed to define more complex state structures.

Modelling unit states in MCS is relatively simple for a two-state unit and achieved by generating a random number U in the range (0, 1). This value of U is compared with the FOR. If $U <$ FOR, then the unit is deemed to be in the down state; otherwise the unit is deemed to be available.

This principle can be extended to any number of states. Consider a unit with one derated state and let P(down) and P(derated) be the probabilities of residing in the totally down and derated states respectively. A random number U is again generated and:

- if $U < P$(down), the unit is deemed to be in the totally down state;

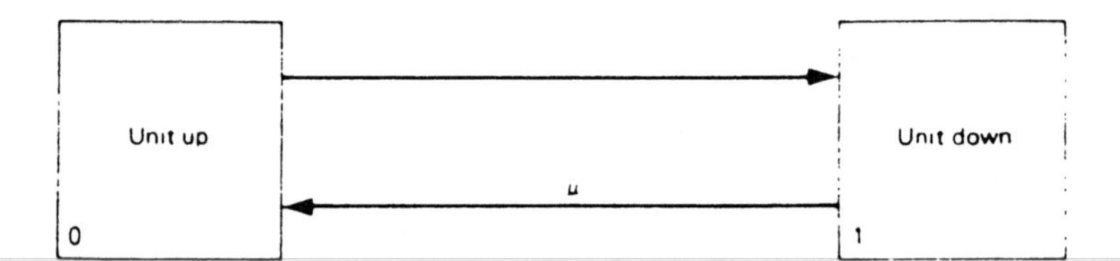

Fig. 12.3 Two-state model of generating unit

- if $P(\text{down}) < U < [P(\text{down}) + P(\text{derated})]$, the unit is deemed to be in the derated state;
- otherwise the unit is deemed to be available.

This depth of modeling is sufficient if the analysis is limited to random state modeling. However, if the chronology of the process is to be considered (i.e., in sequential MCS), then state durations are also needed. This modeling process is described below.

(b) *Duration of States*

If the duration of a state is to be sampled, then the random number generated must be transformed into time. In the case of an exponential distribution, this is given by Equation (12.2) and discussed in detail in *Engineering Systems*.

Consequently, for a two-state generating unit described by the model shown in Fig. 12.3, random values of time to failure (TTF) and time to repair (TTR) are given by

$$\text{TTF} = -\frac{1}{\lambda}\ln U_1 \tag{12.5}$$

$$\text{TTR} = -\frac{1}{\mu}\ln U_2 \tag{12.6}$$

where U_1 and U_2 are two random numbers (0, 1). A typical sequence of up-down or operating-repair cycles can be deduced by sequentially sampling a value of TTF, then TTR, then TTF, etc. This produces a sequence typically illustrated in Fig. 12.4.

This basic procedure can be extended to include any number of states [3, 4]. For instance, consider the derated state model shown in Fig. 12.5. If the sequence starts in the up state (state 1), two random numbers are generated, one determines the time to enter state 2 from 1 (TT_{12}) and the other the time to enter state 3 from 1 (TT_{13}). If $\text{TT}_{12} < \text{TT}_{13}$, then the system enters the derated state (state 2) after a time TT_{12}, but if $\text{TT}_{12} > \text{TT}_{13}$ then the system enters the down state (state 3) after a time TT_{13}. Depending on whether the system is simulated to enter state 2 or state 3, random numbers are subsequently generated to determine how long the system remains in the state, and the state which the system next encounters. This principle [3, 4] can be applied to any system containing any number of states, each one of

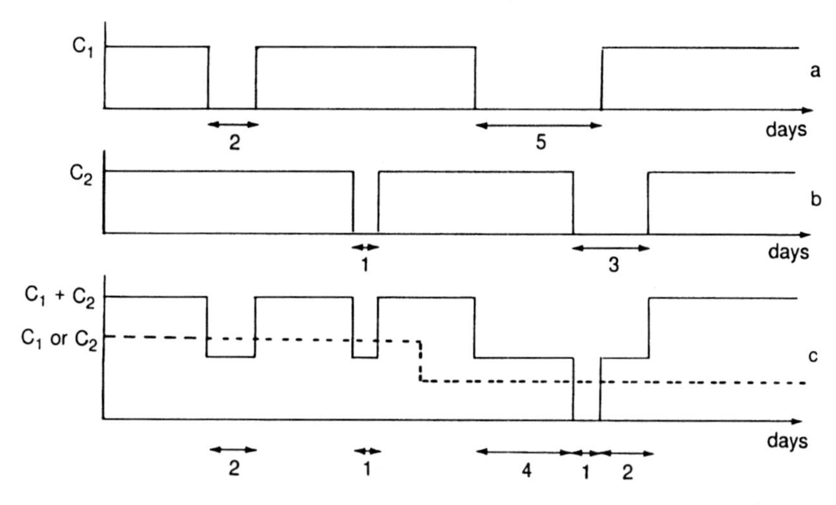

Fig. 12.4 Typical up/down sequence for two identical units; (a) unit 1, (b) unit 2, (c) combined units; —— = variation of generating capacity, - - - - - - - = variation of load

which can have any number of departure transitions. The number of random numbers that need to be generated to determine the departure from each state is equal to the number of departure transition rates from that state. This could produce typically the sequence shown in Fig. 12.6 for the state diagram shown in Fig. 12.5.

(c) *Load Model*

There are two main ways of representing the variation of load: chronological and nonchronological. Both are used in MCS.

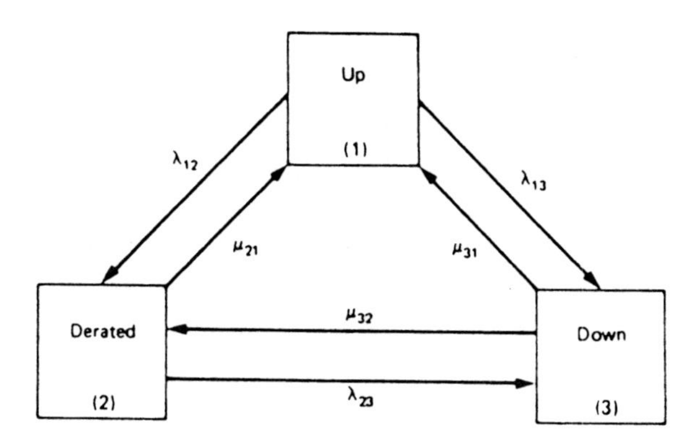

Fig. 12.5 Three-state unit model

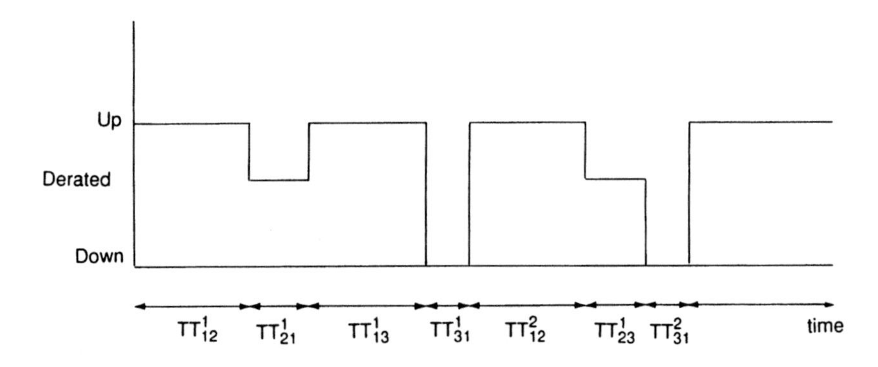

Fig. 12.6 Typical sequence of unit with derated state

The first enumerates the load levels in the sequential or chronological order in which they occur or are expected to occur. This can be on an annual basis or for any other continuous time period. This load model can be used to represent only the daily peaks, giving 365 values for any given year, or to represent the hourly (or half-hourly) values, giving 8760 (or 17,520) individual values for a given year.

The second way is to numerate the load levels in descending order to form a cumulative load model (see Section 2.3.1). This load model is known as a daily peak load variation curve (DPLVC), if only daily peaks are used, or as a load duration curve (LDC), if hourly or half-hourly loads are used. It produces characteristics similar to that of Fig. 2.4.

These models can be used in MCS using the following approaches. The chronological model is the simplest to describe since this model as defined is superimposed on the simulated generation capacity to obtain knowledge of deficiencies. The following examples illustrate this procedure. The nonchronological model is treated differently. Several methods exist, but the following is probably the most straightforward and is directly equivalent to the generating unit model described in (a).

The DPLVC or LDC is divided into a number of steps to produce the multistep model shown in Fig. 12.7. This is an approximation: the amount of approximation can be reduced by increasing the number of steps. The total time period, d_i, for which a particular load level L_i can exist in the period of interest T determines the likelihood (or probability) of L_i and an estimate for this probability is given by $d_i/T\,(=p_i)$. The cumulative values of probability are easier to use than the individual ones. These cumulative values are

$$P_1 = p_1$$

$$P_2 = p_1 + p_2$$

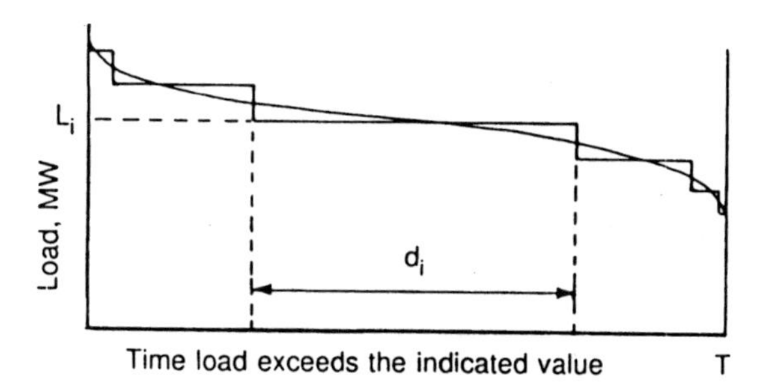

Fig. 12.7 Load model

$$P_i = \sum_{j=1}^{i} p_j$$

$$P_n = 1$$

The simulation process is as follows. A random number $U_k(0, 1)$ is generated:

- if $U_k \leq P_1$, then load level L_1 is deemed to occur;
- if $P_1 < U_k \leq P_2$, then load level L_2 is deemed to occur;
- if $P_{i-1} < U_k \leq P_i$, then load level L_i is deemed to occur;
- if $P_{n-1} < U_k \leq 1.0$, then load level L_n is deemed to occur.

This procedure is also illustrated in the following examples.

12.6.3 LOLE assessment with nonchronological load

(a) *Objective*

The present example is based on random sampling of generation and load states and therefore does not take into account the sequential variation of the load with time or the duration of the generation states. As in Section 2.6.2, the example is also based on the DPLVC and not the LDC. Consequently, neither frequency, duration, nor energy indices can be evaluated; only LOLP and LOLE (in days/year) can be assessed. This replicates the analysis performed in Chapter 2 using the analytical approach.

(b) *System studied*

The system studied is the same as that used in Section 2.3.2, i.e., a system containing five 40-MW units each with an FOR of 0.01. The system load is represented by the DPLVC shown in Fig. 2.5 having a forecast maximum daily peak load of 160 MW, a minimum daily peak load of 64 MW, and a study period of 365 days (one year). It should be noted that the straight-line model is used for illustrative purposes only and would not generally occur in practice.

The analytical LOLE result for this system is shown in Section 2.3.2 as 0.150410 days/yr.

(c) *Simulation procedure*

The process of simulation is discussed in detail in *Engineering Systems*. This process can be translated into the following steps:

Step 0 Initialize $D = 0$, $N = 0$.

Step 1 Generate a uniform random number U_1 in the interval (0, 1).

Step 2 If $U_1 <$ FOR(0.01), then unit 1 is deemed to be in the down state ($C_1 = 0$); otherwise unit 1 is available with full capacity ($C_1 = 40$). (See Section 12.6.2(a)).

Step 3 Repeat Steps 1–2 for units 2–5 (giving C_2 to C_5).

Step 4 Cumulate available system capacity, $C = \Sigma_{i=1}^{5} C_i$.

Step 5 Generate a uniform random number U_2 in the interval (0, 1).

Step 6 Compare the value of U_2 with the cumulative probabilities representing the DPLVC (see Section 12.6.2(c)). If $P_{i-1} < U_2 \leq P_i$, then load level $L = L_i$.
(In this example, the load model is a straight line and it is more convenient to calculate the load level from $L = 64 + (160 - 64)U_2$.)

Step 7 If $C < L$, then $D = D + 1$.

Step 8 $N = N + 1$.

Step 9 Calculate LOLP = D/N.

Step 10 Calculate LOLE = LOLP × 365.

Step 11 Repeat Steps 1–10 until acceptable values of LOLP/LOLE or stopping rule is reached.

(d) *Results*

The LOLE results for this example are shown in Table 12.1 and Fig. 12.8. It should be noted that this is a trivial example, and, if only the value of LOLE is required, there is no benefit to using MCS since the analytical approach is far superior. However the results are extremely useful since the two approaches can be compared very easily. Specific comments are as follows.

(i) The trend in the MCS results is dependent on the random number generator being used, and hence it is not likely that the reader will repeat these results exactly.

(ii) The results fluctuate considerably for small sample sizes but eventually settle to a value around 0.15 day/yr. An acceptable value for LOLE of 0.150482 day/yr is reached after 342,000 samples. This compares favorably with the analytical value of 0.150410 day/yr.

(iii) The number of samples required may seem very large for a small example. This is due to the small value of FOR used; as the value of FOR increases, the required number of samples decrease. In the present example, a sample size of 300,000 produced only 120 days on which a deficiency occurred.

Table 12.1 Variation of LOLE with number of simulations

Simulation	*LOLE (day/yr)*
2000	0.000000
3000	0.121626
4000	0.009120
5000	0.218956
6000	0.182470
9000	0.243306
15000	0.194654
16000	0.228111
23000	0.174558
24000	0.212908
46000	0.150758
47000	0.147550
50000	0.153297
51000	0.150291
60000	0.145998
80000	0.155123
100000	0.153300
200000	0.138700
300000	0.146000
375000	0.150420

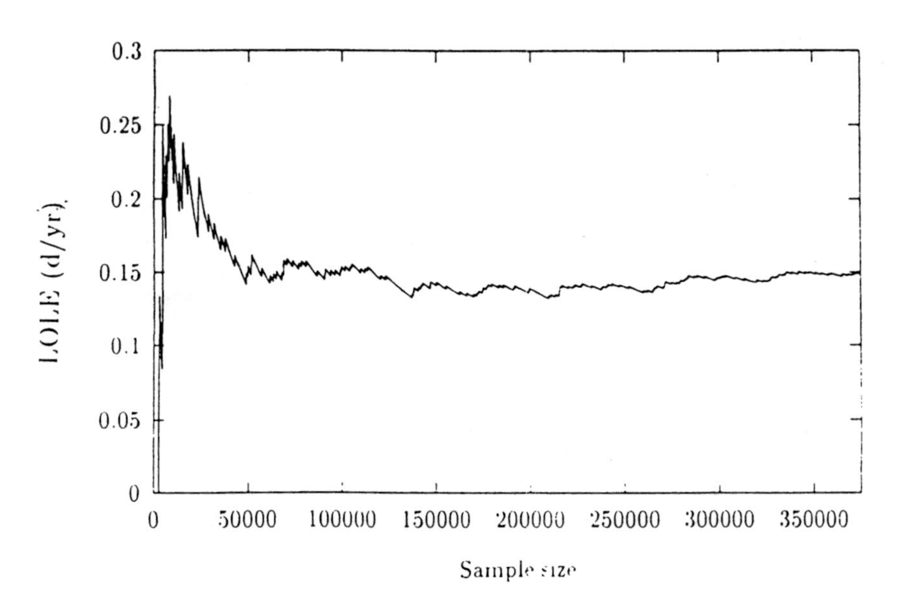

Fig. 12.8 Variation of LOLE with number of simulations

(iv) It can be seen from Table 12.1 that the MCS produced results very close to the analytical value at relatively small sample sizes (e.g., 46000, 51000, etc.). Without knowing the analytical results, this feature is also unknown and cannot be used until after the complete MCS has been done—hindsight is of no benefit during the analysis.

(v) This simple application demonstrates that MCS is a straightforward procedure, produces results that are comparable with the analytical approach, but may take significantly longer computational times to converge on an acceptable result. If only the basic assessment of LOLE is required, then the MCS approach offers no advantages over the analytical methods.

12.6.4 LOLE assessment with chronological load

(a) *Objective*

One of the main disadvantages of the analytical approach is that it is not conducive to determining frequency histograms or probability distributions, only average or expected values. Also the basic and most widely used analytical approach cannot evaluate frequency and duration indices. The present example demonstrates how the example described in Section 12.6.3 can be extended to give not only the expected number of days on which a deficiency may occur but also the distribution associated with this expected value. This was not, and could not be, done with the analytical approach and therefore illustrates one real benefit that can be achieved from MCS.

(b) *System studied*

The generating system is the same as that used in Sections 2.3.2 and 12.6.3. However, in order to generate a sequential capacity model, values of λ and μ are also required. These were chosen as $\lambda = 1$ f/yr and $\mu = 99$ rep/yr, giving the same value of FOR (0.01) as used in Section 2.6.3. Also a chronological load has not been defined for this system. Therefore a load model based on that of the IEEE Reliability Test System (IEEE-RTS) [5] has been used; this is defined in Appendix 2. A chronological or sequential daily peak load model was developed using a maximum daily peak load of 160 MW and the weekly and daily variations shown in Tables A2.2 and A2.3.

(c) *Simulation procedure*

The simulation procedure consisted of the following steps:

Step 0 Initialize $N = 0$ (N = number of years sampled).

Step 1 Consider sample year $N = N + 1$.

Step 2 Generate an up-down sequence in sample year i for each unit using the approach described in Section 12.6.2(b).

Step 3 Combine these sequences to give the generating capacity sequence for the system in sample year i.

Step 4 Superimpose the chronological load model on this sequence.

Step 5 Count the number of days d_i on which the load model exceeds the available generating capacity (d_i is then the LOLE in days for sample year i).

Step 6 Update the appropriate counter which cumulates number of sample years (frequency) in which no days of trouble (capacity deficiency), one day of trouble, two days, etc., are encountered; e.g., if $d_i = 2$, this means that two days of trouble are encountered in sample year i and the counter for number of sample years in which two days are encountered is increased by 1. (This enables frequency and hence probability distributions to be deduced.)

Step 7 Update total number of days of trouble $D = D + d_i$.

Step 8 Calculate updated value of LOLE = D/N (this gives the average value of LOLE).

Step 9 Repeat Steps 1–8 until acceptable value of LOLE or stopping rule is reached.

A very simple illustrative example is shown in Fig. 12.4. This indicates a typical up-down sequence for two units, the combined sequence of available generating capacity and the effect of superimposing a load model having a daily peak load level L_1 for x days and L_2 for y days. The result is that the system encounters four days of trouble. If a further nine sample years produced 1, 0, 2, 0, 0, 1, 3, 0, 1 days,

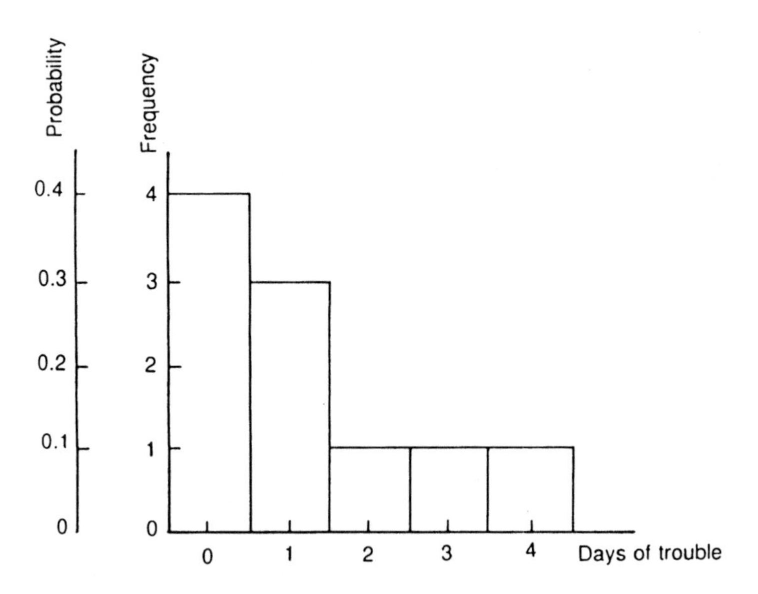

Fig. 12.9 Typical frequency histogram/probability distribution

then LOLE = 13/10 = 1.3 days/yr, and the frequency histogram/probability distribution is shown in Fig. 12.9.

(d) *Results*

The results for this example are shown in Figs. 12.10 and 12.11 and in Table 12.2.

Figure 12.10 illustrates the variation of LOLE as the number of sample years is increased. This indicates that LOLE settles to an acceptably constant value of 0.1176 day/yr after about 5000 sample years. It is interesting to note that in the previous case (Fig. 12.8) the value of LOLE first rapidly increased followed by a gradual decay, whereas in the present case (Figure 12.10) the value of LOLE essentially follows an increasing trend. These differences have little meaning and are mainly an outcome of the random number generating process.

Table 12.2 and Fig. 12.11 both show the frequency (histogram) or probability distribution of the days on which trouble may be encountered. Several important points can be established from, and should be noted about, these results.

(i) The results do not imply that 5000 years have been studied because this would have no physical or real meaning; instead the same year has been sampled 5000 times, thus creating an understanding of not only what may happen to the real system in that time but also the likelihood of these alternative scenarios.

(ii) On this basis, the results indicate that the most likely outcome (prob = 4723/5000 = 0.9446) is to encounter no trouble, but at the other extreme

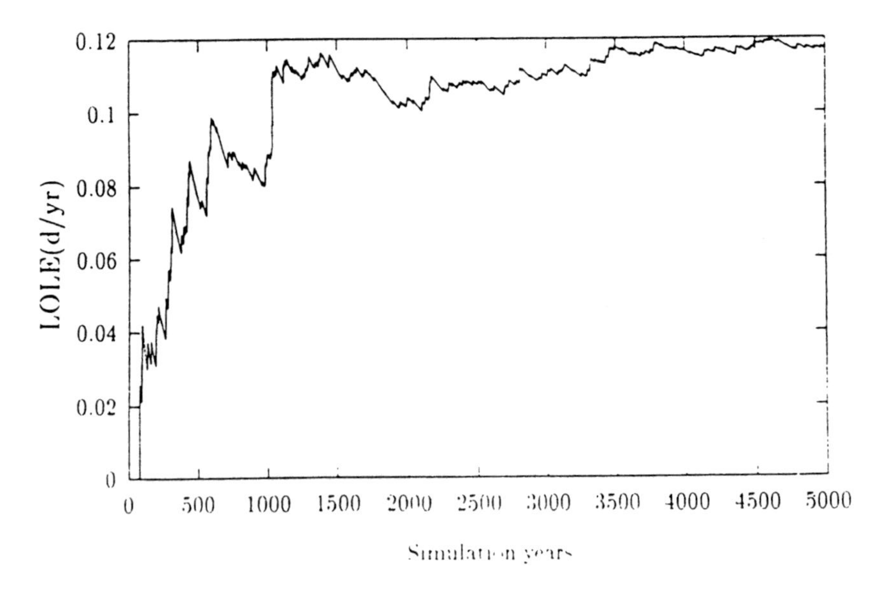

Fig. 12.10 Variation of LOLE with number of simulations

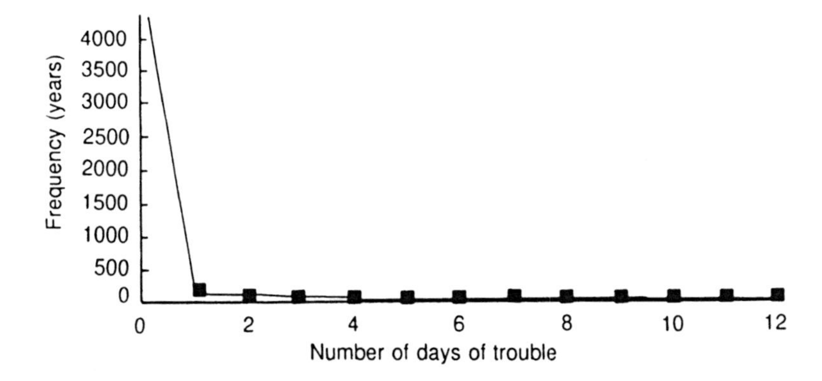

Fig. 12.11 Frequency histogram/probability distribution of days of trouble

it is possible to encounter nine days of trouble with a probability of 1/5000 = 0.0002.

(iii) An alternative interpretation of the same results could indicate that, if 5000 identical systems were operated under the same conditions. 4723 would experience no trouble, 123 would experience one day of trouble, etc.

(iv) The results are typical of many power system reliability problems. Because the system is "very" reliable, the probability distribution is very skewed, the average value is very close to the ordinate axis, and the very large extreme values are masked by the high degree of skewness. Average values only can give a degree of comfort to system planners and operators which may not be warranted. It also makes it very difficult to compare

Table 12.2 Frequency of number of days of trouble

Days of trouble	*Frequency (years)*
0	4723
1	123
2	83
3	29
4	16
5	16
6	5
7	3
8	1
9	1
10	0

calculated reliability results with specified deterministic criteria. For instance, an average value may be less than a specified criterion, but this does not necessarily mean that a system performs satisfactorily.

12.6.5 Reliability assessment with nonchronological load

(a) *Objective*

This example is intended to indicate how the basic example considered in Section 12.6.3 can be extended to produce energy-based indices in addition to load-based ones, i.e., LOEE as well as LOLE. Again a nonchronological load model is considered, but this time it is assumed that the model is a LDC and not a DPLVC. This is essential because the area under a LDC represents energy, while that under a DPLVC does not.

(b) *System studied*

The system is the same as that used in Section 12.6.3 except that the load model is a LDC with a maximum peak load of 170 MW and a minimum load of 68 MW (=40% as before). In this case the results for LOLE are in hr/yr and it is assumed that the load remains constant during each hour simulated. The last assumption enables energy to be evaluated when the magnitude of the deficiency in MW is known.

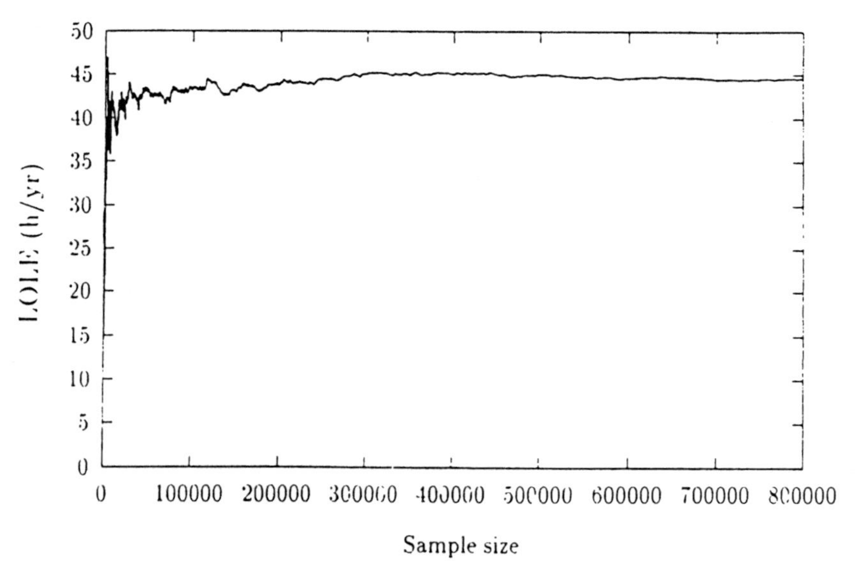

Fig. 12.12 Variation of LOLE with number of simulations

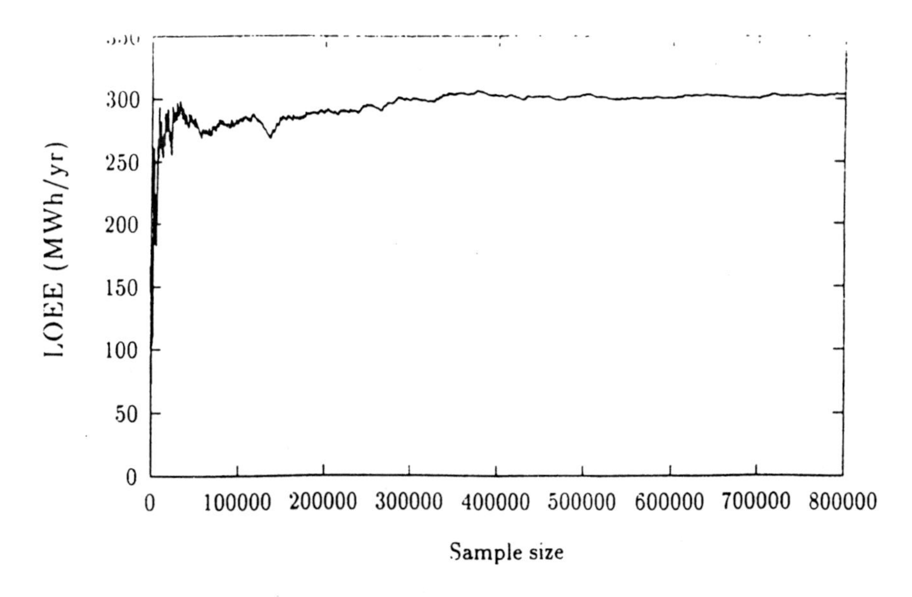

Fig. 12.13 Variation of LOEE with number of simulations

(c) *Simulation procedure*

The procedure is essentially the same as that described in Section 12.6.3(c), with the following steps modified.

Step 0 Initialize $H = 0$, $N = 0$, $E = 0$ (H = hours of trouble, N = hours simulated, E = energy not supplied).

Step 7 If $C < L$, then $H = H + 1$ and $E = E + (L - C)$.

Step 9 Calculate LOLP = H/N.

Step 10 Calculate LOLE = LOLP × 8760 and LOEE = $E \times 8760/N$.

(d) *Results*

The results for LOLE and LOEE are shown in Figs. 12.12 and 12.13 respectively. The values of LOLE and LOEE reached acceptably constant levels after about 800,000 simulations of 44.59 hr/yr and 303.8 MWh/yr respectively.

12.6.6 Reliability assessment with chronological load

(a) *Objective*

This example is intended to indicate how energy-based indices can be evaluated using the chronological load model, including average values and the frequency histograms and probability distributions.

(b) *System studied*

The system is the same as that used in Section 12.6.4 except that the peak load is 170 MW and the chronological load model is represented on an hourly basis using Table A2.4 in addition to Tables A2.2 and A2.3.

(c) *Simulation procedure*

The procedure is essentially the same as that described in Section 12.6.4(c), except that energy as well as time should be counted and cumulated. In addition, the frequency of encountering trouble can be assessed, together with indices expressed on a per-interruption basis as well as on an annual basis. These additional concepts are better described by way of an example rather than simply as a step-by-step algorithm.

Consider the simple chronological example illustrated in Fig. 12.8. This represents one particular sample year and is redrawn in Fig. 12.14 to include additional information such as magnitude of load levels and energy not supplied. For this particular year i:

(i) Annual system indices

Frequency of interruption FOI

$$= \text{number of occasions load exceeds available capacity}$$
$$= 3$$
$$\text{LOLE} = \Sigma \text{ individual interruption durations}$$
$$= 96 \text{ hr}$$
$$\text{LOEE} = \Sigma \text{ energy not supplied during each interruption}$$
$$= 2160 \text{ MWh}$$

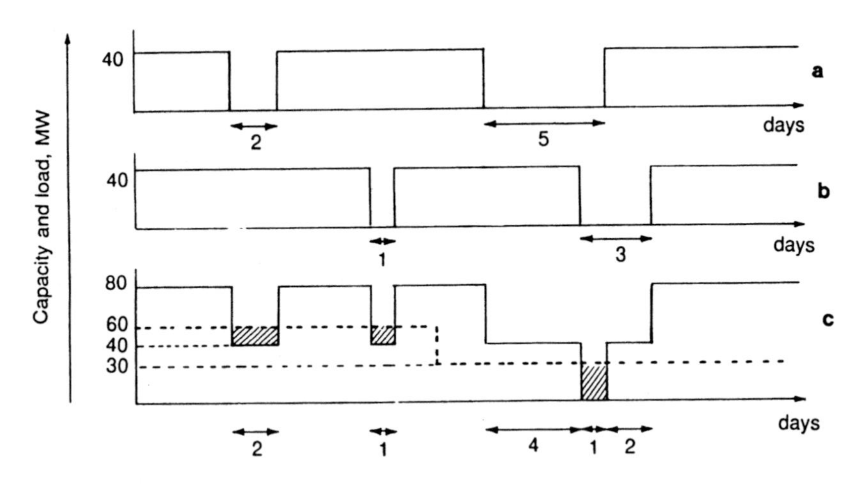

Fig. 12.14 Typical up/down sequence showing energy not supplied: (a) unit 1, (b) unit 2, (c) combined units; shaded areas = energies not supplied

Other indices such as EIR, EIU, system minutes, etc., can be determined from these system indices.

(ii) Interruption indices

Duration of interruption DOI = LOLE/FOI hr/int
= 32 hr/int

Energy not supplied ENSI = LOEE/FOI MWh/int
= 720 MWh/int

Load-curtailed LCI = LOEE/LOLE MW/int
= 22.5 MW/int

These system and interruption indices can be calculated individually for each sample year to give the frequency histograms and probability distributions. They can also be cumulated and then divided by the number of sample years (*N*) to give average or expected values. This process, by way of example, replaces Steps 5–8 of the procedure described in Section 12.6.4.

(d) *Results*

It is evident from the above description that this sequential or chronological approach to MCS can produce an extensive set of indices compared with the restricted set given by the state sampling approach. Whether these are required depends on the application, and it is not appropriate to be prescriptive in this

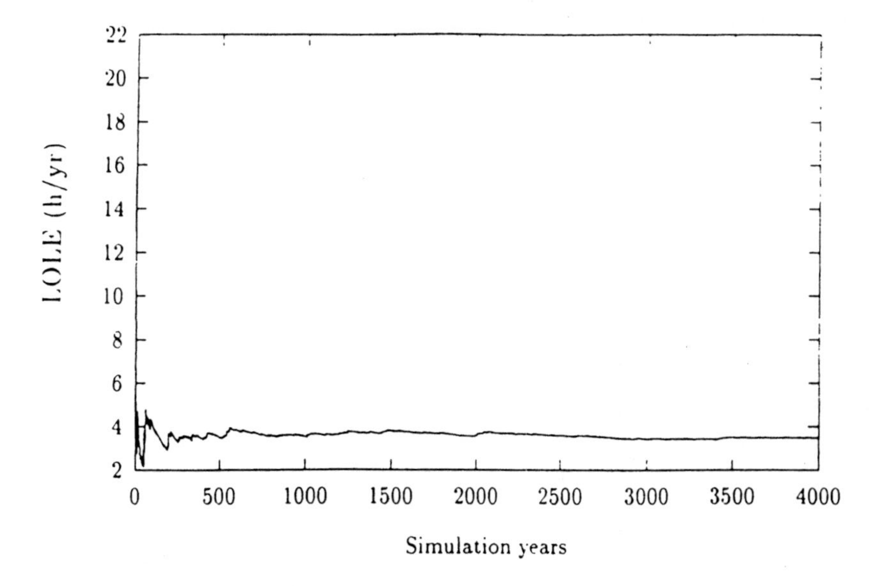

Fig. 12.15 Variation of LOLE with number of simulations

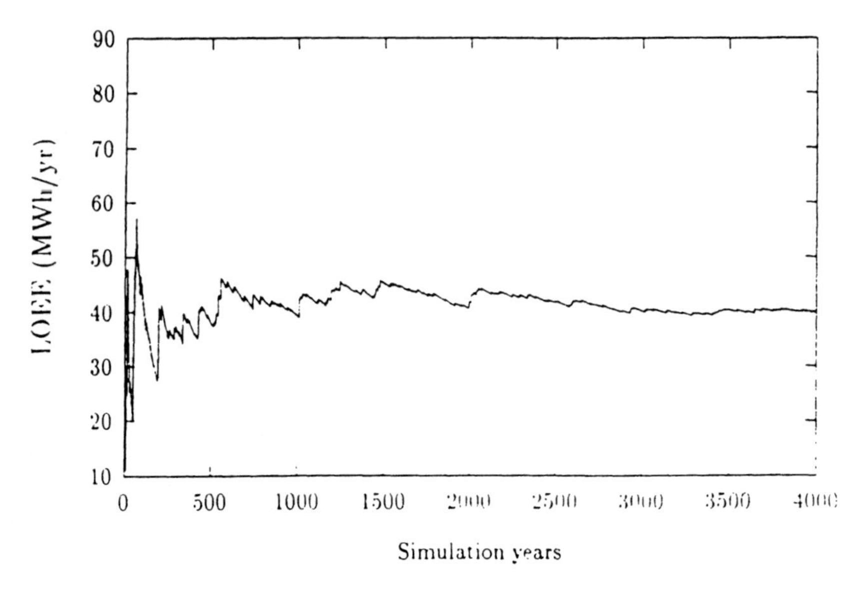

Fig. 12.16 Variation LOEE with number of simulations

teaching text. Many examples exist in which both the state sampling and sequential sampling approaches have been applied [6–8].

In the present case, typical results are shown in Figs. 12.15 and 12.16 for the variation of LOLE and LOEE respectively as a function of simulation years, in Table 12.3 and Fig. 12.17 for the frequency distribution of LOLE, and in Table 12.4

Table 12.3 Frequency of loss of load interval

Loss of load interval (hr)	*Frequency (years)*
0–10	3558
10–20	241
20–30	82
30–40	47
40–50	26
50–60	15
60–70	11
70–80	9
80–90	3
90–100	1
100–110	2
110–120	2
120–130	3
130–140	0

Table 12.4 Frequency of loss of energy intervals

Loss of energy interval (MWh)	*Frequency (years)*
0–100	3732
100–200	58
200–300	55
300–400	36
400–500	26
500–600	21
600–700	17
700–800	10
800–900	12
900–1000	4
1000–1100	5
1100–1200	4
1200–1300	6
1300–1400	1
1400–1500	0
1500–1600	2
1600–1700	0
1700–1800	1
1800–1900	2
1900–2000	2
2000–2100	4
2100–2200	2
2200–2300	0

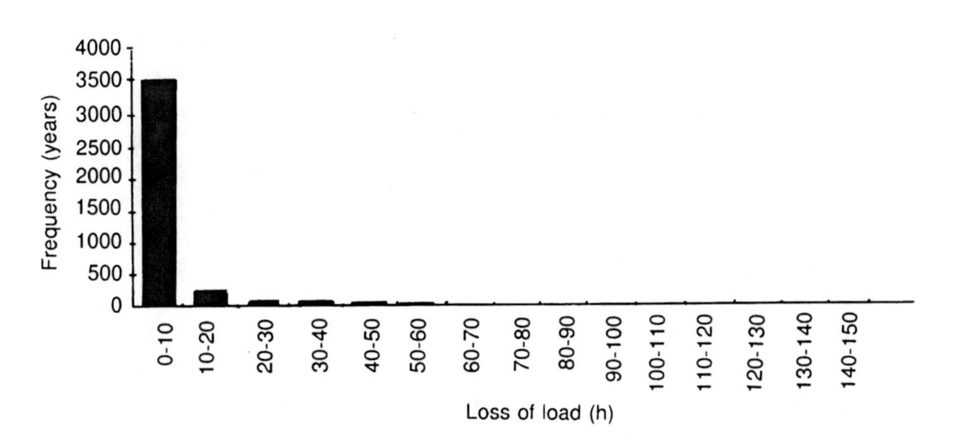

Fig. 12.17 Frequency histogram for LOLE

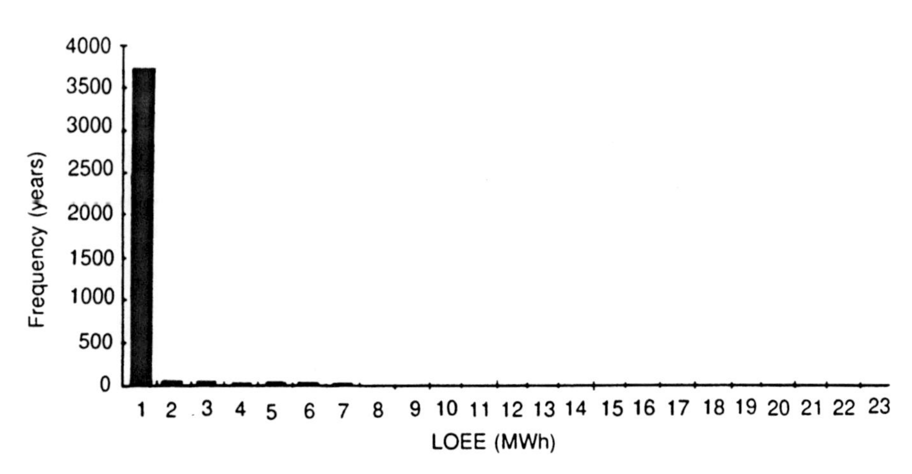

Fig. 12.18 Frequency histogram for LOEE

and Fig. 12.18 for the frequency distribution of LOEE. The latter two histograms should be compared with the average values of LOLE and LOEE of 3.45 hr/yr and 39.86 MWh/yr respectively. In both cases, it is seen that the maximum values encountered were 40–50 times the average values.

12.7 Application to composite generation and transmission systems

12.7.1 Introduction

The application of analytical techniques to composite systems is described in detail in Chapter 6. The approach examines the adequacy of system states; the basic principle of this being achieved by selecting a state deterministically, evaluating the probability, frequency, and duration of the state to give state indices using rules of probability, assessing the adequacy or inadequacy of the state to give severity indices using an appropriate load flow or state assessment algorithm, and combining probabilistically the state indices with the severity indices to give the overall system and load point reliability indices.

In principle, this logical process does not change when using the MCS approach. The major difference is that, with the analytical approach, states are selected generally on the basis of increasing level of contingency (i.e., first-order outages, second-order, etc.), and each state is selected at most only once. With the MCS approach, states are selected using random numbers similar to the procedure described in Section 12.6.2 for generating systems, and each state may be selected and analyzed several times: in fact the likelihood of a state is calculated on the basis of the number of times it is selected by the random number process since the most likely events are selected more frequently. While in one of the system states selected by the simulation process, the adequacy assessment is frequently identical or similar to that used in the analytical approach.

The foregoing discussion relates to the basic principle of the assessment procedure. However, it should be noted that this principle can be significantly extended and that any system parameter can be treated as a random variable, the value of which can be selected using a random number generator. Some of these extended considerations are commented on in Section 12.7.4, but the reader is referred to other in-depth considerations [9–18] for more details and applications. Instead this section is intended to illustrate the basic principles and concepts of applying MCS to the assessment of adequacy in composite systems.

12.7.2 Modelling concepts

The concepts used to model the system components of a composite system (i.e., generators, lines, transformers, etc.) are essentially the same as those used for generators in Section 12.6.2. Two specific approaches can be used as in the previous case: random state sampling and sequential simulation.

The random state sampling procedure is identical to that described in Section 12.6.2(a). The data required are the availabilities and unavailabilities of each two-state system component to be modeled or the availabilities of all states of a multistate component. Sufficient random numbers are generated in order to deduce the states in which each component resides at random points of time. The contingency order depends on the number of components found to be in a failure state on that occasion. This approach has the same merits and demerits as before; it enables load-based indices to be assessed but not frequency, duration, or energy-based ones.

The sequential simulation approach uses the same procedure as described in Section 12.6.2(b). A particular sequential behavior for each system component (up-down cycles) is deduced for a convenient period of time, perhaps one year, or even longer if it is necessary to include some events which are known to occur less frequently, such as scheduled maintenance. The frequency and duration of single and multiple contingencies can be deduced by combining the sequential behavior of all system components and identifying single and overlapping events.

Finally the load can be modeled using the procedures described in Section 12.6.2(c); the DPLVC or LDC for nonchronological analyses and the daily or hourly sequence of loads for chronological analyses. The only difference in this case is that a suitable load model is required for each load point in the system rather than the global or pooled load used in generating capacity assessment.

12.7.3 Numerical applications

A limited number of numerical examples are considered in this section in order to illustrate the application of MCS to composite systems. Only the basic applications are considered, and the reader is referred to the extensive literature on the subject to ascertain knowledge of more detailed applications. Further discussion of this is given in Section 12.7.4.

The concepts are applied to the three-bus, three-line system shown in Fig. 6.2 and analyzed using analytical techniques in Section 6.4. Several individual case studies have been made. The reader is referred to Chapter 6 for the generation, network, and load data.

Case A. A DPLVC is considered having a peak load of 110 MW and a straight-line load curve from 100% to 60% as defined in Section 6.4. Using the random-state sampling procedure, the variation of LOLE with sample size is shown in Fig. 12.19. A reasonable final value for LOLE is found to be 1.2885 day/yr, which compares with the analytical result (Section 6.4) of 1.3089 day/yr.

Case B. A similar study was made but this time considering a LDC having a peak load of 110 MW and a straight-line load curve from 100% to 40%. The variation of LOEE with sample size is shown in Fig. 12.20. This shows that a reasonable final value of LOEE is 265.4 MWh/yr compared with the analytical result (Section 6.4) of 267.6 MWh/yr.

Case C. This study replicates that used to obtain the results given in Tables 6.10 and 6.11; i.e., the load remains constant at 110 MW, transmission constraints are not considered, frequency calculations consider line departures only, and the generation is considered as a single equivalent unit. The MCS results obtained after 300,000 samples are compared with the analytical results (Tables 6.10 and 6.11) in Table 12.5.

Case D. This study replicates that used to give the results in Table 6.13; i.e., the load remains constant at 110 MW, transmission constraints are not considered,

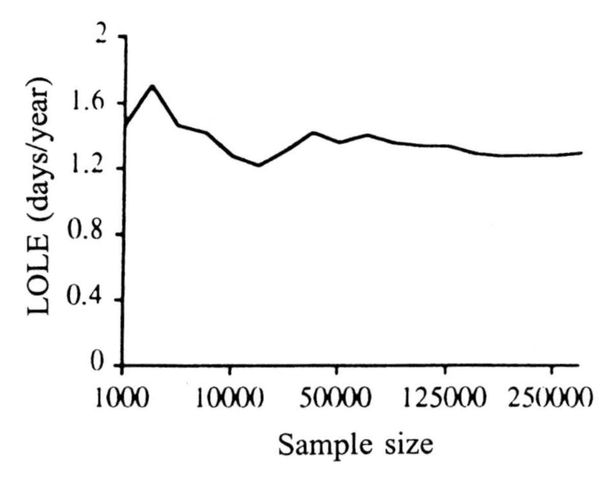

Fig. 12.19 Variation of LOLE with number of simulations

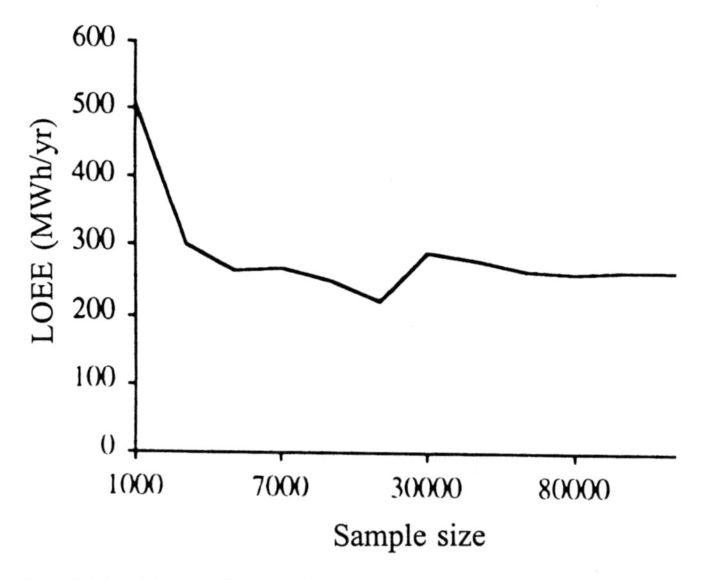

Fig. 12.20 Variation of LOEE with number of simulations

generating unit and line departures are considered, and generating units and lines are considered as separate components. The MCS results obtained after 300,000 samples are also compared with their equivalent analytical results (Table 6.13) in Table 12.5.

12.7.4 Extensions to basic approach

The results provided in Section 12.7.3 illustrate the application of MCS to composite systems at the basic level. In reality, this approach can be and has been extended in a number of ways [3, 4, 19–26]. These include the following considerations and features:

(a) An increased number of indices can be calculated, including all the load point and system indices defined and described in Section 6.6.

Table 12.5 Comparison of MCS and analytical results

	MCS results		*Analytical results*	
Case study	*Probability*	*Frequency (occ/yr)*	*Probability*	*Frequency (occ/yr)*
C	0.09789	2.4795	0.09807	2.4259
D	0.09859	9.3681	0.09783	9.1572

(b) Not only average values but also the underlying probability distributions can be assessed. These generally cannot be evaluated using the analytical approach.
(c) Various stopping rules can be applied to determine when to cease the simulation process as indicated in *Engineering Systems*.
(d) Variance reduction techniques can be applied in order to obtain convergence of the results with fewer iterations [8, 20].
(e) An increased number of system effects can be assessed, including common mode failures [3, 4] and weather-related effects [22, 26].
(f) The examples given in Section 12.7.3 only considered random state sampling. Extensive use has been made of sequential simulation [24, 25] in order that additional indices such as frequency, duration, and energy can be assessed together with their probability distributions, and the effect of chronological events can be realistically considered. The latter is particularly important in the case of hydro-systems in which reservoir capacity is limited, rainfall is very variable, hydro is the dominant source of energy, and/or pumped storage is used.

12.8 Application to distribution systems

12.8.1 Introduction

The application of analytical techniques to distribution systems and to electrical networks when the generation sources are neglected is described in detail in Chapters 7–10. It will be recalled that the techniques are mainly based on a failure modes and effects analysis (FMEA), using minimal cuts sets and groups of equations for calculating the reliability indices of series and parallel systems. Generally only expected values of these indices can be calculated. If only the expected values are required, then there is little or no benefit to using any method other than this analytical approach for analyzing distribution systems. However, there are several instances when this is not the case, and a short discussion of these instances is useful.

In many cases, a decision regarding the benefit between alternative planning or operational decisions can be easily made if the expected or average values of a parameter are known. In some cases, this is not possible, and knowledge of the distribution wrapped around the average value is of great benefit. One particular case is when target performance figures are set. Knowledge that the average value is less than such a target figure is of little significance; instead knowledge of how probable it is that the target figure will be exceeded is of much greater importance. This can only be deduced if the probability distribution is calculated. The objective would then be to minimize this value of probability commensurate with the cost of achieving it. An example of this is in the United Kingdom where, depending on their peak demand, customers should be reconnected following an interruption within 15 min, 3 h, or 24 h [1]. Failure to do so in the latter case involves penalty payments to the affected customers. Knowledge of the outage time distribution then becomes important.

A second instance is in the case of evaluating customer outage costs. This is described in Chapter 13, where it is seen that these costs are a function of the outage time. A knowledge of the outage time distribution is therefore a valuable piece of information needed to calculate these outage costs. Using the average value of outage time only can produce erroneous estimates of these outage costs since the cost is a nonlinear function of duration.

There are many other examples that can be quoted, including number of interruptions greater than certain specified levels, amount of energy not supplied greater than specified values, etc. In addition, all of these parameters may be required at individual load points, for groups of load points, for complete feeders, areas, etc. Although analytical techniques can be used to evaluate these distributions under certain assumptions and conditions [27, 28], the most suitable method of analysis is based on simulation approaches [2, 29].

12.8.2 Modeling concepts

The basic indices required to assess the reliability of distribution networks are the failure rate (λ), average down time (r), and annual outage time or unavailability (U) of each load point. These can then be extended to evaluate load and energy indices at each load point, and all indices for groups of load points, complete feeders, areas, etc. These concepts are described in Chapters 7–9.

The principle for calculating the same sets of indices using a simulation approach is described in Chapter 13 of *Engineering Systems*. This principle centers on randomly sampling up times and down times of each component to produce a simulated sequence of component up times and down times. Sufficient sequences are simulated to produce a representative picture of the overall system behavior. Although the underlying concepts are the same for radial and meshed (or parallel) networks, small differences exist in the simulation processes for each because the failure of a single component causes problems in radial systems, whereas overlapping failures are generally dominant in meshed or parallel systems. These differences are highlighted in the following.

(a) Radial systems

The types of systems being considered are those previously discussed in Chapter 7, i.e., simple radial systems feeding several load points via a main feeder consisting of several sections and lateral distributors. The following description can be easily extended to cover more general radial networks consisting of an increased number of branches.

The basic algorithm is described in Section 13.5.11 of *Engineering Systems*, i.e.:

Step 1 Generate a random number.

Step 2 Convert this number into a value of up time using a conversion method on the appropriate times-to-failure distribution of the component.

Step 3 Generate a new random number.

Step 4 Convert this number into a value of repair time using a conversion method on the appropriate times-to-repair distribution.

Step 5 Repeat Steps 1–4 for a desired simulation period. In order to obtain distributions this should be for a period of time which is able to capture the outage events to be considered. For radial systems, a period of one year is usually sufficient. However, consider a general period of n years.

Step 6 Repeat Steps 1–5 for each component in the system.

Step 7 Repeat Steps 1–6 for the desired number of simulated periods.

These steps create the scenarios from which the load point reliability indices can be deduced. The principles of the subsequent procedure are as follows:

Step 8 Consider the first simulated period lasting n years.

Step 9 Consider the first component (feeder section or lateral distributor).

Step 10 Deduce which load points are affected by a failure of this component.

Step 11 Count the number of times this component fails during this period. Let this be N. The failure rate (λ) is approximately equal to N/n. [This is strictly frequency and a better estimate would be given by dividing N by the total up time (i.e., Σ TTF used in *Engineering Systems*) rather than n, but the difference is generally negligible.]

Step 12 Evaluate the total down time of the load point. This will be equal to the total down time (repair time) of the component if it cannot be isolated and the load point restored to service by switching. If the component can be isolated, the down time of the load point is the total time taken to restore the load point by switching. This latter value can be considered either as a deterministic value of time or itself sampled from an assumed switching time distribution. In either case, define the value as Σ TTR, total time taken to restore the load point in the n-year period. Then the average down time (r) is Σ TTR/N.

Step 13 The annual unavailability (U) is given by the product λr.

Step 14 Steps 8–13 creates one row in the FMEA tables shown in Chapter 7 (e.g., row 1 of Table 7.8) for one simulated period.

Step 15 Repeat Steps 9–14 for each system component to produce the complete FMEA table.

Step 16 From this FMEA table, calculate the values of reliability indices at each load point and for the system for one simulated period using Equations (7.1)–(7.14). This set of indices represents one point on each of the probability distributions.

Step 17 Repeat Steps 8–16 for each of the simulated periods. This produces a series of individual points from which the complete probability distributions can be determined. Generally these distributions are calculated and plotted as frequency histograms or probability distributions using the principles of classes and class intervals described in Section 6.15 of *Engineering Systems*.

Step 18 The average value, standard deviations, and any other desired statistical parameter of these distributions can then be evaluated.

There are various alternative ways in which the foregoing procedure can be implemented, but the essential concepts remain. The point to note, however, is that the procedure allows distributions to be deduced. If only overall average values are required, then only one simulated period is needed, but then n (number of simulated years) must be great enough to give confidence in the final results. However, such average values can generally be best evaluated using the analytical approach.

(b) *Meshed or parallel systems*

The systems considered in this section are of the form discussed in Chapters 8–10, and can include substations and switching stations. The principle is very similar to that described for radial systems except that there is now a need to consider overlapping outages in the FMEA procedure. Therefore combinations of components and their effect on a load point must be considered in addition to the first-order events described before. The steps in the previous algorithm can therefore be modified to include not only first-order events leading to an outage of a load point but also combinations. The principle of this was described in Section 13.5.11 of *Engineering Systems* and, in particular, the sequence shown in Fig. 12.21 (reproduction of Fig. 13.13 of *Engineering Systems*).

An alternative approach, and one that is similar in concept to that used in Chapters 8–10, is to consider one load point at a time, deduce the minimal cut sets (failure events) and simulate the times-to-failure and times-to-repair (or restoration) for each of these, one at a time. The main difference between this approach and the previous algorithm is that previously one component and all load points are considered simultaneously, while this latter approach considers one load point and all components simultaneously. The final result should be the same.

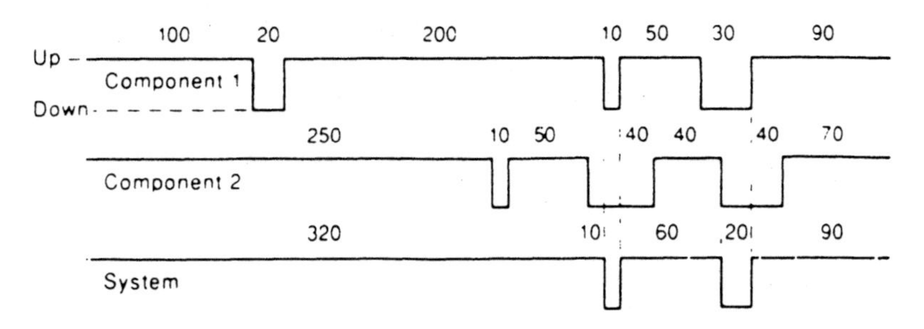

Fig. 12.21 Typical operation/repair sequences of a two-component system

It will be recalled that different failure modes and restoration procedures were described and analyzed in Chapters 8–10. These have been included in the simulation procedure as well.

12.8.3 Numerical examples for radial networks

In order to illustrate the type of results that can be obtained using MCS and the benefit of the increased information, a series of studies on the systems previously analyzed in Chapter 7 have been performed. Three such case studies have been conducted, namely Case 1, Case 3, and Case 5 described in Table 7.15. In brief these are:

Case 1— base case shown in Fig. 7.4;

Case 3— Case 1, but with fuses in the laterals and disconnects in the feeder as shown in Fig. 7.5;

Case 5— Case 3, but with an alternative supply as shown in Fig. 7.7.

A brief discussion of the results is as follows.

(a) *Case 1*

The network is solidly connected, and therefore any failure affects all load points and repair of the failed component is required before any load point can be restored to service. Therefore all load points behave identically. A set of histograms (λ, r, U, SAIFI, SAIDI, CAIDI, AENS) for this case is shown in Fig. 12.22.

It can be seen that the average values of the indices compare favorably with those obtained using the analytical approach (Table 7.15). However, it can also be seen that considerable dispersion exists around these average values, with most distributions being skewed quite significantly to the right. The implication of this skewness is that, although an average value may seem to be acceptable as a value in its own right, there is a significant probability in many cases for values much greater than that to occur. Relying only on the average value can therefore be delusory with respect to the system performing as required and against specified targets.

In practice, these distributions can be used to determine the likelihood of a parameter being greater than a particular value. A decision would then be made whether a particular reinforcement scheme reduces this likelihood to an acceptable level and at what cost. The effect on likelihood is illustrated in Cases 3 and 5. The effect on customer outage cost is illustrated in Chapter 13.

(b) *Case 3*

The reliability of the system is improved by including fuses in the laterals and disconnects in the feeder sections. This is demonstrated in Chapter 7 on the basis of average values only, where it was observed that these reinforcements had a significant effect on all load points, with the greatest benefit derived by load point A and less so by load point D (see Table 7.15). This effect is clearly seen by

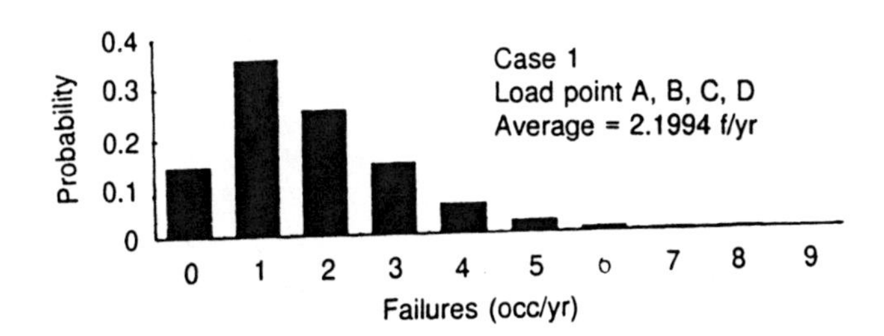

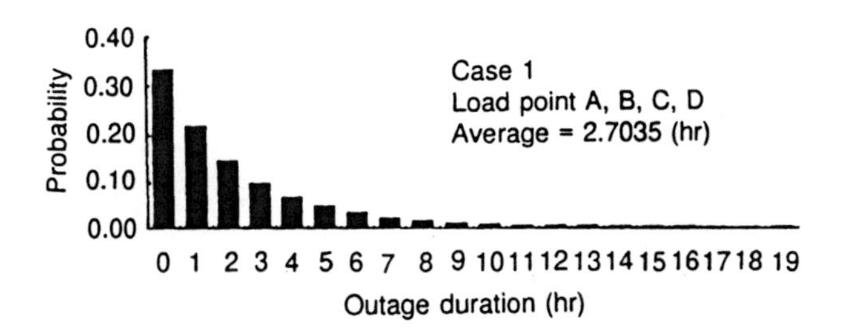

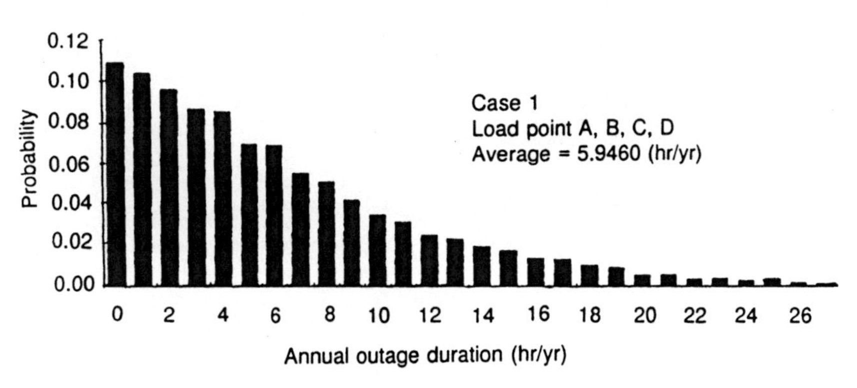

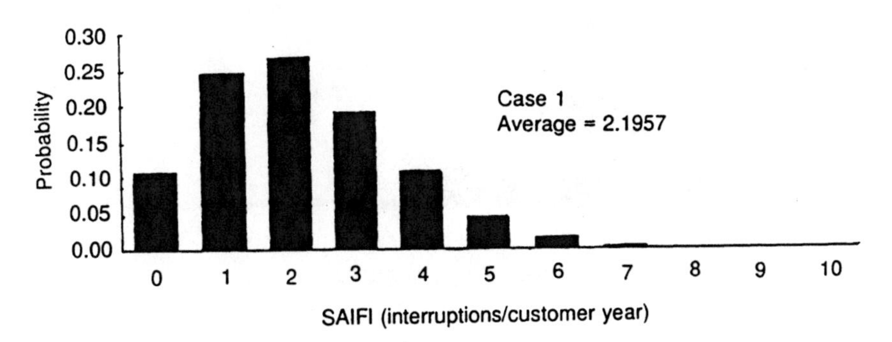

Fig. 12.22 Results for Case 1

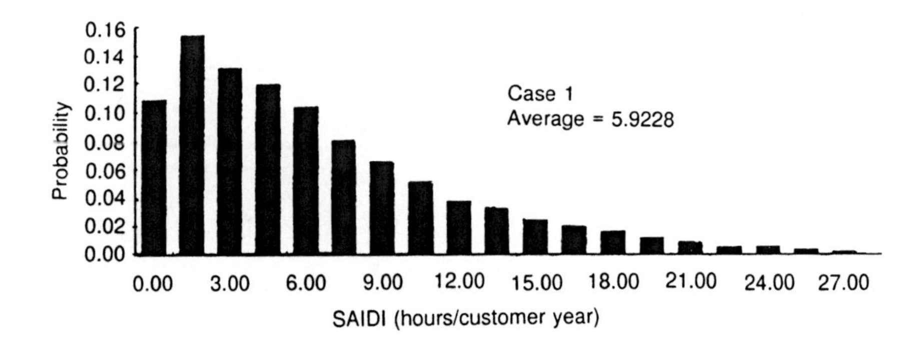

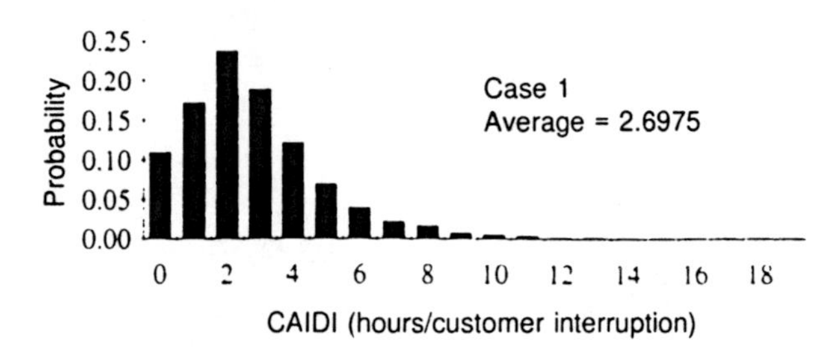

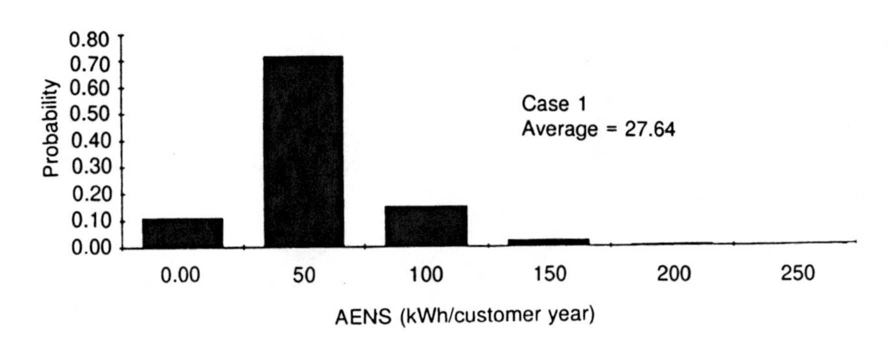

Fig. 12.22 Continued

comparing the histograms of average outage time (r) and annual outage duration (U) for load points A and D shown in Fig. 12.23. This figure also includes the histograms of some of the system indices, from which the effect of reduced load point outage durations on SAIDI in particular is very evident.

(c) *Case 5*

In Chapter 7 it was observed that the effects of an alternative supply or backfeeding had a considerable benefit on the outage time of load point D but no effect on that of load point A. This is also reflected in the relevant histograms, some of which are shown in Fig. 12.24, in which those associated with the outage time of load point A are the same as for Case 3, but those for load point D and the system duration indices of CAIDI and SAIDI change.

12.8.4 Numerical examples for meshed (parallel) networks

The dual-transformer feeder network shown in Fig. 8.1 is used in Chapter 8 to illustrate the application of reliability assessment to meshed networks using analytical techniques. A range of studies starting from basic assessments to the incorporation of multi-failure and outage modes and the effect of weather were discussed. It is possible to replicate the same range of studies using MCS and produce similar average values as well as probability distributions. There is no benefit to be derived from this since the merits of evaluating probability distributions have been demonstrated in Section 12.8.3 using the radial networks. Therefore the only benefits at this point are to demonstrate that the theoretical concepts described in Section 12.8.2 can be used to obtain expected values and probability distributions and to produce an illustrative set of examples to indicate the shape of these distributions.

In order to do this, consider the dual transformer feeder shown in Fig. 8.1 and the reliability data shown in Table 8.1. Using the principles described in the algorithm of Section 12.8.2, the following results are obtained for two case studies, assuming:

Case 1— the repair times are exponentially distributed with the average values shown in Table 8.1;

Case 2— the repair times are lognormally distributed with the average values shown in Table 8.1 and standard deviations equal to one sixth of the average value.

The load point indices (λ, r, U) obtained using 10,000 simulation years are shown in Table 12.6, together with the equivalent analytical values (previously given in Table 8.2). In addition, the probability distributions for the load point down times are shown in Fig. 12.25.

It can be seen from Table 12.6 that the average values of the load point indices obtained from MCS compare very favorably with those obtained from the analytical approach.

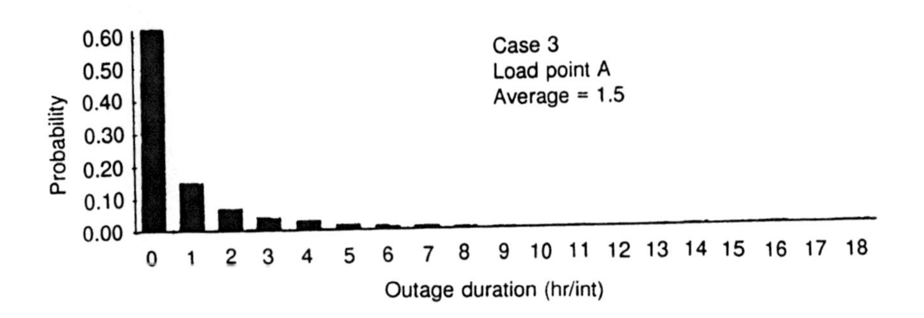

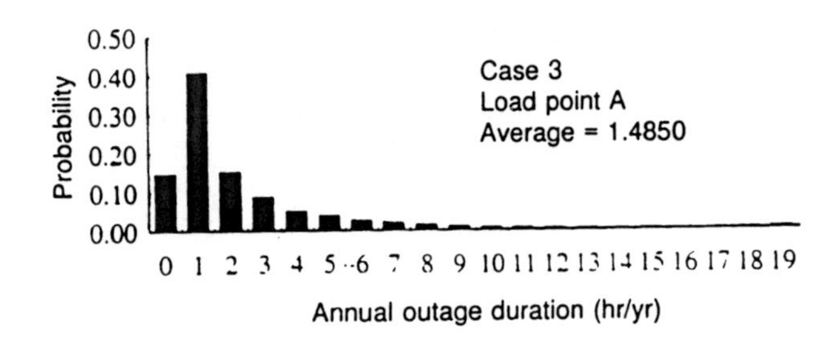

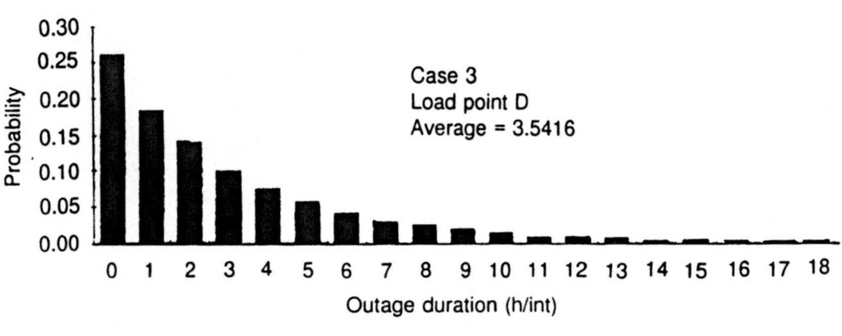

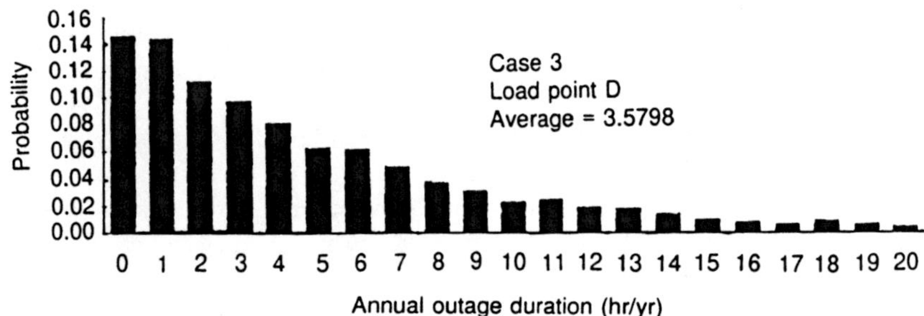

Fig. 12.23 Results for Case 3

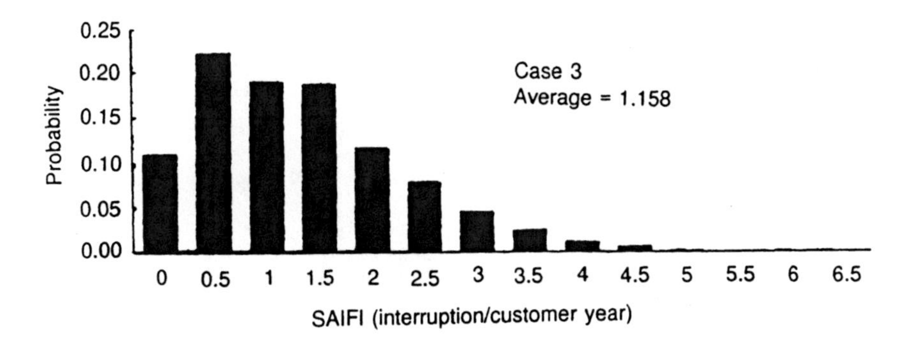

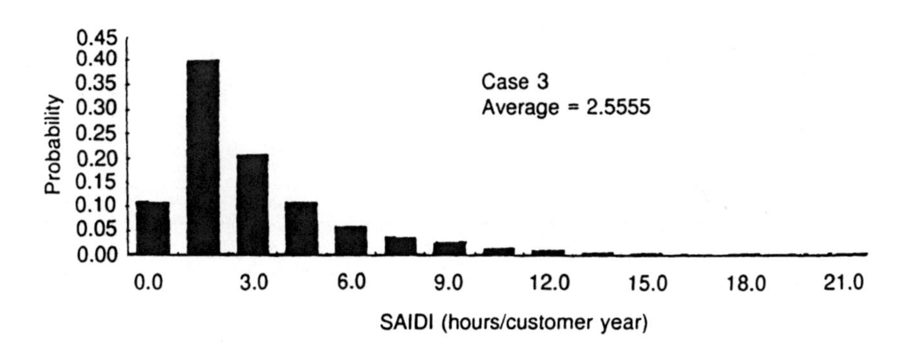

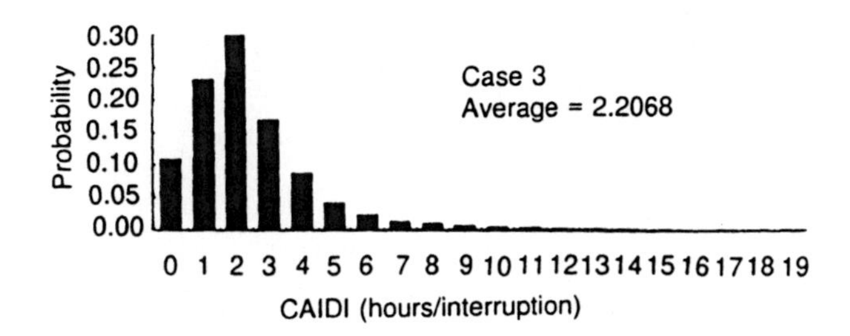

Fig. 12.23 Continued

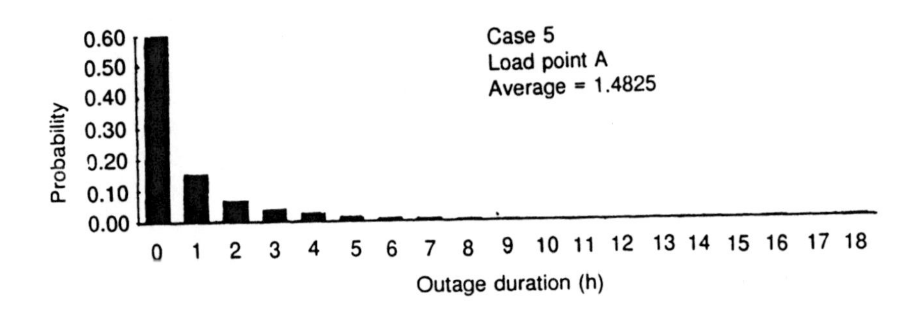

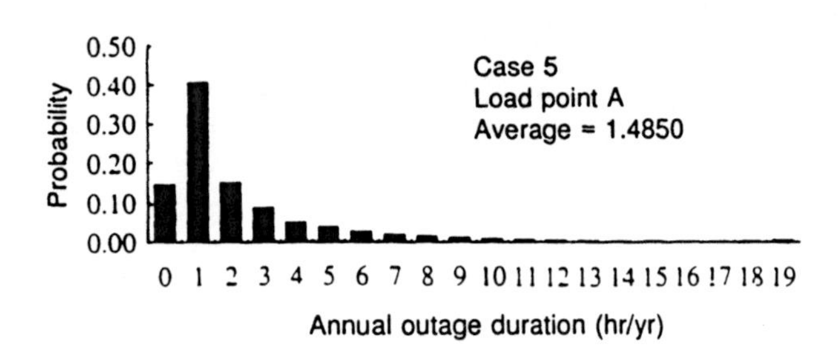

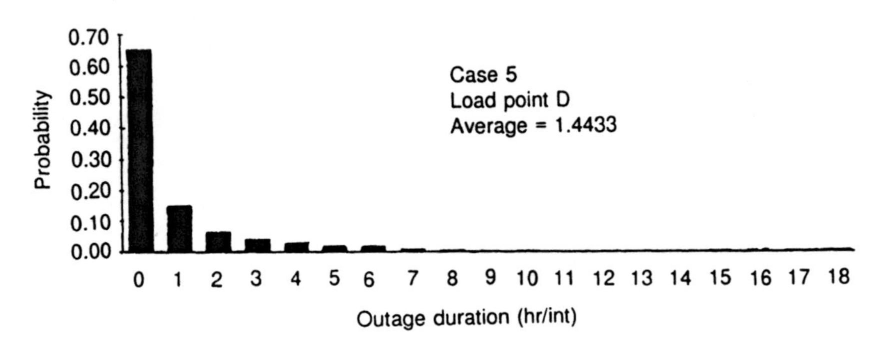

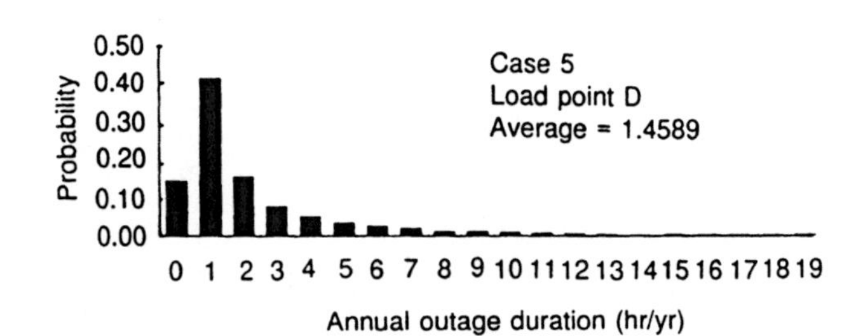

Fig. 12.24 Results for Case 5

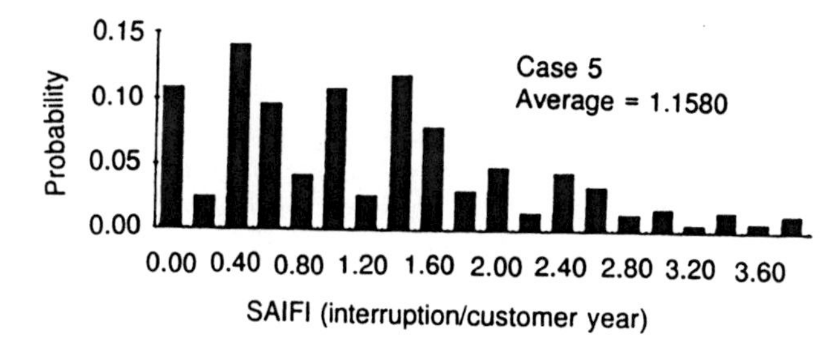

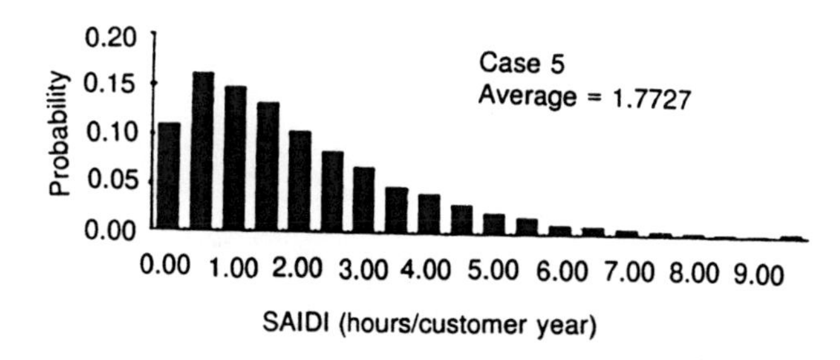

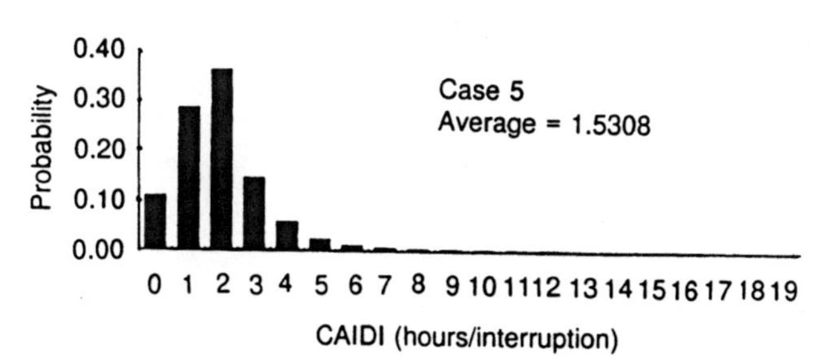

Fig. 12.24 Continued

Table 12.6 Average values for dual-transformer feeder

		Simulation approach	
Reliability index	*Analytical approach*	*Exponential distribution*	*Lognormal distribution*
λ (f/yr)	6.99×10^{-4}	7.0×10^{-4}	6.6×10^{-4}
r (hr)	5.88	5.77	6.03
U (hr/yr)	4.11×10^{-3}	4.04×10^{-3}	3.98×10^{-3}

Two distributions are shown in Fig. 12.25; the outage duration characteristics assuming times to repair are exponentially distributed and lognormally distributed respectively. Several observations can be made.

The first major observation is that the shapes are significantly different. The one assuming the exponential distribution itself exhibits a negative exponential shape and therefore is very skewed to the right, with a peak at very small values of outage duration. This general shape will not change as the average value changes; as the average down time increases, the characteristic becomes more skewed to the right with the likelihood of long outage durations increasing. The distribution

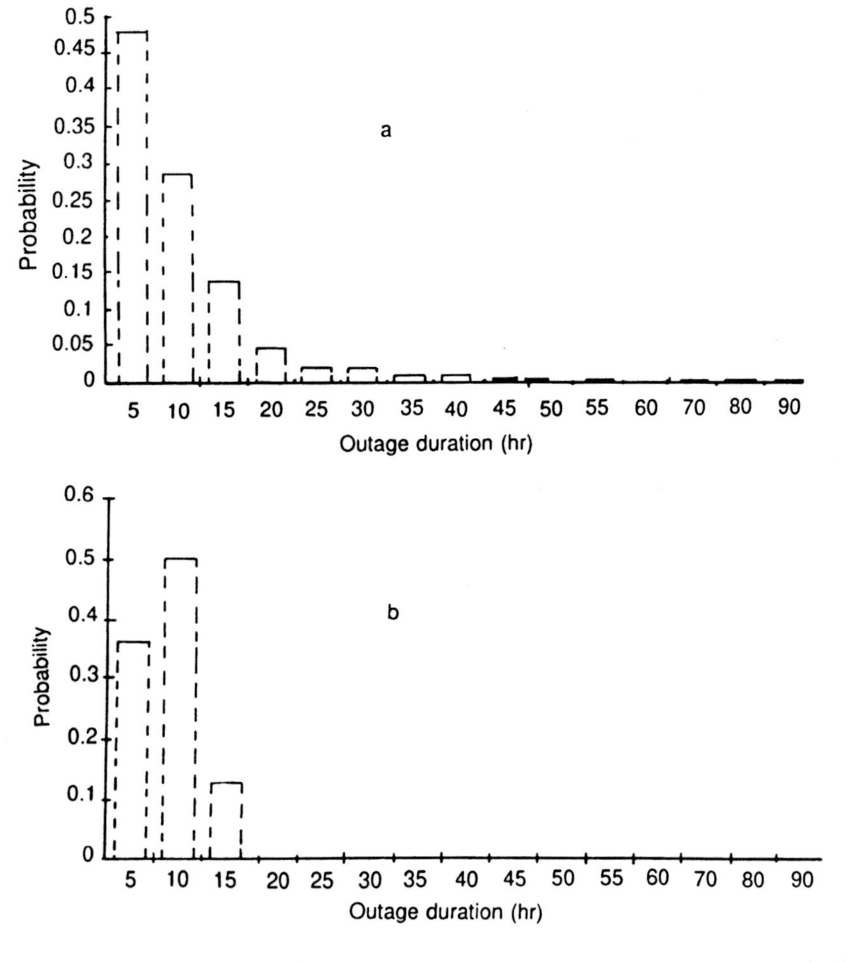

Fig. 12.25 Results for parallel feeders: (a) for exponential distribution; (b) for lognormal distribution

produced assuming a lognormal distribution is more bell-shaped. As discussed in Chapter 6 of *Engineering Systems*, the shape of the lognormal distribution changes with the value of standard deviation, whereas the exponential distribution has no shape parameter. Therefore the shape indicated in Fig. 12.25 for the lognormal case is not general and will reflect the value of standard deviation assumed.

The second major observation is that the average value of down time is about 6 hr, yet it is seen from both characteristics that the actual value for any given interruption can be significantly greater than this. This is particularly the case for the exponential distribution for which outage durations greater than 30 hr occur, albeit infrequently. The upper values for the lognormal case are less excessive but still significant. The result of this outcome is that, although the average time may be less than an acceptable target value, this provides little or no information to how likely it is for the target value to be exceeded either in duration or frequency. If this is important, and it is becoming increasingly so, the need for the distributions becomes evident.

12.8.5 Extensions to the basic approach

The examples provided in this section are purely illustrative and do not pretend to be an exhaustive discussion of all applications. However, they do provide an insight into the types and shapes of distributions. In principle, the modeling concepts in Section 12.8.2 can be extended to take other factors into account. For instance, if weather is being modeled, this would also need to be treated as a random variable and the actual duration of any weather state sampled using a random number generator, the assumed distribution representing state times, and the appropriate location and shape parameters. Similarly, if two different failure modes are being considered, two random values would be generated, one for each failure mode, and that for which the sampled time is smallest would be the assumed failure mode on that occasion. This principle can be extended to any number of failure modes. In general, the concept is generally simple, but the application including keeping track of the numerous number of events, consequences, and indices is the complex and time-consuming part of the analyses.

12.9 Conclusions

This chapter has described the basic application of MCS to the reliability analysis of a wide range of power system applications. More extensive discussion and applications can be found in a wide range of papers and publications [9–18]. The present chapter complements the developments in previous chapters which were concerned only with analytical techniques. The principles described at the end of all previous chapters also apply in the case of this one because the general purpose for using MCS is the same as that for using the analytical approach, i.e., to achieve objective measures for all parts of the system so that better and more informed decision making is possible. This basic concept does not change.

The most important additional principle associated with the MCS approach is that it can provide the probability distributions associated with the various reliability indices which, in consequence, can provide additional and more informative data about the behavior of a system. In particular, it can indicate the likelihood of various target levels being either complied with or violated.

One further merit of the MCS approach which has not been stressed so far is that it is often used as an independent means for assessing the accuracy of an analytical model or technique, or the degree of inaccuracy introduced by an approximation that may be inbuilt into another set of models or techniques.

A final point is one of warning. Although the MCS approach can generally be used to model any system parameter or function, it is not a mindless approach and cannot be used to analyze any system without a sound understanding of the real problem and system behavior: it is therefore not a substitute for engineering appreciation and thought.

12.10 Problems

1. Problem 1 of Chapter 2
2. Problem 2 of Chapter 2
3. Problem 7 of Chapter 2
4. Problem 1 of Chapter 4
5. Problem 2 of Chapter 6
6. Problem 1 of Chapter 7
7. Problem 1 of Chapter 8
8. Problem 1 of Chapter 10

12.11 References

1. Office of Electricity Regulation (OFFER), 'Report on customer services', Annual Report, OFFER, Birmingham, UK.
2. Allan, R. N., da Guia da Silva, M., 'Evaluation of reliability indices and outage costs in distribution systems', *IEEE Trans. Power Systems*, **PWRS-10** (1995), pp. 413–19.
3. Billinton, R., Li, W., Consideration of multi-state generating unit models in composite system adequacy assessment using Monte Carlo simulation, *Canadian Journal of Electrical and Computer Engineering*, **17** (1) (1992).
4. Bhuiyan, M. R., Allan, R. N. 'Modelling multistate problems in sequential simulation of power system reliability studies', *Proc IEE*, **142** part C (1995), pp. 343–49.

5. IEEE Committee Report, 'IEEE reliability test system', *IEEE Trans. on Power Apparatus System*, **98** (1979), pp. 2047–54.
6. Billinton, R., Li, W., 'A Monte Carlo method for multi-area generation system reliability assessment', *IEEE Trans. on Power Systems*, **7** (4), (1992), pp. 1487–92.
7. Allan, R. N., Roman, J., 'Reliability assessment of generation systems containing multiple hydro plant using simulation techniques', *IEEE Trans. on Power Systems*, **PWRS-4** (1989), pp. 1074–80.
8. Allan, R. N., Ubeda, J. Román, 'Reliability assessment of hydro thermal generation systems containing pumped storage plants', *Proc IEE*, **138** part C (1991), pp. 471–78.
9. Billinton, R., *Bibliography on Application of Probability Methods in the Evaluation of Generating Capacity Requirements*, IEEE/PES Winter Meeting (1966), Paper 31 CP 66-62.
10. Billinton, R., 'Bibliography on the application of probability methods in power system reliability evaluation', *IEEE Trans. on Power Apparatus Systems*, **91** (1972), pp. 649–60.
11. IEEE Committee Report, 'Bibliography on the application of probability methods in power system reliability evaluation, 1971–1977', *IEEE Trans. on Power Apparatus Systems*, **97** (1978), pp. 2235–42.
12. Allan, R. N., Billinton, R., Lee, S. H., 'Bibliography on the application of probability methods in power system reliability evaluation, 1977–1982', *IEEE Trans. on Power Apparatus Systems*, **103** (1984), pp. 275–82.
13. Allan, R. N., Billinton, R., Shahidehpour, S. M., Singh, C., 'Bibliography on the application of probability methods in power system reliability evaluation, 1982–1987', *IEEE Trans. on Power Systems*, **3** (4) (1988), pp. 1555–64.
14. Allan, R. N., Billinton, R., Briepohl, A. M., Grigg, C. H., 'Bibliography on the application of probability methods in power system reliability evaluation, 1987–1991', *IEEE Trans. on Power Systems*, **PWRS-9** (1994).
15. Billinton, R., Allan, R. N., *Reliability Assessment of Large Electric Power Systems*, Kluwer, Boston (1988).
16. CIGRE Working Group 38.03, *Power System Reliability Analysis*: Vol. 1, *Application Guide*, CIGRE Publications, Paris (1988).
17. CIGRE Working Group 38.03, *Power System Reliability Analysis*: Vol. 2, *Composite Power System Reliability Evaluation*, CIGRE Publications, Paris (1992).
18. Billinton, R., Li, W., *Reliability Assessment of Electric Power Systems Using Monte Carlo Simulation*, Plenum, New York (1994).
19. Noferi, P., Paris, L., Salvaderi, L., *Monte Carlo Methods for Power System Reliability Evaluation in Transmission and Generation Planning*, Proceedings 1975 Annual Reliability and Maintainability Symposium, Washington, DC (1975), Paper 1294, pp. 449–59.

20. Oliveira, G. C., Pereira, M. V. F., Cunha, S. H. F., 'A technique for reducing computational effort in Monte Carlo based composite reliability evaluation', *IEEE Trans. on Power Systems*, **4** (3) (1989), pp. 1309–15.
21. Billinton, R., Li, W., 'Hybrid approach for reliability evaluation of composite generation and transmission systems using Monte Carlo simulation and enumeration technique', *IEE Proc.*, **138** (3) Pt. C (1991), pp. 233–41.
22. Billinton, R., Li, W., 'A novel method for incorporating weather effects in composite system adequacy evaluation', *IEEE Trans. on Power Systems*, **6** (3) (1991), pp. 1154–60.
23. Billinton, R., Li, W., *Composite System Reliability Assessment Using a Monte Carlo Approach*, The Third International Conference on Probabilistic Methods Applied to Electric Power Systems, London, 3–5 July, 1991.
24. Ubeda, J. Román, Allan, R. N., 'Sequential simulation applied to composite system reliability evaluation', *Proc. IEE*, **139** Part C (1992), pp. 81–6.
25. Ubeda, J. Román, Allan, R. N., 'Reliability assessment of composite hydro thermal generation and transmission systems using sequential simulation', *Proc. IEE*, **141** Part C (1994), pp. 257–62.
26. Bhuiyan, M. R., Allan, R. N., 'Inclusion of weather effects in composite system reliability evaluation using sequential simulation', *Proc. IEE*, **141** Part C (1994), pp. 575–84.
27. Patton, A. D., *Probability Distribution of Transmission and Distribution Reliability Performance Indices*, 1979 Reliability Conference for the Electric Power Industry, pp. 120–23.
28. Billinton, R., Goel, R., 'An analytical approach to evaluate probability distributions associated with the reliability indices of electric distribution system', *IEEE Trans. on Power Distribution*, **1** (3) (1986), pp. 245–51.
29. Billinton, R., Lian, G., 'Station reliability evaluation using a Monte Carlo approach', *IEEE Trans. on Power Distribution*, **8** (3) (1993), pp. 1239–45.

13 Evaluation of reliability worth

13.1 Introduction

As discussed in Chapter 1, the basic function of an electric power system is to satisfy the system load and energy requirements as economically as possible and with a reasonable assurance of continuity and quality [1]. The two aspects of relatively low cost electrical energy at a high level of reliability are often in direct conflict and present power system managers, planners, and operators with a wide range of challenging problems. Electric power utilities are also facing increasing uncertainty regarding the economic, political, societal, and environmental constraints under which they operate and plan their future systems. This has created increasing requirements for extensive justification of new facilities and increased emphasis on the optimization of system costs and reliability. An integral element in the overall problem of allocating capital and operating resources is the assessment of reliability cost and reliability worth [2]. The ability to assess the costs associated with providing reliable service is reasonably well established and accepted [3,4]. In contrast, the ability to assess the worth of providing reliable service is not well established and considerable work will be required before these methodologies can be considered to be completely acceptable. Establishing the worth of service reliability is a difficult and subjective task, as direct evaluation does not appear to be feasible at this time. A practical alternative, which is being widely utilized, is to evaluate the impacts and the monetary losses incurred by customers due to electric power supply failures. Customer interruption costs provide a valuable surrogate for the actual worth of electric power supply reliability [2]. This chapter illustrates the application of reliability worth assessment to HLI, HLII, and to the distribution functional zone.

13.2 Implicit/explicit evaluation of reliability worth

The general planning problem in an electric power system consists traditionally of a comparison between various alternatives for system development made on the basis of system cost. There are two fundamental approaches to the system cost [1]. The first approach is one that has been used for many years, and it can be argued to have resulted in the high level of reliability enjoyed by electrical energy consumers in developed countries. In this approach, system investment is driven by deterministic criteria or by fixed quantitative reliability indices selected on the basis of experience and judgment. The capital cost of the proposed facilities plus the cost of operating and maintaining them are compared under the assumption that each alternative provides the same reliability based on whatever deterministic or probabilistic technique is used. This approach implies that an implicit socioeconomic cost is associated with the selection of the reliability criterion. The deterministic or probabilistic criteria adopted by a utility are therefore presumed to be based on a perception of public need and shaped by economic and/or regulatory forces to implicitly include recognition of the socioeconomic costs. Utilization of such criteria should therefore reflect the optimum trade-offs between the cost of achieving the required reliability and the benefits derived by society.

The second approach, known as the explicit cost technique, incorporates reliability in the costing process by comparing the overall costs, including the societal costs of unreliability. In both methods, the costs are those incurred by the utility in each year of the timespan considered and equated using present worth analysis. The explicit cost approach uses subjective and objective measures of customer monetary losses arising from electric energy supply curtailments. The LOEE, sometimes expressed as the expected energy not supplied (EENS), is usually used as the index to link system unreliability with reliability worth. The unit cost of losses due to energy not supplied is a composite parameter formed from the various classes of customers affected by a given interruption. Considerable work has been done on developing procedures for assessing customer monetary losses due to electric supply failures, and there is a wide range of available literature [5,6].

The explicit cost approach to reliability worth assessment can be used to provide valuable information in two major ways. It can be used to quantify the fundamental electric utility requirement of what is a reasonable level of reliability at all three hierarchical levels. It can be used in a more direct and practical fashion in a wide range of utility decision-making processes. The basic concepts associated with the explicit cost approach to reliability worth assessment are illustrated in Chapter 1. As shown in Figure 1.3, the utility cost which includes capital investment, operating and maintenance increases as the reliability level increases. The socioeconomic losses in the form of customer costs decrease as the reliability increases. The total societal cost is the sum of the utility and customer costs. The optimum level of reliability therefore occurs at the point of minimum total cost. The concepts illustrated in Figure 1.3 are quite general and can be applied within each functional zone and hierarchical level.

13.3 Customer interruption cost evaluation

A variety of methods has been utilized to evaluate customer impacts due to interruptions [5–7]. These methods can be grouped, based on the methodological approach used, into three broad categories: various indirect analytical evaluations, case studies of blackouts, and customer surveys. While a single approach has not been universally adopted, utilities appear to favor customer surveys as the means to determine specific information for their purposes [2].

A necessary preliminary step in the determination of interruption costs is an understanding of the nature and variety of customer impacts resulting from electric service interruptions. Impacts may be classified as direct or indirect, economic or otherwise (social), and short term or long term. Direct impacts are those resulting directly from cessation of supply while indirect impacts result from a response to an interruption. Hence, direct economic impacts include lost production, idle but paid for resources (raw materials, labor, capital), process restart costs, spoilage of raw materials or food, equipment damage, direct costs associated with human health and safety, and utility costs associated with the interruption. Direct social impacts include inconvenience due to lack of transportation, loss of leisure time, uncomfortable building temperatures, and personal injury or fear. Indirect losses usually arise as spin-off consequences and it may be difficult to categorize them as social or economic. Examples of such costs are civil disobedience and looting during an extended blackout, or failure of an industrial safety device in an industrial plant, necessitating neighboring residential evacuation. The final distinction between short-term and long-term impacts relates to the immediacy of the consequence. Specifically, long-term impacts are often identified as adaptive responses or mitigation undertaken to reduce or avoid future outage costs. Installation of protective switchgear, voltage regulation equipment, and cogeneration or standby supplies would be included in this category, as would the relocation of an industrial plant to an area of higher electric service reliability.

Broadly speaking, the cost of an interruption from the customer's perspective is related to the nature of the degree to which the activities interrupted are dependent on electrical supply. In turn, this dependency is a function of both customer and interruption characteristics. Customer characteristics include type of customer, nature of the customer's activities, size of operation, and other demographic data, demand, and energy requirements, energy dependency as a function of time of day, etc. Interruption characteristics include duration, frequency, and time of occurrence of interruptions; whether an interruption is complete or partial; if advance warning or duration information is supplied by the utility; and whether the area affected by the outage is localized or widespread. Finally, the impact of an outage is partially dependent on the attitude and preparedness of customers, which in turn is related to existing reliability levels.

13.4 Basic evaluation approaches

Many of the approaches devised to evaluate interruption costs can be broadly categorized as indirect analytical methods which infer interruption cost values from associated indices or variables. Examples of such substitution or proxy approaches include the following:

(i) Electrical supply rates or tariffs are used to derive value of service reliability estimates [8]. The minimum estimate of customers' willingness to pay is based on electrical rate structures and the maximum is based on cost of standby plant.

(ii) The value of foregone production is determined by taking the ratio of the annual gross national product to the total electrical consumption ($/kWh) and ascribing to it the value of service reliability [9]. A similar value-added approach has been used to evolve an analytical model which, with appropriate adjustments, was applicable to different customer categories [10]. The approach made use of detailed and specific data (sales data, value-added data, employee data) and numerous assumptions and derivations of average consumption, price, and price elasticity.

(iii) The value of foregone leisure time based on customers' wage rates has been used in several residential interruption cost assessments. This is based on the notion that consumers can and do make optimum labor/leisure decisions and that earnings are equal. Some derivations are based on estimates, minisurveys, or discussions and are presumed to include actual losses, household activities, and leisure time [11]. Others make simplifying assumptions and base their results principally on lost leisure time [12].

(iv) The hourly depreciation rates of all electrical household appliances unavailable because of an outage have been used as the basis of residential outage costs [13].

The advantages of these and other similar methods are that they are reasonably straightforward to apply, make use of readily available data, and, consequently, are inexpensive to implement. Their disadvantages are that most are based on numerous and severely limiting assumptions. Most generate global rather than specific results and consequently do not reveal variations in cost with specific parameters as required by the utilities. Therefore, the usefulness of the results to the utilities for planning purposes is significantly reduced.

A second category of outage cost assessment is to conduct an after-the-fact case study of a particular outage. This approach has been limited to major, large-scale blackouts such as the 1977 New York blackout [14]. The study attempted to assess both direct and indirect short-term costs. Direct costs included food spoilage, wage loss, loss of sales, loss of taxes, etc. Indirect costs included emergency costs, losses due to civil disorder (looting, rioting, and arson), and losses by governments and insurance companies resulting from social disorder. The study also considered a wide range of societal and organizational impacts. Such impacts are significant but difficult to evaluate in monetary terms. While specific data obtained were based on assumptions and were incomplete in many respects, some important conclusions resulted. In particular, the results indicated that the indirect costs were much higher than the direct costs.

The third methodological approach that has been used to assess direct, short-term customer interruption costs is that of customer surveys [15–19]. With this method, customers are asked to estimate their costs or losses due to supply outages of varying duration and frequency at different times of the day and year. The strength of this method lies in the fact that the customer is probably in the best position to assess the losses. Direct costs are relatively easy to determine for some customer categories (e.g., industrial), but users' opinions are particularly important in assessing less tangible losses, such as inconvenience, for other categories (e.g., residential). Another advantage is that the method can readily be tailored to seek particular information as related to the specific needs of the utility. Obviously, this method is beset with all the problems of questionnaire surveys, and the cost and effort of undertaking surveys is significantly higher than using the other approaches outlined earlier. Nevertheless, this approach appears to be the method favored by utilities which require outage cost data for planning purposes.

13.5 Cost of interruption surveys

13.5.1 Considerations

Costs of interruption surveys are usually undertaken with specific objectives in mind, such as system expansion/upgrading or major rate revision. Typically, the customer pool is broken down into appropriate major customer categories or sectors, such as residential, industrial, commercial, agricultural, etc., so that category-specific survey instruments can be used. The Standard Industrial Classification (SIC) system of customer identification is commonly utilized because of its wide general acceptance by industry and government, and often it has already been adopted by the utility for other reasons. Development of survey instruments for each of the customer sectors is a major and important step in the process. Questionnaire preparation and the attendant survey procedures require an understanding of the many difficulties which can be encountered in conducting surveys, such as representative sample selection, questionnaire bias, nonresponse bias, compromising questionnaire content with length to ensure satisfactory response rates, etc. Additionally, the nature and approach of the survey instrument should reflect a sound theoretical basis and a clear statement of objectives [2].

13.5.2 Cost valuation methods

Perhaps the most important questionnaire design consideration is the choice of interruption cost valuation methodology, since determining this "cost" is the primary objective of the survey. It is in this regard that the greatest variations in approach exist. If one accepts that indirect analytical methods are inadequate and that the customer is the best source of the desired information as discussed earlier, the problem still remains: In what manner is the information solicited from the customer? There appears to be concurrence that some methods are more suitable than others for particular sectors, but there is no universal agreement as to what those methods are.

The most obvious approach is a direct solicitation of the customers' interruption costs for given outage conditions. Guidance can be offered as to what should and should not be included in the cost estimate so that the meaning of the result is not ambiguous. This approach provides reasonable and consistent results in those situations where most losses tend to be tangible, directly identifiable, and quantifiable. Independent researchers have derived valuations which are reasonably similar in magnitude [17,19,20,21]. The approach is applicable for the industrial sector, most large users, and for the commercial sector (retail trades and services). It has also been used for large institutions and office buildings [15]. Its major weakness lies in those areas where the impacts tend to be less tangible and the monetary loss is not directly identifiable.

Another approach is to ask respondents what they would be willing to pay to avoid having the interruptions, or conversely what amount they would be willing to accept for having to experience the outage. The basis of this approach is that incremental willingness to pay (willingness to accept) constitutes a valuation of corresponding marginal increments (decrements) in reliability. Theory would suggest that incremental "willingness to pay" amounts should be nearly equal to "willingness to accept" valuations. However, actual valuations consistently yield willingness to pay values significantly less than willingness to accept values. This result is believed to support the earlier argument that electric service and its reliability do not perform as normal "markets," though other factors may be at work. Nevertheless, valuations based on willingness to pay and accept are worthwhile measures, possibly as outside bounds, if the limitations are recognized.

A third approach is that of indirect worth evaluation. If direct valuation is not possible, customer-selected alternatives or responses to indirect method questions may be used to derive a value. The intent is to devise a suitable approach so as to decrease the problems associated with rate-related antagonism and the lack of experience in rating the worth of reliability. This is achieved by asking questions which the respondents can relate to in the context of their experience. A limitation is the possibility that the derived value is not an estimate of the worth but some other entity associated with the indirect approach. Possible question forms that have been used or considered include the following:

(i) cost of hypothetical insurance policies to compensate for possible interruption effects, and the appropriate compensation payable in the event of an interruption claim;
(ii) respondents' opinions as to the appropriate interruption cost figures utilities should use in planning;
(iii) respondents' predictions of what preparatory actions they might take in the event of recurring interruptions;

(iv) respondents' selection of interruptible or curtailable options with reduced rates, which are, in effect, self-predictions of willingness to accept decreased rates for reductions in reliability;

(v) respondents' rank ordering of a set of reliability/rate alternatives and choosing an option that is most suitable to their needs.

It should be noted that several of the options use a form of substitution either in services or in monetary terms. While the substitution concept is similar to that discussed earlier, the difference here is that the substitution is reasonably direct and, more importantly, the selection is being made by the customer rather than by the analyst. A matter of concern for most of the approaches cited is the question of how closely would customers' actions match with their prior prediction of their actions? Put another way, how valid is the customers' perception? Perhaps the strongest rejoinder to this issue is that it is the customer's perception that is sought, and that there are markets where customers' selections are based more on their perception than on factual evidence. An aside at this juncture is that most of the approaches cited attempt to establish a market or at least an inferred market for reliability. It is believed that the "market" responses for small variations around the current reliability value are reasonable, though obviously it can only be as "accurate" as the particular substitution can accomplish. Attempting considerable variations from the current reliability level, however, may not yield meaningful results, mainly because service reliability may not respond as a true market as discussed earlier. Additionally, the customer's perception is doubtful in extreme situations because of lack of experience, and the useful range for most of the substitutions is questionable.

Considerable use has been made of the preparatory action approach as it appears to secure reasonable results. The approach has been used in Canadian residential and agricultural surveys [22,23] and in the U.K. residential surveys [20,21]. With this approach, the respondent is presented with a list of actions that one might conceivably take in preparation against recurring interruptions. A reasonable cost figure for purchase and application of each action is assigned and included in the list. The list ranges from making no preparations through to buying a self-starting standby generator capable of supplying the entire load. Respondents are then asked to indicate what action or actions they would take for different failure scenarios. During analysis, the cost(s) of the chosen action(s) are used as an estimate of the expenditure respondents are willing to undertake on their own behalf so as to prevent or nullify the full effects of the interruption. This represents an indirect estimate of reliability worth in that the derived expenditures are considered to be the user's perception of the value of avoiding the interruption consequences. Respondent interviews during questionnaire development are essential to ensure that respondents accept the overall approach and that they consider the choice of actions adequate and the quoted costs reasonable.

13.6 Customer damage functions

13.6.1 Concepts

One convenient way to display customer interruption costs is in the form of customer damage functions (CDF). The CDF can be determined for a given customer type and aggregated to produce sector customer damage functions for the various classes of customers in the system. Table 13.1 shows a series of sector CDF expressed in kilowatts of annual peak demand. These values were obtained from the Canadian survey [24–26]. Similar values were obtained from the U.K. survey [20,21].

The data shown in Table 13.1 is expressed graphically in Fig. 13.1.

Table 13.1 Sector interruption cost estimates (CDF) expressed in kilowatts of annual peak demand ($/kW)

	Interruption duration				
User sector	*1 min*	*20 min*	*1 hr*	*4 hr*	*8 hr*
Large users	1.005	1.508	2.225	3.968	8.240
Industrial	1.625	3.868	9.085	25.163	55.808
Commercial	0.381	2.969	8.552	31.317	83.008
Agricultural	0.060	0.343	0.649	2.064	4.120
Residential	0.001	0.093	0.482	4.914	15.690
Govt. & inst.	0.044	0.369	1.492	6.558	26.040
Office & bldg.	4.778	9.878	21.065	68.830	119.160

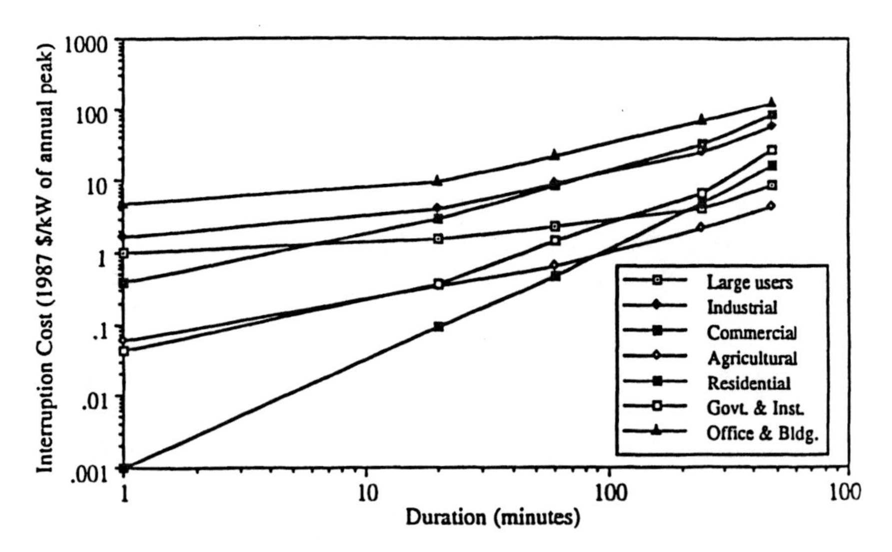

Fig. 13.1 Sector interruption cost estimates

The sector CDF can be aggregated at any particular load point in the system to produce a composite customer damage function (CCDF) at that load point. The assumption in this case is that all load curtailments will be distributed proportionally across all the customer sectors. The weighting used to produce a CCDF is usually done in terms of the per-unit energy for each sector. The per-unit peak demand for each sector is sometimes used for short interruption durations. The weighting procedure is illustrated in the following section.

13.6.2 Reliability worth assessment at HLI

The system model at HLI is shown in Fig. 13.2 (see Section 2.1).

Assume that the load composition in terms of the annual peak demand and energy consumption is as shown in Table 13.2.

The system CCDF obtained using the data in Tables 13.1 and 13.2 is shown in Table 13.3 and presented graphically in Fig. 13.3. The weighting procedure used to obtain the CCDF is as follows. The CCDF value at the 1-min duration is obtained using the 1-min values in Table 13.1 and the sector peak % values in Table 13.2.

$$\begin{aligned}\text{CCDF (1-min duration)} &= 0.30(1.005) + 0.14(1.625) \\ &\quad + 0.10(0.381) + \ldots (0.02)(4.778) \\ &= 0.668040 \\ &= 0.67\ \$/\text{kW}\end{aligned}$$

The CCDF (20-min duration) was also obtained using the sector peak % values. The 2-, 4-, and 8-hr values were obtained using the sector energy % values.

The CCDF can be converted into an extended index that links system reliability with customer interruption costs. One suitable form being used in Canada is known as the interrupted energy assessment rate (IEAR) expressed in $/kWh of unsupplied energy. A detailed description of the concepts involved in calculating an IEAR using a basic frequency and duration (F & D) approach or Monte Carlo simulation is presented in Ref. 25. A brief description of the F & D approach is given here to illustrate the salient features. The estimation of the IEAR at HLI involves the generation of a capacity margin model which indicates the severity, frequency, and duration of the expected negative margin states. The basic approach to developing the margin model is described in detail in Chapter 3. This model can be used in

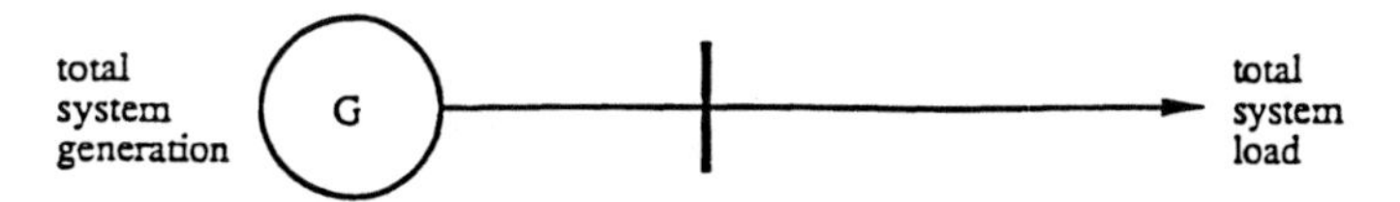

Fig. 13.2 Basic HLI system model

Table 13.2 Load composition for the assumed service area, based on annual peak demand and annual energy consumption

User sector	*Sector peak (MW)*	*Sector peak (%)*	*Sector energy (%)*
Large users	55.5	30.0	31.0
Industrial	25.9	14.0	19.0
Commercial	18.5	10.0	9.0
Agricultural	7.4	4.0	2.5
Residential	62.9	34.0	31.0
Govt. & inst.	11.1	6.0	5.5
Office & bldg.	3.7	2.0	2.0
Total	185.0	100.0	100.0

conjunction with the CCDF for the given service area to estimate the IEAR. The generation model is developed from the capacities, forced outage rates, failure rates, and repair rates of the generating units. The exact-state load model described in Chapter 3, which represents the actual daily system load cycle by a sequence of discrete load levels is utilized. The total LOEE for the estimated loss of load events within the period of study is

$$\text{Total LOEE} = \sum_{i=1}^{N} m_i f_i d_i \text{ kWh/day} \tag{13.1}$$

where m_i is the margin state capacity for load loss event i (kW), f_i is frequency of load loss event i (occ/day), d_i is the duration of load loss event i (hr), and N is the total number of load loss events.

The cost $c_i(d_i)$ of the energy not supplied during load loss event i can be obtained from the duration d_i and the CCDF for the given service area. The total expected cost of all the system load curtailment events is

$$\text{Total cost} = \sum_{i=1}^{N} m_i f_i c_i(d_i) \text{ \$/day} \tag{13.2}$$

Table 13.3 System CCDF ($/kW) calculated from the sector CDFs

Interruption duration				
1 min	*20 min*	*2 hr*	*4 hr*	*8 hr*
0.67	1.56	3.85	12.14	29.41

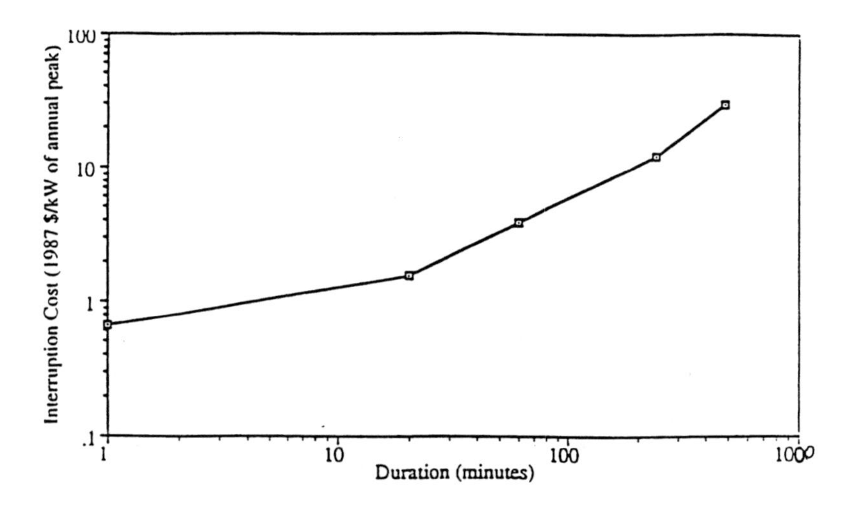

Fig. 13.3 System composite customer damage function

The IEAR for the service area is calculated as the ratio of the total cost and total LOEE:

$$\text{Estimated IEAR} = \frac{\sum_{i=1}^{N} m_i f_i c_i(d_i)}{\sum_{i=1}^{N} m_i f_i d_i} \ \$/\text{kWh} \tag{13.3}$$

The procedure can be illustrated using the generating system example presented in Chapter 3. Table 3.14 shows the margin array for the three-unit generating system and the simple load model. The average duration of each margin state is obtained by dividing the probability of the state by the frequency of encountering it. Table 13.4 shows the negative margin array including the duration, expected energy not supplied (EENS), and the expected interruption cost at each negative margin.

The total EENS using Equation (13.1) is 1.946136 MWh/day and the total cost, using Equation (13.2) is 7.454387 k$/day. The IEAR obtained using Equation (13.3) is 3.83 $/kWh.

The average interruption cost can be calculated for different generation and load compositions, or a single IEAR value can be used for a range of studies. Sensitivity analyses [25] show that the IEAR is reasonably stable and does not vary significantly with peak load or other operating conditions. Using a single IEAR

Table 13.4 Negative margin array from Table 3.14 extended by including ECOST and EENS

Margin state (MW)	*Probability*	*Frequency (occ/day)*	*Duration (hr)*	*ECOST (k$/day)*	*EENS (MWh/day)*
−5	0.001960	0.004938	9.526124	0.907426	0.2352
−15	0.003920	0.009877	9.525160	5.444404	1.4112
−21	0.000078	0.000234	8.0	0.144521	0.039312
−25	0.000078	0.000234	8.0	0.172048	0.04680
−30	0.000078	0.000234	8.0	0.206458	0.056160
−40	0.000157	0.000469	8.034116	0.554737	0.150720
−46	0.000001	0.000003	8.0	0.004059	0.001104
−50	0.000001	0.000003	8.0	0.004411	0.0012
−55	0.000001	0.000003	8.0	0.004853	0.00132
−65	0.000002	0.000006	8.0	0.01147	0.00312

allows the analyst to simply determine the expected energy not supplied or LOEE for a given generation-load situation and then convert it to expected interruption costs or reliability worth using Equation (13.4). It should be appreciated that the expected interruption cost (ECOST) can be calculated using the method shown in Table 13.4 for each change in the system configuration. The use of a single index such as the IEAR for a wide range of studies considerably simplifies the process without introducing great inaccuracies:

$$\text{Expected customer interruption cost (ECOST)} = (\text{IEAR})(\text{LOEE}) \qquad (13.4)$$

The application of a single IEAR in reliability worth evaluation at HLI is illustrated using the simple generating system example in Section 2.10.1. The system has 5–40 MW generating units and a straight-line load duration curve. Assume that the forecast system peak load is 170 MW and the IEAR is $3.83/kW. Additional capacity is in the form of 10-MW gas turbine units with an annual fixed

Table 13.5 Case 1 analysis

Situation	*Total capacity (MW)*	*Reserve margin (%)*	*EENS (MWh/yr)*	*ECOST ($M/yr)*	*Fixed cost ($M/yr)*	*total cost ($M/yr)*
original	200	17.66	313.854	1.20206	0.00000	1.20206
add 1 × 10 MW	210	23.53	74.278	0.28449	0.50000	0.78449
add 2 × 10 MW	220	29.41	40.865	0.15651	1.00000	1.15651
add 3 × 10 MW	230	35.29	19.498	0.07468	1.50000	1.57468
add 4 × 10 MW	240	41.18	6.294	0.02411	2.00000	2.02411
add 5 × 10 MW	250	47.06	1.176	0.00450	2.50000	2.50450

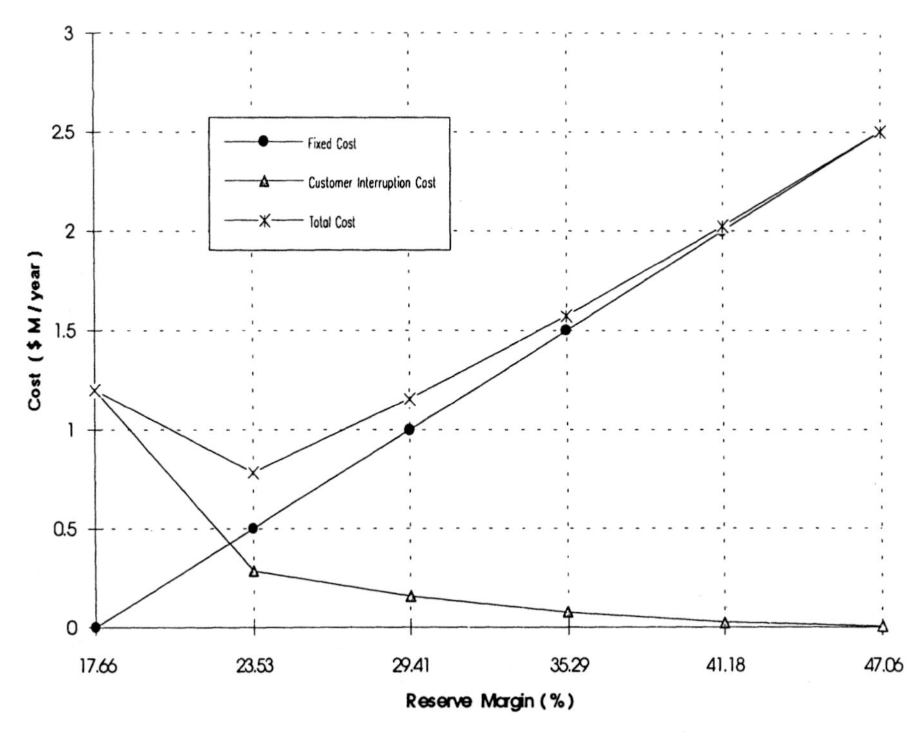

Fig. 13.4 Case 1: Change in fixed, customer, and total costs with reserve margin

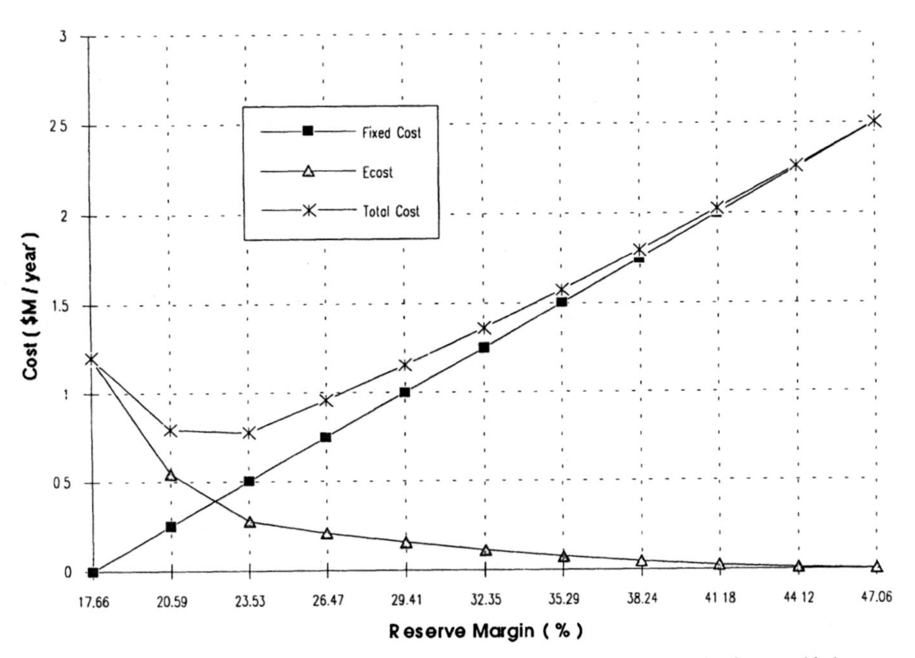

Fig. 13.5 Change in fixed, customer, and total costs with reserve margin as 5-MW units are added

Table 13.6 Case 2 analysis

Situation	*Total capacity (MW)*	*Reserve margin (%)*	*EENS (MWh/yr)*	*ECOST ($M/yr)*	*Fixed cost ($M/yr)*	*Total cost ($M/yr)*
original	200	17.66	313.854	2.40412	0.00000	2.40412
add 1 × 10 MW	210	23.53	74.278	0.56897	0.50000	1.06897
add 2 × 10 MW	220	29.41	40.865	0.31303	1.00000	1.31303
add 3 × 10 MW	230	35.29	19.498	0.14935	1.50000	1.64935
add 4 × 10 MW	240	41.18	6.294	0.04821	2.00000	2.04821
add 5 × 10 MW	250	47.06	1.176	0.00901	2.50000	2.50901

cost of $50/kW. The annual fixed costs associated with the existing 5–40 MW units and the annual production costs are not included in the analysis.

Case 1

Table 13.5 shows the EENS and ECOST for the original system and with the subsequent addition of 5–10 MW units. The results given in Table 13.5 are shown graphically in Fig. 13.4, where it can be seen that the customer costs decrease

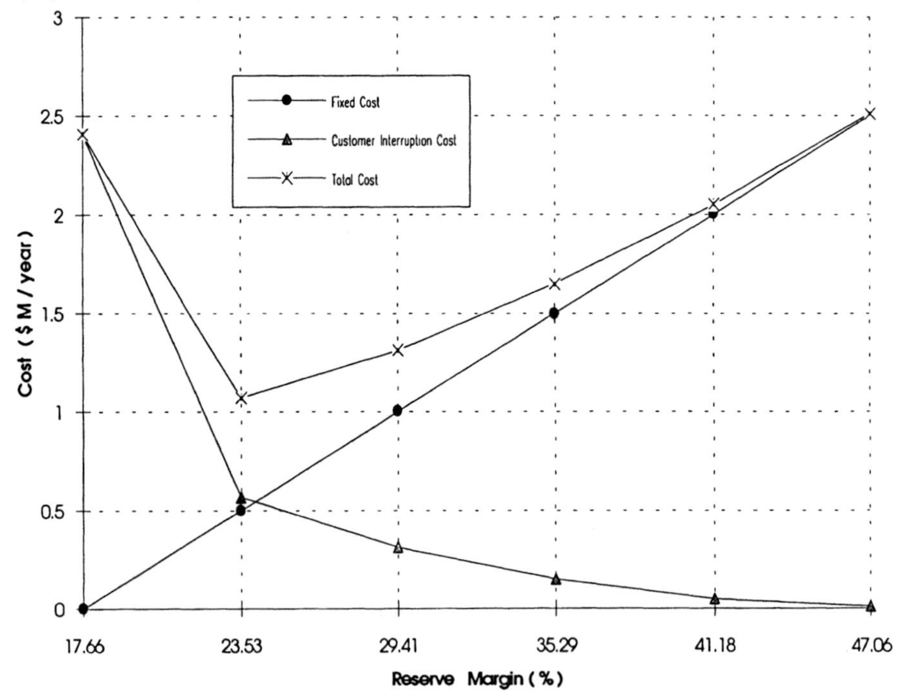

Fig. 13.6 Case 2: Change in fixed, customer, and total costs with reserve margin

Table 13.7 Case 3 analysis

Situation	*Total capacity (MW)*	*Reserve margin (%)*	*EENS (MWh/yr)*	*ECOST ($M/yr)*	*Fixed cost ($M/yr)*	*Total cost ($M/yr)*
original	200	17.66	313.854	1.20206	0.00000	1.20206
add 1 × 10 MW	210	23.53	74.278	0.28448	1.00000	1.28448
add 2 × 10 MW	220	29.41	40.865	0.15651	2.00000	2.15651
add 3 × 10 MW	230	35.29	19.498	0.07468	3.00000	3.07468
add 4 × 10 MW	240	41.18	6.294	0.02411	4.00000	4.02411
add 5 × 10 MW	250	47.06	1.176	0.00450	5.00000	5.00450

rapidly as additional capacities are added to the system and the fixed costs increase. The least cost reserve margin occurs with the addition of one 10-MW unit and at a reserve margin of 23.53%.

Figure 13.5 shows the variation in the fixed, customer, and total costs with reserve margin as 5-MW units are added rather than the 10-MW units used in Case 1. The optimum reserve margin is again 23.53%.

Case 2

Table 13.6 and Fig. 13.6 show the variation in costs when the IEAR used in Case 1 is doubled. The customer costs increase, but the optimum reserve margin is still 23.53%.

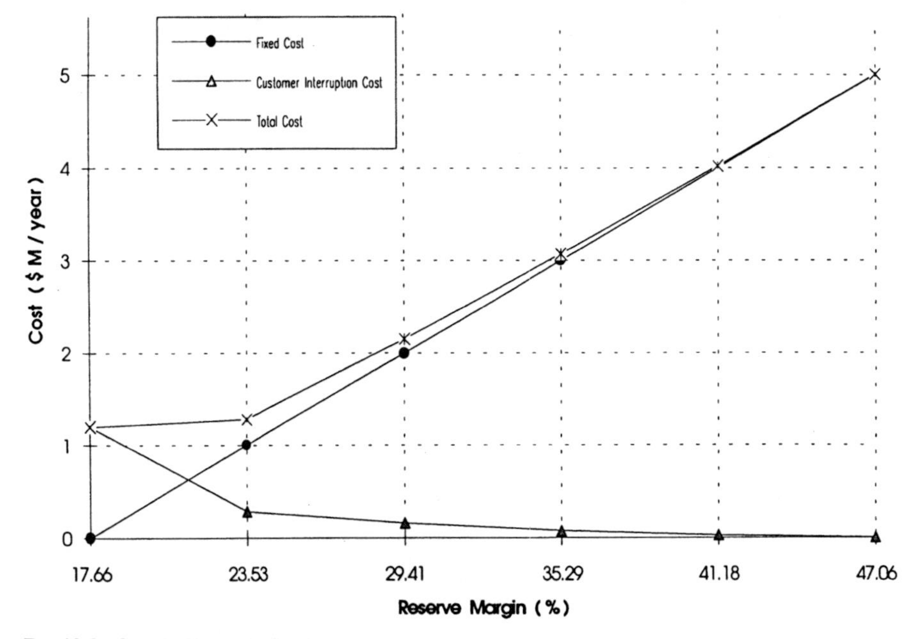

Fig. 13.7 Case 3: Change in fixed, customer, and total costs with reserve margin

Table 13.8 Case 4 analysis

Situation	Total capacity (MW)	Reserve margin (%)	EENS (MWh/yr)	ECOST ($M/yr)	Fixed cost ($M/yr)	Total cost ($M/yr)
original	200	11.11	942.826	3.54208	0.00000	3.54208
add 1 × 10 MW	210	16.67	308.986	1.18342	0.50000	1.68342
add 2 × 10 MW	220	22.22	74.929	0.28698	1.00000	1.28698
add 3 × 10 MW	230	27.78	39.322	0.15060	1.50000	1.65060
add 4 × 10 MW	240	33.33	18.833	0.07213	2.00000	2.07213
add 5 × 10 MW	250	38.89	6.202	0.02375	2.50000	2.52375

Case 3

Table 13.7 and Fig. 13.7 show the variation in costs when the additional unit fixed costs are doubled. In this case, the original system reserve margin of 17.66% is the least cost value.

Case 4

Table 13.8 and Fig. 13.8 show the variation in costs when the peak load in Case 1 is increased to 180 MW. The optimum reserve margin in this case is 22.22%.

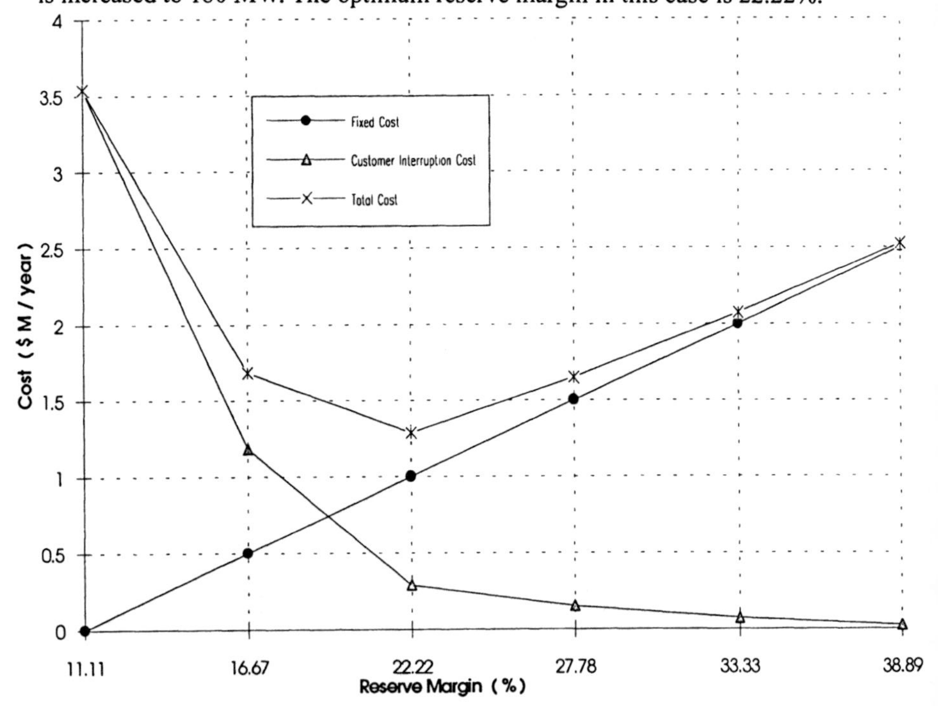

Fig. 13.8 Case 4: Change in fixed, customer, and total costs with reserve margin

The total costs as a function of reserve margin for the four cases are shown in Fig. 13.9. As might be expected in this small system example, the optimum reserve margin occurs at a reserve of 40 MW, the capacity of the large units in the system. The optimum reserve margin is clearly dependent on the data used in the system evaluation. Figure 13.10 illustrates the variation in total cost with percent reserve margin for the situation in which 5-MW units are added. The FOR of the units in the original system have been doubled. The optimum reserve margin in this case increases from 23.53% to 26.47%.

13.6.3 Reliability worth assessment at HLII

Quantitative evaluation of system reliability at HLI is a straightforward procedure, and therefore it is relatively simple to extend the concept to HLI reliability worth assessment. This is not the case at HLII where few electric power utilities apply quantitative reliability assessment techniques. There is, however, a large body of published material available on HLII reliability evaluation which describes various techniques and computer programs. Chapter 6 describes in detail the basic elements of the contingency enumeration approach to HLII evaluation. This technique can be extended to include reliability worth considerations.

The basic contingency enumeration approach considers all component outages up to a specified level. For each outage contingency, the system state is scrutinized and if necessary appropriate corrective actions are taken. A system failure is recorded when corrective actions, other than curtailing customer loads, are unable to eliminate the system problem. The severity of the failure is evaluated by calculating the frequency, duration, magnitude, and location of the load curtailment. For each contingency j that leads to load curtailment at the load bus k, the variables generated in the contingency evaluation are the magnitude L_{kj} (MW) of load curtailment, the frequency f_j (occ/year), and the duration d_j (hours) of the contingency j.

The procedure used to calculate the expected customer interruption cost (ECOST) at each bus is very similar to that used in Section 13.6.2 for HLI analysis and is given by Equations (13.5)–(13.7).

The EENS at bus k due to all the contingencies that lead to load curtailment is given by Equation (13.5). The cost $c_j(d_j)$ of an outage of duration d_j can be obtained from the CCDF of bus k. The expected total cost (ECOST) of power interruptions to customers at bus k for all contingencies is given by Equation (13.6). The IEAR at bus k is evaluated using Equation (13.7) and the aggregate system IEAR is calculated using Equation (13.8).

$$\text{EENS}_k = \sum_{j=1}^{\text{NC}} L_{kj} f_j d_j \quad \text{MWh/yr} \tag{13.5}$$

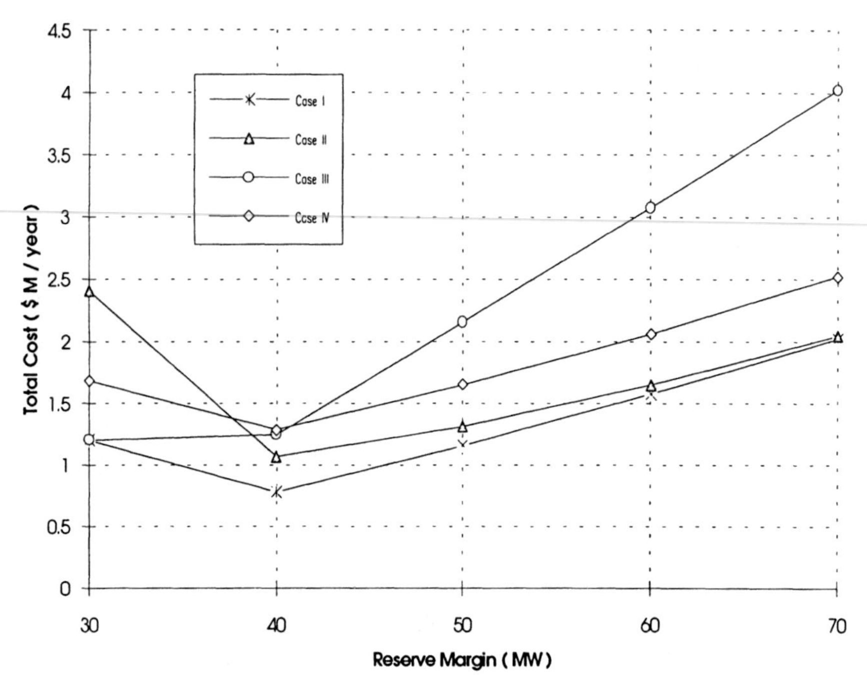

Fig. 13.9 Total cost comparison for Cases 1–4

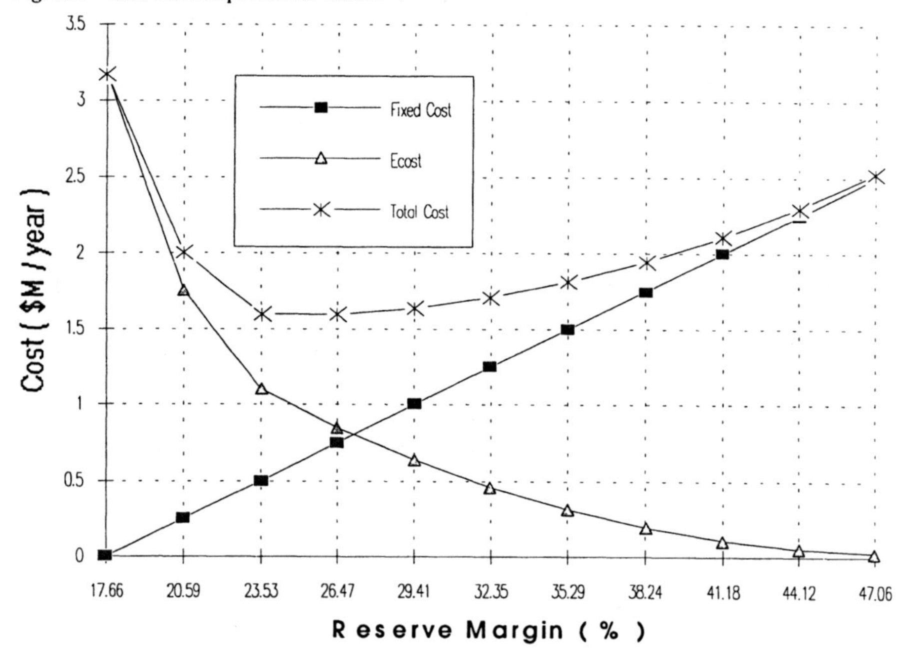

Fig. 13.10 Change in fixed, customer, and total costs with reserve margin as 5-MW units are added. Original system FOR is doubled

Table 13.9 Sector load distribution

Sector	*Energy and peak*
Industrial	25%
Commercial	35%
Residential	40%
	100%

$$\mathrm{ECOST}_k = \sum_{j=1}^{\mathrm{NC}} L_{kj} f_j c_j(d_j) \quad \mathrm{MWh/kW\,yr} \tag{13.6}$$

$$\mathrm{IEAR}_k = \frac{\sum_{j=1}^{\mathrm{NC}} L_{kj} f_j c_j(d_j)}{\sum_{j=1}^{\mathrm{NC}} L_{kj} f_j d_j} \quad \$/\mathrm{kWh} = \frac{\mathrm{ECOST}_k}{\mathrm{EENS}_k} \tag{13.7}$$

$$\text{Aggregate system IEAR} = \sum_{k=1}^{\mathrm{NB}} \mathrm{IEAR}_k \, q_k \ \$/\mathrm{kWh} \tag{13.8}$$

Here NC is the total number of outages that lead to power interruption at bus k, NB is the total number of load buses in the system, and q_k is the fraction of the system load utilized by the customers at bus k.

Equations (13.5)–(13.8) are illustrated using the system shown in Fig. 6.2 and the load distribution data in Table 13.9.

The sector CCDF are given in Table 13.1. The CCDF at the bus can be obtained by weighting the individual sector CDF by the bus sector energy distribution. Assuming that, in this case, the energy and peak load in per unit are equal, the CCDF is given in Table 13.10. Interruption costs for durations longer than 8 hr were estimated using the same slope as the straight line joining the 4- and 8-hr values.

The individual contingencies for the system shown in Figure 6.2 are listed in Table 6.15. The fixed load level in this analysis in 110 MW. Table 13.11 shows the extension of the contingency enumeration approach provided by Equations (13.5)

Table 13.10 Load point CCDF ($/kW)

Duration				
1 min	*20 min*	*1 hr*	*4 hr*	*8 hr*
0.54	2.04	5.45	19.22	49.28

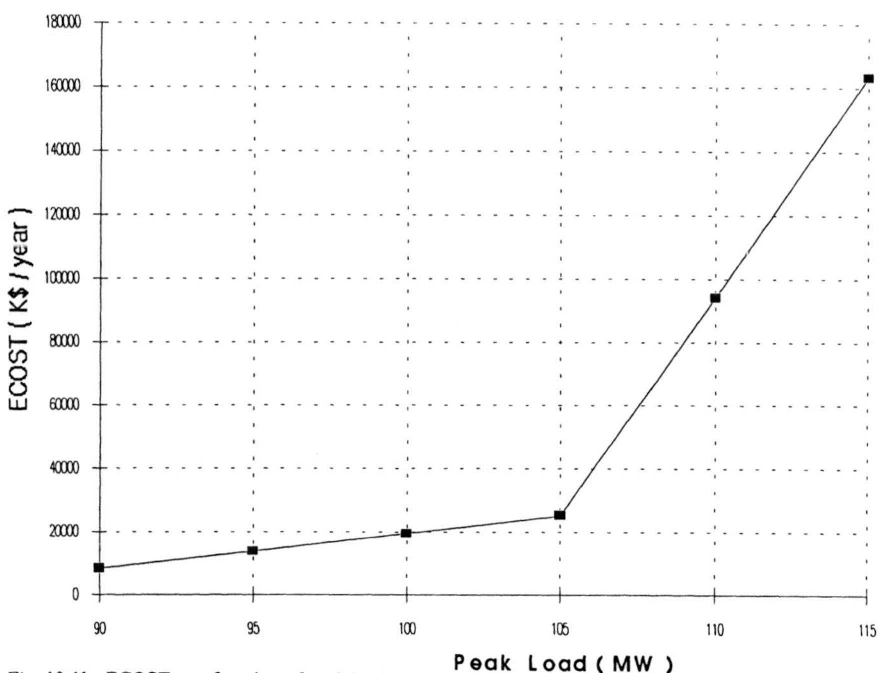

Fig. 13.11 ECOST as a function of peak load

and (13.6) for a range of annual load levels. The system ECOST at a load of 110 MW is 94,242.4 k$/yr, and the EENS is 7310.65 MWh/yr. The IEAR is therefore 12.89 $/kWh.

Figure 13.11 shows the system ECOST as a function of peak load. The ECOST increases rapidly at peak loads greater than 105 MW, at which point the loss of a 30-MW generating unit results in load curtailment (see Section 6.6.2).

The results in Table 13.11 can be used to find an annual ECOST. Assume that the system load duration curve can be approximated by the three-step model in Table 13.12. Using the data in Tables 13.11 and 13.12, the annual ECOST is

$$(94242.40)(0.1) + (19427.22)(0.3) + (8263.79)(0.6) = 20210.68 \text{ k\$/yr}$$

The annual EENS is

$$(7310.65)(0.1) + (2082.09)(0.3) + (841.33)(0.6) = 1860.49 \text{ MWh/yr}$$

The system and load point IEAR is 10.86 $/kW. This IEAR can be used in a wide range of studies to evaluate reliability worth at the load point.

13.6.4 Reliability worth assessment in the distribution functional zone

The estimation of distribution system reliability worth in terms of customer interruption costs proceeds in a similar manner to that used in HLII evaluation and involves five basic steps [1, 27].

Table 13.11 ECOST calculation at selected load levels

			Peak load = 90		Peak load = 95		Peak load = 100		Peak load = 105		Peak load = 110		Peak load = 115	
State	Frequency (occ/yr)	Duration (hr)	L_{ki} (MW)	ECOST (K$/yr)	L_{ki} (MW)	ECOST (K$/yr)	L_k (MW)	ECOST (K$/yr)	L_{ki} (MW)	ECOST (K$/yr)	L_{ki} (MW)	ECOST (K$/yr)	L_k (MW)	ECOST (K$/yr)
1	18.85227476	398.18	—	—	—	—	—	—	—	—	—	—	—	—
2	4.15477080	73.00	—	—	—	—	—	—	—	—	—	—	—	—
3	0.11436062	40.18	—	—	—	—	5	252.37	10	504.75	15	757.12	20	1009.49
4	0.63414996	50.34	5	1900.85	10	3801.71	15	5702.56	20	7603.42	25	9504.27	30	11405.13
5	0.15329376	7.23	—	—	—	—	—	—	—	—	—	—	—	—
6	0.19145910	7.23	9	74.01	14	115.13	19	156.24	24	197.36	29	238.47	34	279.59
7	0.11774001	8.82	—	—	5	33.12	10	66.25	15	99.37	20	132.49	25	165.62
8	6.85537252	115.26	—	—	—	—	—	—	—	—	5	63312.76	10	126625.52
9	0.30858620	67.38	15	4123.36	20	5497.82	25	6872.27	30	8246.73	35	9621.18	40	10995.64
10	0.38783327	7.44	—	—	—	—	—	—	—	—	5	86.59	10	173.18
11	0.48438029	7.45	9	195.02	14	303.36	19	411.71	24	520.05	29	628.40	34	736.74
12	0.29315559	9.23	—	—	5	87.72	10	175.44	15	263.17	20	350.89	25	438.61
13	3.48402390	7.87	—	—	—	—	—	—	—	—	—	—	—	—
14	0.03150290	3.84	35	20.42	40	23.34	45	26.25	50	29.17	55	32.09	60	35.01
15	0.02128992	4.41	15	7.01	20	9.34	25	11.68	30	14.02	35	16.35	40	18.69
16	4.35112256	7.88	9	1890.59	14	2940.92	19	3991.25	24	5041.58	29	6091.91	34	7142.24
17	0.02659900	4.41	90	52.53	95	55.45	100	58.37	105	61.29	110	64.21	115	67.12
18	2.62652070	9.79	—	—	5	851.42	10	1702.83	15	2554.25	20	3405.67	25	4257.09
Total				8263.79		13719.33		19427.22		25135.16		94242.40		163349.67

Table 13.12 Three-step load model

Load level (MW)	*Probability*
100	0.1
100	0.3
90	0.6

1. For load point p connected to the network, obtain the indices of λ_j, r_j, and U_j for each outage event j contributing to its outage.
2. Evaluate the cost of interruption C_{jp} (in \$/kW) using the corresponding CDF and outage duration r_j.
3. Evaluate the corresponding CIC and EENS due to event j using Equations (13.8) and (13.9):

$$\text{CIC}_{jp} = C_{jp} L_p \lambda_j \tag{13.8}$$

$$\text{EENS}_{jp} = L_p U_j \tag{13.9}$$

4. Repeat Steps 2 and 3 for each outage event contributing to load point p. The total load point EENS and CIC are then evaluated using Equations (13.10) and (13.11), where N is the number of outage events within the radial segments which cause the isolation of load point p.

$$\text{CIC}_p = \sum_{j=1}^{N} \text{CIC}_{jp} \quad \text{\$/yr} \tag{13.10}$$

$$\text{EENS}_p = \sum_{j=1}^{N} \text{EENS}_{jp} \quad \text{MWh/year} \tag{13.11}$$

The calculation of expected interruption cost in a radial distribution system can be illustrated using the simple configuration given in Fig. 7.4. The data for this system are given in Tables 7.7 and 7.9. Assume that at each bus there is a 40%

Table 13.13 Load point CCDF

Duration	*\$/kW*
1 min	0.1530
20 min	1.2434
1 hr	3.7100
4 hr	15.4752
8 hr	42.6172

Table 13.14 Failure mode effect analysis for the system in Table 7.8 including EENS and ECOST

Component failure	*Load pt A*					*Load pt B*					*Load pt C*					*Load pt D*				
Section	λ	*r*	*u*	*EENS*	*ECOST*	λ	*r*	*u*	*EENS*	*ECOST*	λ	*r*	*u*	*EENS*	ECOST	λ	*r*	*u*	*EENS*	*ECOST*
1	0.2	4	0.8	4	15.4752	0.2	4	0.8	3.2	12.3802	0.2	4	0.8	2.4	9.2851	0.2	4	0.8	1.6	6.1901
2	0.1	4	0.4	2	7.7376	0.1	4	0.4	1.6	6.1901	0.1	4	0.4	1.2	4.6426	0.1	4	0.4	0.8	3.0950
3	0.3	4	1.2	6	23.2128	0.3	4	1.2	4.8	18.5702	0.3	4	1.2	3.6	13.9277	0.3	4	1.2	2.4	9.2851
4	0.2	4	0.8	4	15.4752	0.2	4	0.8	3.2	12.3802	0.2	4	0.8	2.4	9.2851	0.2	4	0.8	1.6	6.1901
Distributor																				
a	0.2	2	0.4	2	7.5771	0.2	2	0.4	1.6	6.0617	0.2	2	0.4	1.2	4.5463	0.2	2	0.4	0.8	3.0308
b	0.6	2	1.2	6	22.7313	0.6	2	1.2	4.8	18.1850	0.6	2	1.2	3.6	13.6388	0.6	2	1.2	2.4	9.0925
c	0.4	2	0.8	4	15.1542	0.4	2	0.8	3.2	12.1234	0.4	2	0.8	2.4	9.0925	0.4	2	0.8	1.6	6.0617
d	0.2	2	0.4	2	7.5771	0.2	2	0.4	1.6	6.0617	0.2	2	0.4	1.2	4.5463	0.2	2	0.4	0.8	3.0308
Total	2.2	2.73	6.0	30	114.9405	2.2	2.73	6.0	24	91.9525	2.2	2.73	6.0	18	68.9644	2.2	2.73	6.0	12	45.9761

Table 13.15 Failure mode effect analysis for the system in Table 7.10 including EENS and ECOST

Component failure	*Load pt A*					*Load pt B*					*Load pt C*					*Load pt D*				
Section	λ	r	u	*EENS*	*ECOST*	λ	r	u	*EENS*	*ECOST*	λ	r	u	*EENS*	*ECOST*	λ	r	u	*EENS*	*ECOST*
1	0.2	4	0.8	4	15.4752	0.2	4	0.8	3.2	12.3802	0.2	4	0.8	2.4	9.2851	0.2	4	0.8	1.6	6.1901
2	0.1	4	0.4	2	7.7376	0.1	4	0.4	1.6	6.1901	0.1	4	0.4	1.2	4.6426	0.1	4	0.4	0.8	3.0950
3	0.3	4	1.2	6	23.2128	0.3	4	1.2	4.8	18.5702	0.3	4	1.2	3.6	13.9277	0.3	4	1.2	2.4	9.2851
4	0.2	4	0.8	4	15.4752	0.2	4	0.8	3.2	12.3802	0.2	4	0.8	2.4	9.2851	0.2	4	0.8	1.6	6.1901
Distributor																				
a	0.2	2	0.4	2	7.5771															
b						0.6	2	1.2	4.8	18.1850										
c											0.4	2	0.8	2.4	9.0925					
d																0.2	2	0.4	0.8	3.0308
Total	1.0	3.6	3.6	18	69.4779	1.4	3.14	4.4	17.6	67.7057	1.2	3.33	4.0	12	46.2330	1.0	3.6	3.6	7.2	27.7911

Table 13.16 Failure mode effect analysis for the system in Table 7.11 including EENS and ECOST

Component failure	*Load pt A*					*Load pt B*					*Load pt C*					*Load pt D*				
Section	λ	*r*	*u*	*EENS*	*ECOST*	λ	*r*	*u*	*EENS*	*ECOST*	λ	*r*	*u*	*EENS*	*ECOST*	λ	*r*	*u*	*EENS*	*ECOST*
1	0.2	4	0.8	4	15.4752	0.2	4	0.8	3.2	12.3802	0.2	4	0.8	2.4	9.2851	0.2	4	0.8	1.6	6.1901
2	0.1	0.5	0.05	0.25	0.9307	0.1	4	0.4	1.6	6.1901	0.1	4	0.4	1.2	4.6426	0.2	4	0.4	0.8	3.0950
3	0.3	0.5	0.15	0.75	2.7921	0.3	0.5	0.15	0.6	2.2337	0.3	4	1.2	3.6	13.9277	0.3	4	1.2	2.4	9.2851
4	0.2	0.5	0.1	0.5	1.8614	0.2	0.5	0.1	0.4	1.4891	0.2	0.5	0.1	0.3	1.1168	0.2	4	0.8	1.6	6.1901
Distributor																				
a	0.2	2	0.4	2	7.5771															
b						0.6	2	1.2	4.8	18.1850										
c											0.4	2	0.8	2.4	9.0925					
d																0.2	2	0.4	0.8	3.0308
Total	1.0	1.5	1.5	7.5	28.6365	1.4	1.89	2.65	10.6	40.4781	1.2	2.75	3.3	9.9	38.0647	1.0	3.6	3.6	7.2	27.7911

Table 13.17 Failure mode effect analysis for the system in Table 7.13 including EENS and ECOST

Component failure	*Load pt A*					*Load pt B*					*Load pt C*					*Load pt D*				
Section	λ	*r*	*u*	*EENS*	*ECOST*	λ	*r*	*u*	*EENS*	*ECOSt*	λ	*r*	*u*	*EENS*	*ECOST*	λ	*r*	*u*	*EENS*	*ECOST*
1	0.2	4	0.8	4	15.4752	0.2	0.5	0.1	0.4	1.4891	0.2	0.5	0.1	0.3	1.1168	0.2	0.5	0.1	0.2	0.7446
2	0.1	0.5	0.05	0.25	0.9307	0.1	4	0.4	1.6	6.1901	0.1	0.5	0.05	0.15	0.5584	0.1	0.5	0.05	0.1	0.3723
3	0.3	0.5	0.15	0.75	2.7921	0.3	0.5	0.15	0.6	2.2337	0.3	4	1.2	3.6	13.9277	0.3	0.5	0.15	0.3	1.1168
4	0.2	0.5	0.1	0.5	1.8614	0.2	0.5	0.1	0.4	1.4881	0.2	0.5	0.1	0.3	1.1168	0.2	4	0.8	1.6	6.1901
Distributor																				
a	0.2	2	0.4	2	7.5771															
b						0.6	2	1.2	4.8	18.1850										
c											0.4	2	0.8	2.4	9.0925					
d																0.2	2	0.4	0.8	3.0308
Total	1.0	1.5	1.5	7.5	28.6365	1.4	1.39	1.95	7.8	29.5870	1.2	1.88	2.25	6.75	25.8122	1.0	1.5	1.5	3.0	11.4546

Table 13.18 Summary of ECOST for the four cases studied

	ECOST (k$/yr)				
Case	*Load point A*	*Load point B*	*Load point C*	*Load point D*	*Total*
1 (Table 7.8)	114.9405	91.9525	68.9644	45.9761	321.8335
2 (Table 7.10)	69.4779	67.7057	46.2330	27.7911	211.2077
3 (Table 7.11)	28.6365	40.4781	38.0647	27.7911	134.9704
4 (Table 7.13)	28.6365	29.5870	25.8122	11.4546	95.4903

commercial and 60% residential customer mix for both energy and peak load. Using the residential and commercial CDF data in Table 13.1, the CCDF at the load points in Fig. 7.4 is shown in Table 13.13.

Table 7.8 shows the calculation of the load point failure rate, average outage duration, and annual outage time using the failure mode and effect analysis approach. These calculations are extended in Table 13.14 by including the EENS and ECOST for each contingency. This is designated as Case 1.

The analysis in Chapter 7 considers several alternative configurations and their effects on the customer load point reliability. The reliability worth calculations have been extended for the following cases.

Case 2. Section 7.5 describes the effect of including lateral distributor protection. The analysis is shown in Table 7.10.

Case 3. Section 7.6 describes the effect of adding disconnects in the main feeder. The analysis is shown in Table 7.11.

Case 4. Section 7.8 describes the effect of transferring load to an adjacent feeder. The analysis is shown in Table 7.13.

The extended FMEA for Cases 2, 3, and 4 are shown in Tables 13.15, 13.16, and 13.17 respectively and summarized in Table 13.18.

Table 13.18 clearly shows how the customer interruption costs decrease with subsequent modifications and investment in the system. The ECOST figures in each case can be considered in conjunction with the annual costs associated with making the modification in order to decide on the optimum alternative.

13.6.5 Station reliability worth assessment

The extended FMEA approach can be applied to the reliability worth evaluation of stations using the approach described for distribution system evaluation. This approach can also be used for networked transmission and distribution systems. Table 10.8 shows the calculation of load point failure rates, average outage durations, and annual outage times for the alternatives in Fig. 10.10. The analysis is extended in Tables 13.19–13.23 by including the ECOST for each failure event. The load point CCDF is given in Table 13.13.

Table 13.19 ECOST analysis for Case A

Failure event	λ_{pp} *(f/yr)*	r_{pp} *(hr)*	U_{pp} *(hr/yr)*	*L* *(MW)*	*ECOST* *(k$/yr)*
Total outage					
7	2.40×10^{-2}	2.00	4.80×10^{-2}	15	2.7278
1 + 4	1.36×10^{-5}	3.67	4.97×10^{-5}	15	0.0029
1 + 5	5.89×10^{-5}	6.39	3.77×10^{-4}	15	0.0271
1 + 6	2.12×10^{-6}	2.13	4.52×10^{-6}	15	0.0003
2 + 4	5.89×10^{-5}	6.39	3.77×10^{-4}	15	0.0271
2 + 5	1.14×10^{-4}	25.00	2.85×10^{-3}	15	0.3853
2 + 6	1.21×10^{-5}	2.83	3.42×10^{-5}	15	0.0020
3 + 4	2.12×10^{-6}	2.13	4.52×10^{-6}	15	0.0003
3 + 5	1.21×10^{-5}	2.83	3.42×10^{-5}	15	0.0020
3 + 6	2.74×10^{-7}	1.50	4.11×10^{-7}	15	0.0000
Active failure					
3A	1.00×10^{-2}	1.00	1.00×10^{-2}	15	0.5565
6A	1.00×10^{-2}	1.00	1.00×10^{-2}	15	0.5565
Active + stuck					
1A + 3S	4.50×10^{-3}	1.00	5.40×10^{-3}	15	0.3005
2A + 3S	6.00×10^{-3}	1.00	6.00×10^{-3}	15	0.3339
4A + 3S	5.40×10^{-3}	1.00	5.40×10^{-3}	15	0.3005
5A + 6S	6.00×10^{-3}	1.00	6.00×10^{-3}	15	0.3339
Total	6.71×10^{-2}	1.41	9.45×10^{-2}	—	5.5566

Table 13.20 ECOST analysis for Case B

Failure event	λ_{pp} *(f/yr)*	r_{pp} *(hr)*	U_{pp} *(hr/yr)*	*L* *(MW)*	*ECOST* *(k$/yr)*
Total outage					
7	2.40×10^{-2}	2.00	4.80×10^{-2}	15	2.7278
1 + 4	1.36×10^{-5}	3.67	4.97×10^{-5}	15	0.0030
1 + 5	5.89×10^{-5}	6.39	3.77×10^{-4}	15	0.0271
2 + 4	5.89×10^{-5}	6.39	3.77×10^{-4}	15	0.0271
2 + 5	1.14×10^{-4}	25.00	2.85×10^{-3}	15	0.3853
Active failure					
1A	9.00×10^{-2}	1.00	9.00×10^{-2}	15	5.0085
2A	1.00×10^{-1}	1.00	1.00×10^{-1}	15	5.5650
4A	9.00×10^{-2}	1.00	9.00×10^{-2}	15	5.0085
5A	1.00×10^{-1}	1.00	1.00×10^{-1}	15	5.5650
Total	4.04×10^{-1}	1.07	4.32×10^{-1}	—	24.3173

Table 13.21 ECOST analysis for Case C

Failure event	λ_{pp} *(f/yr)*	r_{pp} *(hr)*	U_{pp} *(hr/yr)*	*L (MW)*	*ECOST (k\$/yr)*
Load L1					
7	2.40×10^{-2}	2.00	4.80×10^{-2}	7.5	1.3639
1	9.00×10^{-2}	7.33	6.60×10^{-1}	7.5	25.3147
2	1.00×10^{-1}	50.00	5.00	7.5	465.3794
Load L2					
8	2.40×10^{-2}	2.00	4.80×10^{-2}	7.5	1.3639
4	9.00×10^{-2}	7.33	6.60×10^{-1}	7.5	25.3147
5	1.00×10^{-1}	50.00	5.00	7.5	465.3794
Total	4.28×10^{-1}	26.70	11.42	—	984.1160

The results presented in Tables 13.19–13.23 are summarized in Table 13.24. The ECOST is different in each case. These values can be used in conjunction with the investment and operation/maintenance costs associated with each alternative to select the optimum configuration.

Table 13.22 ECOST analysis for Case D

Failure event	λ_{pp} *(f/yr)*	r_{pp} *(hr)*	U_{pp} *(hr/yr)*	*L (MW)*	*ECOST (k\$/yr)*
For load L1 (load L2 identical)					
7	2.40×10^{-2}	2.00	4.80×10^{-2}	7.5	1.3639
1 + 4	1.36×10^{-5}	3.67	4.97×10^{-5}	7.5	0.0014
1 + 5	5.89×10^{-5}	6.39	3.77×10^{-4}	7.5	0.0136
1 + 8	2.30×10^{-6}	1.57	3.61×10^{-6}	7.5	0.0001
1 + 9	2.12×10^{-6}	2.13	4.52×10^{-6}	7.5	0.0001
2 + 4	5.89×10^{-5}	6.39	3.77×10^{-4}	7.5	0.0136
2 + 5	1.14×10^{-4}	25.00	2.85×10^{-3}	7.5	0.1926
2 + 8	1.42×10^{-5}	1.92	2.74×10^{-5}	7.5	0.0008
2 + 9	1.21×10^{-5}	2.83	3.42×10^{-5}	7.5	0.0010
1A	9.00×10^{-2}	1.00	9.00×10^{-2}	7.5	2.5043
2A	1.00×10^{-1}	1.00	1.00×10^{-1}	7.5	2.7825
9A	1.00×10^{-2}	1.00	1.00×10^{-2}	7.5	0.2783
4A + 9S	5.40×10^{-3}	1.00	5.40×10^{-3}	7.5	0.1503
5A + 9S	6.00×10^{-3}	1.00	6.00×10^{-3}	7.5	0.1670
8A + 9S	1.44×10^{-3}	1.00	1.44×10^{-3}	7.5	0.0401
Total (L1 & L2)	4.74×10^{-1}	1.12	5.30×10^{-1}	—	15.0192

Table 13.23 ECOST analysis for Case E

Failure event	λ_{pp} (f/yr)	r_{pp} (hr)	U_{pp} (hr/yr)	L (MW)	ECOST (k$/yr)
Load L1					
7	2.40×10^{-2}	2.00	4.80×10^{-2}	7.5	1.3639
1	9.00×10^{-2}	1.00	9.00×10^{-2}	7.5	2.5043
2	1.00×10^{-1}	1.00	1.00×10^{-1}	7.5	2.7825
9A	1.00×10^{-2}	1.00	1.00×10^{-2}	7.5	0.2783
Load L2					
8	2.40×10^{-2}	2.00	4.80×10^{-2}	7.5	1.3639
4	9.00×10^{-2}	1.00	9.00×10^{-2}	7.5	2.5043
5	1.00×10^{-1}	1.00	1.00×10^{-1}	7.5	2.7825
9A	1.00×10^{-2}	1.00	1.00×10^{-2}	7.5	0.2875
Total	4.48×10^{-1}	1.11	4.96×10^{-1}	—	13.8580

Table 13.24 ECOST, EENS, and IEAR comparison

Case	*A*	*B*	*C*	*D*	*E*
ECOST (k$/yr)	5.5566	24.3173	984.1160	15.0192	13.8580
EENS (MWh/yr)	1.4175	6.4800	85.6500	3.9750	3.7200
IEAR (k$/yr)	3.9200	3.7527	11.4900	3.7784	3.7253

13.7 Conclusions

This chapter illustrates the extension of the quantitative reliability evaluation techniques presented earlier in this book to the assessment of reliability worth. The ability to quantify the worth associated with reliability of electrical energy supply systems provides the opportunity to explicitly consider reliability in an economic appraisal of alternative plans, designs, and operating philosophies. Customer interruption costs serve as a functional surrogate for reliability worth and can be used in a wide range of studies. The monetary effect on consumers of interruptions in electrical energy supply has been examined using a wide range of techniques [2]. While a single approach has not been universally adopted, it appears that utilities favor the survey approach as a practical vehicle to obtain the required data [15–24,28,29].

This chapter briefly reviews the approaches devised to evaluate interruption costs, with particular emphasis on the survey technique. The results obtained can be portrayed in the form of customer damage functions (CDF) which provide estimates of customer outage costs for interruptions of different durations. These data can be customer-specific or aggregated to provide sector CDF. These values can be aggregated at any particular load point in the system to produce a composite customer damage function (CCDF) at that load point.

This chapter illustrates the utilization of sector CDF to create CCDF at HLI, HLII, and distribution load points. These CCDF are then combined with the quantitative reliability procedures and indices described earlier in this book to assess reliability worth at HLI, HLII, and at distribution system load points.

The approach utilized in this chapter to incorporate customer interruption costs in the assessment of reliability worth is a basic extension of the contingency enumeration technique. This procedure can also be applied to the Monte Carlo simulation (MCS) approach to reliability evaluation [30]. Chapter 12 illustrates the determination of failure states in HLI, HLII, and distribution system load point evaluation using the MCS sequential approach. Each time a failure state occurs, the duration and impact of that state can be combined with an appropriate CCDF to generate a customer interruption cost. The costs can be aggregated to estimate the expected annual cost over the sampling period. The application of both the analytical and MCS techniques to HLI are illustrated in [25] and the results compared. Distributions of annual customer outage costs can also be generated using the concepts described in Chapter 12.

Reliability worth evaluation is an important extension of predictive reliability assessment as it permits reliability to be explicitly considered in system economic analysis.

13.8 References

1. Billinton, R., 'Evaluation of reliability worth in an electric power system', *Reliability Engineering and System Safety*, **46** (1994), pp. 15–23.
2. Wacker, G., Billinton, R., 'Customer cost of electric interruption', *IEEE Proc.*, **77** (6) (1989), pp. 919–30.
3. Billinton, R., Allan, R. N., *Reliability Assessment of Large Electric Power Systems,* Kluwer, Boston (1988).
4. Billinton, R., Goel, L., 'Overall adequacy assessment of an electric power system', *IEE Proc C*, **139**(1) (1992).
5. Billinton, R., Wacker, G., Wojczynski, E., 'Comprehensive bibliography on electric service interruption costs', *IEEE Trans. Power Systems* **102** (1983), pp. 1831–8.
6. Tollefson, G., Billinton, R., Wacker, G., 'Comprehensive bibliography on reliability worth and electric service consumer interruption costs, 1980–1990', *IEEE Trans. Power Systems*, **6**(4) (1991), pp. 1508–14.
7. *The value of service reliability to consumers.* EPRI EA-4494, Res. Project 1104-6, Proceedings, May (1966).
8. Webb, M. G., 'The determination of reserve generating capacity criteria in electricity supply systems', *Applied Economics*, **9** (1977), pp. 19–31.

9. Shipley, R., Patton, A., Denison, J., 'Power reliability cost vs. worth', *IEEE Trans. Power App. Syst.*, **PAS-91**, (1972), pp. 2204–12.
10. Myers, D., *The Economic Effects to a Metropolitan Area of Power Outage Resulting from an Earthquake,* Earthquake Engineering Systems Inc., San Francisco (1978).
11. Lundberg, L., *Report of the Group of Experts on Quality of Service from the Customer's Points of View,* UNIPEDE, Rep. 60/D.l, 1972.
12. Munasinghe, M., 'The costs incurred by residential electricity consumers due to power failures', *J. Consumer Res.*, **6** (1980).
13. Markel, L., Ross, N., Badertscher, N., *Analysis of Electric Power System Reliability,* California Energy Resources Conservation and Development Commission (1976).
14. Corwin, J., Miles, W., *Impact Assessment of the 1977 New York City Blackout,* U.S. Department of Energy, Washington, DC (1978).
15. Skof, L. V., *Ontario Hydro Surveys on Power System Reliability Summary of Customer Viewpoints,* Ontario Hydro Rep. R&MR 80-12, EPRI Seminar (1983).
16. Wacker, G.. Wojczynski, E., Billinton, R., 'Interruption costs methodology and results—a Canadian residential survey', *IEEE Trans. Power App. Syst.*, **102** (1983), pp. 3385–91.
17. Wojczynski, E., Billinton, R., Wacker, G., 'Interruption cost methodology and results—a Canadian commercial and small industrial survey', *IEEE Trans. Power App. Syst.*, **103** (1984), pp. 437–43.
18. Wacker, G., Billinton, R., 'Farm losses resulting from electric service interruptions—a Canadian survey', *IEEE Trans. Power Syst.*, **4**(2) (1989), pp. 472–78.
19. IEEE Committee, 'Report on reliability survey of industrial plants. Part II. Cost of power outage, plant restart time, critical service loss, duration, time and type of loads lost versus time of power outages', *IEEE Trans. Industry Appl.*, **IA-10**, (1976), pp. 236–41.
20. Billinton, R., Wacker, G., Wojczynski, E., *Customer Damage Resulting from Electric Service Interruptions,* Canadian Electrical Association, R&D Project 907 U 131 Rep. (1982).
21. Wacker, G., Billinton, R., Brewer, R., *Farm Losses Resulting from Electric Service Interruptions,* Canadian Electric Association R&D Research Project 3409 U 403, May (1987).
22. Subramaniam, R. K., Billinton, R., Wacker, G., 'Factors affecting the development of an industrial customer damage function', *IEEE Trans. Power App. Syst.*, **104** (1985), pp. 3209–15.
23. Billinton, R., Wacker, G., Subramaniam, R. K., 'Factors affecting the development of a commercial customer damage function', *IEEE Trans. Power Syst.*, **1** (1986), pp. 28–33.

24. Billinton, R., Wacker, G., Subramaniam, R. K., 'Factors affecting the development of a residential customer damage function', *IEEE Trans. Power Syst.*, **2** (1987), pp. 204–9.
25. Billinton, R., Oteng-Adjei, J., Ghajar, R., 'Comparison of two alternate methods to establish an interrupted energy assessment rate', *IEEE Trans Power Systems*, **2** (1987), pp. 751–7.
26. Billinton, R., *Composite System Adequacy Assessment—the Contingency Enumeration Approach*. IEEE Tutorial Text, Reliability assessment of composite generation and transmission systems No. 90, EHO311-1-PWR (1990).
27. Goel, L., Billinton, R., Gupta, R., *Basic Data and Evaluation of Distribution System Reliability Worth*, Proc. IEEE Wescanex 91 Conference, Regina, Canada (1991).
28. Kariuki, K. K., Allan, R. N., *Reliability Worth in Distribution Plant Replacement Programmes*, 2nd IEE Conference on Reliability of Transmission and Distribution Equipment, Coventry (1995), IEE Conf. Publ. 406, pp. 162–7.
29. Kariuki, K. K., Allan, R. N., Palin, A., Hartwight, B., Caley, J., *Assessment of Customer Outage Costs due to Electricity Service Interruptions*, CIRED, Brussels (1995), Vol. I/2, pp. 2.05.1–2.05.6.
30. Billinton, R., Li, W., *Reliability Assessment of Electric Power Systems Using Monte Carlo Methods,* Plenum Press, New York (1994).

14 Epilogue

This book is concerned with the quantitative reliability evaluation of power systems. It describes many of the available modeling and evaluation techniques and the various indices that can be deduced and used in practical applications.

As discussed in Chapter 1 and consolidated in the subsequent chapters, the evaluation process considers the system, not as a whole, but as a set of interrelated subsystems, now generally understood as hierarchical levels. This process allows the quantitative assessment of each subsystem or hierarchical level to be done separately and combined appropriately to give indices of relevant systems or subsets of systems. The various sets of HL indices can be used in the managerial decision-making process at that hierarchical level in order to determine the most appropriate expansion and reinforcement schemes, operating policies, and maintenance strategies. A managerial decision is required in order to decide to which part of the system limited investment capital should be directed. Quantitative reliability evaluation does not remove this decision-making process from the engineer or manager, but simply enhances the quality of the decision by adding quantitative measures to the decision process. It is therefore worth reviewing the present applications of reliability assessment at these different hierarchical levels.

The use of reliability evaluation techniques at HLI varies considerably around the world. For example, in the United Kingdom there is no central body since privatization (1990) that is responsible for deciding when additional capacity is required or how much is required. These decisions are left to individual private generators in response to market forces. The values of LOLP and VOLL (value of lost load) embedded in the Pool (energy trading) pricing mechanism are intended to be indicators that encourage or discourage the installation of additional generation by these private generators. The process, however, is very different to that in Canada where utilities operate on a provincial basis. The regulatory bodies require decisions relating to capacity expansion plans to be made on objective bases, which have encouraged all Canadian utilities to use some form of probabilistic reliability assessments. The criteria and indices vary considerably but the basic concept is the same. Several utilities have incorporated a system IEAR (interrupted energy assessment rate) in their planning process and selected capacity reserve criterion based on reliability cost/worth principles.

At present, HLII studies are not used extensively in practice. However, interest has changed dramatically in recent times, and there is a widening awareness of the need to develop relevant computer tools and applications. Such studies are likely to become of significant importance in the near future. It should be appreciated that these studies are extremely valuable in comparing alternatives such as reinforcements, maintenance schedules, operating strategies, etc. It is worth noting that while individual utilities or regulatory bodies need alternative indices in order to reflect particular system conditions and requirements, they may require only a few, even one, for their decision-making process.

Many distribution systems are still designed according to deterministic standards. These views are also changing quite significantly and there is now a positive awareness of the need to assess system design alternatives in a probabilistic sense. There is also a rapidly growing appreciation, inside and outside the industry, of the need to account for customers' expectations and their assessment of the worth of supply. Since the latter cannot be objectively assessed without adequate and objective reliability measures, it is expected that the two aspects, reliability and worth of supply, will become of significant importance in the very near future.

In conclusion, it should be stressed that, although the book considers a very wide range of power system problem areas, it does not purport to cover every conceivable technique, model, or evaluation process. This is simply not practical within the scope available. It does, however, provide readers with a wealth of information which should enable them to consider most of the problems likely to be encountered on a day-to-day basis. After assimilating the models and techniques described in this book, readers should be able to widen and deepen their knowledge and understanding of power system reliability by reading the available relevant technical papers and publications dealing with new techniques which will continue to be published in the foreseeable future.

Appendix 1 **Definitions**

The following terms and associated definitions have been extracted from the IEEE Standard 346-1973 part 2: *Terms for Reporting and Analyzing Outages of Electrical Transmission and Distribution Facilities and Interruptions to Customer Service.* A revised Standard (IEEE Standard 859-1987: *Terms for Reporting and Analyzing Outage Occurrences and Outage States of Electrical Transmission Facilities*) has been published which contains similar terms with slightly different definitions. A reader may wish to refer to this new Standard in addition to the following.

(i) *Component.* A piece of equipment, a line, a section of line, or a group of items which is viewed as an entity for purposes of reporting, analyzing, and predicting outages.

(ii) *System.* A group of components connected or associated in a fixed configuration to perform a specified function.

(iii) *Outage.* Describes the state of a component when it is not available to perform its intended function due to some event directly associated with that component. An outage may or may not cause an interruption of service to consumers depending on system configuration.

  (a) *Forced outage.* An outage that results from emergency conditions directly associated with a component requiring that it be taken out of service immediately, either automatically or as soon as switching operations can be performed, or an outage caused by improper operation of equipment or human error.

  (b) *Scheduled outage.* An outage that results when a component is deliberately taken out of service at a selected time, usually for purposes of construction, preventive maintenance, or repair.

  (c) *Transient forced outage.* An outage whose cause is immediately self-clearing so that the affected component can be restored to service either automatically or as soon as a switch or circuit breaker can be reclosed or a fuse replaced. An example of a transient forced outage is a lightning flashover which does not permanently disable the flashed component.

  (d) *Permanent forced outage.* An outage whose cause is not immediately self-clearing but must be corrected by eliminating the hazard or by repairing or replacing the component before it can be returned to service. An example of a permanent forced outage is a lightning flashover which shatters an insulator, thereby disabling the component until repair or replacement can be made.

(iv) *Weather conditions*

(a) *Normal weather*. Includes all weather not designated as adverse or major disaster.

(b) *Adverse weather*. Designates weather conditions which cause an abnormally high rate of forced outages for exposed components while such conditions persist, but do not qualify as major storm disasters. Adverse weather conditions can be defined for a particular system by selecting the proper values and combinations of conditions reported by the weather bureau: thunderstorms, tornadoes, wind velocities, precipitation, temperature, etc.

(c) *Major storm disaster*. Designates weather which exceeds design limits of plant and which satisfies all of the following:
—extensive mechanical damage to plant;
—more than a specified percentage of customers out of service;
—service restoration times longer than a specified time.

(v) *Exposure time*. The time during which a component is performing its intended function and is subject to outage.

(vi) *Outage rate*. For a particular classification of outage and type of component, the mean number of outages per unit exposure time per component.

(a) *Adverse weather permanent forced outage rate*. For a particular type of component, the mean number of outages per unit of adverse weather exposure time per component.

(b) *Normal weather permanent forced outage rate*. For a particular type of component, the mean number of outages per unit of normal weather exposure time per component.

(vii) *Outage duration*. The period from the initiation of an outage until the affected component or its replacement once again becomes available to perform its intended function.

(a) *Permanent forced outage duration*. The period from the initiation of the outage until the component is replaced or repaired.

(b) *Transient forced outage duration*. The period from the initiation of the outage until the component is restored to service by switching or fuse replacement.

(c) *Scheduled outage duration*. The period from the initiation of the outage until construction, preventive maintenance, or repair work is completed.

(viii) *Switching time*. The period from the time a switching operation is required due to a forced outage until that switching operation is performed.

(ix) *Interruption*. The loss of service to one or more consumers. An interruption is the result of one or more component outages.

(a) *Scheduled interruption*. An interruption caused by a scheduled outage.

(b) *Forced interruption*. An interruption caused by a forced outage.

(x) *Interruption duration*. The period from the initiation of an interruption to a consumer until service has been restored to that consumer.

(a) *Momentary interruption*. An interruption of duration limited to the period required to restore service by automatic or supervisory controlled switching operations or by manual switching at locations where an operator is immediately available.

(b) *Sustained interruption*. A sustained interruption is any interruption not classified as a momentary interruption.

Appendix 2 Analysis of the IEEE Reliability Test System

A2.1 Introduction

The concepts and algorithms presented in Chapters 2 and 3 for evaluation of generating capacity systems were illustrated in these chapters by application to some relatively small system examples. These examples can all be evaluated using a desk calculator and in many cases by basic concepts which do not require the use of a recursive algorithm. This is not the case with practical system studies which require efficient algorithms and a suitable digital computer. In order to provide the opportunity to develop and test a digital computer program which can be used in practical studies, this chapter presents a test system and a set of calculated results for a range of studies. The IEEE Subcommittee on the Application of Probability Methods has developed a Reliability Test System [1] (RTS) which includes both generation and major transmission facilities. The main objective was to provide a basic model which could be used to test or compare methods for reliability analysis of power systems. The generating capacity and load model data are presented in the following section.

A2.2 IEEE–RTS

The total installed capacity in the RTS is 3405 MW. The system capacity composition is given in Table A2.1.

Table A2.2 presents the annual load model in terms of the weekly peak loads as a percentage of the annual peak load. If Week 1 is taken as the first week in January then the load model represents a winter peaking system. A summer peaking system can be created by taking a suitable time for Week 1.

Table A2.3 presents the daily peak load cycle, as a percentage of the weekly peak. The same weekly peak load cycle is assumed to apply for all times of the year. The data in Tables A2.2 and A2.3 defines a daily peak load model of 364 days with Monday as the first day of the year.

Table A2.4 gives weekday and weekend hourly load data for each of three seasons. Combining the data given in Tables A2.2–4 defines an hourly load model of 8736 hours.

Additional data are given in Refs. 1 and 2.

Table A2.1 Generating unit reliability data

Unit size (MW)	*Number of units*	*Forced outage rate*	*MTTF (hours)*	*MTTR (hours)*	*Scheduled maintenance (wks/year)*
12	5	0.02	2940	60	2
20	4	0.10	450	50	2
50	6	0.01	1980	20	2
76	4	0.02	1960	40	3
100	3	0.04	1200	50	3
155	4	0.04	960	40	4
197	3	0.05	950	50	4
350	1	0.08	1150	100	5
400	2	0.12	1100	150	6

Table A2.2 Weekly peak load as a percentage of annual peak

Week	*Peak load (%)*	*Week*	*Peak load (%)*
1	86.2	27	75.5
2	90.0	28	81.6
3	87.8	29	80.1
4	83.4	30	88.0
5	88.0	31	72.2
6	84.1	32	77.6
7	83.2	33	80.0
8	80.6	34	72.9
9	74.0	35	72.6
10	73.7	36	70.5
11	71.5	37	78.0
12	72.7	38	69.5
13	70.4	39	72.4
14	75.0	40	72.4
15	72.1	41	74.3
16	80.0	42	74.4
17	75.4	43	80.0
18	83.7	44	88.1
19	87.0	45	88.5
20	88.0	46	90.9
21	85.6	47	94.0
22	81.1	48	89.0
23	90.0	49	94.2
24	88.7	50	97.0
25	89.6	51	100.0
26	86.1	52	95.2

Table A2.3 Daily peak load as a percentage of weekly peak

Day	*Peak load (%)*
Monday	93
Tuesday	100
Wednesday	98
Thursday	96
Friday	94
Saturday	77
Sunday	75

Table A2.4 Hourly peak load as a percentage of daily peak

	Winter weeks 1–8 & 44-52		*Summer weeks 18–30*		*Spring/Fall weeks 9–17 & 31–43*	
Hour	*Wkdy*	*Wknd*	*Wkdy*	*Wknd*	*Wkdy*	*Wknd*
12–1am	67	78	64	74	63	75
1–2	63	72	60	70	62	73
2–3	60	68	58	66	60	69
3–4	59	66	56	65	58	66
4–5	59	64	56	64	59	65
5–6	60	65	58	62	65	65
6–7	74	66	64	62	72	68
7–8	86	70	76	66	85	74
8–9	95	80	87	81	95	83
9–10	96	88	95	86	99	89
10–11	96	90	99	91	100	92
11–Noon	95	91	100	93	99	94
Noon–1pm	95	90	99	93	93	91
1–2	95	88	100	92	92	90
2–3	93	87	100	91	90	90
3–4	94	87	97	91	88	86
4–5	99	91	96	92	90	85
5–6	100	100	96	94	92	88
6–7	100	99	93	95	96	92
7–8	96	97	92	95	98	100
8–9	91	94	92	100	96	97
9–10	83	92	93	93	90	95
10–11	73	87	87	88	80	90
11–12	63	81	72	80	70	85

Wkdy = Weekday, Wknd = Weekend.

A2.3 IEEE–RTS results

A2.3.1 Single system

The total installed capacity of the RTS is 3405 MW. The complete capacity model with no rounding increment used will have a large number of states. If the model is truncated at a cumulative probability of 1×10^{-8}, the system has 1872 states. Table A2.5 presents a set of representative results for comparison purposes at selected capacity levels. In practice, these tables are truncated and rounded, which, when convolved with load models, which may also be approximated, can give results with varying degrees of inaccuracy.

In order to identify the effect of rounding, a series of studies was conducted [2] in which no approximations were made in the evaluation process. This provides a set of exact indices against which the results from alternative and approximate methods can be compared. These results are:

using daily peak loads, LOLE = 1.36886 day/yr

using hourly loads, LOLE = 9.39418 hr/yr

This was followed by assessing the effect of approximations in both the generation and load models on the system indices. These results [2] are shown in Table A2.6.

In addition to these studies, Ref. 2 outlines restrictions which exist in the generation data of the RTS, extends the RTS by including more factors and system conditions, and provides the results with these additional factors included. The reader is referred to Ref. 2 for the full details; the following is a summary of some of the information:

(a) The 350 MW and 400 MW units were given derated states, producing the results shown in Table A2.7.

Table A2.5 Representative generation model data

State	*Cap. out (MW)*	*Individual probability*	*Cumulative probability*	*Cumulative frequency/day*
1	0	0.23639495	1.0	0.0
31	100	0.02999154	0.54760141	0.14607832
90	200	0.00128665	0.38132840	0.12174396
153	265	0.00001312	0.33556693	0.09192086
288	400	0.06572832	0.26187364	0.06434489
444	556	0.00000345	0.08457820	0.05360552
488	600	0.00035769	0.06211297	0.04291001
838	950	0.00006431	0.00749197	0.00712004
1088	1200	0.00002413	0.00079125	0.00104271
1388	1500	0.00000030	0.00004043	0.00006923

Table A2.6 Effect of rounding

Capacity model rounding interval (MW)	*Load model (no. of points)*	*LOLE (d/yr)*
20	exact	1.38587
40	exact	1.37978
60	exact	1.39806
80	exact	1.37687
100	exact	1.41622
exact	10	1.74649
exact	100	1.42843
exact	200	1.38993
exact	364	1.37256
20	100	1.43919
20	200	1.39869
20	364	1.38967
40	100	1.45041
40	200	1.41514
40	364	1.39415

(b) The load model was associated with varying degrees of uncertainty, giving the results shown in Table A2.8.

(c) The peak load was varied between 0.84 and 1.10 of that specified in the RTS, giving the results shown in Table A2.9.

(d) The generation was expanded to include a varying number of additional 25-MW gas turbines, giving the results shown in Table A2.10.

Table A2.7 Effect of derated states

Units derated	*LOLE (d/yr)*
1 × 400 MW	1.16124
2 × 400 MW	0.96986
2 × 400 + 1 × 350 MW	0.88258

Table A2.8 Effect of load forecast uncertainty

Uncertainty (%)	*LOLE (d/yr)*
2	1.45110
5	1.91130
10	3.99763
15	9.50630

Table A2.9 Effect of peak load

Multiplying factor p.u.	*Peak load (MW)*	*LOLE (d/yr)*
1.10	3135	6.68051
1.06	3021	3.77860
1.04	2964	2.67126
1.00	2850	1.36886
0.96	2736	0.65219
0.92	2622	0.29734
0.88	2508	0.12174
0.84	2394	0.04756

(e) Finally the exact energy indices of the basic RTS were calculated and found to be LOEE = 1.176 GWh and EIR = 0.999923.

A2.3.2 Interconnected systems

A study was also conducted on two identical IEEE Reliability Test Systems connected through a completely reliable tie line of variable capacity to illustrate how the LOLE is affected by the tie line capacity. Under the assumptions stated previously for two interconnected systems, the variation of the LOLE of one system as a function of the tie line capacity is shown in Table A2.11.

A2.3.3 Frequency and duration approach

The IEEE–RTS provides an excellent vehicle for comparing the results obtained from the two different load models available in the F&D approach. A series of studies has been performed using the IEEE–RTS to provide a comprehensive set of results which can be used in program testing.

Table A2.10 Effect of adding gas turbines

No. of gas turbines	*LOLE (d/yr)*
1	1.18293
3	0.86372
5	0.62699
8	0.38297
10	0.27035
12	0.18709
15	0.10674
16	0.08850

Table A2.11

Tie line capacity (MW)	*LOLE (days/year) (no rounding)*
0	1.3689
100	0.7500
200	0.4633
300	0.3413
400	0.2934
500	0.2771
600	0.2740
700	0.2740

These studies were conducted using the daily peak loads over the period of a year consisting of 365 days unless otherwise specified.

(1) Generation

All generators regardless of unit size were assumed as binary units. The daily generation model was developed by converting the hourly data into daily data. The capacity outage table was truncated at a cumulative probability less than 10^{-8}.

(2) Load

A winter peaking system was adopted by taking Week 1 as January and Monday as the first day of the year. Since the test system provides only 364 daily peak loads in a year, it was assumed that the daily peak load on 31 December was the same as that on 1 January.

(a) *Exact-state load model*

The daily peak loads were assumed to exist for 12 hours giving an exposure factor of 0.5. The 365 daily peaks were arranged in ascending order and then grouped in

Table A2.12 Load occurrence data table

Level j	*Load level L_j (MW)*	*No. of Occurrences*	*Prob. $p(L_j)$*	*$+(L_j)$/day*	*$-(L_j)$/day*
1	2687	12	0.01643836	0	2
2	2454	82	0.11232877	0	2
3	2188	108	0.14794520	0	2
4	1953	116	0.15890411	0	2
5	1593	47	0.06438356	0	2
6	1485	365	0.50000000	2	0

Table A2.13

Exposure factor	*Cumulative probability*	*Cumulative frequency/day*
0.2	0.64202767 E-2	0.37328424 E-2
0.5	0.16050218 E-2	0.45170295 E-2
0.6	0.19260199 E-2	0.47784252 E-2

class intervals of 1450–1750, 1750–2050, 2050–2350, 2350–2650, 2650 and greater. The mean of each class was taken as the load level and the class frequency as the number of occurrences of that load level as shown in Table A2.12. After rounding off the decimal places, the exact-state load model was obtained as follows:

No. of load levels (including low load) = 6

Annual peak load (MW) = 2850

Exposure factor, e = 0.5

Period of study (days) = 365

The first negative margin was taken as the load loss situation and therefore the calculated indices are the cumulative probability and frequency associated with this margin. Table A2.13 also shows results for exposure factors between 0.2 and 0.6.

(b) *Cumulative state load model*

This load model is formulated using the chronological input data, and therefore depends on which peaking system (winter or summer) is used. The load levels in this case (Table A2.14) were assumed to be equally spaced between the annual peak and annual low loads. The following expression was used to compute the load level increment between any two load levels:

$$\text{load level increment} = \frac{(\text{annual peak load}) - (\text{annual low load})}{(\text{no. of load levels}) - 2}$$

The cumulative state load model was obtained as follows:

No. of load levels = 8

Annual peak load (MW) = 2850

Annual low load (MW) = 1485.56

Period of study (days) = 365

The margin of 0 MW was taken as the load loss situation and therefore the calculated indices are the cumulative probability and frequency associated with this margin.

The number of load levels is an imbedded parameter built into the cumulative state load model which affects the cumulative probability and frequency of the load loss situation. The selection of a large number of load levels will give accurate

Table A2.14 Load level data

Level i	*Load level L (MW)*	*Prob. P(L)*	*Freq. F(L)/day*
1	2850.00	0.	0.
2	2622.59	0.038356	0.005479
3	2395.19	0.202740	0.030137
4	2167.78	0.419178	0.104110
5	1940.38	0.704110	0.098630
6	1712.97	0.879452	0.087671
7	1485.56	0.997260	0.065753
8	1485.56	1.000000	0.

Table A2.15

No. of load levels	*Cumulative probability*	*Cumulative frequency/day*
8	0.90109048 E-2	0.75496321 E-2
10	0.73679441 E-2	0.65175359 E-2
15	0.54434502 E-2	0.56968695 E-2
20	0.49097231 E-2	0.52101543 E-2
30	0.45568782 E-2	0.50016633 E-2
40	0.42683125 E-2	0.48351036 E-2
50	0.41478655 E-2	0.46945866 E-2

results, but on the other hand it demands more execution time and computer memory. A question which arises is 'what is the optimal number of load levels?'

The cumulative probability and frequency of the load loss situation were computed as a function of the number of load levels. The results are shown in Table A2.15.

The frequency and duration approach can be used in interconnected system evaluation. Consider two identical IEEE–RTS Systems A and B and assume that they are interconnected by a finite capacity interconnection and the requirement is

Table A2.16 Exact-state load model

Tie capacity (MW)	*Cumulative probability*	*Cumulative frequency/day*
0	0.16511638 E-2	0.46503918 E-2
200	0.36053659 E-3	0.10929590 E-2
400	0.65338427 E-4	0.22418916 E-2
600	0.14944406 E-4	0.68437925 E-2
800	0.86339767 E-5	0.47665535 E-2
1000	0.81682158 E-5	0.45980385 E-4

Table A2.17 Cumulative state load model

Tie capacity (MW)	*Cumulative probability*	*Cumulative frequency/day*
0	0.88971496 E-2	0.74702451 E-2
200	0.23814005 E-2	0.23080875 E-2
400	0.49108965 E-3	0.69563516 E-3
600	0.12680453 E-3	0.27716927 E-3
800	0.60721801 E-4	0.20183909 E-3
1000	0.53263615 E-4	0.19192208 E-3

to evaluate the F&D reliability indices in System A. System B is considered to help System A as much as it can without curtailing its own load. System B has the same capacity model as System A and an exact-state load model. The peak loads in both systems are assumed to be uncorrelated.

The tie capacity was varied from 0 to 1200 MW and the tie lines were assumed to be 100% reliable. The cumulative probability and frequency of the load loss situation for different load models in System A as a function of the tie capacity are given in Tables A2.16 and A2.17.

A2.4 Conclusion

This chapter has presented the basic generation–load data from the IEEE–RTS given in Refs. 1 and 2. It has also presented the calculated indices for a range of studies. It is hoped that these results will provide useful reference values for those interested in applying the algorithms given in Chapters 2 and 3 to practical system studies.

A2.5 References

1. Reliability Test System Task Force of the IEEE Subcommittee on the Application of Probability Methods, 'IEEE Reliability Test System', *IEEE Transactions*, **PAS-98** No. 6, Nov/Dec (1979), pp. 2047–54.
2. Allan, R.N., Billinton, R., Abdel–Gawad, N.M., 'The IEEE reliability test system—extensions to and evaluation of the generating system', *IEEE Trans. on Power Systems*, **PWRS-1** (4) (1986), pp. 1–7.

Appendix 3 Third-order equations for overlapping events

A3.1 Introduction

The equations presented in this appendix relate to third-order overlapping events. They can be applied either to a three-component parallel system or to a third-order minimal cut set (failure event). All the equations can and have been deduced logically using the concepts of overlapping outages described in Chapter 8. The equations are presented using this logical and sequential structure so that the reader may rededuce them if desired. It should be noted, however, that the assumptions and conditions described in Chapter 8 are embedded in these derivations. Consequently, the following equations should not be used indiscriminately and the reader should first check that these assumptions and conditions are applicable to the system he is analyzing. If they are not applicable, suitable modifications should be made, using the same basic logic, and a similar set of appropriate equations should be deduced.

A3.2 Symbols

The following symbols are used in the equations presented in the following sections.

λ_i { = permanent failure rate of component i (if single weather state)
= permanent failure rate of component i in normal weather (if two weather states)

λ_{ti} = temporary/transient failure rate of component i

λ'_i = permanent failure rate of component i in adverse weather

λ''_i = maintenance outage rate of component i

λ_{ij} = common mode failure rate of components i and j

λ_{ijk} { = common mode failure rate of components i, j and k (if single weather state)
= common mode failure rate of components i, j and k in normal weather (if two weather states)

λ'_{ijk} = common mode failure rate of components i, j and k in adverse weather

$\hat{\lambda}_{ijk}$ = average common mode failure rate of components i, j and k

r_i = repair time of component i

r_{ti} = reclosure time of component i
r''_i = maintenance time of component i
r_{ij} = reciprocal of repair rate representing simultaneous repair of components i and j
r_{ijk} = reciprocal of repair rate representing simultaneous repair of components i, j and k
N = average duration of normal weather
S = average duration of adverse weather

A3.3 Temporary/transient failure overlapping two permanent failures

The following equations relate to a third-order event in which one component suffers either a transient or a temporary failure and the two other components are forced out of service as a result of permanent failure. The equations neglect any contribution due to two transient failures, two temporary failures or one of each. If necessary, the equations can be adapted to include these effects although the probability of such an event is generally negligible.

$$\lambda_{pt} = \lambda_{t1}\lambda_2 r_{t1}\lambda_3 \frac{r_{t1}r_2}{r_{t1}+r_2} + \lambda_{t1}\lambda_3 r_{t1}\lambda_2 \frac{r_{t1}r_3}{r_{t1}+r_3}$$

$$+ \lambda_2\lambda_{t1}r_2\lambda_3 \frac{r_2 r_{t1}}{r_2+r_{t1}} + \lambda_2\lambda_3 r_2\lambda_{t1} \frac{r_2 r_3}{r_2+r_3}$$

$$+ \lambda_3\lambda_{t1}r_3\lambda_2 \frac{r_3 r_{t1}}{r_3+r_{t1}} + \lambda_3\lambda_2 r_3\lambda_{t1} \frac{r_3 r_2}{r_3+r_2}$$

+ twelve similar terms involving temporary/transient failures of components 2 and 3

$$= \lambda_{t1}\lambda_2\lambda_3(r_{t1}r_2 + r_2 r_3 + r_3 r_{t1})$$

+ two similar terms involving temporary/transient failures of components
2 and 3

$$\lambda_{pt} = \lambda_a + \lambda_b + \lambda_c$$

$$r_{pt} = (\lambda_a r_a + \lambda_b r_b + \lambda_c r_c)/\lambda_{pt}$$

where

$$\lambda_a = \lambda_{t1}\lambda_2\lambda_3(r_{t1}r_2 + r_2 r_3 + r_3 r_{t1})$$

$$\lambda_b = \lambda_1\lambda_{t2}\lambda_3(r_1 r_{t2} + r_{t2}r_3 + r_3 r_1)$$

$$\lambda_c = \lambda_1\lambda_2\lambda_{t3}(r_1r_2 + r_2r_{t3} + r_{t3}r_1)$$

$$r_a = r_{t1}r_2r_3/(r_{t1}r_2 + r_2r_3 + r_3r_{t1})$$

$$r_b = r_1r_{t2}r_3/(r_1r_{t2} + r_{t2}r_3 + r_3r_1)$$

$$r_c = r_1r_2r_{t3}/(r_1r_2 + r_2r_{t3} + r_{t3}r_1)$$

If the reclosure time is assumed negligible compared with the repair time, the above equations reduce to

$$\lambda_{pt} \simeq \lambda_{t1}\lambda_2\lambda_3r_2r_3 + \lambda_1\lambda_{t2}\lambda_3r_1r_3 + \lambda_1\lambda_2\lambda_{t3}r_1r_2$$

$$r_{pt} \simeq (\lambda_{t1}\lambda_2\lambda_3r_{t1}r_2r_3 + \lambda_1\lambda_{t2}\lambda_3r_1r_{t2}r_3 + \lambda_1\lambda_2\lambda_{t3}r_1r_2r_{t3})/\lambda_{pt}$$

A3.4 Temporary/transient failure overlapping a permanent and a maintenance outage

The following equations relate to a third-order event in which one component is out on scheduled maintenance, one component suffers a transient or temporary failure and one component is forced out of service as the result of a permanent failure. As discussed in Chapter 8, it is assumed that the first component in each sequential outage event is on scheduled maintenance, i.e. a component is not taken out of service for scheduled maintenance if a forced outage in the related event has already occurred.

$$\lambda_{pmt} = \lambda_1''\lambda_2 r_1''\lambda_{t3}\frac{r_1''r_2}{r_1''+r_2} + \lambda_1''\lambda_{t3}r_1''\lambda_2\frac{r_1''r_{t3}}{r_1''+r_{t3}}$$

$$+\lambda_1''\lambda_3 r_1''\lambda_{t2}\frac{r_1''r_3}{r_1''+r_3} + \lambda_1''\lambda_{t2}r_1''\lambda_3\frac{r_1''r_{t2}}{r_1''+r_{t3}}$$

+ eight similar terms involving maintenance of components 2 and 3

$$= \lambda_1''\lambda_2\lambda_{t3}(r_1'')^2\left(\frac{r_2}{r_1''+r_2} + \frac{r_{t3}}{r_1''+r_{t3}}\right)$$

$$+\lambda_1''\lambda_{t2}\lambda_3(r_1'')^2\left(\frac{r_3}{r_1''+r_3} + \frac{r_{t2}}{r_1''+r_{t2}}\right)$$

+ four similar terms involving maintenance of components 2 and 3.

Therefore

$$\lambda_{pmt} = \lambda_a + \lambda_b + \lambda_c + \lambda_d + \lambda_e + \lambda_f$$

$$r_{pmt} = (\lambda_ar_a + \lambda_br_b + \lambda_cr_c + \lambda_dr_d + \lambda_er_e + \lambda_fr_f)/\lambda_{pmt}$$

where

$$\lambda_a = \lambda_1'' \lambda_2 \lambda_{t3} (r_1'')^2 \left(\frac{r_2}{r_1'' + r_2} + \frac{r_{t3}}{r_1'' + r_{t3}} \right)$$

$$\lambda_b = \lambda_1'' \lambda_{t2} \lambda_3 (r_1'')^2 \left(\frac{r_3}{r_1'' + r_3} + \frac{r_{t2}}{r_1'' + r_{t2}} \right)$$

$$\lambda_c = \lambda_2'' \lambda_1 \lambda_{t3} (r_2'')^2 \left(\frac{r_1}{r_2'' + r_1} + \frac{r_{t3}}{r_2'' + r_{t3}} \right)$$

$$\lambda_d = \lambda_2'' \lambda_{t1} \lambda_3 (r_2'')^2 \left(\frac{r_3}{r_2'' + r_3} + \frac{r_{t1}}{r_2'' + r_{t1}} \right)$$

$$\lambda_e = \lambda_3'' \lambda_1 \lambda_{t2} (r_3'')^2 \left(\frac{r_1}{r_3'' + r_1} + \frac{r_{t2}}{r_3'' + r_{t2}} \right)$$

$$\lambda_f = \lambda_3'' \lambda_{t1} \lambda_2 (r_3'')^2 \left(\frac{r_2}{r_3'' + r_2} + \frac{r_{t1}}{r_3'' + r_{t1}} \right)$$

$$r_a = \frac{r_1'' r_2 r_{t3}}{r_1'' r_2 + r_2 r_{t3} + r_{t3} r_1''}$$

$$r_b = \frac{r_1'' r_{t2} r_3}{r_1'' r_{t2} + r_{t2} r_3 + r_3 r_1''}$$

$$r_c = \frac{r_1 r_2'' r_{t3}}{r_1 r_2'' + r_2'' r_{t3} + r_{t3} r_1}$$

$$r_d = \frac{r_{t1} r_2'' r_3}{r_{t1} r_2'' + r_2'' r_3 + r_3 r_{t1}}$$

$$r_e = \frac{r_1 r_{t2} r_3''}{r_1 r_{t2} + r_{t2} r_3'' + r_3'' r_1}$$

$$r_f = \frac{r_{t1} r_2 r_3''}{r_{t1} r_2 + r_2 r_3'' + r_3'' r_{t1}}$$

If the reclosure time is negligible compared with the repair and maintenance times, the above equations reduce to

$$\lambda_a \simeq \lambda_1'' \lambda_2 \lambda_{t3} (r_1'')^2 / (r_1'' + r_2) \qquad \lambda_b \simeq \lambda_1'' \lambda_{t2} \lambda_3 (r_1'')^2 r_3 / (r_1'' + r_3)$$

$$\lambda_c \simeq \lambda_2'' \lambda_1 \lambda_{t3} (r_2'')^2 / (r_1 + r_2'') \qquad \lambda_d \simeq \lambda_2'' \lambda_{t1} \lambda_3 (r_2'')^2 r_3 / (r_2'' + r_3)$$

$$\lambda_e \simeq \lambda_3'' \lambda_1 \lambda_{t2} (r_3'')^2 /(r_1 + r_3'') \; \lambda_f \simeq \lambda_3'' \lambda_{t1} \lambda_2 (r_3'')^2 r_2/(r_2 + r_3'')$$

$$r_a = r_c \simeq r_{t3} \qquad r_b = r_e \simeq r_{t2} \qquad r_d = r_f \simeq r_{t1}$$

A3.5 Common mode failures

A3.5.1 All three components may suffer a common mode failure

The following equations relate to a third-order event in which all three components may suffer a common mode failure. These equations are presented assuming permanent failures only, but the structure can be adapted to accommodate transient, temporary and maintenance outages. Two sets of equations are included. The first set relates to the situation in which independent failures and common mode failures lead to the same down state. The system therefore has a single down state similar in concept to Fig. 8.13. The second set relates to the situation in which independent failures and common mode failures lead to distinctly separate down states. This is conceptually similar to Fig. 8.14.

(a) *Single down state*

$$\lambda = \lambda_1\lambda_2\lambda_3(r_1r_2 + r_2r_3 + r_3r_1) + \lambda_{123}$$

$$r = \frac{r_1r_2r_3r_{123}}{r_1r_2r_3 + r_1r_2r_{123} + r_1r_3r_{123} + r_2r_3r_{123}}$$

(b) *Separate down states*

$$\lambda = \lambda_1\lambda_2\lambda_3(r_1r_2 + r_2r_3 + r_3r_1) + \lambda_{123}$$

$$r = \frac{\lambda_1\lambda_2\lambda_3r_1r_2r_3 + \lambda_{123}r_{123}}{\lambda}$$

A3.5.2 Only two components may suffer a common mode failure

The following equations relate to a third-order event in which only two of the components (2 and 3) may suffer a common mode failure and the other component (1) can only fail independently. As discussed in Section A3.5.1, there are two sets of equations—one for a single down state and the other for separate down states.

(a) *Single down state*

$$\lambda = \lambda_a + \lambda_b$$

$$r = (\lambda_a r_a + \lambda_b r_b)/\lambda$$

where

$$\lambda_a = \lambda_1\left(\lambda_2 r_1 \lambda_3 \frac{r_1 r_2}{r_1+r_2} + \lambda_3 r_1 \lambda_2 \frac{r_1 r_3}{r_1+r_3} + \lambda_{23} r_1\right) + [\lambda_2\lambda_3(r_2+r_3) + \lambda_{23}]\lambda_1 r_e$$

$$\lambda_b = \lambda_2\lambda_1 r_2 \lambda_3 \frac{r_1 r_2}{r_1+r_2} + \lambda_3\lambda_1 r_3 \lambda_2 \frac{r_1 r_3}{r_1+r_3}$$

i.e.

$$\lambda = \lambda_1\lambda_2\lambda_3(r_1 r_2 + r_1 r_3 + r_2 r_e + r_3 r_e) + \lambda_1\lambda_{23}(r_1 + r_e)$$

and

$$r_a = r_1 r_e/(r_1 + r_e)$$

$$r_b = r_1 r_2 r_3/(r_1 r_2 + r_2 r_3 + r_3 r_1)$$

$$r_e = r_2 r_3 r_{23}/(r_2 r_3 + r_3 r_{23} + r_{23} r_2)$$

(b) *Separate down states*

$$\lambda = \lambda_a + \lambda_b$$

$$r = (\lambda_a r_a + \lambda_b r_b)/\lambda$$

where

$$\lambda_a = \lambda_1\lambda_2\lambda_3(r_1 r_2 + r_2 r_3 + r_3 r_1)$$

$$\lambda_b = \lambda_1\lambda_{23}(r_1 + r_{23})$$

$$r_a = r_1 r_2 r_3/(r_1 r_2 + r_2 r_3 + r_3 r_1)$$

$$r_b = r_1 r_{23}/(r_1 + r_{23})$$

A3.6 Adverse weather effects

The following equations relate to a third-order event in which the three components share a common environment which can vary between normal weather and adverse weather. These equations are presented assuming permanent failures only, but the structure can be adapted to accommodate transient, temporary and maintenance outages. Two sets of equations are given as Tables A3.1 and A3.2. The first set relates to the situation where repair can be performed in adverse weather. The second set relates to the situation where repair can only be done in normal weather. The concepts and assumptions discussed in Chapter 8 are again used in the formulation of these equations.

Table A3.1 Repair is possible in adverse weather

The system indices are given by combining the tabulated contributions ($i = 1$ to $i = 8$) as follows: $\lambda = \Sigma_{i=1}^{8} \lambda_i$; $r = r_1 r_2 r_3/(r_1 r_2 + r_2 r_3 + r_3 r_1)$.

No. (i)	Failure mode 1st	2nd	3rd	Contribution to the system failure rate (λ_i)
1	N	N	N	$\frac{N}{N+S}[\lambda_1(\lambda_2 R_7)(\lambda_3 R_5) + \lambda_1(\lambda_3 R_7)(\lambda_2 R_6)$ + similar terms for components 2 and 3]
2	N	A	N	$\frac{N}{N+S}\left[\lambda_1 \frac{r_1}{N}(\lambda_2 R_3)(\lambda_3 R_5) + \lambda_1 \frac{r_1}{N}(\lambda_3 R_3)(\lambda_2 R_6)\right.$ + similar terms for components 2 and 3]
3	A	N	N	$\frac{S}{N+S}[\lambda_1(\lambda_2 R_7)(\lambda_3 R_5) + \lambda_1(\lambda_3 R_7)(\lambda_2 R_6)$ + similar terms for components 2 and 3]
4	A	A	N	$\frac{S}{N+S}[\lambda_1(\lambda_2 R_3)(\lambda_3 R_5) + \lambda_1(\lambda_3 R_3)(\lambda_2 R_6)$ + similar terms for components 2 and 3]
5	A	A	A	$\frac{S}{N+S}[\lambda_1(\lambda_2 R_3)(\lambda_3 R_8) + \lambda_1(\lambda_3 R_3)(\lambda_2 R_9)$ + similar terms for components 2 and 3]
6	A	N	A	$\frac{S}{N+S}\left[\lambda_1(\lambda_2 R_7)\frac{R_1}{N}(\lambda_3 R_8) + \lambda_1(\lambda_3 R_7)\frac{R_2}{N}(\lambda_2 R_9)\right.$ + similar terms for components 2 and 3]
7	N	A	A	$\frac{N}{N+S}\left[\lambda_1 \frac{r_1}{N}(\lambda_2 R_3)(\lambda_3 R_8) + \lambda_1 \frac{r_1}{N}(\lambda_3 R_3)(\lambda_2 R_9)\right.$ + similar terms for components 2 and 3]
8	N	N	A	$\frac{N}{N+S}[\lambda_1(\lambda_2 R_7)\frac{R_1}{N}(\lambda_3 R_8) + \lambda_1(\lambda_3 R_7)\frac{R_2}{N}(\lambda_2 R_9)$ + similar terms for components 2 and 3]

N represents the normal weather state.
A represents the adverse weather state.

$$R_1 = \frac{r_1 r_2}{r_1 + r_2} \qquad R_2 = \frac{r_1 r_3}{r_1 + r_3} \qquad R_3 = \frac{S r_1}{S + r_1} \qquad R_5 = \frac{N r_1 r_2}{N r_1 + N r_2 + r_1 r_2}$$

$$R_6 = \frac{N r_1 r_3}{N r_1 + N r_3 + r_1 r_3} \qquad R_7 = \frac{N r_1}{N + r_1} \qquad R_8 = \frac{S r_1 r_2}{S r_1 + S r_2 + r_1 r_2}$$

$$R_9 = \frac{S r_1 r_3}{S r_1 + S r_3 + r_1 r_3}$$

Table A3.2 Repair is not done during adverse weather

The system indices are given by combining the tabulated contributions ($i=1$ to $i=8$) as follows: $\lambda = \lambda_a + \lambda_b$, $r = (\lambda_a r_a + \lambda_b r_b)/(\lambda_a + \lambda_b)$, where

$$\lambda_a = \sum_{i=1}^{4} \lambda_i \qquad \lambda_b = \sum_{i=5}^{8} \lambda_i$$

$$r_a = \frac{r_1 r_2 r_3}{r_1 r_2 + r_2 r_3 + r_3 r_1} \qquad r_b = \frac{r_1 r_2 r_3}{r_1 r_2 + r_2 r_3 + r_3 r_1} + S$$

No.	*Failure mode*			
(i)	*1st*	*2nd*	*3rd*	*Contribution to the system failure rate* (λ_i)
1	N	N	N	$\frac{N}{N+S}[\lambda_1(\lambda_2 r_1)(\lambda_3 R_1) + \lambda_1(\lambda_3 r_1)(\lambda_2 R_2)$ + similar terms for components 2 and 3]
2	N	A	N	$\frac{N}{N+S}\left[\lambda_1 \frac{r_1}{N}(\lambda_2' S)(\lambda_3 R_1) + \lambda_1 \frac{r_1}{N}(\lambda_3' S)(\lambda_2 R_2)\right.$ + similar terms for components 2 and 3]
3	A	N	N	$\frac{S}{N+S}[\lambda_1'(\lambda_2 r_1)(\lambda_3 R_1) + \lambda_1'(\lambda_3 r_1)(\lambda_2 R_2)$ + similar terms for components 2 and 3]
4	A	A	N	$\frac{S}{N+S}[\lambda_1'(\lambda_2' S)(\lambda_3 R_1) + \lambda_1'(\lambda_3' S)(\lambda_2 R_2)$ + similar terms for components 2 and 3]
5	A	A	A	$\frac{S}{N+S}[\lambda_1'(\lambda_2' S)(\lambda_3' S) + \lambda_1'(\lambda_3' S)(\lambda_2' S)$ + similar terms for components 2 and 3]
6	A	N	A	$\frac{S}{N+S}\left[\lambda_1'(\lambda_2 r_1)\frac{R_1}{N}(\lambda_3' S) + \lambda_1'(\lambda_3 r_1)\frac{R_2}{N}(\lambda_2' S)\right.$ + similar terms for components 2 and 3]
7	N	A	A	$\frac{N}{N+S}\left[\lambda_1 \frac{r_1}{N}(\lambda_2' S)(\lambda_3' S) + \lambda_1 \frac{r_1}{N}(\lambda_3' S)(\lambda_2' S)\right.$ + similar terms for components 2 and 3]
8	N	N	A	$\frac{N}{N+S}\left[\lambda_1(\lambda_2 R_1)\frac{R_1}{N}(\lambda_3' S) + \lambda_1(\lambda_3 r_1)\frac{R_2}{N}(\lambda_2' S)\right.$ + similar terms for components 2 and 3]

$N, S, R_1, R_2, R_3, R_4, R_5, R_6, R_7, R_8, R_9$, as in Table A3.1.

A3.7 Common mode failures and adverse weather effects

The following equations relate to a third-order event in which all three components may suffer a common mode failure as well as sharing a common environment. They are therefore derived from those given in Sections A3.5.1 and A3.6, using the concepts described in Chapter 8. In addition it is assumed that a single down state model is applicable, i.e. similar in concept to Fig. 8.13. The equations can be extended, however, to the case in which only two of the components can suffer a common mode failure and the case in which separate down states are applicable.

A3.7.1 Repair is possible in adverse weather

The contributions to the system indices by the independent failure modes are identical to those given in Table A3.1. Therefore

$$\lambda = \sum_{i=1}^{8} \lambda_i + \frac{N}{N+S}\lambda_{123} + \frac{S}{N+S}\lambda'_{123}$$

$$= \sum_{i=1}^{8} \lambda_i + \hat{\lambda}_{123}$$

where λ_i ($i = 1$ to $i = 8$) is given in Table A3.1:

$$r = \frac{r_1 r_2 r_3 r_{123}}{r_1 r_2 r_3 + r_1 r_2 r_{123} + r_1 r_3 r_{123} + r_2 r_3 r_{123}}$$

A3.7.2 Repair is not done during adverse weather

The contributions to the system indices by the independent failure modes are identical to those given in Table A3.2. Therefore

$$\lambda = \lambda_c + \lambda_d$$

$$r = (\lambda_c r_c + \lambda_d r_d)/(\lambda_c + \lambda_d)$$

where

$$\lambda_c = \lambda_a + \frac{N}{N+S}\lambda_{123} \qquad \lambda_d = \lambda_b + \frac{S}{N+S}\lambda'_{123}$$

λ_a, λ_b are given in Table A3.2;

$$r_c = \frac{r_1 r_2 r_3 r_{123}}{r_1 r_2 r_3 + r_1 r_2 r_{123} + r_1 r_3 r_{123} + r_2 r_3 r_{123}}$$

$$r_d = r_c + S.$$

Solutions to problems

Chapter 2

1. (a)

Peak load	LOLE
150	0.085719 (d/yr)
160	0.120551
170	0.151276
180	0.830924
190	2.057559
200	3.595240

(b)

200	0.105548 (d/yr)
210	0.159734
220	0.210936
230	0.257689
240	1.259372
250	2.641976
260	4.245280

(c) Increase in peak load carrying capability = 45 MW

(d)

Peak load	LOLE
150	0.085349 (d/yr)
160	0.119730
170	0.281061
180	0.940321
190	2.119768
200	3.565910

200	0.106232 (d/yr)
210	0.159137
220	0.210046
230	0.448675
240	1.335556
250	2.686116
260	4.221469

(e) Increase in peak load carrying capability = 43 MW

2. LOLE = 0.7819 d/period
   EIR = 0.998386

3. EIR = 0.983245
   EES (50 MW unit) = 112 800 MWh
   EES (Unit C) = 13570 MWh
4. (a) LOLE (Gen Bus) = 3.32 days/year
       LOLE (Load Bus) = 7.96 days/year
   (b) LOLE (Gen Bus) = 5.62 days/year
       LOLE (Load Bus) = 10.22 days/year
   (c) EENS = 1288.1 MWh
       EIR = 0.997936
5. (a) LOLE = 0.028363 days/day
   (b) LOLE = 0.030041 days/day
6. (a) EIR = 0.990414
       EES (Unit D) = 3560 MW days
       EES (Unit C) = 1646
       EES (Unit B) = 274.5
       EES (Unit A) = 65.8
7. LOLE = 4.17 days/year

Chapter 3

1. Cumulative frequency = 7.435 occurrences/yr.
   Cumulative probability = 0.173906
   Average duration = 204.9 hr.
2. Cumulative frequency = 0.013003/day
   Cumulative probability = 0.016546
   Average duration = 1.273 days
3.

Capacity	*Probability*	*Cumulative probability*	*Cumulative frequency*
120	0.796466	1.000000	—
100	0.172555	0.203534	19.114
80	0.015577	0.030979	7.906
70	0.014566	0.015402	5.509
60	0.000761	0.000836	0.235
40	0.000021	0.000075	0.046
0	0.000054	0.000054	0.039

4.

Capacity margin (MW)	*Probability*	*Frequency*
200	0.02476499	0.05076823
160	0.00252704	0.00639341
140	0.00619124	0.01269204
120	0.01000912	0.02061218
100	0.00682300	0.01428736
80	0.00101291	0.00256480

Capacity margin (MW)	*Probability*	*Frequency*
60	0.00065755	0.00167584
40	0.00251778	0.00520085
20	0.00002632	0.00007948
0	0.00025354	0.00064220
–20	0.00000054	0.00000189
–40	0.00001032	0.00003107
–60	0.00000001	0.00000004
–80	0.00000021	0.00000073
–100	0.0	0.0
–120	0.0	0.0
–160	0.0	0.0

Capacity margin (MW)	*Cumulative probability*	*Cumulative frequency*
200	0.05479457	—
160	0.03002958	0.05081388
140	0.02750254	0.05473079
120	0.02121130	0.04265788
100	0.01130218	0.02345378
80	0.00447918	0.00983507
60	0.00346627	0.00735963
40	0.00280872	0.00568379
20	0.00029094	0.00073295
0	0.00026462	0.00065504
–20	0.00001108	0.00003309
–40	0.00001054	0.00003122
–60	0.00000022	0.00000077
–80	0.00000021	0.00000073
–100	0.0	0.0
–120	0.0	0.0
–160	0.0	0.0

Chapter 4

1. LOLE = 0.010720 days/day
2. $LOLE_A$ = 0.007491 days/day
   $LOLE_B$ = 0.045925 days/day
3. Interconnection 100% available
   $LOLE_A$ = 0.1942 days, $ELOL_A$ = 2.475 MW
   $LOLE_B$ = 0.0585 days, $ELOL_B$ = 0.659 MW
   Interconnection availability included
   $LOLE_A$ = 0.1943 days, $ELOL_A$ = 2.478 MW
   $LOLE_B$ = 0.0585 days, $ELOL_B$ = 0.661 MW

4. LOLE = 0.319015 days/period
5. Interconnection 100% available
   LOLE_A = 1.162 days/year
   LOLE_X = 0.994 days/year
   Interconnection 95% available
   LOLE_A = 1.502 days/year
   LOLE_X = 1.196 days/year

Chapter 5

1. (a) 0.011370, 0.000058 (b) 0.316789, 0.009485
   (c) 0.006829, 0.000021 (d) 0.009105, 0.000037
   (e) 0.000187, $<10^{-6}$
2. (a) 0.0004761 (b) 0.0000001 (c) 0.0000477
3. 0.001972, 0.000003
4. (a) 8 (b) 45 MW each (c) 0.0003805
5. 0.000011, 0.005126

Chapter 6

1. 4.17 days/year
2. *Generation indices*: LOLE (4.37 d/yr), ELL (0.3168 MW), EENS (2755 MWh/yr)

Annualized load point indices:

Index	*Load A*			*Load B*		
Q	0.012124			0.016201		
F (occ./yr)	2.8595			7.4495		
	Total	*Isolated*		*Total*	*Isolated*	
ENC (occ./yr)	2.8595	—		7.4495	0.0284	
ELC (MW)	36.81	—		84.13	1.70	
EENS (MWh/yr)	1356.5	—		1712.3	7.88	
EDLC (h)	106.2	—		141.9	0.13	
	Value	*Outage*	*Prob.*	*Value*	*Outage*	*Prob.*
max LC (MW)	30	$L1, L2$	0.000014	60	$L2, L3$	0.000015
max energy curt. (MWh/yr)	773.7	$G1, G2$	0.007066	773.7	$G1, G2$	0.007066
max DLC (h)	61.9	$G1, G2$	0.007066	61.9	$G1, G2$	0.007066
aver. LC (MW/curt.)	12.87			11.29		
aver. ENS (MWh/curt.)	474.4			229.9		
aver. DLC (h)	37.14			19.05		

Annualized system indices:

bulk power interruption index (MW/MWyr)	0.86
bulk power supply average MW curt./disturbance	11.73
bulk power energy curt. index (MWh/MWyr)	21.92
modified bulk power energy curt. index	0.002502

aver. NC/load point (occ./yr)	5.155		
aver. LC/load point (MW)	60.47		
aver. DLC/load point (h)	124.06		
max system LC (MW)	60	*L*2, *L*3	0.000015
max system ENS (MWh)	1547.5	*G*1, *G*2	0.007066

3. (a) *P*(capacity deficiency) = 0.016059
   F(capacity deficiency) = 1.97 occ./year
   LOLE = 5.86 days/year
   ELL = 0.3623 MW
   EENS = 3174 MWh/yr

   (b)

Bus	*Load*	*Probability*	*Frequency*
3	1	0.024426	5.49 occ./yr
4	2	0.016032	0.50 occ./yr

   (c) *Annualized load point indices*:

Index	*Load 1*			*Load 2*		
Q	0.023509			0.016412		
F (occ./yr)	7.9258			3.1141		
	Total	*Isolated*		*Total*	*Isolated*	
ENC (occ./yr)	7.9258	0.0894		3.1141	0.0699	
ELC (MW)	80.58	5.36		34.64	2.80	
EENS (MWh/yr)	2199.9	37.34		1561.9	19.63	
EDLC (h)	205.9	0.62		143.8	0.49	
	Value	*Outrage*	*Prob.*	*Value*	*Outrage*	*Prob.*
max LC (MW)	60	*L*2, *L*3	0.000037	40	*L*2, *L*4	0.000034
max energy curt. (MWh/yr)	726.96	*G*1, *G*2	0.008299	726.96	*G*1, *G*2	0.008299
max DLC (h)	72.7	*G*1, *G*2	0.008299	72.7	*G*1, *G*2	0.008299
aver. LC (MW/curt.)	10.17			11.12		
aver. ENS (MWh/curt.)	277.6			501.6		
aver. DLC (h)	25.98			46.17		

Annualized system indices:

bulk power interruption index (MW/MWyr)	1.15		
bulk power supply average MW curt./disturbance	10.44		
bulk power energy curt. index (MWh/MWyr)	37.62		
modified bulk power energy curt. index	0.004294		
aver. NC/load point (occ./yr)	5.52		
aver. LC/load point (MW)	57.61		
aver. DLC/load point (h)	174.86		
max system LC (MW)	100	*L*2, *L*4	0.000034
max system ENS (MWh)	1453.9	*G*1, *G*2	0.008299

(d) *Annualized load point indices*:

Index	*Load 1*	*Load 2*
Q	0.015105	0.015070
F(occ./yr)	2.3906	2.3516

	Total	*Isolated*	*Total*	*Isolated*
ENC (occ./yr)	2.3906	—	2.3516	0.0288
ELC (MW)	25.11	—	25.58	1.15
EENS (MWh/yr)	1430.2	—	1433.0	7.71
EDLC (h)	132.3	—	132.0	0.19

	Value	*Outrage*	*Prob.*	*Value*	*Outrage*	*Prob.*
max LC (MW)	12.5	$G2, G2$	0.005843	40	$L3, L4$	0.000022
max energy curt. (MWh/yr)	720.77	$G1, G2$	0.008228	720.77	$G1, G2$	0.008228
max DLC (h)	72.1	$G1, G2$	0.008228	72.1	$G1, G2$	0.008228
aver. LC (MW/curt.)	10.50			10.88		
aver. ENS (MWh/curt.)	598.4			609.4		
aver. DLC (h)	55.36			56.14		

Annualized system indices:

bulk power interruption index (MW/MWyr)	0.51		
bulk power supply average MW curt./disturbance	10.69		
bulk power energy curt. index (MWh/MWyr)	28.63		
modified bulk power energy curt. index	0.003268		
aver. NC/load point (occ./yr)	2.37		
aver. LC/load point (MW)	25.35		
aver. DLC/load point (h)	132.17		
max system LC (MW)	40	$L3, L4$	0.000022
max system ENS (MWh)	1441.5	$G1, G2$	0.008228

(e) *Annualized load point indices*:

Index	*Load 1*	*Load 2*
Q	0.015419	0.015050
F (occ./yr)	2.7918	2.3568

	Total	*Isolated*	*Total*	*Isolated*
ENC (occ./yr)	2.7918	—	2.3568	0.0294
ELC (MW)	29.11	—	25.65	1.18
EENS (MWh/yr)	1457.6	—	1431.2	7.71
EDLC (h)	135.1	—	131.8	0.19

	Value	*Outage*	*Prob.*	*Value*	*Outage*	*Prob.*
max LC (MW)	12.5	$G2, G2$	0.005836	40	$L3, L4$	0.000022
max energy curt. (MWh/yr)	719.90	$G1, G2$	0.008218	719.90	$G1, G2$	0.008218
max DLC (h)	71.99	$G1, G2$	0.008218	71.99	$G1, G2$	0.008218
aver. LC (MW/curt.)	10.43			10.88		
aver. ENS (MWh/curt.)	522.1			607.3		
aver. DLC (h)	48.38			55.94		

Annualized system indices:			
bulk power interruption index (MW/MWyr)	0.55		
bulk power supply average MW curt./disturbance	10.64		
bulk power energy curt. index (MWh/MWyr)	28.89		
modified bulk power energy curt. index	0.003298		
aver. NC/load point (occ./yr)	2.57		
aver. LC/load point (MW)	27.38		
aver. DLC/load point (h)	133.46		
max system LC (MW)	40	$L3$, $L4$	0.000022
max system ENS (MWh)	1439.8	$G1$, $G2$	0.008218

Chapter 7

		SAIFI (*int / cust. yr*)	*SAIDI* (*h / cust. yr*)	*CAIDI* (*h / cust. int*)	*ASAI*	*AENS* (*kWh / cust. yr*)
1.	(i)	2.20	3.76	1.71	0.999571	16.97
	(ii)	2.20	3.10	1.41	0.999646	14.17
	(iii)	0.85	2.43	2.86	0.999723	10.98
	(iv)	0.94	2.47	2.63	0.999718	11.17
2.(a)	(i)	2.20	2.32	1.05	0.999735	10.78
	(ii)	2.20	2.32	1.05	0.999735	10.78
	(iii)	0.85	1.65	1.94	0.999812	7.60
	(iv)	0.94	1.69	1.80	0.999808	7.78
2.(b)	(i)	2.20	2.55	1.16	0.999708	11.80
	(ii)	2.20	2.55	1.16	0.999708	11.80
	(iii)	0.85	1.88	2.21	0.999785	8.62
	(iv)	0.94	1.92	2.04	0.999781	8.80

Chapter 8

1. (a) (i) 4.57×10^{-4} f/yr, 4 hours, 1.83×10^{-3} hours/yr
1.00×10^{-1} f/yr, 2.01 hours, 2.02×10^{-1} hours/yr
(ii) 2.13×10^{-7} f/yr, 2.64 hours, 5.63×10^{-7} hours/yr
3.44×10^{-5} f/yr, 1.68 hours, 5.76×10^{-5} hours/yr
(b) (i) 2.46×10^{-3} f/yr, 1.56 hours, 3.83×10^{-3} hours/yr
1.03×10^{-1} f/yr, 1.99 hours, 2.05×10^{-1} hours/yr
(ii) 2.00×10^{-3} f/yr, 1 hour, 2.00×10^{-3} hours/yr
2.03×10^{-3} f/yr, 1.01 hours, 2.06×10^{-3} hours/yr
2. (a) (i) 1.37×10^{-3} f/yr, 4 hours, 5.48×10^{-3} hours/yr
1.01×10^{-1} f/yr, 2.04 hours, 2.06×10^{-1} hours/yr
(ii) 9.07×10^{-7} f/yr, 2.62 hours, 2.38×10^{-6} hours/yr
3.55×10^{-4} f/yr, 3.43 hours, 1.22×10^{-3} hours/yr
3. 1.11×10^{-1} f/yr, 8.68 hours, 9.62×10^{-1} hours/yr
4. 1.14×10^{-1} f/yr, 8.54 hours, 9.70×10^{-1} hours/yr

5. 4.64×10^{-3} f/yr, 0.92 hours, 4.26×10^{-3} hours/yr
6. (i) (a) 2.28×10^{-3} f/yr, 5 hours, 1.14×10^{-2} hours/yr
   (b) 1.17×10^{-2} f/yr, 5 hours, 5.84×10^{-2} hours/yr
   (c) 4×10^{-2} f/yr, 5 hours, 2×10^{-1} hours/yr
   (ii) (a) 2.28×10^{-3} f/yr, 5 hours, 1.14×10^{-2} hours/yr
   (b) 1.32×10^{-2} f/yr, 6.36 hours, 8.42×10^{-2} hours/yr
   (c) 4.6×10^{-2} f/yr, 6.5 hours, 3×10^{-1} hours/yr
7. (a) 5.04×10^{-3} f/yr, 4.69 hours, 2.37×10^{-2} hours/yr
   (b) 1.60×10^{-2} f/yr, 6.17 hours, 9.85×10^{-2} hours/yr
   (c) 4.88×10^{-2} f/yr, 6.49 hours, 3.16×10^{-1} hours/yr
8. (a) 0.645 f/yr (b) 61.6%
   (c) 1.54×10^{-2} f/yr, 4 hours, 6.16×10^{-2} hours/yr
9. 7.98×10^{-2} f/yr, 4 hours, 3.20×10^{-1} hours/yr

Chapter 9

1. (i) 0.04 f/yr, 7.5 hours, 0.30 hours/yr, 3.75 MW, 1.126 MWh/yr.
   (ii) 0.046 f/yr, 7.8 hours, 0.356 hours/yr, 1.41 MW, 0.503 MWh/yr.
   (iii) 0.039 f/yr, 7.4 hours, 0.287 hours/yr, 1.96 MW, 0.562 MWh/yr.
   (iv) 0.04 f/yr, 7.5 hours, 0.30 hours/yr, 1.98 MW, 0.594 MWh/yr.

Chapter 10

1. (a) 8.81×10^{-2}, 0.51, 4.47×10^{-2}, 0.671
   (b) 9.01×10^{-2}, 0.51, 4.58×10^{-2}, 0.687
   (c) 7.21×10^{-2}, 0.51, 3.67×10^{-2}, 0.551
   (d) 8.21×10^{-2}, 0.51, 4.14×10^{-2}, 0.621
   (e) 1.04×10^{-1}, 0.50, 5.24×10^{-2}, 0.786
   (f) 1.20×10^{-1}, 0.50, 6.04×10^{-2}, 0.906
   (g) 7.41×10^{-2}, 0.52, 3.82×10^{-2}, 0.573
2. (a) 5.00×10^{-2}, 10.4, 5.20×10^{-1}
   (b) 4.01×10^{-2}, 0.51, 2.05×10^{-2}

Chapter 11

1. 117.09
2. 116.15, 118.03
3. 118.06
4. 116.15
5. 116.82, 116.82, 116.96
6. 0.994447(100), 0.001824(70), 0.001371(60)
   0.001216(50), 0.001142(0)
7. 0.793425(100), 0.048767(90), 0.045479(80)
   0.036712(70), 0.033425(60), 0.042192(0) 0.0805
8. 0.977674, 0.009826, 0.012500

0.967898, 0.029380, 0.002722
0.965477, 0.034220, 0.000303
9. 0.972786, 0.014664, 0.000075, 0.012475

Index